COMPOSITION, CHEMISTRY, AND CLIMATE OF THE ATMOSPHERE

COMPOSITION, CHEMISTRY, AND CLIMATE OF THE ATMOSPHERE

▼

Edited by

Hanwant B. Singh

VAN NOSTRAND REINHOLD

I⟨T⟩P™   A Division of International Thomson Publishing Inc.

New York • Albany • Bonn • Boston • Detroit • London • Madrid • Melbourne
Mexico City • Paris • San Francisco • Singapore • Tokyo • Toronto

Copyright © 1995 by Van Nostrand Reinhold
I(T)P™ A division of International Thomson Publishing Inc.
The ITP logo is a trademark under license

Printed in the United States of America
For more information, contact:

Van Nostrand Reinhold
115 Fifth Avenue
New York, NY 10003

International Thomson Publishing GmbH
Königswinterer Strasse 418
53227 Bonn
Germany

International Thomson Publishing Europe
Berkshire House
168–173 High Holborn
London WC1V 7AA
England

International Thomson Publishing Asia
221 Henderson Road #05-10
Henderson Building
Singapore 0315

Thomas Nelson Australia
102 Dodds Street
South Melbourne, 3205
Victoria, Australia

International Thomson Publishing Japan
Hirakawacho Kyowa Building, 3F
2-2-1 Hirakawacho
Chiyoda-ku, 102 Tokyo
Japan

Nelson Canada
1120 Birchmount Road
Scarborough, Ontario
Canada M1K 5G4

International Thomson Editores
Campos Eliseos 385, Piso 7
Col. Polanco
11560 Mexico D.F. Mexico

1 2 3 4 5 6 7 8 9 10 BKR 01 00 99 98 97 96 95

Library of Congress Cataloging-in-Publication Data
Composition, chemistry, and climate of the atmosphere/edited by
 Hanwant B. Singh.
   p.   cm.

Includes bibliographical references and index.
ISBN 0-442-01264-0
1. Atmospheric physics.   2. Atmospheric chemistry.
3. Air-Pollution.   I. Singh, H. B.
QC 861.2.C66   1995   94-45885
551.5′11-dc20   CIP

Contents

List of Contributors

H. Berresheim
Georgia Institute of Technology
Atlanta
GA 30332

present address:
DWD/MOHp
Albin-Schwaiger-Weg 10
82383 Hohenpeissenberg
Germany

P. Brimblecombe
School of Environmental Sciences
University of East Anglia
Norwich NR4 7TJ
UK

P. J. Crutzen
Max Planck Institute for Chemistry
D-55020 Mainz
Germany

D.D. Davis
Georgia Institute of Technology
Atlanta
GA 30332

J.M. Hales
ENVAIR
Kennewick
WA 99337

H.E. Jeffries
University of North Carolina
Chapel Hill
NC 27599–7400

M.A.K. Khalil
Oregon Graduate Institute
Portland
OR 97291

P. Middleton
ACT/ASRC/SUNYA
c/o University for Atmospheric Research
Box 300
Boulder
CO 80307

present address:
Science and Policy Associates, Inc.
3445 Penrose Place, Suite 140
Boulder
CO 80301

M.J. Molina
Department of EAPS and Chemistry
Massachusetts Institute of Technology
Cambridge
MA 02139

R.F. Pueschel
NASA Ames Research Center
Moffett Field
CA 94035

R.A. Rasmussen
Oregon Graduate Institute
Portland
OR 97291

J.M. Roberts
NOAA/ERL Aeronomy Laboratory
Boulder
CO 80303

D.A. Salstein
Atmospheric and Environmental Research
   Inc.
Cambridge
MA 02139

T.-L. Shen
Department of Earth, Atmospheric, and
   Planetary Sciences
Massachusetts Institute of Technology
Cambridge
MA 02139

H.B. Singh
NASA Ames Research Center
Moffett Field
CA 94035

P.H. Wine
Georgia Institute of Technology
Atlanta
GA 30332

P.J. Wooldridge
Department of Earth, Atmospheric, and
   Planetary Sciences
Massachusetts Institute of Technology
Cambridge
MA 02139

D.J. Wuebbles
Global Climate Research Division
Lawrence Livermore National Laboratory
7000 East Avenue, L-262
Livermore
CA 94550

present address:
Department of Atmospheric Sciences
University of Illinois
105 South Gregory Avenue
Urbana
IL 61801

Preface

The notion that human activities may endanger the Earth's environment has emerged as a leading societal concern in the post-industrial era. Under the ever increasing pressures of population growth and industrialization, the problems of local air pollution have now become matters of both local and global concern. Sources of man-made pollution have become competitive with nature's own cycles and are threatening to perturb long-established ecological balances at a rapid pace. Smog, toxic chemicals, acid rain, ozone depletion, and greenhouse warming have become household words, and an intense public policy debate about the cost and benefits of environmental protection continues. There is a growing realization that the consequences of air pollution can be felt in unpredictable ways in near and faraway places. Unpopulated regions of the world such as the Arctic now suffer from Arctic haze, and ozone depletions are the largest in the Antarctic stratosphere. Over the last three decades many countries have instituted ambient air quality standards designed to mitigate problems of health and welfare associated with the release of chemicals such as sulfur dioxide, nitrogen oxides, hydrocarbons, carbon monoxide, and particulate matter. Global agreements to prevent the depletion of the ozone layer and to slow down climatic warming are being actively debated and formulated.

The atmosphere is a complex mixture of more than a thousand trace chemicals that are constantly reacting and redistributing. The need to understand the sources and distribution of these chemicals, along with the mechanisms by which they are transformed, transported, and ultimately removed from the atmosphere, has grown in parallel with the increased concern about air pollution and its consequences. The subject dealing with the atmospheric fate and impact of chemicals has largely fallen under the purview of an emerging discipline termed "atmospheric chemistry". Within this discipline the traditional techniques of chemistry, physics, and meteorology have been applied to the study of the atmosphere. The science behind the issues of air pollution and global change is highly complex and interdisciplinary. Efforts are under way to integrate these concepts into the undergraduate and graduate science curricula. This book attempts to present a synthesis of the science behind most of the important issues in atmospheric chemistry in a simple yet accurate fashion. Local and global air pollution phenomena are integrated into a scientifically comprehensible theme with a clear distinction between "natural" and "man-made" effects. In all cases man's impact is measured as a perturbation of the natural cycle, and the understanding of the latter is a central theme in this book. It is hoped that this book will be useful both to the expert atmospheric scientist as well as to the student at the undergraduate and higher level.

The science of air pollution, air chemistry, and climate change is explored in 13 chapters. Chapters 1–4 are instructive chapters that provide a basic understanding of the Earth's atmosphere. Chapter 1 presents a historical perspective on air pollution problems, evolution of policy, legislation, and scientific progress. Chapter 2 defines the mean physical and dynamical characteristics of the atmosphere, Chapter 3 provides direct scientific evidence to demonstrate that atmospheric composition has indeed changed on a global scale coincident with the growth of population and industrialization, while Chapter 4 deals with attempts to quantify the sources of pollution on regional and global scales. Chapter 5 is devoted to the subject of particles in the atmosphere, with the recognition that particle and gas phase phenomena are more intertwined than hitherto believed. Nitrogen, halogen, and sulfur cycles, which are responsible for important atmospheric perturbations, are discussed in Chapters 6–8 respectively. The science behind the consequences of air pollution resulting in phenomena such as photochemical smog, tropospheric ozone change, stratospheric ozone depletion (including the Antarctic ozone hole), acidic precipitation, and global greenhouse warming is treated in Chapters 9–13 respectively. An appendix summarizes useful numerical values and symbols. In-depth references are provided enabling each subject to be explored fully.

Thanks are due to all the chapter authors for their diligent efforts in the completion of this work. I am grateful to many colleagues for their advice and critical review of some chapters and to VNR editors for persistence in the efforts required to bring this volume to completion. Special thanks are due to my wife Raman for her unyielding service in the review and organization of this volume.

Hanwant B. Singh
Palo Alto, California

1

History of air pollution

Peter Brimblecombe

1.1. INTRODUCTION

The history of air pollution has until recently been of interest mainly to scientists, who approach history rather as amateurs. It has resulted in a history rich in anecdote and mythology but often lacking in primary references and perceptive interpretation of wider social contexts. Historians have often left environmental history to footnotes and seem uncomfortable with its use of geochemical and geophysical proxy data in parallel with the documentary record. Fortunately a growing group of environmental historians is now beginning to approach the subject with considerable sympathy and ability.

This chapter describes the main features of air pollution history. There are quite a few papers on the development of air pollution in given locations

(London: Brimblecombe, 1987; York: Brimblecombe and Bowler, 1992; Pittsburgh: Davidson, 1979; Rome: Camuffo, 1993; Ruhr: Bruggemeier, 1990). Hence the aim here is to write around a theme more central to this book and attempt to trace the history of air pollution from localized microenvironments (indoors and urban areas) through to regional and global scales.

1.2. EARLY HISTORY OF AIR POLLUTION

1.2.1. Indoors

Humans have been exposed to smoke, the archetypical air pollutant, since the earliest use of fire (Rossotti, 1993). Although some have imagined indoor air pollution to be a relatively new condition in the history of civilization, it is likely that it predated general urban air pollution. There is paleopathological evidence of the effects of indoor air pollutants on ancient health. Anthracosis, a blackening of the lung, appears quite general in the samples of mummified human lung tissue that survive from both Arctic and dry low-latitude climates, and no doubt arose from high concentrations of indoor smoke. There is also some evidence that indoor smoke may have induced an increased incidence of sinusitis in some populations exposed to high concentrations of indoor smoke in the past. The Roman poet Horace had already written of his irritation with interior smoke.

We know from modern measurements in the developing world that the concentrations of air pollutants in dwelling places where biofuels are used can be extremely high (see Smith, 1987). Experiments in Iron Age dwellings reconstructed at a Swedish site also suggested that poor ventilation could lead to highly polluted interiors. Indoor smoke has always been a problem, and the design of chimneys has vexed architects and builders over the centuries. In classical writing there is evidence of the damage caused in smoky interiors. In the *Odyssey* Homer writes of the damage that smoke caused to the armor that hung upon the walls. Later, when coal was used as a fuel the problem became even more severe. We know that in one medieval Northumbrian household wood had to be specially purchased to prevent the Christmas arras from being damaged. Hangings were also said to be uncommon in London houses of the 18th century because they were so readily damaged by coal smoke.

1.2.2. Smell and Habitation

When people gathered in large groups, rubbish often accumulated in massive amounts. Taboos, social practices, and then laws tended to govern what it was appropriate to do with refuse. Aristotle and a range of Hebrew and Islamic writers dealt with the question of disposal of rubbish beyond the city walls and good sanitary practices that would lead to freedom from miasmatic odors. In ancient times strange smells and odors were often linked to disease (miasmatic theories). The perceived effects of odor pollution could sometimes be quite dramatic. The *Victory Stela of King Pi(ankhy)* records the Nubian siege of Hermopolis (Un) in Egypt *ca* 734 B.C. Although the meaning is not entirely clear, military resistance collapsed when the town became too foul to be habitable (Lichtheim, 1980). "Days passed, and Un was a stench to the nose, for lack of air to breathe. Then Un threw itself on its belly, to plead before the king." In medieval London the monks at White Friars claimed that the air was so malodorous that some of their brethren had died from the stench.

The stench from rotting organic matter was an ever-present problem in cities. It was said to be so bad in Paris that silverware rapidly tarnished. Culinary odors that so troubled Seneca in Imperial Rome were also to be found in the Papal City (Andrieux, 1962). Police regulations in Papal Rome discouraged cooking, as there seemed to be an objection to the odors. Tripe and cabbage were not prepared in the home for fear of prosecution, so tripe was cooked in enormous cauldrons set up in front of the church of San Marcello and cabbages boiled near the Piazza Colonna. Coffee was roasted at the foot of the column of Marcus Aurelius. However, this was an atypical solution to the problem of odors in cities. In 16th century England, for instance, legal action was taken as a result of continuous objections to the smells from beef boilers. In modern times the production of pet food, chocolate, beer, and coffee brings frequent complaint.

1.2.3. Primary Urban Air Pollution

Smoke problems existed in great cities of antiquity. We know that it was bad enough in Rome to blacken temples and to have doctors argue that their sensitive patients should leave for the cleaner air of the country. There were some attempts at zoning in ancient Rome which kept smoky industries away from the wealthy parts of the city, and occasional court cases show that neighbors were thought to be responsible for smoke. However, despite Rome's size and problems, it was 13th century London where air pollution became particularly evident. Wood shortages caused a number of industries (particularly cement making) to adopt coal as a fuel. The peculiar smell of coal smoke allowed it to be considered as a health hazard and rapidly resulted in a series of attempts to prevent its use in the late 1200s. These were unsuccessful and coal became widely used, shifting to the domestic sector with the adoption of the household chimney by the 16th century. The problems of a coal burning city were fully developed in London of the 17th century, with effects on buildings, health, plants, visibility, etc., all well described in John Evelyn's small book *Fumifugium* of 1661.

In other cities of the early modern period coal use remained small. Inevitably there were local regulations that governed specific processes in cities as they grew in size and density, but air pollution did not seem to concern central government. The development of the steam engine drew particular attention to the problem of smoke. The early engines were noisy, dirty, and dangerous and allowed the public to focus particular attention on point sources that were especially objectionable (Brimblecombe, 1987). Given their small number, they probably represented only a tiny fraction of the domestic smoke produced by large cities. Engineers were concerned about smoke because it represented a loss of fuel. Nevertheless, the earliest modern concerns about air pollution control, particularly those of central government, were stimulated by this growth of furnaces and steam engines at the opening of the 19th century. The regulations were largely ineffective, as general thoughts centered about the notion of "burning your own smoke" that had been outlined by Benjamin Franklin. Smoke reduction was a function of careful stoking, so much pressure was put to bear on underpaid and often unskilled stokers, although some enlightened regulators placed emphasis on training for stokers.

The first half of the 19th century saw an enormous interest in improving the sanitary conditions of cities in Europe and North America. Smoke abatement was usually seen as an element of sanitary reform. In England smoke abatement clauses are scattered through the sanitary regulations of the 19th century. There were also a number of laws relating to specific industries. Perhaps most significant were the UK Alkali Acts that regulated hydrogen chloride emissions during the production of alkali for soap making. The hydrogen chloride devastated vast tracts of land until the law demanded that it be absorbed. The Acts were successful in removing the worst effects of the industry. They were not opposed by the industry as strongly as might be expected, because (i) the industry was growing tired of facing numerous damage cases under nuisance law, (ii) the laws were nation-wide and affected all manufacturers equally, and (iii) the hydrochloric acid produced could be marketed. The Alkali Acts were important in establishing the role of the scientific civil servant. Angus Smith, the first Alkali Inspector, brought great energy and thought to the task. He established many of the principles (e.g., best practical means) that have influenced thinking about pollution control ever since.

The early smoke abatement clauses failed to work partly because administrators tended to be sympathetic to the needs of industry. This was not always true, but local variability was a problem, as was the lack of carefully considered frameworks for the legislation. In the UK it was not until the 1875 Public Health Acts and the requirements for the creation of a sanitary inspectorate, distinct from the police, within cities that smoke abatement became more manageable. Under the harsher application of laws in colonial cities, where stokers were summarily dismissed or beaten, smoke abatement measures were more effective (Anderson, 1992). Still, it is probable that even where there was a will to improve the conditions of the air, the appropriate technology for abating smoke became available only slowly. It does seem that gradual improvements were realized in some cities of the 20th century where smoke control regulations slowly forced industry to be careful in its choice of replacement boilers and furnaces with regard to their smoke emissions (e.g., Brimblecombe and Bowler, 1992).

Early measurements, proxy data such as fog frequency and simple model estimates of pollutant concentration (London: Brimblecombe, 1977; East Coast American cities: Sherwood and Bambaru, 1991) suggest that concentrations of smoke and, to a lesser extent, primary pollutants such as SO_2, often decreased in conurbations of highly industrial societies during large parts of the 20th century. This is a little odd, because increases in urban air pollution concentrations are expected with increasing energy use in cities. However, concomitant increases in air pollution can be counteracted by factors such as the following.

1. Dilution: cities increase in size, especially once they develop effective transport networks.
2. Fuel change: there is often a transition from one fuel to another as the result of shortages or technological changes. Where these changes are from a high sulfur or sooty fuel such as coal to gas or petrol then reductions in SO_2 or smoke emissions are to be expected.
3. Electrification: a switch to electricity, especially where the generation takes place well away from urban areas, can lead to substantial improvement in urban air quality.
4. Legislation: the regulation of emissions can be an important driving force in reducing pollutants in urban air.

In cities which achieved a high fossil fuel use in the 19th century or early 20th century the importance of these factors probably arose in the order listed above. In London, for instance, dilution was probably significant in the late 19th century, while fuel change and electrification were controlling influences throughout the first half of the current century. There may have been a subtle effect through the introduction of smoke abatement clauses within the early public health acts, but the clearest indication of a direct effect of legislation has probably been the reduction in lead emissions that followed reductions in its addition to petrol. In cities which have experienced their most striking growth this century, the temporal sequence of the factors listed above appears to have been much compressed.

1.3. PHOTOCHEMICAL SMOG

Photochemical smog, often just called "smog", is a relatively new problem for cities. Photochemical episodes cannot really form without high concentrations of volatile organic matter, nitrogen oxides, and sunlight. Clearly such a situation will most readily be found in sunny locations with high automobile usage. Los

Angeles not only has sunlight and a substantial automobile population, but also gentle sea breezes, a backdrop of mountains, and subsidence inversions, which create near-perfect conditions for the formation of smogs.

It was sometime in the early 1940s when it became apparent that there was something rather new in the intensity of air pollution experienced in Los Angeles. The pollution was even mistaken for Japanese gas attacks. Much of the blame for the smog was directed toward the oil refineries and a butadiene plant, but the desire to penalize specific installations was accompanied by a growing awareness that it was a rather generalized problem. Professor Raymond Tucker, a St Louis-based air pollution expert, investigated the smog and recommended more controls on industry in 1947, but absolved automobiles from responsibility. Nevertheless, the oil industry became increasingly aware that regulation was very close, and the unpalatable complicity of the motor vehicle itself was much in evidence by the mid 1950s.

The early mistake in believing the smogs of Los Angeles to be similar to the primary air pollution episodes found elsewhere is perhaps understandable on the grounds of the complexity of the smog forming process. In the early 1950s the group of John Middleton at the Citrus Experimental Station of the University of California, Riverside had recognized that traditional primary pollutants such as SO_2 were not responsible for plant damage. In a nearby laboratory the biochemist Arie Jan Haagen-Smit was also convinced that the smog was not smoke and SO_2. It simply had the wrong smell; it reminded him of a terpene laboratory and led him to consider smog to be derived from the ozone-induced cleavage of olefins. Experiments quickly showed that ozonolysis of gasoline vapors had similar effects on plants to those of the smog. He then tried NO_2/gasoline vapor mixtures with the same result, which led to the controversial assertion that NO_2 and petroleum vapors produce ozone on irradiation. Within a short time, long-path infrared spectroscopic methods could be used to confirm the original suggestion of Haagen-Smit (Stephens, 1987).

Three hundred years after Evelyn's classic description (1661) of the London smog, Leighton (1961) wrote his work on the photochemistry of the Los Angeles smog. This book shows the enormous progress made in the 1950s, and after the identification of the route for formation of the hydroxyl radical in the atmosphere (Levy, 1971), the basis of our present understanding of atmospheric photochemistry was firmly established. The unique conditions that favor photochemical smog formation in Los Angeles may have prevented its early detection elsewhere, even though it was obvious that the precursors would be present in the air of many industrialized cities (Leighton, 1961). In the late 1960s measurements began to be made in the US and Europe which soon identified its presence, and in the 1970s it could be argued that photochemical smog had been observed in most major cities of the world. In Europe high ozone concentrations took on an almost continental dimension in the summers of the 1970s (Guicherit and Van

Dop, 1977). The heat wave of 1976 gave rise to an impressive episode across southern England that particularly captured the public imagination. The increasing frequency of ozone episodes started a debate about its role in forest damage. Re-examination of the Montsouris data set and comparison of this 19th century record with modern data sets further increased awareness of increasing ozone in the troposphere (Volz and Kley, 1988).

The more complex issues that photochemical smog raised meant that legislative approaches had to be broader. Hence the development of the US Clean Air Act (1970) and its subsequent amendments have been very different from the narrowly defined UK Clean Air Act (1956, 1968), which is essentially concerned with smoke control. The US legislation has been concerned with setting both health and welfare standards for a wide range of pollutants and regulating their emission. A personal review by Schulze (1993) discusses the evolution of the air pollution control regulations in the US in response to difficulties with older and wider ranges of sources. There have been great legislative advances and improvements in air quality within the US. However, many cities do not meet the air quality standard for ozone, an issue which is addressed specifically within Title I of the US Clean Air Act amendments of 1990. Within Europe the widening concerns inherent in air quality management are handled by a growing list of EC directives.

1.4. MAJOR EVENTS

The 20th century has seen a number of major air pollution incidents which have had great impact on policy. It often seems that the policy impacts have far exceeded the physical change caused by the event. The development of the media as a tool for informing and influencing the public has meant that events can trigger social pressure for cleaner air. However, levels of concern rarely hold widespread popular interest for long enough to carry over from one incident to the next (Lowenthal, 1990). The following examples illustrate some of the incidents that have generated interest in the 20th century.

1.4.1. Smelters

The late 19th and early 20th centuries saw a number of incidents related to air pollution from smelters in North America and Europe. Over many years these caused great damage to livestock and crops. The Trail incident (Dean and Swain, 1944) had special ramifications, because a smelter in one country (Canada) caused damage in another (US) and invoked transnational pollution issues. Perhaps the most important incidents were those that affected human health in the areas around large industrial centers, e.g., the Meuse Valley in Belgium in

1930 (some 60 deaths) and Donora, Pennsylvania in 1948 (some 17 deaths). Historians find that a long period of local concern and activism actually preceded the Donora incident.

1.4.2. Industrial Accidents

There have been numerous releases of air pollutants through industrial accidents (Marshall, 1988). Two with particularly important repercussions were those at Seveso and Bhopal. In 1976 an explosion at a plant in Seveso, Italy released a mix of chemicals that contained as much as 3 kg of "dioxin". The effects have fortunately been less severe than initially anticipated (Bertazzi, 1991), and there is evidence that humans may be rather resistant to dioxin. Regardless of any revisions to the toxicity of dioxin, the accident and other problems concerning dioxin have had an enduring effect on policy. This is perhaps most evident in the Seveso Directive of the European Community, which formalizes the public's right to information on nearby industrial plants.

The accident at Seveso raised particular concerns, because it was seen to involve a parent company from another country with more stringent environmental standards. This incident and the more recent one at Bhopal in India gave substance to claims of environmental imperialism and exploitation. More than 2000 people died after the release of methylisocyanide from a storage tank in Bhopal. Union Carbide claimed that this was not an accident but sabotage by a disgruntled worker. Environmentalists see this as simply an attempt by the parent company to lessen their moral and financial responsibility for the incident. Such argument over the origins of the disaster and the lengthy court proceedings have little relevance to the uncompensated citizens who suffered so much. Bhopal will inevitably be seen as a turning point for environmental attitudes in the developing world (Lowenthal, 1990).

1.4.3. London Smog

The London smog of December 1952 was actually one of many that had brought excess deaths to the city since the late 19th century. Indeed there is evidence of increases in the death rate during fogs back in the 17th century (Brimblecombe, 1987). In 1952 there were probably as many as 4000 excess deaths, and public and political reaction was strong enough to result in an inquiry that led to the Clean Air Act of 1956 (Brimblecombe, 1987). The smog and its subsequent legislation are often viewed by many as a turning point in dealing with the problems of urban pollution. However, the effectiveness of the Act has often been questioned. Some have felt that the Act simply reinforced shifts to less smoky fuels that were already under way, while others were critical of its inability to cover anything but smoke emissions.

1.4.4. Chernobyl

The nuclear reactor accident at Chernobyl on April 26, 1986 is regarded as the worst civil accident ever known, even by groups sympathetic to energy generation (Lewins, 1991). The fire in the nuclear reactor allowed radioactive core material to be released. Shortly after the accident more than a hundred people present at the site were hospitalized with radiation sickness and many subsequently died. The longer-term consequences of the dispersed radiation were the cause of great public anxiety. Some 90 PBq ^{137}Cs was released in the accident (compared with 1500 PBq from decades of weapons testing). The radiation was distributed over many parts of Europe and was ultimately detected around the northern hemisphere.

Official and scientific studies have been numerous, with thousands of papers available. In the area of most general concern, the health effects, official reviews have generally taken the position that these have not been as severe as expected. This view has been dismissed as far too optimistic by Byelorussian and Ukrainian representatives and as "whitewash" by environmental groups. Public dissatisfaction with information supplied and the suppression of opposition to official views have bred beliefs in a cover-up. The accident has transcended national differences and created a near global public disenchantment with nuclear power and a growing realization of the potentially hemispheric impact of accidents.

1.4.5. Kuwait Oil Fires

As Allied forces grouped to end the Iraqi occupation of Kuwait in late 1990, King Hussein of Jordan suggested that an environmental disaster was imminent. Iraqi troops had mined the oil wells, and it was argued that the resulting fires would bring about an environmental disaster. Comparisons with the "nuclear winter" scenario were widely reported, although many scientists felt these visions exaggerated. The rapidity of events and active media interest have left a confused chronology. However, the issue was much discussed, with both political and scientific disagreements influencing predicted outcomes.

It has been generally accepted in the scientific literature that the global impacts of the oil fires were much exaggerated by the earliest media reports (Small, 1991; Marshall, 1991), although some have seen little reason to modify their views on the magnitude of the catastrophe (Vidal, 1991). As predictions of a global apocalypse faded, the specter of local pollution grew. However, early predictions of "people dropping like flies" failed to receive much support from the ground-based observations or the measurements of trace gases made by government and environmental groups (Hunt, 1992).

Some argued that total particulate levels were high enough to affect respiratory health in the region (Okita et al., 1994); however, given the magnitude of the emissions, ground-level pollutant gas concentrations were often surprisingly low. Concern shifted to longer-term effects on pregnancy outcome and cancer incidence, but it is not likely that there will be sufficient data to establish these with any certainty.

Accidental pollution of the atmosphere is no longer restricted to local contamination. A global perspective became publicly apparent with Chernobyl and nuclear-winter-type predictions. Large-scale accidents have been important in forging public opinion. The media are generally successful in disseminating predictions of catastrophic impacts. However, these are often revised downward after the events, which more often evokes distrust than a sense of relief, causing widespread cynicism about official studies. Human error and bad industrial relationships remain a worrying aspect to many of these environmental accidents.

1.5. ACID RAIN

Acid rain undoubtedly became a key environmental issue in the 1980s and to some extent replaced public and media interest in issues such as nuclear power and lead in gasoline. Although acid rain achieved considerable international prominence at the UN Conference on the Human Environment in Stockholm in 1972, a high level of public concern really became more apparent in the 1980s.

A number of papers have treated a history far longer than this (e.g., Cowling, 1982; Brimblecombe, 1992; Camuffo, 1992). The earliest observations of rainfall composition relate to "blood rain", rain containing reddish dust, often used in prognostication (Ehrenberg, 1847; Britton, 1937; Loye-Pilot et al., 1986). Classical writers (Pliny's *Natural History XXXI.29*) comment on damage from saline rains, as do observers of the early modern period (Derham, 1704). In the late 16th century there were complaints that smoke from extensive agricultural burning in England damaged the vine crops in France (Evelyn, 1661). Camuffo (1992) gives extensive reference to early accounts of acid rain damage from volcanic eruptions in the Mediterranean area. In the north of England the terms "moorgrime" and "moorgroime" arose in the dialects of Yorkshire and Lancashire to describe soot transported long distances into upland areas, where it accumulated in the fleeces of sheep.

Early writers were certainly aware of the long-range transport of pollutants (Brimblecombe and Wigley, 1978). By the 19th century, descriptions of contaminated rain and snow were widespread. The acid rain issue features in the Norwegian playwright Ibsen's play *Brand* (1865) and other more factual writings (e.g., Brogger, 1881). In Scotland of the 1860s the Reverend James Rust wrote a book on black rain (Rust, 1864). Air trajectories for the black rain incidents recorded by Rust indicate that the air probably passed over the industrial areas of England. The analytical determination of chloride in rain, as distinct from simple descriptions was a significant development. There were some 17th century determinations (Brimblecombe, 1992), but perhaps the first notable use of such analysis was John Dalton's (1766–1844) observation that inland chloride concentrations were higher when the wind blew more directly from the sea. The detection of "acid rain" tends to be attributed to Angus Smith in the 1870s (Smith, 1872), but it needs to be remembered that Ducros described "pluie acide" in 1845. Some lakes and tarns in southern Norway began to lose their fish populations by 1890 owing to acidification, although in Britain much of the declining fish stock is blamed on the discharge of effluents directly into rivers. In Norway in the 1920s there were measurements of surface water pH, its relationship to trout production, and even some experiments in adding lime to control acidity (Brimblecombe, 1992).

Sporadic attempts to monitor rainfall composition were made in mid-19th century Britain, but the only major program to make a lasting contribution was begun at the agricultural experimental station at Rothamsted, where a large rain gauge gave many years' data on chloride, ammonia, and nitrate in rainfall. Sulfate analyses are more limited, although the few that are available are interesting to compare with modern determinations. Between 1881 and 1887 some 7.8 kg of sulfur was deposited on each hectare at Rothamsted. In the measurements made between 1955 and 1966 this had increased to $12.2 \, kg \, ha^{-1}$, probably owing to the increased amounts of sulfur present in the air over the British Isles. By the end of the century there were agricultural stations all round the world undertaking rainfall analyses, but interest waned until after the Second World War, when there was a reawakening of interest in the contribution that rain made to agricultural nutrients, particularly in Sweden (Eriksson, 1952a, b). In England Gorham (1958) began his studies of precipitation input to lakes and found that the acids introduced into precipitation by industrial processes caused a progressive loss of alkalinity in surface waters and a rise in the free acidity of soils. In North America the atmospheric chemist Junge (1963), made a series of studies of the composition of precipitation.

Concern about the "acid rain" problem in its modern sense was initiated by Oden (1968), who argued that the long-term increases in the deposit of acidity could have profound ecological consequences. His claims initiated many major studies of the problem. Work suggested that acid rain changed the pH of surface waters, led to a decline in fish stocks, leached toxic minerals from soils, decreased forest growth, and accelerated damage to materials. Norway, for instance, showed its concern in 1972 by establishing the SNSF Project to examine the effects of acid precipitation on forests and fish.

1.6. RECOGNITION OF CHANGE IN CLIMATE AND ATMOSPHERIC COMPOSITION

The variability of climate has been the subject of much discussion. Aristotle provided evidence in *Meteorologica* that the whole of Egypt had become continually drier. As for ancient Greece, he argued that the Argive was marshy at the time of the Trojan War, supporting only a small population, with Mycenae in excellent condition. By Aristotle's time Mycenae had become dry and the Argive fertile. Alarm at sudden climate change in Europe at the end of the 18th century seems to have been a catalyst for the development of early observational networks in France in 1775 and in Prussia in 1817 (Lamb, 1965). In Britain there was a particular interest in climate change over the later part of the 18th century driven by an interest in its effects on agriculture (Williams, 1806). By the mid-19th century Koppen (1873) attempted to use the instrumental data gathered by H.W. Dove to assess likely global temperature variations since 1750.

Despite observations within classical literature and the concerns of some observers about long-term change, they tended not to be the dominant view as meteorology developed into a science. Lamb has argued that climatologists of the early part of the present century were convinced that climate was essentially static and could be well described by long records averaged to produce norms for the various parameters. This view gained support through the influential textbook of Hahn and the studies on lengthy climate records by Mossmann (Lamb, 1965). However, the increasing temperatures observed between 1900 and 1940 were so striking that in the end climate change could no longer be ignored or denied.

In his textbook of 1972 Lamb lists some observed causes of climatic change: variations in sea ice, in ocean temperature, volcanic activity and other types of dust and pollution, and solar disturbance, in addition to orbital variations of the Earth for longer-term changes. This perspective seems rather different from the increasing focus on greenhouse gases that has come to dominate intellectual study over the last decades of the century. It is also interesting to note that the 1960s and 1970s were not entirely dominated by thoughts that global temperatures would rise. There were parallel concerns that we might be on the brink of an ice age, as ice coverage in the North Atlantic was the most extensive for 60 years. Thus it was easy to wonder whether we faced an "ice age" or "heat death".

The idea that atmospheric carbon dioxide concentrations could affect temperature was suggested by Tyndall in 1861. At the end of the century Arrhenius (1896) was able to calculate that a doubling of CO_2 would give rise to a 5 K increase in temperature. Callendar (1940) gathered together the most reliable measurements of carbon dioxide he could find and concluded that the modern measurements showed an increase of 30 ppm over those taken in the 19th century. This was insufficient to account for the amount released, so he suggested that absorption by sea-water played an important role in controlling atmospheric CO_2. He rather underplayed the importance of increasing concentrations by arguing that "There is no danger that the amount of CO_2 in the air will become uncomfortably large, because as soon as excess pressure in the air becomes appreciable, say about 0.0003 atmos., the sea will be able to absorb the gas as fast as it is likely to be produced." In a later article he addressed the issue of climate change more directly. Questions about the statistical certainty of a CO_2 increase in the data record were still being raised in the 1950s. Accurate measurements of CO_2 from a number of sites throughout the 1960s removed doubts about the secular increase. The whole story was already well documented in the more popular literature (e.g., Cook, 1955), so it had the potential for a wide early acceptance by the scientifically interested lay public. Since this time, increases in a range of gases have been detected, so the impact of human activities on the global concentrations of trace gases has now become widely accepted.

1.7. OZONE DEPLETION

Detection and preservation of the Earth's ozone layer has engaged scientists intensively in the 20th century. Chapman was able to explain the origin of the Earth's ozone layer in the 1930s by proposing a simple reaction scheme for the photochemical production of ozone from oxygen. This work, despite its incredible degree of completeness (Chamberlain and Hunten, 1987), was ultimately discovered to be insufficient to explain the profile of ozone in the upper atmosphere. Bates and Nicolet (1950) had suggested that catalytic removal of O_3 was possible through the interaction of water vapor. In the 1970s interest in stratospheric transport and the role of nitric oxide in ozone depletion came under intense scrutiny. The decision not to proceed with a fleet of commercial supersonic aircraft was no doubt influenced by these environmental considerations (Johnston, 1971).

Soon after the supersonic transport debate, public concern was again aroused about the potentially damaging effect of chlorofluorocarbon (CFC) release on the ozone layer (Molina and Rowland, 1974). These chemicals were thought to be completely harmless and were routinely released into the environment. Interest in this issue declined somewhat as models became more complex, and any changes in stratospheric ozone concentrations were both difficult to predict and even more difficult to detect. The discovery of the ozone hole over the Antarctic by Farman et al. (1985) reawakened interest. This was a dramatic and unexpected development that had been overlooked by observers and modelers alike. In the late 1980s international agreements (Montreal Protocol) were hatched to phase out the production and release of CFCs over a period of about 10 years. The phase-out of CFC release proposed under the Montreal Protocol has led to the development of new replacement compounds that are expected to be more benign to the environment (Manzer, 1990).

1.8. CONCLUSIONS

Parallel with increased understanding of the atmosphere there have been changes in the human perception of the most tenuous of the global reservoirs. Air, in a primitive sense, has been associated with breath and life. Miasmatic theories have invariably been a common part of disease etiology from the earliest of times and may have led to the practical health-related element of the early response to air pollution that in many ways continues up to the present (Brimblecombe and Nicholas, 1993). While it is possible to identify ways in which religions oppose the pollution of our environment, a number of writers have laid particular blame for the environmental crisis on Judeo-Christian beliefs (e.g., White, 1967). Parallel with this there has been an enthusiasm for the cultures of people who are perceived to live close to nature.

Various developments of the late 18th and 19th centuries seem important to add to the background of our environmental thinking. The re-evaluation of nature with the rise of the Romantic movement in Europe and to a lesser extent North America is perhaps the most obvious (e.g., Clayre, 1977). An increasing penetration of aspects of government into parts of our lives, hitherto matters of personal choice, caused great changes. There were expressions of outrage, largely moralistic, at the increasingly poor sanitary conditions within Victorian cities. These outbursts spread to criticisms of life styles of native peoples living in colonial settings, with a desire for the imposition of European sanitary regulation abroad. Sanitary regulation unfortunately often placed much of its burden on the poor, while the benefits were often restricted to the rich, though that is not to say that individuals working within the growing legislative framework did not have a genuine concern to increase urban health as a whole.

The administrative changes in 19th century environmental regulation were profound but, one could argue, rather practical, focusing on human health in the main. Much air pollution legislation continues to focus on health, but there are extensions into the realm of protection of vegetation and amenity (e.g., prevention of significant deterioration as found in the 1977 amendments to the US Clean Air Act; Schulze, 1993).

In contrast with officials, there were some early advocates of more radical views of our relationship with the environment. "Back to nature" movements were fairly common, as were outspoken critics of the contemporary way of life (e.g., Walter, 1990). In the current century ecological writers such as Aldo Leopold (Flader, 1974) and Rachel Carson (1963) have particularly stimulated debate. The 1970s saw numerous publications with apocalyptic visions of the future (e.g., Ehrlich, 1970; Meadows et al., 1972; Goldsmith et al., 1972). Contemporary books of gloom about the future do not receive quite the breadth of readership that they did in the 1970s. In the 1990s environmental interest appears more focused on specific issues, with the public both aware of and concerned about the problems of species diversity, AIDS, acid rain, the greenhouse effect, and the ozone hole.

An influential synthesis of environmental ideas to emerge from atmospheric science has been the Gaia hypothesis (the suggestion that the entire Earth is an entity which maintains homeostasis) of James Lovelock (1979) and its geophysiological notions (Charlson et al., 1987). His ideas have been variously interpreted by environmental writers, but have probably had more impact on popular environmental thought than on the sciences. I have always found these ideas of more use as a metaphor than a testable hypothesis, and others have argued likewise (e.g., Kirchner, 1989).

Twentieth century atmospheric science has done much to convince the world of the global nature of human effects on the environment in a graphic way (e.g., as satellite images of the Antarctic ozone hole or the dramatic oil fires in Kuwait) that can influence both political and public perspectives. This input has been seen at the major conferences: Stockholm, Montreal, London, and Rio. While there is much complaint about the weakness of the international protocols that emerge from these conferences, there is no indication that the problems themselves are not viewed as global. The awareness of human impact on the environment is certainly broad, and the conquest of nature is now viewed as deplorable by all sides in the environmental debate. However, the development of sensitive environmental policy is not necessarily a simple function of increasing environmental awareness (e.g., see the balanced review of Lowenthal, 1990). Despite the wide sympathy for "green" ideas, the links between idealism and action remain difficult to weld.

References

Anderson, M.R. (1992). The English town in a colonial context. *Environ. History Newslett.* **4**, 48–49.

Andrieux, M. (1962). *La Vie Quotidienne dans la Rome Pontificale*, Hatchette, Paris.

Arrhenius, S. (1896). On the influence of carbonic acid in the air upon the temperature near the ground. *Philos. Mag.*, **41**, 237–276.

Bates, D.R. and Nicolet, M. (1950). The photochemistry of atmospheric water vapour. *J. Geophys. Res.*, **55**, 301.

Bertazzi, P.A. (1991). Long-term effects of chemical disasters: lessons and results from Seveso. *Sci. Total Environ.*, **106**, 5–20.

Brimblecombe, P. (1977). London air pollution, 1500–1900. *Atmos. Environ.*, **11**, 1157–1162.

Brimblecombe, P. (1987). *The Big Smoke*, Methuen, London.

Brimblecombe, P. (1992). History of atmospheric acidity. In Radojevic, M. and Harrison, R.M., Eds, *Atmospheric Acidity*, Elsevier Applied Science, London, pp. 267–304.

Brimblecombe, P. and Bowler, C, (1992). The history of air pollution in York, England. *J. Air Waste Manag. Assoc.*, **42**, 1562–1566.

Brimblecombe, P. and Nicholas, F.M. (1993). History and ethics of clean air. In Berry, R.J., Ed, *Ethical Dilemmas*, Chapman and Hall, London, pp. 72–85.

Brimblecombe, P. and Wigley, T.M.L. (1978). Early observations of London's urban plume. *Weather*, **33**, 215–220.

Britton, C.E. (1937). *A meteorological chronology to AD 1450*. Geophysical Memoirs No. 70.

Brogger, W.C. (1881). Mindre meddelelser. *Naturen*, **47**.

Bruggemeier, F.-J. (1990). The Ruhr Basin 1850–1980: a case of large scale environmental pollution. In Brimblecombe, P. and Pfister, C., Eds, *The Silent Countdown*, Springer Verlag, Heidelberg, pp. 210–227.

Callendar, G.S. (1940). Variations in the amount of carbon dioxide in different air currents. *Q. J. R. Meteorol. Soc.*, **66**, 395–400.

Camuffo, D. (1992). Acid rain and the deterioration of monuments: how old is the problem? *Atmos. Environ.*, **26B**, 241–247.

Camuffo, D. (1993). Reconstructing the climate and the air pollution of Rome during the life of Trajan Column. *Sci. Total Environ.*, **128**, 205–226.

Carson, R. (1963). *Silent Spring*, Hamish Hamilton, London.

Chamberlain, J.W. and Hunten, D.M. (1987). *Theory of Planetary Atmospheres*, Academic Press, New York, NY.

Charlson, R.J., Lovelock, J.E., Andreae, M.O. and Warren, S.G. (1987). Oceanic phytoplankton, atmospheric sulfur, cloud albedo and climate. *Nature*, **326**, 655–661.

Clayre, A. (1977). *Nature and Industrialization*, Oxford University Press, Oxford.

Cook, J.G. (1955). *Our Astonishing Atmosphere*, Scientific Book Club, London.

Cowling, E.B. (1982). Acid precipitation in historical perspective. *Environ. Sci. Technol.*, **16**, 110A–121A.

Davidson, C.I. (1979). Air pollution in Pittsburgh: a historical perspective. *J. Air Poll. Control Assoc.*, **29**, 1035–41.

Dean, R.S. and Swain, R.E. (1944). *US Bureau of Mines Report No. 453*.

Derham, J. (1704). Observations of the late great storm. *Philos. Trans. R. Soc. Lond.*, **abV**, 60.

Ducros, M. (1845). Observation d'une pluie acide. *J. Pharm. Chim.*, **3**, 273–77.

Ehrenberg, C.G. (1847). Passat-Staub und Blut-Regen. *Phys. Abhand. K. Akad. Wiss. Berlin*.

Ehrlich, P. (1970). *The Population Bomb*, Ballantine Books, New York, NY.

Eriksson, E. (1952a). Composition of atmospheric precipitation I Nitrogen compounds. *Tellus*, **4**, 215–232.

Eriksson, E. (1952b). Composition of atmospheric precipitation II Sulphur chloride iodide compounds. *Tellus*, **4**, 280–303.

Evelyn, J. (1661). *Fumifugium, or The Inconvenience of the Aer and Smoak of London Dissipated. . .*, Gabriel Bedel and Thomas Collins, London.

Farman, J.C., Gardener, B.G. and Shanklin J.D. (1985). Large losses of total ozone in Antarctica reveal seasonal ClO_x/NO_x interaction. *Nature*, **315**, 207.

Flader, S.L. (1974). *Thinking Like a Mountain*, University of Nebraska Press, Lincoln, NE.

Goldsmith, E., Allen, R., Allaby, M., Davoll J. and Lawrence, S., Eds (1972). A blueprint for survival, *The Ecologist*, **2(1)**.

Gorham, E. (1958). The influence and importance of daily weather conditions in the supply of chloride, sulphate and other ions to fresh waters from precipitation. *Philos. Trans. R. Soc. B.*, **241**, 147–178.

Guicherit, R. and Van Dop, H. (1977). Photochemical production of ozone in Western Europe (1971–1975) and its relation to meteorology, *Atmos. Environ.*, **11**, 145–156.

Hunt, W.F. (1992). The impact of the Kuwait oil fires – an overview. *Air and Waste Management Association 85th Ann. Meet.*, Preprint 92–76.06.

Johnston, H.S. (1971). Reduction of stratospheric ozone by nitrogen oxide catalysis from supersonic transport exaust. *Science*, **173**, 517–522.

Junge, C.E. (1963). *Air Chemistry and Radioactivity*, Academic Press, New York, NY.

Kirchner, J.W. (1989). The Gaia hypothesis: can it be tested? *Rev. Geophys.*, **27**, 223–235.

Koppen, W. (1873). Uber mehrjahriger Perioden der Witterung, insbesondere über 11 Jährige Periode der Temperatur. *Z. Osterreich. Ges. Meteorol.*, **8**, 241–248.

Lamb, H.H. (1965). Britain's changing climate. In Johnson, C.G. and Smith L.P., Eds, *The Biological Significance of Climate Change in Britain*, Academic Press, New York, NY, pp. 3–34.

Lamb, H.H. (1972). *Climate Past Present and Future*, Vol. 1, Methuen, London.

Leighton, P.A. (1961). *Photochemistry of Air Pollution*. Academic Press, New York, NY.

Levy, H. (1971). Normal atmosphere: large radical and formaldehyde concentrations predicted. *Science*, **173**, 141–143.

Lewins J. (1991). *Five Years After Chernobyl: 1986–1991*, Watt Committee on Energy, Savoy Hill, London.

Lichtheim, M. (1980). *Ancient Egyptian Literature*, Vol. III, *The Late Period*, University of California Press, Berkeley, CA.

Lovelock, J.E. (1979). *Gaia: A New Look at Life on Earth*, Oxford University Press, Oxford.

Lowenthal, D. (1990). Awareness of human impacts: changing attitudes and emphases. In Turner, B.L., *The Earth as Transformed by Human Action*, Cambridge University Press, Cambridge, pp. 121–135.

Loye-Pilot, M.D., Martin, J.M. and Morelli, J. (1986). Influence of Saharan dust on the rain acidity and atmospheric input to the Mediterranean. *Nature*, **321**, 427–428.

Manzer, L.E. (1990). The CFC–ozone issue: progress on the development of alternatives to CFCs. *Science*, **249**, 31–35.

Marshall, E. (1991). Nuclear winter from Gulf War discounted. *Science*, **251**, 372.

Marshall, V.C. (1988). A perspective of individual disasters. *Atom*, **376**, 2–7.

Meadows, D.H., Meadows, D.L., Randers, J. and Behrens III, W.W. (1972). *The Limits to Growth*, Signet, New York, NY.

Molina, M.J. and Rowland, F.S. (1974). Stratospheric sink for chlorofluoromethanes: chlorine atom catalyzed destruction of ozone. *Nature*, **249**, 810–812.

Oden, S. (1968). *The Acidification of the Atmosphere and Precipitation and Its*

Consequences in the Natural Environment, SNSR, Stockholm.

Okita, T., Yanagihara, M., Yoshida, K., Iwata M., Tanabe, K. and Hara, H. (1994). Measurements of air pollution associated with oil fires in Kuwait by a Japanese research team. *Atmos. Environ.* **28**, 2255–2259.

Rossotti, H. (1993). *Fire*, Oxford University Press, Oxford.

Rust, Rev. J. (1864). *The Black Scottish Rain Showers and Pummice Shoals of the Years 1862 and 1863*, Blackwood, Edinburgh.

Schulze, R.H. (1993). The 20-year history of environmental air pollution control legislation in the USA. *Atmos. Environ.*, **27B**, 15–22.

Sherwood, S.I. and Bambaru, D., Eds. (1991). Final touch. *J. Preserv. Technol.*, **23**, 72.

Small, R.D. (1991). Environmental impact of fires in Kuwait. *Nature*, **350**, 11–12.

Smith, A.R. (1872). *Air and Rain*, London.

Smith, K.R. (1987). *Biofuels, Air Pollution and Health*, Plenum Press New York, NY.

Stephens, E.R. (1987). Smog studies of the 1950s. *EOS*, **68** 89–93.

Tyndall, J. (1861). On the absorption and radiation of heat from gases and vapours, and on the physical connection of radiation, absorption and conduction. *Philos. Mag.*, **22**, 169–194, 273–285.

Vidal, J. (1991). Poisoned sand and seas, In Brittain, V., Ed., *The Gulf War Between Us*, Virago, London, pp. 133–142.

Volz, A. and Kley, D. (1988). Evaluation of the Montsuris series of ozone measurements made in the nineteenth century. *Nature*, **332**, 240–242.

Walter, F. (1990). The evolution of environmental sensitivity 1750–1950. In Brimblecombe, P. and Pfister, C., Eds. *The Silent Countdown*, Springer Verlag, Berlin, pp. 231–247.

White, L. (1967). The historical roots of our ecologic crisis. *Science*, **155**, 1203–1207.

Williams, J. (1806). *The Climate of Great Britain*, Baldwin, London.

2

Mean properties of the atmosphere

David A. Salstein

2.1. INTRODUCTION

The atmosphere can be defined as the relatively thin gaseous envelope surrounding the entire planet Earth. It possesses a number of properties related to its physical state and chemical composition, and it undergoes a variety of internal processes and external interactions that can either maintain or alter these properties. Whereas descriptions of the atmosphere's chemical properties form much of the remaining chapters of this book, the present chapter will highlight the atmosphere's physical and dynamical properties, including its temperature, pressure, and motions. Internal energy is contained in the molecular motions of the atmosphere's gases, and these define its temperature structure. In contrast, the larger-scale motions comprise the winds, the global organization of which is often referred to as the general circulation. The framework of the dynamical and thermodynamical laws, including the three principles of conservation of mass, momentum, and energy, are fundamental in describing both the internal processes of the atmosphere and its external interactions. The atmosphere is not a closed system, because it exchanges all three of these internally conservative quantities across the atmosphere's boundary below and receives input from regions outside it. Thus surface fluxes of moisture, momentum, and heat occur to and from the underlying ocean and land. The atmosphere exchanges very little mass and momentum with space, though it absorbs directly a portion of the solar radiational energy received from above.

The total dry mass of the atmosphere, calculated as an annual mean, is estimated to be 5.13×10^{18} kg (Trenberth and Guillemot, 1994). The atmosphere exists in a space that has as its lower boundary the land surface, whose irregular topography results in a varying surface geopotential, as well as the relatively equipotential ocean surface. Thus the local atmospheric surface pressure, and hence the quantity of its mass above an area, varies considerably from one geographical area to another. The extent of topographical features is such that a difference of nearly 3% exists between the mean surface pressure at sea level, near 1013 mbar, and its global mean calculated over both land and ocean, the value of which is estimated to be 984 mbar. However, locally, over the highest mountains, which reach an altitude of over 8000 m, the pressure may be as low as 300 mbar, less than one-third of that at sea level. A representation of the surface pressure of the atmosphere, smoothed to eliminate the smallest features, is given in Figure 2.1. Clear from the figure is the dependence of surface pressure on the orography. In addition, the mean surface pressure depends somewhat on the characteristics of predominant air masses, and hence on temperature as well

Surface Pressure

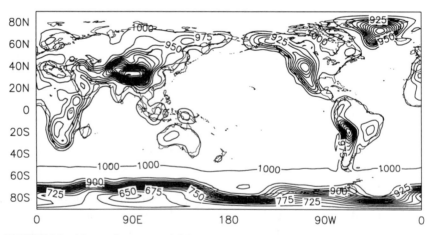

FIGURE 2.1.   Mean surface pressure of the atmosphere, taken from the US National Meteorological Center analyses. The period 1980–1985 was chosen as the period on which to base the figure. Units are millibars, where 10^2 Pa = 1 mbar.

as land surface height. For example, areas at sea level north of 50° S have pressures exceeding 1000 mbar, whereas those south of 50° S are generally under 1000 mbar.

The temperature structure of the atmosphere is such that it is often considered to be stratified in the vertical into several layers. The lowest region, the troposphere, where temperature decreases everywhere with altitude (and hence with decreasing pressure), extends to a level of about 10 km, near the 200 mbar level, although its upper boundary, termed the tropopause, varies with latitude and season. The stability of the atmosphere is a measure of the potential of air to remain static vertically at a constant pressure level and thus resist overturning. This static stability varies considerably in the troposphere, and when it is negative, convection often leads to the production of clouds. At the bottom of the troposphere is the relatively thin boundary layer, where the atmosphere interacts most strongly with the surface below. Above the troposphere is the stratosphere, a very stable region, where temperature increases with altitude. The lower

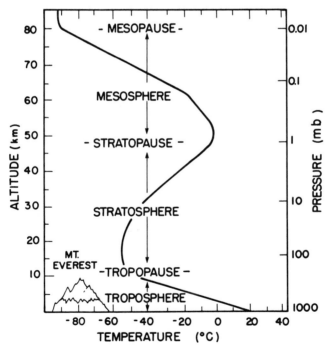

FIGURE 2.2. Schematic picture of the vertical temperature structure of the atmosphere. The commonly defined regions of the atmosphere and their upper limits are noted. The atmospheric pressures approximating the heights are given as well. (Adapted from US Standard Atmosphere, 1976.)

TABLE 2.1   Bulk Mean Mixing Ratios of Major and Minor Constituents of the Global Atmosphere

Gas	Average Mixing Ratio in parts per million (% by volume)[a]	Remarks
N_2	780,840 (78.1%)	Biological
O_2	209,460 (20.9%)	Biological
Ar	9340 (0.9%)	Inert
CO_2	350 (0.03%)	Variable, secular increase
Ne	18	Inert
He	5.2	Inert
CH_4	1.7	Biogenic and anthropogenic
H_2	0.6	Biogenic and anthropogenic
N_2O	0.31	Biogenic and anthropogenic
CO	0.03–0.2	Anthropogenic and chemical
O_3	0.01–0.15	Photochemical
Non-methane hydrocarbons	0.005–0.02	Biogenic and anthropogenic
Halocarbons	0.002	75% man-made; long-lived
Sulfur species	0.0005–0.001	Biogenic and anthropogenic
Nitrogen species (NO_y)[b]	0.0001–0.001	Anthropogenic

[a]On a dry basis. Water vapor is a highly variable physicochemical component of the atmosphere.
[b]High concentrations of O_3 (5–10 ppm) and NO_y (0.02 ppm)) are present in the stratosphere.

stratosphere is hence a minimum level of temperature, and the temperature there is lowest in the tropics. In the stratosphere, interactions with solar radiation have special importance, because molecules there, especially ozone (O_3), absorb the strong ultraviolet solar radiation, shielding the Earth's surface below from its effects. Above the stratosphere are the mesosphere (starting near 47 km, or 1 mbar) and the thermosphere (from about 85 km, or somewhat less than 0.01 mbar). These two high regions have respectively decreasing and increasing vertical lapse rates of temperature. In even higher regions, matter exists in an ionized state and becomes particularly reactive with the solar radiation. A plot of average temperature in these regions is given in Figure 2.2.

The major constituents of air are molecular nitrogen, oxygen, and, to a smaller extent, argon, comprising 99.9% by volume. The remaining portion includes carbon dioxide, whose concentration has been observed to increase significantly in historic times owing to anthropogenic causes. Trace gases include ozone, methane (CH_4), oxides of nitrogen, carbon monoxide (CO), and halocarbons. The proportions of the major dry atmospheric gases are summarized in Table 2.1. Water vapor, a particularly variable atmospheric constituent, resides almost entirely in the lower and middle troposphere; its changing structure will be examined below.

2.2. GOVERNING EQUATIONS

To put the distribution and fluctuations of the mean quantities of the atmosphere in perspective, we rely on a framework based on the conservation principles of physics. In this way, expression of physical laws regarding the atmosphere reveals which atmospheric quantities are important to monitor. Also, they form the basis for the set of equations governing the state and dynamics of the atmosphere, which, together with appropriate boundary and initial conditions, is typically utilized for weather forecasting. These physical principles include the conservation of mass, momentum, and energy.

Conservation of mass is expressible in the atmosphere as the so-called continuity equation, namely, a local change in density, $\partial\rho/\partial t$, is related to the divergence of the (three-dimensional) advection of mass into the area by the vector wind v:

$$\partial\rho/\partial t = -\nabla \cdot \rho\, v \qquad (2.1)$$

This equation leads to the condition in the general circulation that vertical and horizontal flow of the air balance over periods longer than about 1 month, because changes in mass storage on these times scales are not significant. Also, individual constituents such as water vapor, to the extent that they are chemically non-reactive, obey similar conservation laws. In this regard, external sources and sinks of water vapor in the atmosphere, by evaporation and condensation, are solely responsible for changes in the overall quantities of water vapor. Assuming no large change in liquid water storage of clouds over much time, records of precipitation can substitute for condensation.

The principle of conservation of momentum is usually formulated by the atmospheric equations of motion, equivalent to an expression of Newton's second law. In this way, momentum can change locally by advection, and by external forces. When given in the framework of the Earth's rotating coordinate system, local changes in momentum per unit mass are expressible as

$$\partial v/\partial t = -v \cdot \nabla v - 2\Omega \times v - \nabla p/\rho + g + F \qquad (2.2)$$

Here local changes in the three-dimensional velocity vector are accomplished in five ways: by advection of momentum; by the Coriolis force $-2\Omega \times v$ acting normal to the velocity vector; by a pressure gradient body force; by the force of gravity; and by a friction force which can be against the Earth's surface. In equation (2.2), v is the horizontal velocity vector, Ω is the Earth's angular velocity, p is the pressure, ρ is the density, and g and F are the acceleration due to gravity, and surface friction, respectively.

The conservation of energy is expressed as the first law of thermodynamics, in which the quantity of internal energy per unit air mass is written as

$$c_p \, dT/dt = Q + \alpha \, dp/dt \tag{2.3}$$

where c_p is the specific heat of air at constant pressure, Q is the net (diabatic) heating rate per unit mass, and α is the specific volume ($\alpha = 1/\rho$). Use of the equation of state of the atmosphere, the ideal gas law, as well as the concept of potential temperature, allows us to transform this equation to a somewhat more direct form

$$(T/\theta) d\theta/dt = Q \tag{2.4}$$

Here the potential temperature $\theta = T(p_{00}/p)^\kappa$, with $\kappa = 2/7$, is defined as the temperature that a parcel of air would attain if it were brought from its pressure p to a reference level p_{00} adiabatically. In standard meteorological practice this reference level is usually taken to be 1000 mbar.

The energy cycle of the atmosphere may be viewed as a convenient framework in which available potential energy is generated by heating processes, converted to kinetic form by means of baroclinic processes, and dissipated through frictional processes. Monitoring the large-scale aspects of the energy cycle, including measuring these various conversion terms, is a fundamental means for assessing the variability of the atmosphere, particularly on seasonal and longer time scales (Peixoto and Oort, 1974).

2.3. TIME AND SPACE SCALES OF ATMOSPHERIC PROPERTIES

To define the mean of a property in the atmosphere, we must consider both the time period and spatial extent over which to describe it. Because of the overwhelming solar forcing at the annual period, the year is often taken as the convenient unit over which to calculate mean properties. In addition to the mean value of a property over its annual period, the strength of the seasonal cycle, including the amplitudes of the annual and semi-annual components, can be important as well. Furthermore, the atmosphere has peaks of variability on much shorter scales, down, of course, to the daily period, a forcing also related to the solar heating variations. In between these time scales are variations on synoptic scales of about 1 week, related to the motion of weather systems, and intraseasonal scales of 1–2 months (Madden and Julian, 1971). However, when the strengths of atmospheric fluctuations on a whole range of scales are expressed as a kinetic energy spectrum, it is apparent that this spectrum has an especially strong, broad peak on the synoptic and shorter intraseasonal scales, though of less magnitude than that on the annual scale (Peixoto and Oort, 1992).

On much longer, interannual periods, well-known features are the quasi-biennial oscillation (QBO), a reversal of the direction of the winds in the stratosphere occurring on roughly a 26 month cycle. Also prominent, with time scales of several years, is the so-called El Niño/Southern Oscillation (ENSO) phenomenon, a fluctuation of atmospheric mass across the Pacific Ocean, revealed by surface pressure anomalies, with an accompanying signature in the temperature of the ocean surface (Philander, 1989). During an ENSO event, wind and pressure fluctuations in patterns may occur within the year prior to and the 2 years following the warm peak in Pacific sea surface temperatures. The ENSO fluctuations are large enough to affect global measures of variability, including those of the general circulation. Because of the existence of these interannual variations, a measure of the mean state of the atmosphere should cover a multiyear period, approaching at least a decade.

Spatial scales are conveniently discussed first as the zonal averages of quantities over latitudinal bands. For quantities such as temperature, the distribution with latitude zones is considerable as well. In practice, the mean state of the atmosphere may be described on spatial scales that match the scales of common atmospheric forecasting and analysis models, which currently achieve a spectral resolution of order greater than 200 around the globe, where spatial-scale equivalence approaches 100 km at the equator; variations on finer scales are handled by parameterizations and are not explicitly part of the global analysis model.

Transports of quantities, namely advection by winds, are typically divided into those accomplished by mean motions and those due to smaller scale, eddy motions. Furthermore, the eddy terms have typically been divided into their transient and stationary components as follows. The transport by the wind, in, say, the meridional direction, of a quantity A over a period given by $\overline{vA}$ (where the overbar denotes the time mean) can be decomposed into $\overline{vA} = \overline{v}\overline{A} + \overline{v'A'}$ (where primes indicate departures from the time mean). Likewise, the zonal average of the transport of a mean quantity $[\overline{vA}]$ (where brackets denote the zonal mean) is represented as $[\overline{vA}] = [\overline{v}][\overline{A}] \times [v^*A^*]$ (where asterisks indicate departures from the zonal mean). These eddy terms, representing transient and stationary waves, respectively, may be thought of as independent means to transport meridionally the quantity A on a range of time and space scales.

2.4. MEAN METEOROLOGICAL QUANTITIES

As we have seen, fundamental meteorological variables in the atmosphere that appear in descriptions of the governing laws include the pressure, temperature, and wind. Also important are the moisture content of the atmosphere and the proportions of other trace variables. This section will highlight the global distributions of many of these quantities. Also, the heating terms that drive the circulation are particularly important to understand and will be treated in the next section.

For forecast and analysis of weather features, specialized coordinate systems, particularly in the vertical, have been introduced into meteorology. Because the pressure decreases monotonically with height, observations are taken at given pressures, and because certain physical laws, such as the equation of continuity, are conveniently expressed in that system, analyses of many meteorological variables have often been given in pressure coordinates. The lowest such level for the vertical coordinate is usually taken at 1000 mbar. Additionally, though, forecast models of the atmosphere are often performed in other coordinate systems, such as a non-dimensional, terrain-following one. Other atmospheric descriptions of heating rates and studies of dynamics, including so-called diabatic circulations directly influenced by atmospheric heating, have been given in a vertical system based on potential temperature θ, defined above (Johnson, 1989); the θ coordinate can be used because it generally increases with increasing height monotonically. In this chapter, though, we will be confined to descriptions in pressure coordinates.

2.4.1 The Climate Diagnostics Database

Only a few data sets are available on which to base a global description of the mean atmospheric state. One such type is a special analysis of the atmosphere based on the large rawinsonde network comprised of around 1000 stations (Oort, 1983). A second type is derived from atmospheric analyses produced by the daily operational weather forecasting and data assimilation systems of the world's major weather centers. Though differences between operational analyses (Trenberth and Olson, 1988) and between rawinsonde and station analyses (Rosen et al., 1985) have been noted, we have chosen an archived set based on US National Meteorological Center (NMC) analyses (e.g., Kanamitsu, 1989). The vertical extent of these analyses encompasses the troposphere and lower stratosphere, and these regions are thus highlighted in our treatment of the mean quantities. This Climate Diagnostics Data Base (CDDB) has been organized into a set of monthly mean values of parameters and certain cross-products collected on a three-dimensional grid with $2.5° \times 2.5°$ resolution in latitude and longitude and on nine pressure levels in the vertical. From the monthly values, longer averages of one or many years can be formed and taken to represent the mean conditions of the atmosphere, and its variability can be represented as well (Rasmusson and Arkin, 1985).

Here we select one decade, 1980–1989, a period with reasonably reliable reporting, over which to describe atmospheric mean conditions, though it is not as long as the 30 year period sometimes taken for climatological purposes. Although some changes that have taken place in the NMC operational system during the decade may produce inhomogeneities in the records, the use of the CDDB here for long-term mean fields is quite reasonable. Ultimately this inhomogeneity, more serious for studies of interannual fluctuations and trends, will be eliminated with the adoption of the Climate Data Assimilation System/ Reanalysis System (Kalnay and Jenne, 1991). Such a system will produce analyses free from the heterogeneities associated with fluctuations in operational weather analysis procedures.

Besides the decade-long means for the annual period, we will display measures of both seasonal and interannual variability. The seasonal signals, which are often large compared with the long-term mean, are shown by the average of each calendar month for the 10 year period. Conversely, the interannual signal will constitute the mean of all 12 months within each calendar year; because it is typically small, it is given in anomaly form, i.e., the departure from the decade-long signal.

The CDDB fields are given for those levels of the atmosphere whose pressure is less than the mean surface pressure, i.e., above the physical topography. Those areas not plotted are blackened in the figures.

2.4.2 Temperature Structure

Because temperature is related to the heat balance, and hence energetics, of the atmosphere, and is important as well for reaction rates, it is a key parameter to diagnose. Of course, temperature varies with horizontal position as well as elevation. Aspects of its long-term mean, intraseasonal, and interannual variability are shown in Figures 2.3–2.6. In the first of these (Figure 2.3) the mean temperature averaged around a latitude band is given at the nine pressure levels, from 50 to 1000 mbar. Highest values of this zonal mean quantity, of course, occur in the tropics, having a tendency to peak near 10° N rather than at the equator; at the 1000 mbar level the temperature exceeds 295 K. The annual mean temperature structure is largely symmetric about the equator. Temperature generally decreases poleward, with differences between tropics and the highest latitude at the surface approaching 35 °C, and also upward, until near-stratospheric levels, where the meridional temperature gradient starts to reverse. In the tropical stratosphere, at 100 mbar, a minimum is found in the annual mean temperature values, reaching below 200 K; at this stratospheric level, temperature increases poleward by about 20 °C in both hemispheres.

A map of temperature at the lower tropospheric, 850 mbar level (Figure 2.4) expands the distribution of features into the longitudal domain. The 850 mbar level was chosen because it is the first standard meteorological level above much of the topography and is a level at which a considerable quantity of heat energy

Zonal Mean Temperature

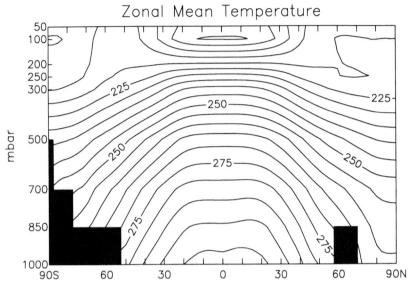

FIGURE 2.3.   Zonal mean temperature for 1980–1989 between the 1000 and 50 mbar levels. Units are kelvins.

Temperature 850 mbar

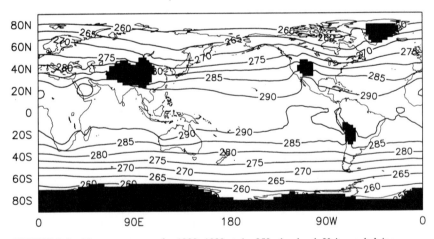

FIGURE 2.4.   Mean temperature for 1980–1989 at the 850 mbar level. Units are kelvins.

is located as well as transported. It is clear that there is much zonal symmetry to the temperature structure of the atmosphere, particularly in the middle latitudes of the southern hemisphere. However, one can note asymmetries too, such as the troughs of lower temperatures dipping southward across North America and Eurasia (265–270 K contours), reflecting the colder, and hence more continental climates in the eastern parts of these land masses. Similar features can be seen over the southern hemisphere continents too, but these are present to a much lesser extent.

The intra-annual variability of the temperature structure at the 850 mbar level is summarized by the mean calendar monthly distribution of zonal mean values over the 10 year period in Figure 2.5. Here a notable difference between the

Temperature 850 mbar

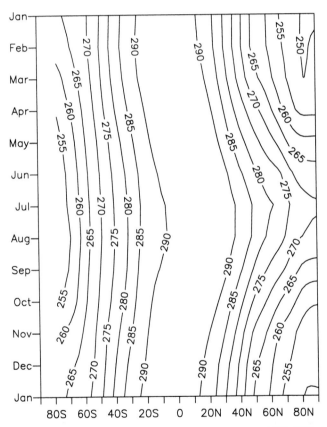

FIGURE 2.5.   Mean calendar monthly zonal mean temperature over 1980–1989 at the 850 mbar level, showing intra-annual variability. Units are kelvins.

hemispheres is evident. Whereas a strong annual cycle exists in the northern hemisphere, especially at middle latitudes, a much weaker one occurs in corresponding southern latitudes. For example, the difference between zonal mean temperatures in the extreme months of January and July is 20 °C near 40° N but only 7 °C near 40° S. Such a difference is due to the contrasting land–ocean distributions of the hemispheres. As was evident in Figure 2.4, the highest temperatures occur near 10° N and are highest for the year during July and August; corresponding temperatures during January and February are not quite as warm near 10° S.

The anomalies from the mean temperature at each latitude over the 10 year period are reasonably small (figure not shown). The zonal mean temperatures do not vary in any year and at any latitude by more than 2 °C, fluctuating weakly during the period. However, the period around 1986–87 is colder than the rest of the decade, particularly in the middle latitudes of the northern hemisphere, but it is warmer again at the end of the decade 1988–1989.

2.4.3 Wind Structure

The zonal mean zonal wind for the 10 year period, given in Figure 2.6, shows a relatively symmetric structure in both hemispheres, with westerly winds

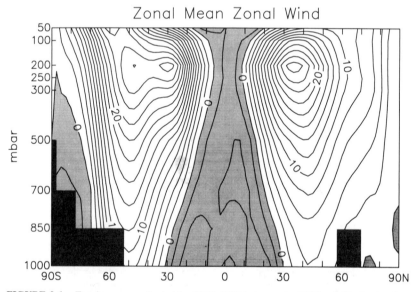

FIGURE 2.6. Zonal mean zonal wind for 1980–1989 between the 1000 and 50 mbar levels. Units are m s^{-1}. Positive values indicate westerly winds, negative values easterly winds; winds are shaded.

predominating throughout much of the globe. However, easterly winds do occur in the lowest latitudes, especially in the lower levels. Indeed, at the surface, easterly winds are observed between 30° N and S, spanning half the area of the globe. Weak easterlies also occur to some extent in the lower altitudes of the polar regions.

Typically the winds become more westerly as they increase with height up to around 200 mb. In this way, in the lower latitudes, winds change from an easterly to a westerly direction and in the middle latitudes, strong westerly jets occur. The major jets have their strongest mean near the 200 mbar level. The annual mean values of the jet maxima reach 27 and 29 m s^{-1} in the northern and southern hemispheres, respectively. This overall increase in winds with altitude is connected by the geostrophic relationship to the meridional temperature gradient. The somewhat stronger winds in the southern hemisphere are consistent with the larger such gradient in that hemisphere.

The mean zonal winds at the 200 mbar level during the decade are given in Figure 2.7. This figure shows the general restriction of the easterlies at this level to the deep tropics in a band with one portion extending from western Africa to the western Pacific, maximizing across the "maritime continent" of Indonesia. Also, there is a smaller portion of easterlies extending westward from South America. The central core of the easterly winds, with values exceeding 10 m s^{-1}, is over Indonesia, a region of intense heating and convection. In general, the wind becomes less easterly and switches to westerly (more positive) the more poleward one goes in latitude. Large values occur off the storm track regions to

Zonal Wind 200 mbar

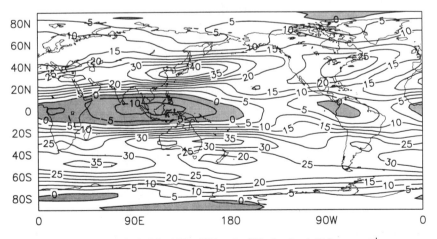

FIGURE 2.7   Mean zonal wind for 1980–1989 at the 200 mbar level. Units are m s^{-1}.

the east of the major northern continents, especially off Japan, where the wind magnitude exceeds 40 m s^{-1}, and off North America. Of particular prominence also are the zonal winds in a band across the middle southern latitudes, maximizing over the southern Indian Ocean.

The seasonal variations in the zonal winds are considerable, as is seen in Figure 2.8, which gives their progression during the calendar year at 200 mbar. Mean easterlies in the tropical band are strongest during July and August, exceeding 10 m s^{-1}; these disappear by December. A strong meridional gradient of wind occurs between tropics and subtropics, especially during the winter months of the respective hemisphere. The amplitude of the annual signal in mean

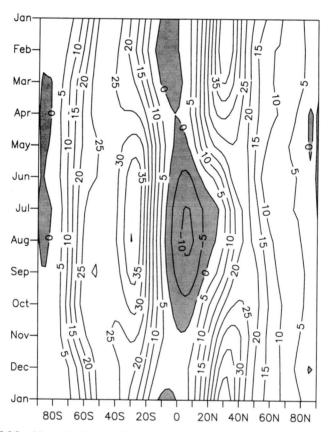

Zonal Wind 200 mbar

FIGURE 2.8. Mean calendar monthly zonal mean zonal wind over 1980–1989 at the 200 mbar level, showing intra-annual variability. Units are m s^{-1}.

zonal wind, strongest in the subtropics, is exceeded in the southern hemisphere by its value in the northern one. Indeed, for this reason, the seasonal signature of the global mean angular momentum of the atmosphere is that of the northern hemisphere (Rosen and Salstein, 1983). The strongest mean value of the zonal jet, at 30° N, exceeds $40 \, \mathrm{m \, s^{-1}}$ in January. Interannual variability during the decade is demonstrated by the anomalies in Figure 2.9, which are strongest in the northern subtropics. Of note are the maxima in 1983 and 1987, years during which major ENSO events occurred; in these two years, winds were particularly strong across the subtropics, especially over the Pacific, consistent with analyses of ENSO behavior.

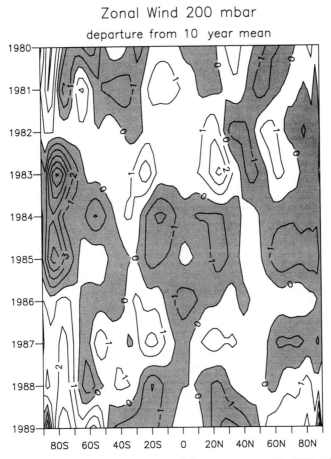

FIGURE 2.9. Mean annual zonal mean zonal wind anomalies over 1980–1989 at the 200 mbar level, showing interannual variaibility. Units are $\mathrm{m \, s^{-1}}$.

The mean meridional wind structure and the mean vertical motions are tied together over long periods through the conservation of mass relation, as noted in Section 2.2. From equation (2.1) a picture of the mean meridional circulation can be formulated in terms of a mass streamfunction, given for the 10 year period, in Figure 2.10. In the figure, the mean flow follows the streamfunction isolines, which are given in units of $10^{10}\,\mathrm{kg\,s^{-1}}$. Furthermore, the strength of the flow is proportional to the cross-isoline gradient. The figure reveals a multicell pattern for the meridional circulation in the atmosphere, which in the tropics contains a direct (Hadley) cell in each hemisphere. The areas of upward motion in these Hadley cells extend between 20° N and 10° S, and the related downward flow occurs in the subtropics between 20° and 40° S and between 20° and 40° N. Because the rising branches are relatively warmer than the sinking ones, the net result is considered a thermally direct circulation. The southern Hadley circulation is stronger than the northern one in this annual mean picture. Because of the mass continuity principle, there is low-level equatorward motion in the low-latitude areas of the Hadley cells and corresponding poleward motion in the upper atmosphere. The centers of the cells are near the 700 mbar level in this analysis, but they have been somewhat higher in analyses based on rawinsonde data alone (Peixoto and Oort, 1992). Poleward of the Hadley cells are the

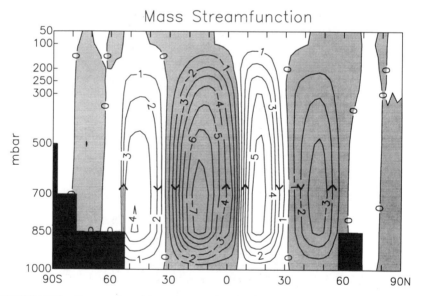

FIGURE 2.10.   Cross-section of the mass streamfunction for the mean meridional circulation for 1980–1989 between the 1000 and 50 mbar levels. Units are $10^{10}\,\mathrm{kg\,s^{-1}}$. Positive values indicate clockwise circulation.

"indirect" Ferrel cells, which have rising motion in their lower-latitude sections, and hence in warmer conditions, than where sinking motion occurs. The meridional motions of the Ferrel cells are reversed from those of the tropical Hadley cells and in fact are driven by different dynamical processes than are the Hadley cells, mostly by synoptic-scale eddies. Lastly, there are weaker direct cells poleward of the Ferrel cells, with corresponding meridional flow aloft.

The meridional circulations for the two individual solsticial seasons, however, are very different from this annual mean picture, as can be seen in Peixoto and Oort (1992). Indeed, the annual mean meridional circulation in Figure 2.10 appears to be a combination of quite different values from all parts of the year. When looking at streamfunctions representing a single season, the twin Hadley circulation reduces to a strong one in the winter hemisphere and a much weaker one in the summer hemisphere. Furthermore, the stronger Hadley cell is displaced meridionally toward the summer hemisphere, following the motion of the sun with the seasons. The Ferrel cells are slightly increased in their respective winter hemispheres, too. Of course the mean meridional winds also follow the flow at upper and lower levels in the latitudes of the Ferrel cell. Fluctuations in the winds within meridional zones tend to be organized with alternations of sign, consistent with zonal wavenumber signals around a hemisphere of up to 6.

2.4.4. Humidity Structure

Water vapor is distributed throughout the lower troposphere and is a rather variable and mobile atmospheric constituent. A number of different parameters may be used to represent humidity: these include mixing ratio, the proportion of water vapor to dry air by mass, specific humidity, the proportion of water vapor to total air, and relative humidity, the fraction of specific humidity to the maximum specific humidity possible for a given temperature and pressure, as determined by a comparison with the saturation vapor pressure. Specific humidity is often the quantity considered for purposes of assessing the role of water vapor in overall water budgets. To a large extent, humidity patterns are similar to those of temperature but are also related to atmospheric pressure level because of the dependence of saturation vapor pressure on these two parameters. Specific humidity varies, though, according to air mass characteristic as well.

The basic pattern of zonal mean specific humidity, given in Figure 2.11, shows a maximum at the 1000 mbar level of $19\,\mathrm{g\,kg^{-1}}$, equivalent to a proportion of nearly 0.2% of atmospheric mass. This value decreases with altitude to $2\,\mathrm{g\,kg^{-1}}$ by the 500 mbar level in the tropics. Also, at the 1000 mbar level, specific humidity decreases with temperature, and hence latitude, to $2\,\mathrm{g\,kg^{-1}}$ by about 65° S and 75° N. The zonal mean structure of water vapor is quite similar in the two hemispheres, although the northern hemisphere tends to hold somewhat more water vapor than does the southern one. A map of specific humidity at

Zonal Mean Specific Humidity

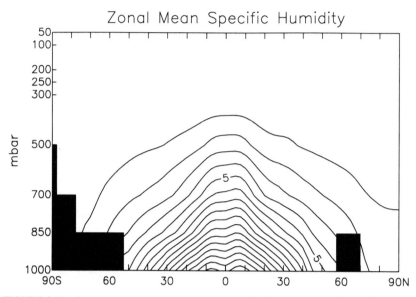

FIGURE 2.11.   Zonal mean specific humidity for 1980–1989 between the 1000 and 50 mbar levels. Units are $g\,kg^{-1}$. Isolines are spaced every $1\,g\,kg^{-1}$.

850 mbars (Figure 2.12) reveals a generally axisymmetric distribution with latitude; however, there is a preference for higher specific humidity over certain regions such as the west tropical Pacific, over Indonesia, and over the Amazon basin. Less moist areas, under $4\,g\,kg^{-1}$, exist for central and eastern Asia,

Specific Humidity 850 mbar

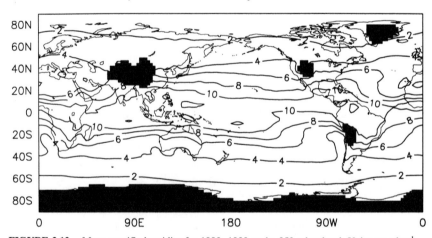

FIGURE 2.12.   Mean specific humidity for 1980–1989 at the 850-mbar level. Units are $g\,kg^{-1}$.

northern Africa, and western Asia, and these are relatively dry for their latitudes. The annual distribution of water vapor, when given as calendar monthly means, has relatively large and small annual cycles in the southern and northern hemispheres, respectively, with temperate latitudes, varying by as much as $5 \, \text{g kg}^{-1}$ from maximum to minimum in middle northern latitudes. January and July tend to be the two extreme months for the distribution of water vapor in both hemispheres.

2.4.5 Meridional Eddy Transport of Heat and Momentum

Analyses of atmospheric energy and momentum budgets, especially for zonal mean quantities (Peixoto and Oort, 1992), require that considerable transports of these quantities exist in the meridional direction. These fluxes have been studied intensively, and, as mentioned above, to help understand the dynamics, such transports are often decomposed into their mean and eddy portions. The mean transport of heat, proportional to the average of the product of meridional wind and temperature, is quite consistent with the mean meridional circulation, reviewed in Section 2.4.3. At low levels, heat is transported southward in the northern hemisphere tropics and southern hemisphere middle latitudes, while northward flow of heat occurs in the other portions of the hemispheres near the ground. The transports aloft in the upper troposphere are reverse, completing the signatures of the mean meridional cells.

The zonal mean values of the transient eddy transport of heat for the decade are displayed in Figure 2.13. This transport is poleward nearly everywhere, at both lower and upper levels in both hemispheres. Especially strong values occur poleward of 20° N and S, however. In this way the atmosphere has arranged itself to carry the excess heat absorbed at the low latitudes into the higher latitudes where there is a relative deficit of heating. These eddy transports, proportional to the temporal covariance of the meridional wind and temperature, denoted by $[\overline{v'T'}]$, maximize between 40° and 50° in both hemispheres and at the 850 mbar level.

In a similar fashion the eddy transport of momentum, proportional to the temporal covariance of the meridional and zonal wind ($[\overline{v'u'}]$), is given in Figure 2.14. The distribution of this momentum transport is toward the poles throughout the low to middle latitudes. However, farther poleward in both hemispheres, transport toward lower latitudes occurs. Thus there is an area of convergence of momentum by the transient eddies in middle to high latitudes, between the maxima of opposite signs. In this way the strong jets in these regions are continually maintained against any dissipation that would occur.

In recognition that both the eddy transport of heat and momentum in middle latitudes act to maintain the whole indirect Ferrel cell, a unified view of the

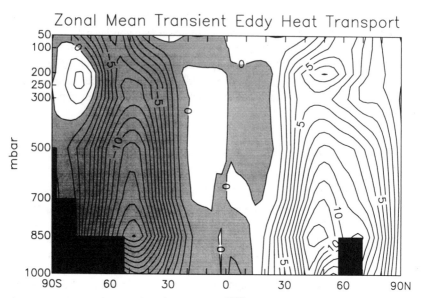

FIGURE 2.13. Zonal mean meridional values of $[\overline{v'T'}]$, proportional to transport by transient eddies of heat, for 1980–1989 between the 1000 and 50 mbar levels. Units are km s^{-1}.

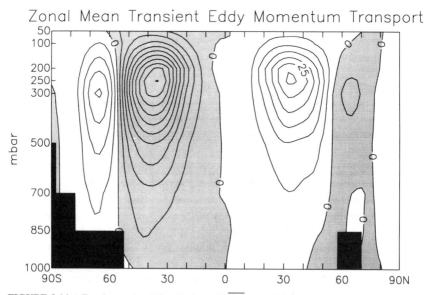

FIGURE 2.14. Zonal mean meridional values of $[\overline{v'u'}]$, proportional to transport by transient eddies of momentum, for 1980–1989 between the 1000 and 50 mbar levels. Units are km^2 s^{-2}.

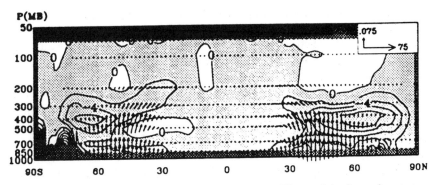

FIGURE 2.15. Meridional cross-section of the transient eddy Eliassen–Palm flux and contours of its divergence, related to the driving of zonal winds by the transient waves, based on seven winters, from meteorological analyses. Horizontal and vertical vector scales (units of $10^7 \, \mathrm{kg \, s^{-2}}$) are displayed in the upper right-hand corner. Contour interval is $2 \, \mathrm{m \, s^{-1} \, day^{-1}}$. (Taken from Black 1993.)

diagnostics of atmospheric wave dynamics has been developed over the last decade. It has as a fundamental invariant quantity, a vector F known as the Eliassen–Palm flux, lying in a meridional plane, whose meridonal component F_y and vertical component F_p involve eddy fluxes in the northward direction of angular momentum and heat, respectively (Edmon et al., 1980). Such a flux may be given by

$$F_y = \rho_o a \cos \phi \, [\overline{-v'_g u'_g}]$$

$$F_p = \rho_o a \cos \phi \, [\overline{v'_g \theta'} \cdot f/\theta_p] \qquad (2.5)$$

where v_g and u_g are the meridional and zonal geostrophic wind, respectively, θ is the potential temperature, θ_p is the static stability, ρ is the density, a is the radius of the Earth, and f is the Coriolis parameter. The divergence of F is a measure of the acceleration of the zonal mean zonal wind by wave activity forcing, and so it has zero divergence for wave-like conservative flow. A measure for mean winter conditions of the Eliassen–Palm flux vector and its divergence is given in Figure 2.15. Upward and equatorward fluxes of wave activity occur from the surface at middle latitudes.

2.5. HEATING OF THE ATMOSPHERE

Although the Sun is ultimately the source of energy reaching the Earth, only about 64% of the energy from the radiation intersected by the Earth is actually absorbed in the atmosphere. The remainder is reflected from the surface and from the clear and cloudy atmosphere. The atmosphere is heated by a number of specific processes, including heating directly by solar radiation, by terrestrial radiation, by sensible processes consisting of conduction and vertical mixing, and by latent processes, which constitute the heat liberated during a change in phase of water substance in the atmosphere.

The radiation consists of short-wave processes, peaking in the visible at wavelength 0.5 µm, whose source is the Sun, and long-wave processes, peaking in the infrared near 10 µm, whose source is radiation emitted from the Earth. The difference in the frequency of radiation results from the vastly different temperatures of the solar and terrestrial surfaces: the wavelength of maximum emission is inversely proportional to the temperature at which it is emitted (Wien's law). Also, radiation is re-emitted by the atmosphere in the long-wave band, based on its temperature. Solar radiation varies strongly with season and time of day in relation to the elevation angle of the Sun. Heat is absorbed in the atmosphere particularly by water vapor molecules in the lower troposphere, with peaks in the visible and near-infrared spectral regions, and is also strongly transmitted to the atmosphere in the absorption bands in the far infrared. Carbon dioxide also has peaks of absorption in the visible and infrared, while ozone, in particular, greatly absorbs ultraviolet radiation.

Sensible heating involves contact with the ground and transfer of heat between levels within the atmosphere. Near the very lowest level the transport from the ground takes place by means of turbulent exchanges in the boundary layer (Section 2.6). Above that layer, sensible exchanges occur but are not nearly as strong.

Latent processes are often negative, providing cooling, near the lowest layer of the atmosphere where evaporation occurs, but condensation to clouds provides positive heating higher in the free atmosphere. Condensation, and hence cloud formation, occurs by means of a number of distinct processes, including those on large scales and on smaller convective scales.

A summary of the heating processes in terms of the zonal mean average of the four heating components is given in Figures 2.16–2.19 for the area above the lowest part of the boundary layer. These results have been obtained from an annual simulation by the global analysis and forecast model of the NASA Goddard Laboratory for Atmospheres. The latent processes within the model, in particular, are treated in a sophisticated manner, involving complex interactions within detailed cloud convection schemes (Arakawa and Schubert, 1974; Sud and Walker, 1993). The heating rates due to clouds are high over the tropical areas, reaching, for mean annual conditions, as much as $3\,\mathrm{K\,day^{-1}}$ as a result of the strong convective activity (Figure 2.16). This maximum occurs at a rather high level near 400 mbar, revealing the very deep convection in this warm region, reaching into the middle and even the upper troposphere. Maps of latent heating reveal particularly strongly convective areas to include the maritime continent

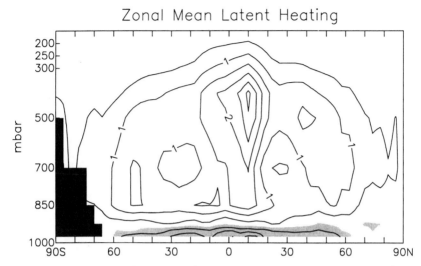

FIGURE 2.16.   Zonal mean latent heating for 1 year based on a simulation using the NASA Goddard Laboratory for Atmospheres global circulation model, with physics modified by Sud and Walker (1993). Units are K day^{-1}, with isoline increments of 0.5 K day^{-1}. Negative values are shaded and indicate cooling.

(Indonesia) and South America. Minima of heating occur in the subtropics, in the region of reduced convection near the descending cells. Farther poleward, maxima in the latent heating rate occur again, with the middle-latitude values resulting from shallower convective activity and large-scale cloudiness, though some deeper convection can occur there too. For individual months the maxima are much stronger than the annual average in Figure 2.16, because the maxima are confined to reasonably narrow latitude bands which move with the solar forcing. For example, in July the middle tropospheric maximum in heating reaches over 4 K day^{-1}, centered at 10° N. Nearest the surface, in a shallow area, particularly in the tropics, cooling due to evaporation occurs. Patterns of latent heating are of smaller scale for much shorter time periods. Such latent heating accompanies convective activity and apparently can move horizontally with weather systems.

Sensible heating (Figure 2.17) is strongest at the surface; in the lowest part of the boundary layer (not shown) it is particularly strong, but values are near 1 K day^{-1} by the top of the boundary layer too. Positive values are seen throughout almost the entire globe for annual means of sensible heating owing to the transfer of heat from the relatively warm surface of the Earth below the atmosphere. In particular, ocean surfaces contain much heat, which is evident in maps of mean low level sensible heating, strong over the warm ocean currents. For example, the air over the extension of the Gulf Stream has much sensible

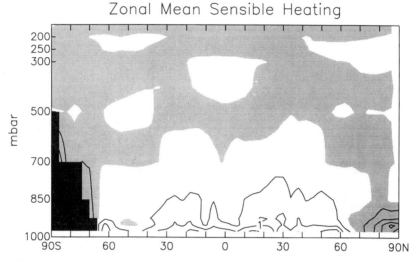

FIGURE 2.17.   Same as Figure 2.15, but for sensible heating.

heating evident even north of 70° N, near Scandinavia. However, in the mean, there is some cooling by sensible processes at the lower-boundary near-polar regions because of heat transfer to the cold continents and ice surfaces below. Sensible heating typically decreases with altitude quite quickly, and by the middle troposphere the magnitude of heating decreases to where regions of small positive and negative values are about equal in size. The annual cycle of sensible heating is similarly dependent on factors such as the temperature of the surface.

Annual mean zonally averaged short-wave radiational heating (Figure 2.18) occurs in a symmetric pattern from pole to pole, with the highest values, as expected, at the equator near the surface, exceeding 2 K day^{-1}. However, near the 900 mbar level a fairly constant mean value of short-wave heating, 1 K day^{-1}, occurs at all latitudes. On the other hand, the long-wave processes (Figure 2.19) contribute to cooling throughout the atmosphere, and do so by as much as 2 K day^{-1} over much of the tropics.

As described above, the means of the sensible, latent, and short-wave radiational heating components each increase from equator to poles and are thus positively correlated with the mean temperature pattern. Because of this distribution, their actions tend to increase the existing meridional temperature gradient and, in so doing, increase the potential energy available for the atmosphere to convert to kinetic energy. Conversely, long-wave radiation is negatively correlated with temperature, cooling in relatively warm areas and vice versa. Long-wave radiation thus acts to reduce this potential energy. Never-

Zonal Mean Shortwave Heating

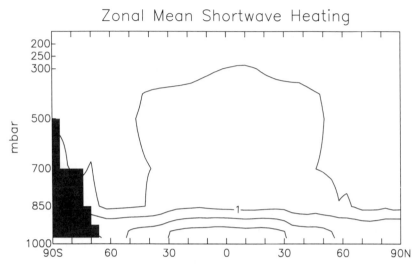

FIGURE 2.18.   Same as Figure 2.15, but for short-wave radiational heating.

theless, the reduction due to the long-wave term is less than those of the other three diabatic heating terms, and so, overall, the meridional distribution of heating helps to maintain the energy cycle which is responsible for the general circulation of the atmosphere.

Zonal Mean Longwave Heating

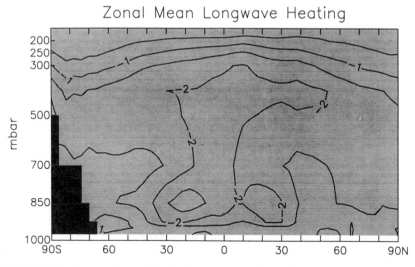

FIGURE 2.19.   Same as Figure 2.15, but for long-wave radiational heating.

2.6. BOUNDARY LAYER OVERVIEW

The atmospheric boundary layer has been defined as that portion of the atmosphere nearest the ground having considerable variations over the course of a day; this is in contrast with the free atmosphere above it, whose diurnal fluctuations are more limited. Also, in the boundary layer the major transfer of momentum, heat, and water vapor with the underlying surface occurs, in the first few centimeters, chiefly by molecular diffusion, but above this area, by turbulent diffusion. Although it may vary considerably with region, season, and time of day, the typical height of the boundary layer is 1 km, or about the 900 mbar level.

The structure of the boundary layer in particular is much dependent on its vertical stability, including the type of air mass in which it is embedded. The basic stability is related to the temperature lapse rate of a parcel of air. If virtual potential temperature (the temperature that the air would achieve were it to release its latent heat and be brought to a reference level) increases with height, the air is unstable and convection readily occurs. Conversely, a stable portion of the air mass is one in which an inversion has formed, limiting the upward motions associated with convection. In a dry air mass without cloud cover, ground cooling leads to strongly stable air, hindering convective activity. In a moist cloudy air mass the usual nighttime ground surface cooling may not occur, tending to keep the air in the subcloud layer unstable. Advection of warmer air above a cooler surface layer may create an inversion as well. A simple picture of an inversion layer below the free unstable portion of the troposphere is given in Figure 2.20.

Treatments of the atmospheric boundary layer have divided it into several sublayers whose properties are very dependent on the time of day and typically change with the diurnal cycle (see review by Stull, 1988). Over land surfaces the

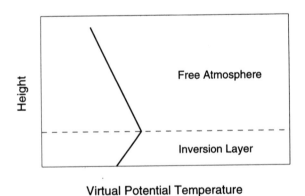

FIGURE 2.20.   Schematic of a low inversion layer in which the atmosphere is stable, thus hindering convection.

boundary layer often has a well-defined structure. Above a very shallow surface layer during the daytime is the mixed layer, developing often as a result of convection related to solar heating of the Earth's surface. During the course of the first few hours after sunrise, turbulent exchanges form an even distribution of heat, momentum, and moisture; such can be seen as constant values of virtual potential temperature, wind, and specific humidity. This mixed layer is at a maximum depth in the late part of the afternoon, and, when convection brings air above the level where condensation takes place, clouds form. After sunset, however, the actions of a radiatively cooled ground change the lowest layer into one with a vertically stable distribution. This stable boundary layer grows between sunset and the first hours after dawn, and its turbulent structures are much reduced compared with those of the mixed layer. The portion of the daytime mixed layer that exists into the night-time hours and is located above the stable boundary layer has been termed the residual layer; in this residual layer the concentration of any passive quantities (tracers) that were placed into the atmosphere in the daytime mixed layer remain relatively constant. Convection no longer occurs in this neutrally stratified layer and the turbulence is isotropic. After sunrise on the next day a new mixed layer forms, taking properties, including entrained moisture, from the night-time residual layer and the stable boundary layer underneath.

Transport in the boundary layer of the quantities occurs from areas of high concentrations to those of lower ones, and so vertical eddy fluxes are often modeled by a relationship proportional to the vertical gradient of the respective quantities. For example, the transient eddy flux of heat, F_H, can be given by

$$F_H = \rho c_p \overline{w'T'} \tag{2.6}$$

and parameterized as

$$F_H = -K_H \rho c_p \partial \overline{\theta}/\partial z \tag{2.7}$$

where K_H is the eddy diffusion term for heat, c_p is the specific heat of air at constant pressure, w is the vertical motion, θ is the potential temperature, and ρ is the density. Alternatively, a drag coefficient dependent on conditions in the lowest layer can be determined, so that the flux of heat is related to the local wind velocity and the difference in potential temperature across the very lowest layers of the atmosphere.

2.7. MAJOR EXCHANGES IN THE ATMOSPHERE

The circulation, as viewed in the mean meridional plane, has been reviewed in Section 2.4.3; especially strong in this regard are the tropical Hadley cells, which transport air between upper and lower levels as well as meridionally. When cross-sections of flow are taken instead in the longitude–height plane, some of the local regions important in vertical transports emerge. One such system has been observed over the Pacific region in low latitudes and has been termed the Walker circulation. The Walker cell typically has rising and descending motions in the eastern and western Pacific areas respectively. However, these motions are variable and their centers move with the annual cycle, related to the predominant convergence zones. Observational evidence exists for strong vertical transport of tropospheric air into the stratosphere, particularly from strong convective activity in the tropics (e.g., Newell and Gould-Stewart, 1981). The very high level of cumulonimbus tops, sometimes to at least 20 km, can be responsible for exchanges of tropospheric and stratospheric air, including deposits of water vapor in the stratosphere (WMO, 1985). In the extratropics, air of tropospheric and stratospheric origins can usually be differentiated by the amounts of potential vorticity P, a quantity defined as

$$P = \alpha \, \nabla\theta \cdot (\nabla \times V + 2\Omega) \tag{2.8}$$

where $\nabla\theta$ is the gradient of potential temperature and Ω is the angular velocity of the Earth. The absolute vorticity, in parentheses, consists of the curl of velocity plus the vorticity of the Earth's rotation, or twice Ω. For adiabatic and frictionless motion, typical to a large extent in the free atmosphere, the amount of P in a parcel of air is conserved and so the origin of an air mass can often be traced to its value of P. Stratospheric air usually has much larger values of P than does tropospheric air. Gradients in P do arise from mixing of tropospheric and stratospheric air masses from a folding of the tropopause, as suggested by Danielson (1968). In such cases, trace gases and aerosols from the lower stratosphere are mixed into the tropospheric air.

To find the horizontal as well as vertical exchanges within major atmospheric compartments, relatively inactive tracers have been utilized in models. Prather et al. (1987), for example, compared results from a three-dimensional chemical tracer model (CTM) with the observed distribution of quantities of chlorofluorocarbons to study air mass exchanges. The place and time of year of maximum air mass transport have been studied in this manner. For example, interhemispheric exchanges are found to occur mainly in the upper troposphere, and horizontal winds in the summer have been found to be more efficient in carrying tracers away from their sources, while in the winter, vertical transports predominate, elevating concentrations higher in the atmosphere.

Modelers have often treated the average exchanges of mass across the boundaries of the hemispheres and of vertical regions as transfers between four compartments (northern troposphere, northern stratosphere, southern troposphere, and southern stratosphere), modeling such exchanges with coefficients κ

TABLE 2.2   Exchange Times Between Atmospheric Compartments

Exchange	Range (years)	Average (years)
Northern troposphere/Southern troposphere	0.7–1.8	1.0
Stratosphere/Troposphere	0.8–2.0	1.4
Northern stratosphere/Southern stratosphere	3–6	4.0

Source: Adapted from Warneck (1988).

whose inverse τ represents the time constant for decay of the compartment's air content (Warneck, 1988). By taking the observed increase in CO_2, for example, as the slope of their trend, and using an equation for the exchange of air between the compartments, one obtains

$$d(m_{NT} - m_{ST})/dt = -2\,\kappa_{NT,ST}\,(m_{NT} - m_{ST}) + 2R \tag{2.9}$$

where m_{NT} and m_{ST} are the mixing ratios of the tracers in the northern and southern tropospheres respectively, $\kappa_{NT,ST}$ is the exchange coefficient (in year^{-1}) between these regions, and R is the annual CO_2 rise. Equation (2.9) has an exponential solution, for which its variable, the difference in mixing ratios, relaxes to a steady state after a time of the order of $\tau_{NT,ST} = 1/\kappa_{NT,ST}$. Estimates of exchange times using such methods are given in Table 2.2, where a range from different estimates is quoted. In general, northern troposphere/southern troposphere exchanges are the fastest, with many values near $\tau = 1$ year, stratosphere/troposphere exchanges are somewhat slower ($\tau \approx 1.4$ years), and intrastratosphere exchanges are the slowest ($\tau \approx 4$ years).

Acknowledgments

The author wishes to give his sincere thanks to Peter Nelson of AER, Inc., who ably produced most of the figures in the chapter. One figure was taken from a report by R.X. Black of AER, Inc. J. Janowiak of NOAA/NMC assisted with the interpretation of the Climate Diagnostics Data Base. Y. Sud of the NASA Goddard Space Flight Center provided the atmospheric modeling results from which I obtained the diabatic heating measurements. R. Rosen of AER, Inc. has reviewed and helped improve the manuscript. Partial support for the writing of the chapter was under contract NAS5–31333 to AER, Inc. from NASA under its Climate Modeling and Analysis Program.

References

Arakawa, A. and Schubert, W. H. (1974). Interaction of a cumulus cloud ensemble with the large-scale environment, Part I. *J. Atmos. Sci.*, **31**, 674–701.

Black, R.X. (1993). Validating seasonal and intraseasonal wave activity in GLA GCM's and data assimilations. *Proc. Seventeenth Ann. Climate Diagnostics Workshop*, US Department of Commerce, NOAA/NWS/CAC/NMC, pp. 236–237.

Danielson E.F. (1968). Stratospheric–tropospheric exchange based on radioactivity, ozone, and potential vorticity. *J. Atmos. Sci.*, **25**, 502–518.

Edmon, H.J., Hoskins, B.J. and McIntyre, M.E. (1980). Eliassen–Palm cross sections for the troposphere. *J. Atmos. Sci.*, **37**, 2600–2616.

Johnson, D. (1989). The forcing and maintenance of global monsoonal circulations: an isentropic analysis. In Saltzman, B., Ed., *Advances in Geophysics*, Academic Press, New York, NY, pp. 43–316.

Kalnay, E. and Jenne, R. (1991). Summary of the NMC/NCAR reanalysis workshop of April 1991. *Bull. Am. Meteorol. Soc.*, **72**, 1897–1904.

Kanamitsu, M. (1989). Description of the NMC global data assimilation and forecast system. *Weather Forecast.*, **4**, 335–342.

Madden, R. and Julian, P.R. (1971). Detection of a 40–50 day oscillation in the zonal wind in the tropical Pacific. *J. Atmos. Sci.*, **28**, 702–708.

Newell, R.E. and Gould-Stewart, S. (1981). A stratospheric fountain? *J. Atmos. Sci.*, **38**, 2789–2796.

Oort, A.H. (1983). Global atmospheric circulation statistics, 1958–1973. *NOAA Professional Paper 14*, US Government Printing Office, Washington, DC..

Peixoto, J. P. and Oort, A.H. (1974). The annual distribution of atmospheric energy on a planetary scale. J Geophys. Res., 79, 2149–2159.

Peixoto, J.P. and Oort, A.H. (1992). *Physics of Climate*, American Institute of Physics, New York, NY.

Philander, S.G.H. (1989). *El Niño, La Niña and the Southern Oscillation*. Academic Press, New York, NY.

Prather, M., McElroy, M., Wofsy, S., Russell, G. and Rind, D. (1987). Chemistry of the global troposphere: fluorocarbons as tracers of air motion. *J. Geophys. Res.*, **92**, 6579–6613.

Rasmusson, E.M. and Arkin, P.A. (1985). Interannual climate variability associated with the El Niño/Southern Oscillation. In *Coupled Ocean Atmosphere Models*, Elsevier, Amsterdam, pp. 697–725.

Rosen, R.D. and Salstein, D.A. (1983) Variations in atmospheric angular momentum on global and regional scales, and the length of day. *J. Geophys. Res.*, **88**, 5451–5470.

Rosen, R.D., Salstein, D.A., Peixoto, J.P., Oort, A.H. and Lau, N.-C. (1985). Circulation statistics derived from level III-b and station-based analyses during FGGE. *Mon. Weather Rev.*, **113**, 65–88.

Stull, R. (1988). *An introduction to Boundary Layer Meteorology*, Kluwer Academic Publishers, Dordrecht.

Sud, Y. and Walker, G. (1993). A rain evaporation and downdraft parameterization to complement a cumulus-updraft scheme and its evaluation using GATE data. *Mon. Weather Rev.*, **11**, 3019–3039.

Trenberth, K.E. and Olson, J.G. (1988). An evaluation and intercomparison of global analyses from the National Meteorological Center and the European Centre for Medium Range Weather Forecasts. *Bull. Am. Meteor. Soc.*, **69**, 1047–1057.

Trenberth, K.E. and Guillemot, C.J. (1994). The total mass of the atmosphere. *J. Geophys.*

Res., **99**, 23079–23088.

US Standard Atmosphere (1976). National Oceanic and Atmospheric Administration.

Warneck, P. (1988). *Chemistry of the Natural Atmosphere*, Academic Press, San Diego, CA.

WMO (World Meteorological Organization) (1985). *Atmospheric Ozone, Report No. 16*, Geneva.

3

The changing composition of the Earth's atmosphere

M.A.K. Khalil and R.A. Rasmussen

3.1. INTRODUCTION

The Earth's atmosphere is 99.9% nitrogen, oxygen, and argon (see Table 2.1 in Chapter 2). Of the remaining 0.1%, roughly half (0.04%) is accounted for by the three transition gases CO_2, Ne, and He. The remaining minute fraction of the atmosphere (0.06%) is composed of a complex mixture of hundreds of trace gases, the most abundant being methane at 1.7 ppm (10^{-6} v/v). The closer we look at the atmosphere, the more gases we find. At present there is no accounting of how many different gases exist in the atmosphere. Many gases, at the parts per trillion level (10^{-12} v/v), are entirely man-made but may still have significant effects on the environment.

Some trace gases control or affect the Earth's climate and habitability. Long-lived gases that are increasing at substantial rates because of human activities are of particular current interest since they may lead eventually to stratospheric ozone depletion (see Chapter 11), global warming (see Chapter 13), and disturbances in atmospheric chemistry (see Chapters 10 and 12) that many believe will be harmful to human life. This chapter is about such gases, their natural levels, origins and fates, and by how much and why they are changing.

If we confine ourselves to global warming and the destruction of the stratospheric ozone layer, along with feedbacks and couplings, then at present there are five notable "source" gases affected by human activities that are involved in global change, namely carbon dioxide (CO_2), methane (CH_4), nitrous oxide (N_2O), trichlorofluoromethane (CCl_3F; CFC-11), and dichlorodifluoromethane (CCl_2F_2; CFC-12). These gases are emitted directly into the atmosphere and have long atmospheric lifetimes. There are many other gases, man-made and natural, that have similar environmental roles, but their concentrations are low and not expected to rise much. Ozone is a trace gas with secondary photochemical sources and its trends are discussed in Chapter 10.

The increase in CO_2 has the greatest effect on global warming, while the chlorofluorocarbons (CFCs) have the greatest effect on destroying stratospheric ozone. Yet all these gases affect both the ozone layer and global temperatures. While the concentrations of CH_4, N_2O, and the CFCs are minute compared with CO_2, they are, molecule for molecule, much more effective in causing global warming. Moreover, there are many such gases; each adds only a little bit, but together they have a substantial effect on the Earth's climate and environment (for current perspectives see Chamberlain and Hunten, 1987; Houghton et al., 1990; Rambler et al., 1989; Warneck, 1988; WMO, 1985, 1988, 1989; Wuebbles and Edmonds, 1991).

Our plan is to discuss the long-term changes in the gases most important for causing global environmental changes. We will, in the next three sections, discuss how the trends are estimated to define the changing atmosphere and then the budgets of trace gases. In the last section we will bring together the connections between these gases and perspectives on the trends that may guide us in forming expectations for the future of the global environment.

3.2. TRENDS OF GASES

3.2.1. Mass Balances and Causes of Trends

During the time between t and $t + dt$, in any volume of the atmosphere, the concentration (C) of a gas is increased by emissions from the sources (S) within the region, or by atmospheric formation of the gas, or when the winds (v) and turbulence (K) transport the gas from outside the region. At the same time the concentration is reduced by transport out of the region and by the chemical or depositional removal (ηC) of the gas within the region. If these processes are not exactly balanced, concentration changes and trends are observed ($\partial C/\partial t \neq 0$). Often the mass balance equation, with n being the number density of air, is written as

$$\partial C(t)/\partial t = S(t) - \{\eta + [\nabla \cdot n(\mathbf{v} - K \cdot \nabla)]/n\}C(t) \tag{3.1}$$

If we make the time scales very short, it is unlikely that a balance will exist, and rapid fluctuations of trace gas concentrations are observed. The trends of most interest to global change science are increases or decreases over time scales longer than a year. However, even these trends are rarely constant for long (more on this later). If equation (3.1) is averaged over the entire atmosphere, we obtain a global average mass balance

$$dC(t)/dt = S(t) - C(t)/\tau(t) \tag{3.2}$$

This simplified equation is useful to discuss the changes that affect trace gases in the environment and provides an insight into the causes of imbalances. In equation (3.2), C is the average mixing ratio, or number of molecules of the gas in the atmosphere, S is the emission rate from all sources (molecules year^{-1}), and $\tau(t)$ is the atmospheric lifetime (years). Increasing trends can occur only when the right-hand side of equation (3.2) is positive, which can happen if the sources or lifetimes are increasing, or if the current source, which may be constant or even declining, is still larger than the annual removal. For example, increasing sources have caused methane concentrations to rise during the last century; increasing combustion of fossil fuels may be tying up oxygen in CO_2, causing an increase in its removal rate and thus a declining trend, although this trend is slow, undetected at present, and not considered important; for CCl_3F the emissions are probably falling, but its lifetime is long and therefore only a small amount is removed every year. This small removal rate is still less than the present annual emissions, so the concentration continues to rise even when the sources do not; see equation (3.2).

3.2.2. Estimating Trends

The foundation for estimating trends of gases is a time series of measurements at one or more locations, usually at the Earth's surface. Often a classical time series decomposition is used to determine the trends:

$$C(t) = \text{trend} + \text{seasonal cycles} + \text{IAV} + \epsilon(t) \tag{3.3}$$

where "trend" is a linear or quadratic function of time, "seasonal cycles" are variations in the gas concentration that repeat every year, "IAV" is the interannual variability, and ϵ is the residual variability, presumed to be random.

The IAV, as we think of it, is a short-term change in the cycle of the gas that occurs every few years. An example is the El Niño events, which may cause aperiodic changes in concentrations and trends of CO_2 (see Bacastow, 1976). Other causes may have a similar effect on the trends. Methods to decompose the time series are discussed elsewhere (Gottman, 1984; Khalil and Rasmussen, 1990b). A trend, whether increasing or decreasing, spans long time scales and shows changes in the concentration of a gas that may eventually cause global environmental change. The existence and causes of the trend are therefore of considerable scientific and practical interest. Indeed, the idea of the changing atmosphere is related fundamentally to the "trend."

It seems simple enough to determine the trend from equation (3.3), but there are several factors that obscure or prevent the detection of trends. Many gases undergo large seasonal variations, especially at middle and higher latitudes. Assuming that there are enough years of data to determine the magnitude of the seasonal variations so that these may be subtracted from the observed concentrations, or if there are no seasonal variations or IAV, the observability of a trend then depends on the noise in the data as represented by ϵ. The random variability ϵ is a combination of several factors, including natural atmospheric variability caused by turbulent transport, variability of sources or sinks between the place where the gas is emitted (source) and the place where it is detected (receptor), and variability introduced by the experimental process. The variability from the experimental process itself has several components, including effects from non-random sampling, storage of samples before analysis, and variability of the response of instruments used to measure the gas. At locations far from sources the atmospheric variability is often inversely proportional to the lifetime, but there is no general rule that determines the experimental variability, except perhaps that qualitatively reactive gases are more likely to be affected by sampling and storage than are less reactive gases.

Natural variability also has many origins. It is, as a composite, often reflected in the atmospheric lifetime as suggested by Junge (1974) (see also Slinn, 1988). The relative variability, defined as the standard deviation of measurements divided by the mean concentration, is inversely proportional to the lifetime. Gases that have a very long atmospheric lifetime have little natural variability. Intuitively this happens because the background concentration of these gases is very large compared with the emission or destruction rate according to equation (3.2). Thus, if a puff happens to reach the detector, its contribution to the concentration is small compared with the accumulated residuum of many years. For such gases the effects of experimental variability may be most significant. For gases with short lifetimes and precise measurement methods the variability is mostly a reflection of the lifetime.

The variability, whether natural or not, affects the detection of trends. For a linear regression calculation of trend, in a time series with evenly spaced measurements, the time needed to detect a trend, in years, at a specified statistical confidence $(Z\alpha)$ is

$$T_D = (12/\delta)^{1/3} [(Z\alpha/b)S_{c.t}]^{2/3} \qquad (3.4)$$

where $S_{c.t}$ (in ppb or other appropriate units) is the standard deviation of the detrended data $C_{measured} - (a + bt)$, δ is the frequency of measurements per year (number per year) so that $1/\delta$ is the time between samples (in years), and b is the trend in units of concentration per year (such as ppb year^{-1}). $S_{c.t}$ may be regarded as a composite of natural (S_N) and experimental (S_E) variabilities, so that $S_{c.t}^2 = S_E^2 + S_N^2$. If $S_E >> S_N$, then the detection of the trend, or narrowing the uncertainties in the trends, depends mostly on improving the experimental techniques to increase precision (as in the case of N_2O). This condition occurs most frequently for very long-lived gases ($\tau > 50$ years). When $S_E << S_N$ and assuming S_N to be λ/τ, or inversely proportional to the lifetime, with the proportionality constant λ dependent on the location of the measurements and possibly other variables, the length of time to detect the trend becomes

$$T_D = (12^{1/2} Z_\alpha \lambda)^{2/3} b^{*-2/3} \tau^{-2/3} \delta^{-1/3} \qquad (3.5)$$

In equation (3.5), b^* is the relative trend equal to b (ppb year^{-1})/C_{av} (ppb), C_{av} being the average concentration (ppb) over the length of the measurements; $b^* \approx (1/C) \, dC/dt$. The equation shows that if the trends are small or the lifetime is short, it takes a long period of measurements to detect a trend, compared with opposite conditions when the lifetime is long or the trends are large. Yet the dependence to the 2/3 power may be regarded as favorable for early detection of trends. For a gas with a lifetime half as long as another, it takes 1.6 times as long to detect the same trend (all other factors being the same). Equations (3.4) and (3.5) also show that increasing the frequency of measurements has diminishing returns, since the time needed to detect the trend goes down only as the 1/3 power of the frequency. For instance, to cut the time to detect the trend in half, the sampling frequency has to be increased by a factor of 8. Increasing sampling frequency is therefore not a substitute for long-term measurements. Furthermore, the frequency cannot be increased arbitrarily to improve the chances of seeing a trend, even if fast response instruments are available, because the concentration does not necessarily change perfectly smoothly. Air masses have persistence, so that a sample may represent the same air mass as other samples collected around the same time and therefore have the same concentration. Under these conditions, not every measured concentration can be regarded as statistically independent as required by equations (3.4) and (3.5). With a high frequency of samples, concentrations have to be averaged over time spans that can make the average values (statistically) independent of each other. This time span is often taken to be 1 month and probably cannot be reduced below 1–2 weeks, which is the average persistence determined for atmospheric variables such as temperature. The times represented by equations (3.4) and (3.5) are also the minimum periods over which we can hope to see changes in the trends that are not caused by random fluctuations. If there are gaps in the time series, the time to detect trends increases accordingly.

By analyzing equations (3.1) and (3.2), it can be shown that the trends b and b^* also depend on the lifetime τ: b^*, the relative trend, has a weaker dependence on the lifetime than does b, but still, the longer the lifetime, the faster the concentrations rise, all else being the same. From this discussion we conclude that there are two reasons why it is easier to detect or document trends of long-lived gases compared with short-lived gases: the longer-lived gases increase faster and have less atmospheric variability. When the lifetime is short, other questions arise regarding the extrapolation of the trend to a global scale and its sustainability. For short-lived gases the trends reflect local imbalances and may not represent global rates of increase; see equation (3.1). Moreover, the trends are most likely to vary from year to year or over time scales of a few years. This is because, to maintain trends, either the sources must increase steadily every year or the sinks must decrease. Any changes in the sources or sinks are immediately reflected in the trends. For long-lived gases the year-to-year changes in sources and sinks are buffered by the high atmospheric concentrations and do not show up as large changes in the trends or concentrations. This effect is another manifestation of the connection between variability and lifetime.

3.2.3. Long-Term Variability of Trends

When there are only a few years of data, it seems appropriate to report a single trend. In time, however, the trend itself (or its rate of increase) changes even when the seasonal variations are taken out. To accommodate the analysis of long time series, we have used a "moving slopes" method (Khalil and Rasmussen, 1990b). In this method we calculate the rate of change dC/dt by a linear regression model or a suitable non-parametric statistical model for the slope over some period T. If there are seasonal variations, these are subtracted before applying the method. The trend as a function of time is denoted $b(t + T/2)$, with $t = 0, 1, \ldots, N$, when $C(t)$ is the concentration time series. The first value of $b(t + T/2) = b(T/2) = (dC/dt)(T/2)$, is the trend of concentrations between times 0 and T; the next value, $b(T/2 + 1)$, is the trend for times between 1 and $T + 1$, and so on.

If the concentrations are increasing steadily, this function is constant; if the "trend" is itself decreasing, it suggests an approach to a new equilibrium or perhaps a limit to the global change that can be caused by the gas; if it is some other, more complex function of time, it may reveal the effects of natural or human events on the global balance of the trace gas. For instance, the trends of CCl_3F have risen and fallen during its presence in the atmosphere, yet its concentration has increased every year during which measurements have been taken and probably since it was first put into use in the 1930s. In this case, defining the trend by a constant value would leave out critical information about the global balance of CCl_3F. Applications of the methods discussed here will be given in Sections 3.3 and 3.4.

3.2.4. Trends from Ice Core Data

At present, the most effective and direct method of measuring the changes in atmospheric composition in pre-industrial and ancient atmospheres is the analysis of bubbles of old air preserved in polar ice. As snow falls and accumulates in polar regions, it undergoes a series of transitions before it becomes solid ice. After the snow falls, it is unpacked ice crystals with air in between. As more snow falls, the weight causes the ice crystals to pack closer in a transition region where bubbles of air form between contiguous regions of ice. In the next stage the air bubbles are sealed from the atmosphere and sink deeper into the ice sheet. Far below the surface of the ice sheet are bubbles that were sealed off from the atmosphere long ago and thus contain the air from ancient atmospheres. The deeper the ice, the older is the air in it (for a review see Raynaud and Chappellaz, 1993).

There are several issues still under investigation that have a bearing on the trends estimated from ice core data. First, as the air is being trapped, it becomes a mixture of air from several years. Second, the core sample analyzed is usually long enough to include many years of air. For shallow ice this may not be significant, but for very deep ice the sample may contain air from several centuries. Perhaps the biggest problem is to date the air. Although it is possible to obtain the date of the ice from various techniques, the age of the air is different from that of the surrounding ice since it takes time for the bubbles to seal. To make matters worse, this firn-to-ice transition time is not the same everywhere on the polar ice sheet. Where the snowfall is greater, resulting in greater accumulation rates, the transition from firn to ice occurs sooner than where the ice accumulation rate is slow. The difference between the age of the air in the bubbles and the age of the ice may be between 20 and 200 years. This matter is unimportant for deep ice but is very important for the shallow ice representing more recent air.

Since the sample does not provide a measure of the concentration during a given year but may be a composite of several years, decades, or even centuries, the trends that may have existed over short periods are lost and the trends observed may be filtered. Besides the issues just mentioned, it is possible that physical and chemical processes affect the trends deduced from ice cores.

Physical processes include diffusion of gases from the bubble to the ice lattice, and chemical processes may involve slow gas phase or heterogeneous chemical reactions that can change the composition of the air in the bubbles. The fact that there is very little air recovered from the ice cores prevents the analysis of many gases that are at low concentrations in the present atmosphere and may have been even rarer in the pre-industrial atmosphere. Experimental procedures often add additional variability to the data. So far the only stable trace gases that have been measured in ice cores are CO_2, CH_4, and N_2O. Having discussed the nature and causes of the trends used to define the changing atmosphere, we can now proceed to look at how the main trace gases have changed during recent times, during the last several thousand years, and over longer times when the climate itself has also undergone significant changes.

3.3. GLACIAL–INTERGLACIAL CHANGES

Early work on glacial–interglacial transitions of CO_2 was reported by Delmas et al. (1980) and Neftel et al. (1982) and on methane concentrations by Raynaud et al. (1988) and Stauffer et al. (1988). The most complete and longest records of carbon dioxide and methane have been compiled from the deep polar ice core drilled at Vostok Station in Antarctica by Russian scientists (see Barnola et al., 1987; Chappellaz et al., 1990; Lorius et al., 1990). These data go back to the last penultimate glaciation and the intermediate warm epochs that span some 160,000 years (Figure 3.1). The data show clearly that both CO_2 and CH_4 concentrations are correlated with the average temperature of the Earth. To further delineate this relationship we show a spectrum of frequencies that make up the temperature record and compare it with the spectrum of CH_4 and CO_2 (Figure 3.2). The cause-and-effect connections between these correlations are still under investigation, but for methane it appears that changes are caused by the climatic variations. Climatic variations, such as a global cooling during an ice age, affect all the sources of methane, mostly in the same direction. During ice ages, populations of wild ruminants and other herbivores decline, soil biological activity is reduced, and temperate wetlands freeze, lowering the net flux of methane to the atmosphere and thus reducing the concentration (see Khalil and Rasmussen, 1989; Chappellaz et al., 1993). At the same time the removal of methane is also changed. The destruction of OH is reduced because of a lack of CH_4 and CO (more on OH in Section 3.4.6). This could cause OH to increase and thus reduce the lifetime of CH_4. However, lower concentrations of O_3 and water vapor reduce the production of OH, compensating the effect of CO and CH_4, and stabilize the concentrations of OH (Pinto and Khalil, 1991; Thompson, 1992). The atmospheric concentrations of methane during an ice age are therefore mostly affected by the reduction of sources; at least, such is the current thinking. There are reasons to believe that N_2O concentrations also decrease during ice

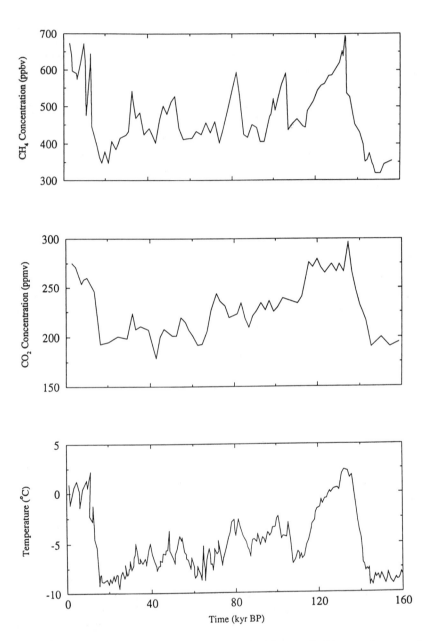

FIGURE 3.1.   Concentrations of methane and carbon dioxide in the Earth's atmosphere over the last 160,000 years. (Data taken from Barnola et al., 1987, Chappellaz et al., 1990, and Jouzel et al., 1987.)

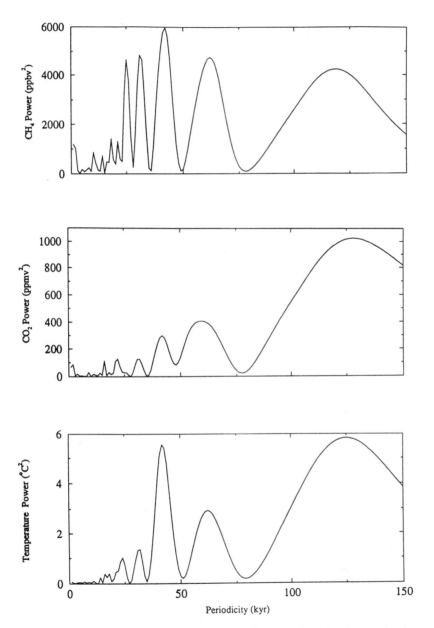

FIGURE 3.2. Spectrum of the average temperature and concentrations of methane and carbon dioxide (from Figure 3.1). The spectrum is given as $P = \Sigma \ a(\omega)^2 + b(\omega)^2$, where C is described as $C_o + a(\omega)\sin(\omega \ t) + b(\omega)\cos(\omega \ t)$. C_o is the average concentration and ω is the frequency of cycles contained in the data.

ages, possibly because of a reduction in global biological activity (see Khalil and Rasmussen, 1989, Zardini et al., 1989).

3.4. CHANGES BETWEEN PRE-INDUSTRIAL AND PRESENT TIMES

Substantial effects of human activities on the global cycles of trace gases have arisen only during the most recent century. Since the middle of the last century the human population has increased from 1 to 5.3 billion, bringing with it vast changes in the surface of the Earth and disturbing the global balances of trace gases. Changes in atmospheric composition that have occurred are unprecedented during the last several thousand years. The rate of change is much faster now than can occur from natural climatic cycles discussed earlier, although it is hard to prove that there were no times when trace gas concentrations did not rise rapidly and later fall. This is because the ice core data, as obtained by current experimental techniques, buffer rapid changes, as we have said before.

The changes that are occurring now have their origins in the industrial revolution and the expansion of technology. These changes are directly related to human events and activities. There is a prevalent belief that the changes man is causing to the atmospheric environment are unnatural and unhealthy. If this is so, then the changes in atmospheric composition during the last century are of the greatest practical significance. These trends also form the basis of plans to stop global pollution and possibly to reverse it. Such plans bring with them enormous social and economic changes that are of immediate interest to everyone, even if global change science is not. Here we review briefly the cycles of the main gases involved in global environmental change.

3.4.1. Carbon Dioxide

Carbon dioxide belongs to the group of gases that lie in the transition between the major gases and the trace gases. It is the most significant gas for causing global warming in the future. This is because it has a long effective lifetime before it is tied down to forms that cannot affect atmospheric concentrations over time scales of our interest and because it has a high concentration compared with the other gases involved in global warming.

Carbon dioxide was the first gas shown to be increasing owing to human activities and was implicated in global warming in the future (see Keeling et al., 1982). The causes of the increase in CO_2 are believed to be dominated by the burning of fossil fuels, which contribute some $6\,PgC\,year^{-1}$ to the atmosphere ($1\,Pg = 10^{15}\,g$) (Marland and Rotty, 1984). Fossil fuels are a reservoir of carbon that would not be released as CO_2 to the atmosphere were it not for the burning of fossil fuels for energy. Deforestation and land use change are also causes of

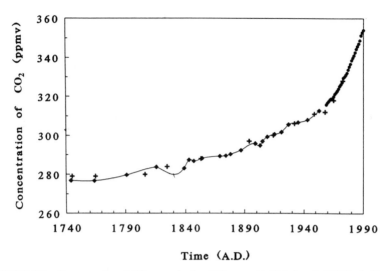

FIGURE 3.3.   Concentration of CO_2 over the last 250 years: ×, Neftel et al. (1985); ◆, Fredli et al. (1986); ●, current. (Data from Friedli et al., 1986.)

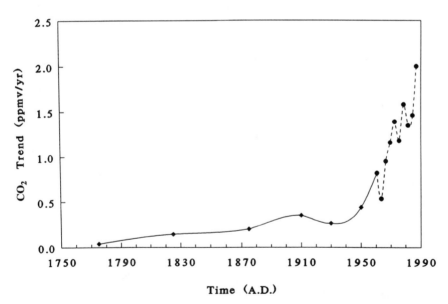

FIGURE 3.4.   Trends of CO_2 during the last two centuries (using the data shown in Figure 3.3). The trends were calculated, using linear regression, over 50-year (non-overlapping) periods between 1750 and 1900, over 10 year periods between 1900 and 1960, and over 3 year periods between 1959 and 1988. The calculated trends are placed at the middle of the span of the data in each calculation. These time scales are determined by the amounts of data available.

increasing CO_2, currently estimated to be about 2 PgC year (Houghton et al., 1987). There are other man-made sources of lesser importance. In addition, human activities have also changed the shorter term cycles of CO_2 and thus its seasonal variations. On more recent time scales, the pre-industrial concentration was about 280 ppm, and now it is 350 ppm, as shown in Figure 3.3. The trend of CO_2 during the recent period of direct global measurements has changed considerably from year to year, reflecting in part the interannual variability (IAV) mentioned earlier. During the years of the El Niño, concentrations of CO_2 are below normal, causing the observed trend to be smaller (Bacastow, 1976). There are other factors that also contribute to changes in trends from year to year (see Figure 3.4).

3.4.2. Methane

The increase of methane in the atmosphere was discovered only recently (Rasmussen and Khalil, 1981). The trends have been verified by several studies since (Blake and Rowland, 1988; Steele et al., 1987). There has been sustained interest in its trends ever since because methane is about 20 times more effective at present concentrations than CO_2 in causing global warming. The lifetime of methane is only 10 years, compared with a 150–200 year lifetime for CO_2. This means that a molecule of methane can cause global warming for a much shorter time than CO_2. On the other hand, the atmospheric oxidation of methane mostly ends by producing CO_2, so that a molecule of methane released to the atmosphere is always likely to be more important for global warming than a molecule of CO_2. During the last decade there has been considerable progress in understanding the origins and fate of atmospheric methane and the role of human activities in causing increasing trends (for compilations of current results see Khalil and Shearer, 1993). We look first at the sources of methane, then at the observed trends during the last several hundred years, and finally at the causes of the increases.

The sources of methane may be classified as major, minor, and very small. Major sources are the natural wetlands, rice agriculture, and domestic cattle (and other animals). Each of these sources probably exceeds 60 Tg year^{-1}. Next are sources from 10–60 Tg year^{-1} including biomass burning, landfills, coal mines, termites, sewage disposal, natural gas leakages, lakes, oceans, and tundra. The very small sources are much less than 10 Tg year^{-1} each and include biogas pits, asphalt, several industrial sources, and possibly others not yet found (see Table 3.1). The reason the minor sources are important is that there are many of them and together they represent a substantial fraction of the annual emissions. It is remarkable that most of the sources now dominated by human activities have natural precedents. The major sources affected by human activities are closely tied to food and energy production and therefore tend to increase with increasing population. Based on data compiled by the United Nations Food and Agriculture

TABLE 3.1   Estimates of Methane Emissions from Various Sources Described in the Following Chapters

	NATO–ARW Methane Budget	
Source	Methane (Tg)	Range (if available)
Natural Sources		
Wetlands[e]	110	
Termites[f]	20	15–35
Open ocean[f]	4	
Marine sediments[f]	[b]	8–65
Geological[f]	10	1–13
Wild fire[g]	2	2–5
Total[a]	150	
Anthropogenic Sources		
Rice[h]	65	55–90
Animals[i]	79	
Manure[j]	15	
Landfills[j]	22	
Waste-water treatment[j]	25	27–80
Biomass burning[g]	50	
Coal mining[k]	46	25–50
Natural gas[c,k]	30	
Other anthropogenic[f,k]	13	7–30
Low temperature fuels[f]		
	17	
Total[a]	360	
[14]C-depleted sources[a,d]	120	
All Sources[a]	510	

[a]Rounded.

[b]The amount of CH_4 generated was estimated, but no estimates of emissions were given.

[c]Beck et al. did not make a final estimate for natural gas emissions. This was inferred from Tables 5–8 in their chapter.

[d][14]C-depleted sources were considered to be the "geological" source, plus emissions from coal mining, natural gas, other anthropogenic (mainly industrial and transportation fossil fuel combustion), and low-temperature fuel combustion.

Source: From Khalil and Shearer (1993). All other references are to various chapters in Khalil, M.A.K., Ed. (1993). *Atmospheric Methane: Sources, Sinks and Role in Global Change*, Springer Verlag, Heidelberg: [e]Matthews; [f]Judd et al.; [g]Levine et al.; [h]Shearer and Khalil; [i]Johnson et al.; [j]Thorneloe et al.; [k] Beck et al.

Organization (FAO), it appears that cattle populations have increased by a factor of 2.5 between the turn of the century and now, while the area of rice harvested has increased twofold during the same time. Similarly, other sources such as natural gas leakage, landfills, and biomass burning have also increased substantially. The larger sources are mostly responsible for the major increase in methane that occurred between a 100 years ago and now.

There are several constraints that define the total emissions and the anthropogenic fraction. These constraints do not apply to emissions from individual sources, however. Although there are many uncertainties about the annual emissions of methane from individual sources, the total budget agrees with both the timing and magnitude of the observed trends during the last 100–200 years including the downturns and increases in the observed trends (Khalil and Rasmussen, 1987). Methane is removed from the atmosphere mostly by reactions with tropospheric OH. Lesser amounts are removed by dry soils and in stratospheric processes.

The concentrations of methane in Figure 3.5 are taken from various sources (Ethridge et al., 1992; Rasmussen and Khalil, 1984; Khalil and Rasmussen, 1987). The time-dependent trends (discussed in Section 3.2.3) calculated by linear regression are shown in Figure 3.6. Several noteworthy features are visible. It seems that the concentrations of methane were nearly constant for several thousand years until recently. Concentrations fell during the Little Ice Age (Khalil and Rasmussen, 1989). For the other times, concentrations have increased, starting about 100 years ago. During this time there are three periods when the trends have slowed. The first, centered at around 1900, is of unknown origins; the second, around the 1930s and 1940s, probably represents the effects of the Great Depression and the Second World War. During these years, data compiled by Mitchell (1980, 1982, 1983) show that cattle populations and rice agriculture both declined. Other sources controlled by human activities may also have been affected during this time. The last slow-down of trend is happening now.

The concentration in the most recent decade is shown in Figure 3.7, while the decreasing trend is shown in Figure 3.8. The causes of this change are not fully understood at present but are likely to be a combination of changes in both sources and sinks. FAO statistics show that cattle populations have risen by only 6% during the decade of the 1980s, while the rice area harvested has increased by a mere 2% during the same time. These slow changes may be contrasted with the post-war boom, when the same sources increased by about 17% and 14%, respectively in the decade of the 1950s. There are many reasons why these major sources controlled by human activities are not increasing as before. Part of the reason is that technological advances have produced higher-yielding rice and, in the US and Europe, more efficient dairy cattle. The global populations of cattle, however, are affected by their numbers in India and the rest of the Third World. There the limitation of cattle populations is often determined by the availability

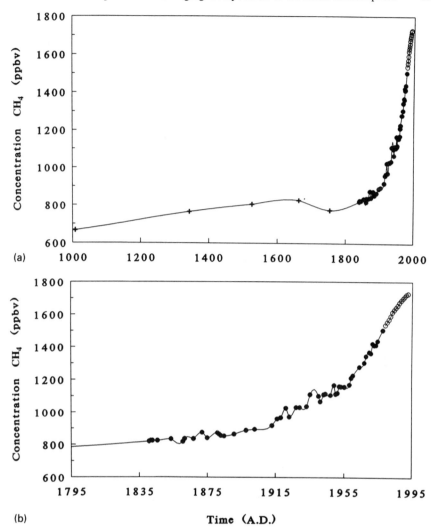

FIGURE 3.5. Concentrations of methane (a) over the last 1000 years and (b) over the last 200 years. Data of Ethridge et al. (1992) (●) and Rasmussen and Khalil (1984) (+) are from ice core samples. Data of Khalil and Rasmussen (○) are global averages from weekly flask samples collected at various latitudes.

of food. Similarly for rice agriculture, the land that is left requires enormous economic resources and effort to farm. One may argue, therefore, that the slow-down in these sources is partly due to natural limits to their continued growth.

It is not known at present whether emissions of methane from other anthropogenic sources such as biomass burning, landfills, or natural gas leakages

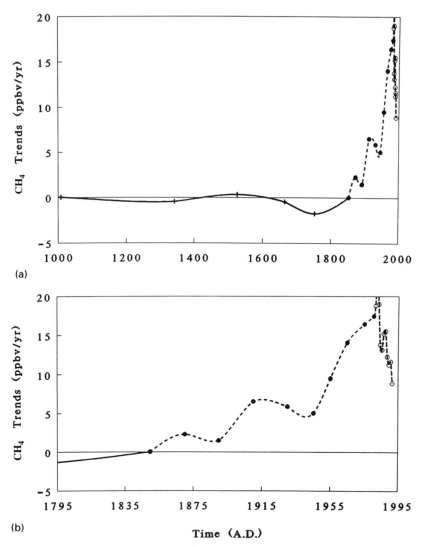

FIGURE 3.6.   Trends of methane concentrations (a) during the last 1000 years and (b) during the last 200 years (using the data shown in Figure 3.5): Data of Ethridge et al. (1992) (●) and Rasmussen and Khalil (1984) (+) are from ice core samples. Data of Khalil and Rasmussen (○) are global averages from weekly flask samples collected at various latitudes. Rapid increases in methane started only about 200 years ago. These are linear regression estimates of trends over various (non-overlapping) periods of time between about 0 A.D. and the present. For data between 1840 and 1940 trends were calculated over 20 year periods, for 1940–1980, over 10 year periods, and for 1980–1992 over 2 year periods. The calculated trends were placed at the middle of the time span in each calculation. The earlier data are more sparse. Trends for the period between 0 and 1800 A.D. were calculated for every 10 data points, and the trends are placed at the average time spanned by the 10 data.

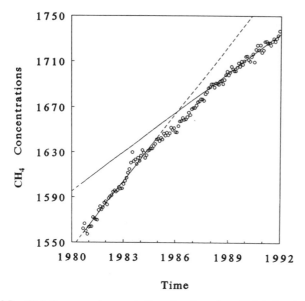

FIGURE 3.7.   Globally averaged concentration of methane (monthly) during the last decade.

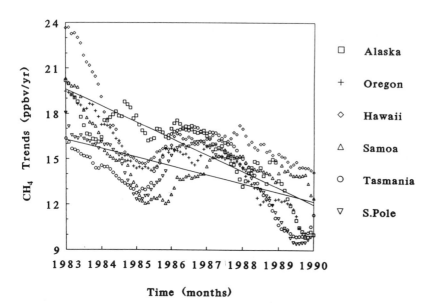

FIGURE 3.8.   Rate of change in methane concentrations during the recent decade. Methane was increasing faster at the beginning of the record than now.

are increasing or decreasing. It is very likely, however, that the total annual emissions are no longer increasing as rapidly as before and may even be constant or declining. Because of the moderately long lifetime of methane (about 10 years), the effect of a slow-down in the trend of sources or a faster removal rate does not immediately show in the atmospheric observations. The other possibility to explain why the trends are slower now is that OH may be increasing. There are several mechanisms by which this can happen, at least during the recent decade (more on this in Section 3.4.6).

3.4.3. Nitrous Oxide

Nitrous oxide is even more potent than methane and, molecule for molecule, about 200 times as effective as CO_2 in causing global warming and is involved directly in the catalytic destruction of stratospheric ozone. The increasing trend of N_2O was reported by Weiss (1981) and by others since. Nitrous oxide has received less

TABLE 3.2   Natural and Identified Anthropogenic Sources of N_2O (in Tg year^{-1})

	Factor[a]		
Source	Middle	Range	Uncertainty
Anthropogenic sources			
Biomass burning[b]	1.6	0.2–3	15
Power plants[c]	0.0	0.0–0.2	20
Nylon manufacture[d]	0.7	NK	NK
Nitrogen fertilizer[e]	1.0	0.4–3	8
Sewage[f]	1.5	0.3–3	10
Cattle–agriculture[g]	0.5	0.3–1	NK
Aquifers–irrigation[h]	0.8	0.8–2	NK
Automobiles[g,i]	0.8	0.1–2	20
Global warming[j]	0.3	0.0–1	NK
Land use change[k]	0.7	NK	NK
Atmospheric formation	NK	NK	
Total	8	5–10	
Natural Sources			
Soils[l]	12	–	–
Oceans[l]	3	–	–
Total	15		

[a]NK, not known; uncertainty factor, max/min of range; some numbers, especially in the ranges, are rounded.

Source: [b]Cofer et al. (1991), lower limits from Crutzen and Andrea (1990); [c]Linak et al. (1990), Sloan and Laird (1990), Yokoyama et al. (1991), Khalil and Rasmussen (1992b); [d]Thiemens and Trogler (1991); [e]Eichner (1990) and primary references therein, Conrad et al. (1983); [f]Kaplan et al. (1978); [g]Khalil and Rasmussen (this work); [h]Ronen et al. (1988); [i]EPA (1986); [j]Khalil and Rasmussen (1989); [k]Matson and Vitousek (1990), Luizao et al. (1989); [l]McElroy and Wofsy (1987), WMO (1985).

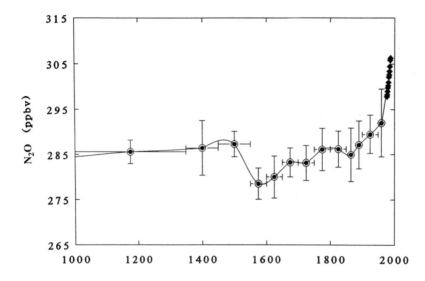

(a)

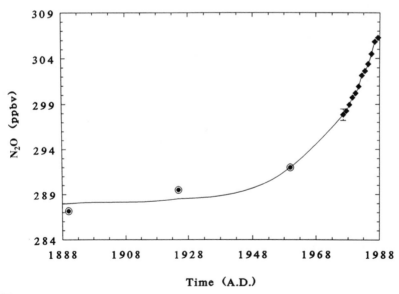

(b)

FIGURE 3.9. Concentrations of nitrous oxide (a) over the last 1000 years and (b) over the last 100 years.

attention than methane and the CFCs, partly because the past rates of increase were slow, but now the trend is escalating (Khalil and Rasmussen, 1992a).

A recent budget of N_2O is given in Table 3.2. It shows that the anthropogenic emissions of N_2O are distributed among many small sources. Emissions from these sources are not well defined at present. N_2O is removed from the atmosphere mostly by photochemical reactions in the stratosphere and by reacting with $O(^1D)$, giving it a long estimated lifetime of between 100 and 150 years. The concentrations and trends of N_2O during the last 1000 years are shown in Figure 3.9. Increases in N_2O started more recently than for methane. An indication that the rate of increase is itself increasing is shown in Figure 3.10, based on detailed measurements taken during the last 15 years. It is therefore possible that N_2O will become much more important in global change than the current assessment suggests. The continued increases in N_2O can arise from increasing uses of nitrogen based fertilizers. For instance, nitrogen fertilizers are being used in more places in rice agriculture where traditionally organic fertilizers have been used. This will cause a shift from methane emissions to nitrous oxide emissions. Since the lifetime of N_2O is long, it can be a more potent gas for causing global warming when compared with methane (see Wang et al., 1976; Lacis et al., 1981; Chapter 13).

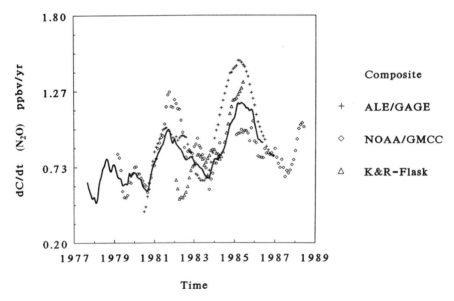

FIGURE 3.10.   Trend of N_2O during the last 15 years. The rate of increase is calculated over 3 year overlapping periods of time as discussed in the text. The trend has been variable but appears to be increasing.

3.4.4. Major Chlorofluorocarbons: CFC-11 and CFC-12

The two chlorofluorocarbons CCl_3F (CFC-11) and CCl_2F_2 (CFC-12) are perhaps the most commonly known gases thought to affect the global environment. These gases were used as propellants in cans of household products ranging from cheese spreads to hair sprays. Partly for this reason, it came as somewhat of a shock to most people that their life styles may change the global environment to a state that may be harmful to the environment and to human health; yet these man-made chemicals were, in many of their uses, unique and otherwise the safest and best suited. The strenuous arguments and political maneuvers have now long been settled, resulting in the Montreal Protocol. The accord and its amendments will eliminate the industrial production of these chlorofluorocarbons and thus reduce or eliminate the threat they may have posed to the global environment (see Chapters 7 and 11 for more details).

The chlorofluorocarbons have no natural sources. If the ice core sample is not contaminated, CFC-11 and CFC-12 concentrations are below detection limits. CFC-11 and CFC-12 have similar categories of uses. CFC-11 is released from refrigerators, rigid polyurethane foams, in CFC-11 manufacturing processes (about 5%), and from spray cans when used as a propellant. How much is released every year is complicated by the variable residence time of CFC-11 in each use. For instance, from spray cans CFC-11 is released perhaps within 1 year after production, but from refrigerators it may be 10–20 years before it is released. The case of rigid foams is different still. When CFC-11 is used to blow foams, a substantial amount is lost in the process (10–20%). The rest is tied up for perhaps as long as 100 years. If the foam is taken out of use, crushed, burned, or put in landfills, the release rate changes again. While it is commonly believed that the annual emissions of CFC-11 and CFC-12 are accurately known, this is not so. Even the amount produced every year has uncertainties because of lack of information from some countries, most notably Russia and China. Lack of data on manufacturing losses and losses during storage increases the uncertainties in estimates of emission. Added to these complexities are the variable and inherently stochastic release patterns from the widely differing uses and the different storage times in applications (for details see Khalil and Rasmussen, 1993). These CFCs are destroyed almost entirely in the stratosphere by photochemical processes, where they release chlorine atoms (CFC + $h\nu \rightarrow$ Cl + products) which catalytically destroy ozone in a series of chain reactions (see Chapter 11 for details). Eventually the chlorine atom reacts with methane and is tied up into HCl or, by another reaction, in a relatively stable reservoir. The HCl then diffuses back down to the troposphere and is removed from the atmosphere. Besides stratospheric destruction, small amounts are removed by soils, and some CFC-11 is stored in the oceans (Khalil and Rasmussen, 1993).

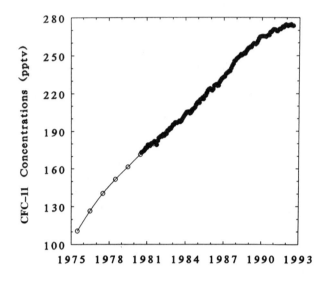

(a)

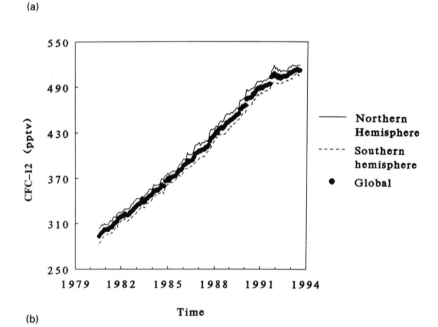

(b)

FIGURE 3.11. Globally averaged concentrations of (a) CCl_3F (CFC-11) and (b) CCl_2F_2 (CFC-12) from the 1970s to the present.

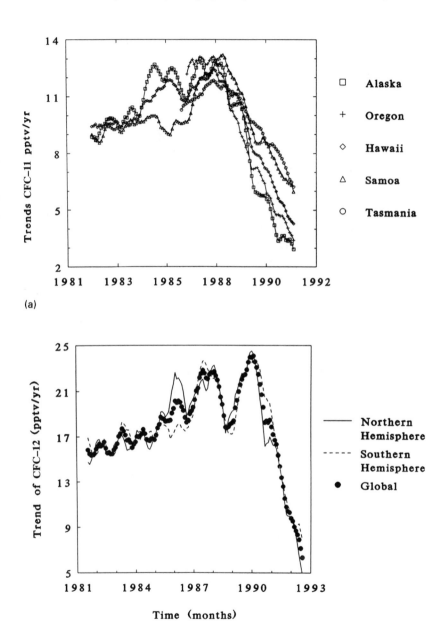

(a)

(b)

FIGURE 3.12.   Trends of (a) CCl$_3$F (CFC-11) and (b) CCl$_2$F$_2$ (CFC-12) during the last decade. Trends are calculated over 3 year overlapping periods for CCl$_3$F and 2 year overlapping periods for CCl$_2$F$_2$.

The concentrations and trends of CFC-11 and CFC-12 are shown in Figures 3.11 and 3.12. Concentrations have increased during the course of the measurements. Since about 1987, however, the rates of increase have plummeted and the concentrations are approaching stability. At present CFC-11 and CFC-12 concentrations are still rising, but perhaps very soon concentrations will start to fall if the Copenhagen Agreement is implemented, which bans production after 1996.

We trace briefly the history of CFC-11 in the atmosphere, since it may be regarded as a prototype representing the behavior of many man-made gases that have an effect on the global environment (more about other gases in the next section). CFC-11 was put into use late in the 1930s. Its use grew steadily and substantially after about 1950. After two decades of increasing usage, CFC-11 (along with CFC-12) was implicated in the depletion of the ozone layer (Molina and Rowland, 1974). The trends were about $14 \, \text{ppt year}^{-1}$ and were at an all time high in 1974. Around this time the trend of CFC-11 started decreasing slowly, probably because of market conditions, and later was affected mostly by the impending ban on inessential uses that was implemented in 1979. Between 1974 and 1982 the rate of accumulation decreased to about $8 \, \text{ppt year}^{-1}$. However, during this time, other uses increased, particularly the blowing of rigid polyurethane foams with CFC-11. After 1982 the rate of increase in CFC-11 picked up again and reached about $11 \, \text{ppt year}^{-1}$ by 1987. Since then the trend of CFC-11 has slowed dramatically as we approach the time for the ban to take effect. The global annual increases in CFC-11 and CFC-12 during the last few years are $3-4 \, \text{ppt year}^{-1}$ and falling; they are the lowest rates ever observed (see Figure 3.12). An accounting of how much

TABLE 3.3   Reservoirs of CFC-11 in 1991

Category	Gg	% of Total Produced
Refrigeration	89	1.0
Foams	1031	11.3
Aerosol + other	24	0.3
Unreported	166	1.8
Stratosphere	741	8.1
Troposphere	5360	58.6
Ocean surface	27	0.3
Ocean deep	6	0.1
Soils	1	0.0
Removed stratosphere	1709	18.7
Total	9152	100

Source: From Khalil and Rasmussen (1993).

CFC-11 was released to the atmosphere since its earliest uses to the present and where it went is given in Table 3.3. It shows that about 67% of all the CFC-11 produced since the 1930s is still in the atmosphere. Only about 19% has been removed by stratospheric processes. The rest is tied up in various reservoirs, the largest at the Earth's surface being polyurethane foams at 11%. This CFC-11 may still make its way into the atmosphere in the future, long after industrial production stops.

3.4.5. Other Long-Lived Gases

There are many man-made trace gases that can affect the global climate and atmospheric chemistry. These gases do not individually rival the potential of the main gases mentioned earlier (namely, CO_2, CH_4, N_2O, CFC-11, and CFC-12). This is either because these gases are not as effective in causing global change or, more likely, because their concentrations are low and expected not to rise to such levels as to be significant. We mention here some particularly noteworthy gases (see Ramanathan et al., 1985 and Chapter 13).

There are several chlorine-containing gases that have effects similar to CFC-11 and CFC-12. The most noteworthy are $CHCl_2F$ (CFC-22), $C_2Cl_3F_3$ (CFC-113), CCl_4, and CH_3CCl_3. CFC-22 is an existing chemical capable of replacing CFC-11 and CFC-12 in some forms of refrigeration and air conditioning and is gaining use accordingly. CFC-113 is a high technology solvent used for cleaning electronic components. CH_3CCl_3 has widespread uses as a degreasing solvent and was until recently found in consumer goods such as household drain cleaners. CCl_4 is also a solvent but is not much used in this capacity in Europe and the US; there is evidence that it is still used in other countries, particularly Russia and China. In western countries it has been used as a starting stock to manufacture CFC-11 and CFC-12, and so a certain fraction is inadvertently released into the atmosphere. The present concentrations of CFC-22 and CFC-113 continue to increase; however, CCl_4 and CH_3CCl_3 concentrations are now actually declining in the atmosphere, so the rate of change is negative.

CF_3Br, CF_2BrCl, and CH_3Br are the main bromine-containing gases from anthropogenic sources that are currently included in plans to protect the ozone layer. The first two are man-made halon fire-extinguishing compounds, while methyl bromide is an agricultural fumigant. All these gases have very low concentrations of 1–3 ppt for the halons and about 10 ppt for CH_3Br. It is estimated that most of the CH_3Br in the atmosphere comes from natural sources, mostly in the oceans. At most, some 3 ppt may come from anthropogenic sources. Moreover, the lifetime of CH_3Br is about 1 year, making it less important for ozone depletion compared with the longer-lived halons and chlorofluorocarbons.

3.4.6. Reactive Gases and the Oxidizing Capacity of the Atmosphere

We have discussed the origins, fates, and cycles of long-lived gases in the atmosphere (lifetimes of 5–200 years). The effects of human activities on these gases are easier to see and document, as discussed in Section 3.2. On the other side of the spectrum are reactive species and gases in the atmosphere that have major roles in the global environment and may be responsible for feedbacks that can amplify or dampen the disturbances of the environment. The most important gases in this category are O_3, OH, and water vapor. CO may be added to this list, although it is of lesser significance. The trends of these gases are not clear and are likely to be variable, depending on the conditions during the year.

O_3 in the stratosphere, which is of photochemical origin, is believed to be decreasing because of the emissions of CFCs, bromine gases, and N_2O from human activities. Decreasing trends have been clearly documented in specific locations, such as the "ozone hole" over Antarctica. Tropospheric ozone is formed by chemical reactions, and some comes from stratospheric–tropospheric exchanges of air. Human activities tend to increase tropospheric ozone, for which there is recent evidence at some locations (WMO, 1988; see Chapter 10).

Hydroxyl radicals are extremely short-lived by the standards of the other gases discussed here, with lifetimes of a few seconds. They are formed by solar radiation at frequencies < 320 nm, which splits tropospheric O_3 into O_2 and $O(^1D)$. The $O(^1D)$ reacts with water vapor to make two OH. Several other mechanisms supplement this production process. The production of OH is balanced by its removal due to reactions with more stable gases, mostly methane and carbon monoxide, but light non-methane hydrocarbons may also be significant on a global scale. The oxidizing capacity of the atmosphere is dependent mostly on OH, since it removes many natural and man-made gases from the atmosphere. If OH decreased, it would reduce the ability of the atmosphere to remove an entire class of gases that can add to global warming and deplete stratospheric ozone. For instance, most of the substitute compounds designed to replace CFC-11 and CFC-12 depend on the reaction with OH to destroy them in the troposphere. Because the concentration of OH can be reduced by high levels of CO and CH_4, it was thought for some time that OH must be decreasing and that there would be a positive feedback on the cycles of methane, CO, and other gases. However, as CO and CH_4 have increased, so has tropospheric O_3, and now possibly even the solar UV, needed to make OH, may be increasing. This can happen, for instance, if stratospheric O_3 is reduced, which allows the additional UV to reach the troposphere. Thus the OH depletion from increased CO and CH_4 is compensated by increased production rates. In recent years it may even be that OH is increasing, thus causing a negative feedback on climate by limiting the concentrations of CH_4 and other gases that can cause

---

global warming (see Madronich and Granier, 1992; Prinn et al., 1992; Thompson, 1992).

At present there is no practical means to directly measure the trends of OH. There is a way to estimate OH trends indirectly by using the mass balance of methyl chloroform (CH_3CCl_3), whose production is confined to a few manufacturers and emissions are well documented. This idea was first suggested by Singh (1977) and Lovelock (1977). The method can produce a benchmark estimate of effective global OH, but there are enough problems and uncertainties that it is not yet sufficient to produce a credible estimate of the trend of OH (see Chapter 7 for an in-depth discussion). In time, however, there is a good chance that, by using select tracers of OH (e.g., CH_3CCl_3), we can probably say whether OH, and therefore the oxidizing capacity of the Earth's atmosphere, is increasing or decreasing on the global scale.

Although it is not yet possible to determine whether OH is increasing or decreasing, and it is not certain whether tropospheric O_3 is increasing on the global scale and by how much, there are sufficient data on CO to detect and quantify trends that are likely to represent the global atmosphere. A budget of CO is given in Table 3.4. It shows that most of the CO in the atmosphere

TABLE 3.4   Sources of Carbon Monoxide (in Tg year^{-1})

Source	Anthropogenic	Natural	Global	Range
Directly from combusion				
Fossil fuels	500	–	500	400–1000
Forest clearing	400	–	400	200–800
Savanna burning	200	–	200	100–400
Wood burning	50	–	50	25–150
Forest fires	–	30	30	10–50
Oxidation of Hydrocarbons				
Methane	300	300	600	400–1000
Non–methane HCs	90	600	690	300–1400
Other Sources				
Plants	–	100	100	50–200
Oceans	–	40	40	20–80
Total	1500	1100	2600	2000–3000

(1) Estimates are expressed to one significant figure. Totals are rounded to two significant digits.
(2) Half the production of CO from the oxidation of CH_4 is attributed to anthropogenic sources and the other half to natural sources based on the budget of CH_4.

Source: Adapted from Logan et al. (1981) and revisions reported by WMO (1985). This version is taken from Khalil and Rasmussen (1990a).

comes from combustion processes and from atmospheric oxidation of hydro-carbons, including methane. The amount from man-made sources, or sources controlled by human activities, is perhaps around 60% of the total. It stands to reason that the concentration of CO must now be much higher than in the pre-industrial atmosphere of 100 or more years past. Because of experimental difficulties, there are no ice core measurements, or any other test, to prove whether CO has increased during the last century or not; it is generally assumed that it has. The idea is supported by the observation that at times CO concentrations have increased in the atmosphere (Khalil and Rasmussen, 1990a; Rinsland and Levine, 1985).

The increasing trends of CO were difficult to quantify by direct measurements and were slow at about 1% per year between 1981 and 1987. During the last few years, however, there is good evidence that CO concentrations are beginning to fall (trend is now negative). With CO concentrations decreasing, it seems more probable that OH is not declining. We show the global CO data in Figure 3.13 and trends in Figure 3.14. The main characteristics of the gases we have discussed in this section are given in Table 3.5. It should be noted, however, that both concentrations and trends are changing and that the data displayed have to be associated with the times over which measurements were taken.

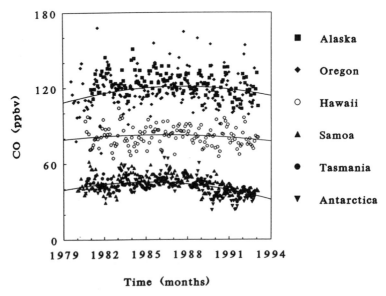

FIGURE 3.13. Atmospheric concentrations of carbon monoxide. (From Khalil and Rasmussen, 1993.)

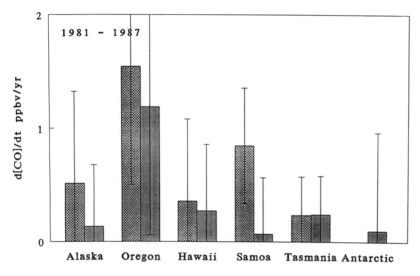

(a)

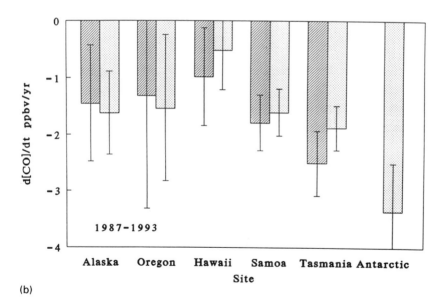

(b)

FIGURE 3.14.   Trends of CO during the last 12 years. (a) During the first 6 years the concentrations increased, while (b) during the last 6 years the concentrations have been decreasing.

3.5. CONCLUSIONS, CONNECTIONS, AND PERSPECTIVES

We have two remaining thoughts that have a bearing on the focus of this chapter. The first is to discuss aspects of the forces behind trends and man-made changes to the atmosphere, and the second is to summarize the many intricate connections between the cycles of various gases that seem to cause stability rather than instability. As we said earlier, human activities can increase the concentrations of gases by directly increasing emissions (e.g., CO_2, CH_4, N_2O, and CO) or by putting gases into the atmosphere that no natural mechanism can produce (e.g., CCl_3F and CCl_2F_2). Alternatively, we can reduce the rate of removal of a gas, which will also cause its concentration to rise. It seems that an increase in emissions is usually the major mechanism for the atmospheric increase in most gases, while changes in the removal rates may play an important but lesser role.

TABLE 3.5   Characteristics of Atmospheric Trace Gases Important in Global Change

Gas		Concentrations[a]		
		C (Present)	C (Pre-industrial)[b]	Gradient (%)[c]
CO_2	(ppm)	350.2 ± 0.3	275	0.47 ± 0.02
CH_4	(ppb)	1686 ± 2	700 ± 19	6.2 ± 0.2
N_2O	(ppb)	302.3 ± 0.2	285 ± 1	0.27 ± 0.03
CCl_3F	(ppt)	198 ± 2	0	6.9 ± 0.3
CCl_2F_2	(ppt)	329 ± 2	0	7.1 ± 0.2
CH_3CCl_3	(ppt)	146.1	0	30 ± 1
CO	(ppt)	90.5	40	250
CCl_4	(ppt)	150	NK	
$CHCl_2F$	(ppt)	120	NK	
$C_2Cl_3F_3$	(ppt)	60	NK	

Gas	Trends[a]			
	C (Quadratic) $= a + b_0 t + b_1 t^2$			
	dC/dt (ppx year^{-1})	a (ppx)[d]	b_0 (ppx year^{-1})	b_1 (ppx year^{-2})
CO_2	1.63 ± 0.05	315 ± 1	0.54 ± 0.01	0.00168 ± 0.00004
CH_4	16.4 ± 0.05	1555 ± 10	21.2 ± 1.7	−0.049 ± 0.017
N_2O	1 ± 0.07	298 ± 1	0.57 ± 0.08	0.020 ± 0.006
CCl_3F	8.9 ± 0.1	156.9₊ 1.1	10.3 ± 0.4	−0.0227 ± 0.0064
CCl_2F_2	16.1 ± 0.2	255.5₊ 1.9	18.7 ± 0.7	−0.0418 ± 0.0112
CH_3CCl_3	5.1 ± 0.8	47 ± 3	11.1 ± 0.3	−0.0236 ± 0.0015
CO	0.9 ± 0.4	–	–	–
CCl_4	2.0	–	–	–
$CHCl_2F$	7.0	–	–	–
$C_2Cl_3F_3$	5.0	–	–	–

TABLE 3.5 (*Continued*)

	Lifetime (years)	Emissions (Tg year^{-1}) S_a^e	Emissions (Tg year^{-1}) S_n^e
		Lifetimes and Source Emissions[a]	
CO_2	200	30000	–
CH_4	9 ± 2	350	200
N_2O	150 ± 30	7	15
CCl_3F	70 ± 20	0.29	0
CCl_2F_2	100 ± 50	0.41	0
CH_3CCl_3	7 ± 2	0.61	0
CO	0.2	1500	1100
CCl_4	50	–	NK
$CHCl_2F$	15	–	NK
$C_2Cl_3F_3$	90	–	0

[a]The periods over which data were taken for all calculations in this table are as follows. For quadratic trends the entire periods are used over which data are available. These periods are: CFC-11 and CFC-12, 1978.58–1983.5; CH_3CCl_3 1975–1988; CO_2, 1958.17–1989; CH_4, 1980.67–1988.67; N_2O, 1976.17–1987.92. The global average concentrations are reported for as recent a time as possible. CO_2 is the average for 1988 and the trend is for the 5 years from 1984 to 1989.0; average CH_4 is for 1987.67–1988.67; average CFC-11 and CFC-12 are for 1982.5–1983.5 and the trends of CH_4, CFC-11 and CFC-12 are for the entire period of the data; average N_2O is for 1987 and the trend is for 1983–1987; average CH_3CCl_3 is for 1988 and the trend is for 1985–1988; CO trend and concentration are for 1980–1988. The emission rates for N_2O are for 1979–1987; CFC-11 and CFC-12 emissions are averages for 1975–1982; CH_3CCl_3 emissions are averaged over 1983–1988. Data for CCl_4, $CHCl_2F$ (CFC-22), and $C_2Cl_3F_3$ (CFC-113) are for 1990 and taken from the IPCC; see Houghton et al. (1990).
[b]Ice core data are taken from Khalil and Rasmussen (1987, 1988), Stauffer et al. (1985), Raynaud et al. (1988) and Friedli et al. (1986). For some halocarbons there are no data from ice cores, although it is likely that there are no natural sources and the concentrations are zero; denoted as NK for "not known".
[c]The % gradient is defined as [(C_n/C_s) – 1] × 100%, where C_n and C_s are the average northern and southern hemispheric concentrations, respectively.
[d]x = m for CO_2, b for CH_4, N_2O, and CO, and t for CH_3CCl_3, CCl_3F, and CCl_2F_2. Trends are correspondingly in units of ppmv year^{-1} or ppbv year^{-1} or pptv year$_{-1}$.
[e]S_a and S_n are anthropogenic and natural source terms, respectively.

Even though the long-term changes in trace gases such as methane may have been caused mostly by increasing sources, the trends during shorter periods, including the last decade, may be more affected by changing sinks.

What then drives the anthropogenic emissions of gases into the atmosphere? For methane, which comes from rice fields and domestic cattle among other sources, an increase in population leads to a proportional increase in agricultural activities, and thus for some time the increase in population is a controlling factor for the trends of emissions, as suggested by Khalil and Rasmussen (1985). After some time, however, other sociological and political factors become more significant. For example, rice agriculture and raising of domestic cattle have been undergoing changes recently that may decouple the relationship with population.

High-yielding rice varieties, for instance, reduce the need to put more land under rice agriculture. There is evidence that it is not the yield that affects methane emissions, but rather the hectares of land under water to produce the rice. New hybrid varieties reduce both the land needed and the length of time the fields are under water, thus reducing methane emissions. To increase yields, the use of nitrogen-based fertilizers replacing organic manure further decreases methane emissions. In this case, then, while the increasing population may continue to require more rice, the per capita methane emissions are much lower than without modern agricultural practices.

Besides these factors, sociological and economic factors may also play a role in the connection between population, agriculture, and methane emissions. For instance, in parts of China, as the country becomes more affluent and the farmers have a greater say in what they grow, they may prefer to grow vegetables and other crops rather than rice because of the higher market value of these crops. We believe that population is only one of the factors influencing the emissions of methane; other factors are political, sociological, and economic, which can enhance or reduce the effects of population increase. These matters relate to all gases emitted from agricultural processes as well as other sources, although we have used methane and rice agriculture as an example.

For gases such as the CFCs, increasing population is even less of a factor than for gases dominated by agriculture. In the case of the CFCs and other industrial gases the determining factors are per capita demand for products that use these materials, the extent to which these chemicals can be used in consumer goods, and the level of market saturation. Methyl chloroform, for instance, when it was first put into use as a degreasing solvent, filled a market left behind by restrictions on other more reactive solvents such as trichloroethylene and perchloroethylene. Once it filled these needs, the growth in demand, and therefore in the sales, slowed considerably. These factors were not much influenced by population increase.

Regulations that ban the worldwide production of a chemical emitted into the atmosphere are relatively recent and affect only the industrially produced gases. These gases are the easiest to control by legislative action. The trends of the CFCs have been affected greatly by these or the impending regulations. Legislative control of gases such as CO_2, CH_4, and N_2O is far more complex, and there are no international agreements to limit the anthropogenic sources. We can conclude that the sources of man-made gases are a major factor in controlling atmospheric concentrations and trends. The factors that drive anthropogenic sources are many and complicated, including population increase, economic conditions, per capita demands, and the level of affluence.

Our second point is that while controlling emissions of man-made gases to prevent global warming or ozone depletion is complicated by the interplay between various social processes, the gases involved are similarly interrelated, so that action on one gas may influence what happens to another. We have discussed some connections already. In review we note the following relationships. CO_2, CH_4, N_2O, CFC-11, and CFC-12 can all cause global warming. CFC-11, CFC-12, and N_2O can deplete the ozone layer. By causing global warming at the Earth's surface, these gases reduce stratospheric temperatures. Lower stratospheric temperatures slow the processes that destroy stratospheric ozone and thus reduce ozone depletion. Global warming is then a negative feedback on ozone depletion (e.g., Groves and Tuck, 1979). An increase in methane emissions leads to more methane in the stratosphere, where it produces H_2O. The additional water vapor destroys ozone, but the higher methane concentration speeds up the removal of chlorine atoms released by the CFCs and all other natural and man-made chlorine-containing compounds. Methane increase therefore reduces the destruction of ozone by the CFCs. The CFCs, as they deplete the ozone layer, may lead to a reduction in the methane concentration by increasing OH in the troposphere. This could reduce global warming from CH_4 and other gases. These series of connections tend to be negative feedbacks that reduce the effects of individual trace gases.

These and other interconnections make it clear that global change cannot be studied as a sum of effects from each gas influenced by human activities. We can observe the trends of gases in the atmosphere and define how the atmosphere is changing. How these changes affect the global environment is a most significant and complex question, especially when the interactions of the effects from some gases produce changes that compensate or exacerbate the effects from other gases. The coupling between atmospheric chemistry and the global climate is therefore the modern frontier of global change science.

Acknowledgments

We thank Martha J. Shearer, Francis Moraes, and Robert MacKay for discussions and comments; Donald Stearns, Robert Dalluge, and James Mohan for laboratory measurements; the NOAA/GMCC and later the NOAA/CMDL programs for collecting flask samples; and Dr Paul J. Fraser of CSIRO, Australia for providing samples from Tasmania. Financial support for this work was provided in part by a grant from the Department of Energy (DOE #DE-FG06–85ER60313). Support for work on trend analyses was provided in part by the US EPA (Order #2D3333NASA) issued to Andarz Co. Additional support was provided from the resources of the Biospherics Research Corporation and the Andarz Company.

References
Bacastow, R.B. (1976). Modulation of atmospheric carbon dioxide by the southern oscillation. *Nature*, **261**, 116–118.
Barnola, J.M., Raynaud, D., Korotkevich, Y.S. and Lorius, C. (1987). Vostok ice core

provides 160,000 year record of atmospheric CO_2. *Nature*, **329**, 408–414.

Blake, D. and Rowland, F. (1988). Continuing worldwide increase in tropospheric methane. *Science*, **239**, 1129–1131.

Chamberlain, J.W. and Hunten, D.M. (1987). *Theory of Planetary Atmospheres: An Introduction to Their Physics and Chemistry*, Academic Press, Orlando, FL.

Chappellaz, J.A., Barnola, J.M., Raynaud, D., Korotkevich, Y.S. and Lorius, C. (1990). Ice core record of atmospheric methane over the past 160,000 years. *Nature*, **345**, 127–131.

Chappellaz, J.A., Fung, I.Y. and Thompson, A.M. (1993). The atmospheric methane increase since the last glacial maximum: 1. Source estimates. *Tellus*, **45B**, 228–241.

Cofer, III, W.R., Levine, J.S., Winstead, E.L. and Stocks, B.J. (1991). New estimates of nitrous oxide emissions from biomass burning. *Nature*, **349**, 689–691.

Conrad, R., Seiler, W. and Bunse, G. (1983). Factors influencing the loss of fertilizer nitrogen into the atmosphere as N_2O. *J. Geophys. Res.*, **88**, 6709–6718.

Crutzen, P.J. and Andreae, M.O. (1990). Biomass burning in the tropics: impact on atmospheric chemistry and biogeochemical cycles. *Science*, **250**, 1669–1678.

Delmas, R.J., Ascencio, J.-M. and Legrand, M. (1980). Polar ice evidence that atmospheric CO_2 20,000 years B.P. was 50% of present. *Nature*, **284**, 155–157.

Eichner, M.J. (1990). Nitrous oxide emissions from fertilized soils: summary of available data. *J. Environ. Qual.*, **19**, 272–280.

EPA (Environmental Protection Agency) (1986). Proceedings of the workshop on N_2O emissions from combustion. *NTIS #PB87–113742*, Washington, DC.

Ethridge, D.M., Pearman, G.I. and Fraser, P.J. (1992). Changes in tropospheric methane between 1841 and 1978 from a high accumulation rate Antarctic ice core. *Tellus*, **44B**, 282–294.

Friedli, H., Lotscher, H., Oeschger, H, Siegenthaler, U. and Stauffer, B. (1986). Ice core record of $^{13}C/^{12}C$ ratio of atmospheric CO_2 in the past two centuries. *Nature*, **324**, 237–238.

Gottman, J.M. (1984). *Time Series Analysis*, Cambridge University Press, New York, NY.

Groves, K.S. and Tuck, A.F. (1979). Simultaneous effects of carbon dioxide and chlorofluoromethanes on stratospheric ozone. *Nature*, **280**, 127–129.

Houghton, J.T., Jenkins, G.J. and Ephraums, J.J. (1990). *Climate Change: The IPCC Scientific Assessment*, Cambridge University Press, New York, NY.

Houghton, R.A., Boone, B.D., Fruli, J.R., Hobbie, J.E., Melillo, J.M., Palm, C.A., Peterson, B.J., Shaver, G.R. and Woodwell, G.M. (1987). The flux of carbon from terrestrial ecosystems to the atmosphere in 1980 due to changes in land use: geographic distribution of global flux. *Tellus*, **316**, 617–620.

Jouzel, J., Lorius, C., Petit, J.R., Genthon, C., Barkov, N.I., Kotlyakov, V.M. and Petrov, V.M. (1987). Vostok ice core: a continuous isotope temperature record over the last climatic cycle (160,000 years). *Nature*, **329**, 403–408.

Junge, C.E. (1974). Residence time and variability of tropospheric trace gases. *Tellus*, **26**, 477–488.

Kaplan, W.A., Elkins, J.W., Kolb, C.E., McElroy, M.B., Wofsy, S.C. and Duran, A.P. (1978). Nitrous oxide in fresh water systems: an estimate of the yield of atmospheric N_2O associated with disposal of human waste. *Pure Appl. Geophys.*, **116**, 424–438.

Keeling, C.D., Bacastow, R.B. and Whorf, T.P. (1982). Measurements of the concentration of carbon dioxide at Mauna Loa Observatory, Hawaii. In Clarke, W.C. Ed., *Carbon Dioxide Review 1982*, Oxford University Press, New York, NY.

Khalil, M.A.K. and Rasmussen, R.A. (1985). Causes of increasing atmospheric methane: depletion of hydroxyl radicals and the rise of emissions. *Atmos. Environ.*, **19**, 397–407.

Khalil, M.A.K. and Rasmussen, R.A. (1987). Atmospheric methane: trends over the last 10,000 years. *Atmos. Environ.*, **21**, 2445–2452.

Khalil, M.A.K. and Rasmussen, R.A. (1988). Nitrous oxide: trends and global mass balance over the last 3000 years. *Ann. Glaciol.*, **10**, 73–79.

Khalil, M.A.K. and Rasmussen, R.A. (1989). Climate-induced feedbacks for the global cycles of methane and nitrous oxide. *Tellus*, **41B**, 554–559.

Khalil, M.A.K. and Rasmussen, R.A. (1990a). Global cycle of CO – trends and mass balance. *Chemosphere*, **20**, 227–242.

Khalil, M.A.K. and Rasmussen, R.A. (1990b). Atmospheric methane: recent global trends. *Environ. Sci. Technol.*, **24**, 549–553.

Khalil, M.A.K. and Rasmussen, R.A. (1992a). The global sources of nitrous oxide. *J. Geophys. Res.*, **97**, 14,651–14,660.

Khalil, M.A.K. and Rasmussen, R.A. (1992b). Nitrous oxide emissions from coal fired power plants. *J. Geophys. Res.*, **97**, 14,645–14,649.

Khalil, M.A.K. and Rasmussen, R.A. (1993). The environmental history and probable future of fluorocarbon-11. *J. Geophys. Res.*, **98**, 23091–23106.

Khalil, M.A.K. and Shearer, M.J. (1993). Sources of methane: an overview. In Khalil, M.A.K., Ed., *Atmospheric Methane: Sources, Sinks and Role in Global Change*, Springer Verlag, Heidelberg, pp. 180–198.

Lacis, A., Hansen, J., Lee, P., Mitchell, T. and Lebedeff, S. (1981). Greenhouse effect of trace gases: 1970–1980. *Geophys. Res. Lett,.* **8**, 1035–1038.

Linak, W.P., McSorley, J.A., Hall, R.E., Ryan, J.V., Srivastava, R.K., Wendt, J.O.L. and Mereb, J.B. (1990). Nitrous oxide emissions from fossil fuel combustion. *J. Geophys. Res.*, **95**, 7533–7541.

Logan, J.A., Prather, M.J., Wofsy, S.C. and McElroy, M.B. (1981). Tropospheric chemistry: a global perspective. *J. Geophys. Res.*, **86**, 7210–7254.

Lorius, C., Jouzel, J., Raynaud, D., Hansen, J. and LeTreut, H. (1990). The ice core record: climate sensitivity and future greenhouse warming. *Nature*, **347**, 139–145.

Lovelock, J.E. (1977). Methylchloroform in the troposphere as an indicator of OH radical abundance. *Nature*, **267**, 32–33.

Luizao, F., Matson, P., Livingston, G., Luizao, R. and Vitousek, P. (1989). Nitrous oxide flux following tropical land clearing. *Global Biogeochem. Cycles*, **3**, 281–285.

McElroy, M.B. and Wofsy, S.C. (1987). Tropical forests: interactions with the atmosphere. In Prance, G.T., Ed., *Tropical Rain Forests and the World Atmosphere*, Westview Publ., Boulder, CO, pp. 33–60.

Madronich, S. and Granier, C. (1992). The impact of recent total ozone changes on tropospheric ozone photodissociation, hydroxyl radicals and methane trends. *Geophys. Res. Lett.*, **19** 465–467.

Marland, G. and Rotty, R.M. (1984). Carbon dioxide emissions from fossil fuels: a procedure for estimation and results from 1950–1982. *Tellus*, **36B**, 232–261.

Matson, P.A. and Vitousek, P.M. (1990). Ecosystems approach to a global nitrous oxide budget. *Bioscience*, **40**, 667–672.

Mitchell, B.R. (1980). *European Historical Statistics*, 2nd Ed., Facts on File, New York.

Mitchell, B.R. (1982). *International Historical Statistics, Africa and Asia*, New York University Press, New York, NY.

Mitchell, B.R. (1984). *International Historical Statistics, the Americas and Australasia*, Gale Research Co., Detroit.

Molina, M.J. and Rowland, F.S. (1974). Stratospheric sink of chlorofluoromethanes: chlorine atom catalyzed sink of ozone. *Nature*, **249**, 810–814.

Neftel, A., Oeschger, H., Schwander, J., Stauffer, B. and Zumbrunn, R. (1982). Ice core measurements give atmospheric CO_2 content during the past 40,000 years. *Nature*, **295**, 220–223.

Neftel, A., Oeschger, H., Schwander, J., Stauffer, B. and Zumbruna, R. (1985). Evidence from polar ice cores for the increase in atmospheric CO_2. *Nature*, **315**, 45–47.

Pinto, J.P. and Khalil, M.A.K. (1991). The stability of tropospheric OH during ice ages, inter-glacial epochs and modern times. *Tellus*, **43B**, 347–352.

Prinn, R.G., Cunnold, D.M., Simmonds, P.G., Alyea, F.N., Boldi, R., Crawford, A.J., Frazer, P., Gutzler, D., Hartley, D., Rosen, R. and Rasmussen, R.A. (1992). Global average concentration and trend for hydroxyl radical deduced from ALE/GAGE trichloroethane (methyl chloroform) data for 1978–1990. *J. Geophys. Res.* **97**, 2445–2461.

Ramanathan, V., Cicerone, R.J., Singh, H.B. and Koehl, J.T. (1985). Trace gas trends and their potential role in climate change. *J. Geophys. Res,*. **96**, 5547–5566.

Rambler, M., Margulis, L. and Fester, R., Eds. (1989) *Global Ecology: Towards a Science of the Biosphere*, Academic Press, New York, NY.

Rasmussen, R.A. and Khalil, M.A.K. (1981). Atmospheric methane: trends and seasonal cycles. *J. Geophys. Res.*, **86**, 9825–9832.

Rasmussen, R.A. and Khalil, M.A.K. (1984). Atmospheric methane in the recent and ancient atmospheres: concentrations, trends, and interhemispheric gradient. *J. Geophys. Res.*, **89**, 11,599–11,605.

Raynaud, D. and Chappellaz, J. (1993). The record of atmospheric methane. In Khalil, M.A.K., Ed., *Atmospheric Methane*, Springer Verlag, Heidelberg, pp. 38–61.

Raynaud, D., Chappellaz, J., Barnola, J.M., Korotkevich, Y.S. and Lorius, C. (1988). Climatic and CH_4 cycle implications of glacial interglacial CH_4 change in the Vostok ice core. *Nature*, **333**, 655–657.

Rinsland, C.P. and Levine, J.S. (1985). Free tropospheric carbon monoxide concentration in 1956 and 1951 deduced from infrared total column air measurements. *Nature*, **318**, 250–254.

Ronen, D., Magaritz, M. and Almon, E. (1988). Contaminated aquifers are a forgotten component of the global N_2O budget. *Nature*, **335**, 57–59.

Singh, H.B. (1977). Preliminary estimation of average tropospheric HO concentrations in the northern and southern hemispheres. *Geophys. Res. Lett.*, **4**, 453–456.

Slinn, W.G.N. (1988). A simple model for Junge's relationship between concentration fluctuations and residence times for tropospheric trace gases. *Tellus*, **40B**, 229–232.

Sloan, S.A. and Laird, C.K. (1990). Measurements of nitrous oxide emissions from p.f.

fired power stations. *Atmos. Environ.*, **24A**, 1199–1206.

Stauffer, B., Fischer, G., Neftel, A. and Oeschger, H. (1985). Increase of atmospheric methane recorded in the Antarctic ice core. *Science*, **229**, 1386–1388.

Stauffer, B., Lochbronner, E., Oeschger, E. and Schwander, J. (1988). Methane concentration in the glacial atmosphere was only half that of the pre-industrial holocene. *Nature*, **332**, 812–814.

Steele, L.P., Fraser, P., Rasmussen, R.A., Khalil, M.A.K., Conway, T.J., Crawford, A.J., Gammon, R.H., Massarie, K.A. and Thoning, K.W. (1987). The global distribution of methane in the troposphere. *J. Atmos. Chem.*, **5**, 125–171.

Thiemens, M. and Trogler, W.C. (1991). Nylon production: an unknown source of atmospheric nitrous oxide. *Science*, **251**, 932–934.

Thompson, A. M. (1992). The oxidizing capacity of the earth's atmosphere: probable past and future changes. *Science*, **256**, 1157–1165.

Wang, W.-C., Yung, Y.L., Lacis, A.A., Mo, T. and Hansen, J.E. (1976). Greenhouse effects due to man-made perturbations of trace gases. *Science*, **194**, 685–690.

Warneck, P. (1988). *Chemistry of the Natural Atmosphere*, Academic Press, Orlando, FL.

Weiss, R.F. (1981). The temporal and spatial distribution of tropospheric nitrous oxide. *J. Geophys. Res.*, **86**, 7185–7195.

WMO (World Meteorological Organization) (1985). *Atmospheric ozone 1985. Global Ozone Research and Monitoring Project – Report #16*, WMO, Geneva.

WMO (1988). *Report of the International Ozone Trends Panel 1988, Global Ozone Research and Monitoring Project – Report #18*, WMO, Geneva.

WMO (1989). *Scientific Assessment of Stratospheric Ozone: 1989, Global Ozone Research and Monitoring Project – Report #20*, WMO, Geneva.

Wuebbles, D.J. and Edmonds, J. (1991). *Primer on Greenhouse Gases, Lewis Publ.*, Chelsea, MI.

Yokoyama, T., Nishinomiya, S. and Matsuda, H. (1991). N_2O emissions from fossil fuel fired power plants. *Environ. Sci. Technol.*, **25**, 347–348.

Zardini, D., Raynaud, R., Scharffe, D. and Seiler, W. (1989). N_2O measurements of air extracted from antarctic ice cores: implication on atmospheric N_2O back to the last glacial–interglacial transition. *J. Atmos. Chem.*, **8**, 189–201.

4

Sources of air pollutants

Paulette Middleton

4.1. INTRODUCTION AND OVERVIEW

Clean air is composed of chemicals that have occurred naturally for thousands of years. Oxygen and nitrogen are the major chemicals that make up air. Clean air is also composed of varying amounts of water vapor (H_2O) and traces of gases such as helium and carbon dioxide (CO_2). This air is referred to as clean because it does not contain *harmful* levels of chemicals that adversely effect human welfare. Polluted air contains other chemical gases and aerosols that can harm people, plants, animals, and materials. Some chemicals have a direct harmful impact. Others cause damage indirectly by forming other harmful chemicals in the air or by changing the amount of incoming sunlight. Changes in incoming sunlight can affect temperature and rainfall at the Earth's surface. Some gases such as CO_2 that are components of clean air become of concern when their concentrations become much higher than those levels found in clean air.

Many of the known major sources and impacts as well as the connections among pollution chemicals are outlined schematically in Figure 4.1. All these major chemicals that act directly or indirectly to produce short- and long-term impacts need to be considered when developing an understanding of air pollution sources and their impacts on human health and the environment. Until about 150 years ago the levels of harmful chemicals in the air were quite low and the sources were mainly natural. As people began to build factories and use automobiles, the levels of pollution increased. This is because most pollution caused by people is associated with the burning of fuels such as coal to run factories and gasoline to operate cars. The burning of such "fossil fuels" produces harmful gases such as sulfur oxides (SO_x), nitrogen oxides (NO_x), and volatile organic compounds (VOCs). The burning also produces small aerosols such as soot and fly ash.

Many of the same human activities are responsible for multiple chemical emissions, and many chemicals come from multiple sources. For example, power plants, industries, and automobiles all produce SO_x, NO_x, VOCs, and aerosols in varying amounts. SO_x comes mainly from power plants and industry, NO_x comes about equally from the same stationary sources as SO_x and from mobile sources, and VOCs come mainly from mobile sources. While SO_x and NO_x are associated mainly with human activities, natural sources are a major contributor of VOCs. Some of the most important individual chemicals included in total VOCs come mainly from vegetation. In parts of the world where there are large numbers of trees and relatively high temperatures, human activities can become less important sources of VOCs.

The sources of chemicals such as CO_2, CH_4, and N_2O that are important for global climate change, the greenhouse effect, as well as ozone depletion are also more equally divided among anthropogenic and natural sources. Other greenhouse gases such as CFCs come exclusively from human activities. The sources of radioactivity, another contributor to atmosphere-related impacts, are mainly natural, involving soil processes under normal conditions.

The effects of air pollution can occur far away from the sources as well as nearby. Chemicals emitted from tall stacks often travel long distances before doing their damage. This is because the gaseous chemicals are too light to deposit quickly from the air. However, chemicals emitted near the ground, such as carbon monoxide (CO), which comes from automobiles and wood burning, do affect people closest to the sources. Because the source heights determine the spatial extent of pollutant impacts, it is important to identify stack characteristics for major point sources.

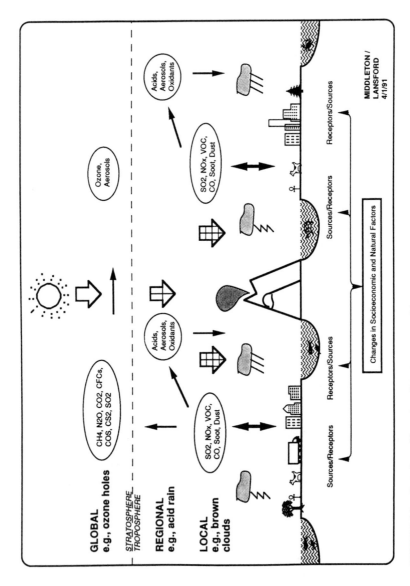

FIGURE 4.1.   Overview of key air pollutant sources and impacts.

Over the last several decades, air pollution has had other increasingly widespread and serious impacts on the natural environment and human society. The health and lives of some city dwellers are threatened by weather episodes that trap stagnant air loaded with carbon monoxide and other toxic gases and aerosols. In recent years, policy makers as well as scientists have given particular attention to damaging effects on aquatic and terrestrial ecosystems of acid deposition from the atmosphere and the effect of oxidants on human health, crops, and forests. In addition, more attention is also being given to the murky skies over our cities and the associated degradation in visibility in important US national parks such as Grand Canyon. Expanding population and industrial growth have made it inevitable that foul air would move beyond city limits, cross national boundaries, create international tensions, and present new difficulties for those attempting to control its harmful effects. Control of certain pollutants in the atmosphere, or finding the best pollution solutions, may be hard, but it is a goal that we can and must achieve.

One major step toward this goal is the development of widely accepted and policy-relevant characterizations of air pollution sources. In this chapter, current information on sources is presented. Much of this source characterization work has been extensively reviewed as part of the Global Emissions Inventory Activity (GEIA) of the International Global Atmospheric Chemistry Program and is reported in several GEIA articles and workshop proceedings (e.g., Graedel et al 1993; Bouwman, 1993; Middleton and Lansford, 1994). In this chapter the processes for characterization of sources are presented as well as source levels themselves. Background on components leading to the estimates of emissions and of uncertainties provides information for assessing the credibility of the estimates and how such estimates can be improved.

4.2. CHARACTERIZING SOURCES

Sources are usually characterized broadly in terms of natural or human-related processes. The human-caused sources are further described as point, area, or mobile sources. Within each group there are many categories, depending on variations in combustion processes and human behaviors, for example. Similarly, natural processes can also be described in terms of multiple attributes. The underlying driving factors for human-related sources include population, land use, and resource use. The sources can usually be described in terms of the emission rate of species over a time interval over a specified area. The rate in turn depends mainly on the number of sources of the type in an area of interest and the time-dependent usage given to each source type in the area.

Source strengths or the magnitude of emissions for a chemical can be described in terms of different areas and time intervals of interest. For example, if sulfur emissions around the world are of interest, then the emissions over the whole globe must be treated, and a way must be found to count (or estimate) the numbers of each type of sulfur source in such distinctly different regions as Northern Europe, the mid-Pacific Ocean, and Antarctica. The result for global descriptions is generally that large land areas and time intervals (probably annual) are used to characterize global emissions. Conversely, if SO_x emissions in a large city are of interest, the variety of sources will not be as large. Time intervals are likely to be daily or hourly, and the metropolitan region will be the only area of interest.

The determination of specific chemicals or groups of individual chemicals is important for characterizing VOC sources. It is not feasible to consider every

TABLE 4.1   Existing Global Primary Databases

Parameter	Spatial Resolution[a]	Temporal Resolution	Effective Year[b]
Atmosphere			
Vector wind	2.5° × 5° × 11L	Monthly	(1958–1973)[c]
Temperature	2.5° × 5° × 11L	Monthly	(1958–1973)[c]
Specific humidity	2.5° × 5° × 11L	Monthly	(1958–1973)[c]
Surface air temperature	2° × 2.5°	Monthly	(1958–1979)[d]
ΔT surface	4° × 5°	Monthly	1880–1990[e]
ΔT surface	Hemisphere	Monthly	1861–1990[f]
Precipitation	2° × 2.5°	Monthly	(1958–1979)[d]
Δprecipitation	"Global"	Seasonal	1880–1990[g]
Clouds	280 km × 280 km	3-Hourly	1983–1990[h]
Clouds	5° × 5°	Monthly	1952–1981[i]
Tropical temperature	280 km × 280 km × 5L	Monthly	1983–1990[h]
Tropical water vapor	280 km × 280 km × 5L	Monthly	1983–1990[h]
Total column O_3	280 km × 280 km	Monthly	1983–1990[h]
Land			
Topography/bathymetry	1° × 1°	None	Present[j]
Topography/bathymetry	5'–10'	None	Present[k]
Albedo	1° × 1°	None	Present[l]
Coal resources	Country	None	Present[m]
Vegetation	1° × 1°	None	Present[l]
Vegetation	0.5° × 0.5°	None	Present[n]
Land cover	1° × 1°	None	Present[o]
Land use	1° × 1°	None	Present[l]
Soils	1° × 1°	None	Present[p]
Soils	1° × 1°	None	Present[o]
Wetlands	1° × 1°	None	Present[q]
Wetlands	2° × 2.5°	None	Present[r]
Drainage basins	2° × 2.5°	None	Present[s]
Vegetation index	1° × 1°	Monthly	1982–1990[t]

TABLE 4.1 (*Continued*)

Parameter	Spatial Resolution[a]	Temporal Resolution	Effective Year[b]
Ocean			
Sea surface temperature	2° × 2°	Monthly	1985–1994[u]
Sea surface temperature	[1024 × 512]	Monthly	1988[v]
Marine climate	2° × 2°	Monthly	1854–1979[w]
Surface wind	2° × 2°	Monthly	1983[x]
Surface wind	[1080 × 540]	Montly	1988[v]
Surface height	[1080 × 540]	Monthly	1988[v]
Temperature	1° × 1° × 33L	Seasonal	(1900–1978)[y]
Salinity	1° × 1° × 33L	Seasonal	(1900–1978)[y]
Oxygen	1° × 1° × 33L	Seasonal	(1990–1978)[y]
Ocean color	[512 × 512]	Seasonal	1978–1986[z]
$\Delta p(CO_2)$	2° × 2°	Seasonal	(1972–1990)[α]

[a]The first and second dimensions indicate grid spacing in latitude and longitude, respectively. The third dimension (followed by L) indicates the number of vertical layers extending through the depth of the troposphere or the depth of the ocean. Numbers in brackets denote the dimension of the global data arrays.
[b]Years in parentheses indicate the period of observations used in the compilation of the statistics or the periods for which the databases are typical. Numbers not in parentheses indicate the years for which gridded data are available.
[c]Oort (1983); [d]Shea (1986); [e]Hansen and Lebedeff (1988); [f]Jones et al. (1986a,b,c); [g]Eischeid et al. (1991); [h]Rossow and Schiffer (1991); [i]Hahn et al. (1988); [j]Gates and Nelson (1975); [k]NGDC (1988)); [l]Matthews (1983); [m]Couch (1988); [n]Olson and Watts (1982); [o]Wilson and Henderson-Sellers (1985); [p]Zobler (1986); [q]Matthews and Fung (1987); [r]Aselmann and Crutzen (1989); [s]Russell and Miller (1990); [t]Tucker et al. (1986); [u]Reynolds (1988); [v]Halpern et al. (1991); [w]Slutz et al. (1985); [x]Woodruff et al. (1987); [y]Levitus (1982); [z]Feldman et al. (1989)); [α]Tans et al. (1990).

individual VOC. The emissions of groups of VOCs need to be considered. For some studies of tropospheric chemistry the hundreds of individual VOCs have been aggregated into two to three dozen chemical classes based on chemical reactivity and emission magnitudes (e.g., Middleton et al., 1990). Once the species and areas of interest have been determined, the sources of chemicals can be identified and counted. For nitrogen oxides in an urban area, for example, these include anthropogenic sources such as motor vehicles, power plants, and industrial operations, and natural sources such as lightning and soil emissions. Motor vehicles need to be further subdivided into automobiles of different types, weights, and fuel efficiencies, as well as trucks, buses, and so forth, because each produces a unique amount of NO_x per unit of fuel.

Once sources are identified and counted, it is then necessary to determine how much and how often they are used. For example, it might be necessary to determine automotive travel frequency for each hour of the day or each day of the week. For natural sources it is important to determine such factors as vegetative growth rates in each season of the year. Understanding all the source types and usages is a complex undertaking. Many assumptions or estimates must be made when information is absent. For regional anthropogenic NO_x, for example, it is common to have the number of sources incompletely known, operating parameter information incomplete, and industrial process emissions unspecified. Notwithstanding these difficulties, unless the analysis is carefully and accurately done, the resulting estimates of source strengths will suffer significantly in accuracy.

The evaluation of the flux terms also is a challenging step. The central question is the rate of emission of species i given a particular source and set of conditions (type of fuel, ambient temperature, etc.). The answer to this question may involve *in situ* sampling of industrial processes, field measurements of natural processes, and the like. Finally, detailed knowledge of the driving factors such as population, resource use, agricultural practices, and energy use is a needed but challenging step. Many of these driving factors and their characteristics globally are summarized in Tables 4.1–4.3. The information is not exhaustive, but it illustrates the wealth of material already available. The primary databases

TABLE 4.2   Existing Global Secondary Databases

Parameter	Spatial Resolution	Effective Year
Agricultural		
Agricultural cultivation	$1° \times 1°$	1980s[a]
Rice cultivation	$1° \times 1°$	1984[b]
Fertilizer application	$1° \times 1°$	1984[c]
Human population	$1° \times 1°$	1980[d]
Cattle population	$1° \times 1°$	1984[e]
Dairy cow population	$1° \times 1°$	1984[e]
Water buffalo population	$1° \times 1°$	1984[e]
Sheep population	$1° \times 1°$	1984[e]
Goat population	$1° \times 1°$	1984[e]
Camel population	$1° \times 1°$	1984[e]
Horse population	$1° \times 1°$	1984[e]
Pig population	$1° \times 1°$	1984[e]
Caribou population	$1° \times 1°$	1984[e]
Indulstrial		
Coal mines	$1° \times 1°$	1980[f]
Coal-fired power stations	Point sources	1985[g]
Oil wells	$1° \times 1°$	1980[f]
Oil refineries	$1° \times 1°$	1980[f]
Natural gas wells	$1° \times 1°$	1980[f]

[a]Matthews (1983); [b]Matthews et al. (1990); [c]GISS (1992, unpublished data); [d]Logan and Dignon (1992, unpublished results); [e]Lerner et al. (1988); [f]*Seydlitz Weltatlas* (1984); [g]Mannini et al. (1990).

TABLE 4.3   Existing Compilations of Emission Factors

Sources Included	Compounds	Effective Year
Numerous	Numerous	1985[a]
Numerous	Numerous	Present[b]
Numerous	Numerous	Present[c]
Numerous	Numerous	None[d]
Road traffice	NO_x, VOCs, CO, TPM	1985[e]
Numerous	Trace elements	1982[f]
Numerous	VOCs (anthro)	1985[g]
Numerous	VOCs (anthro)	1991[h]
Numerous	Ammonia (anthro)	1985[i]
Numerous	Biogenic HC	None[j]
Numerous	Particulate matter (anthro)	1991[k]
Numerous	HCl, HF	1985[l]

[a] EPA (1985, 1987); [b]CEC (1988); [c]CEC (1991); [d]MHPPE (1980, 1983, 1988); [e]CEC (1989); [f] Pacyna (1986); [g]EPA (1986); [h]EPA (1991a); [i]EPA (1990); [j]EPA (1991b); [k]EPA (1991c); [l]Misenheimer et al. (1985).

describe the spatial and temporal variations in atmospheric, terrestrial, and oceanic properties that may control or modulate trace gas and particulate fluxes. The secondary databases describe human activities that are important for emissions. Many of the data sets represent climatology averaged over an observing period. Others are "snapshots" in time or are time series useful for analyzing interannual and longer-term variations or for understanding atmospheric chemistry at a specific point in time.

The primary information on driving factors is categorized by atmosphere, land, and ocean properties. The sampling of the major databases is shown in Table 4.1, while the individual entries are summarized elsewhere (Graedel et al., 1993). Similarly, a selection of the secondary databases is provided in Table 4.2. These data provide the activity values necessary for developing the emission estimates. Keeping these secondary data up to date is important for developing and maintaining accurate emission inventories. As long as political boundaries do not change and as long as the locations of activities within countries remain the same, updating of the gridded databases should be straightforward. Most of the secondary databases pertain to activities in the agricultural or energy sectors, for which the principal source of the data are UN statistics which give production or consumption figures for each country.

Determining emission factors is difficult and exacting. It can be an engineering exercise, as in determining the flux of SO_2 from a specific type of ore smelting operation. It can be an exercise in plant physiology, as in determining the flux of volatile organics from a specific type of vegetation. A selection of factors as described elsewhere (Graedel et al., 1993) is shown in Table 4.3.

4.3. SOURCES OF CHEMICALS

4.3.1. Oxides of Sulfur and Nitrogen

SO_x and NO_x production is mainly associated with human activities. Oxides of both sulfur and nitrogen emanate from major fossil-fuel-burning point sources. Levels of SO_x versus NO_x will vary with the mix of fossil fuels. NO_x is also emitted to a large extent from fossil-fuel-burning mobile sources. Total annual emissions for the world are: $63.7\,Tg\,N\,year^{-1}$ and $22.2\,Tg\,N\,year^{-1}$. Table 4.4 (Benkovitz, 1993) summarizes the status of work on annual emissions of SO_x and NO_x from anthropogenic sources worldwide. The GEIA anthropogenic fossil fuel combustion NO_x inventory gives a global total of 24.5 $Tg\,N\,year^{-1}$, with well over half of the emissions associated with North America and Europe (Muller, 1992; Benkovitz et al. and Scholtz et al., in Middleton and Lansford, 1994; see also Chapter 6). Penner et al. (1991) estimated total global emissions in $Tg\,N\,year^{-1}$ of 22 (fossil fuel combustion), 3 (lightning discharges), 10 (soil microbial activity), and 6 (biomass burning). Estimates of aircraft emissions suggest that NO_x emissions could be significant (Wuebbles, in Middleton and Lansford, 1994). Lightning remains one of the most uncertain and yet an extremely important source of free tropospheric NO_x.

Estimates of natural NO_x on a regional scale (Placet et al., 1990) indicate that the total NO_x produced by lightning is slightly over 5% of the total (1 $Tg\,N\,year^{-1}$ expressed as NO_2), that this NO_x production peaks in the summer, and that the largest rate of production is in the southeastern US. Williams et al. (1992) estimate that the annual NO soil emission in the US is $0.3\,Tg\,N\,year^{-1}$, with strongest emissions occurring during spring and summer and being located predominantly in agricultural areas where large amounts of N fertilizer are used (i.e., the corn-growing counties of Nebraska, Iowa, and Illinois). Because of its strong link to land use practices, part of the NO soil emission has to be considered to be of anthropogenic origin. Considerable uncertainties still exist with respect to the estimates of the lightning and soil NO emission source strength.

More detailed inventories for NO_x are available for North America and Europe (Placet et al., 1990; Lubkert and Zierock, 1989). According to EMEP, $22.4\,Tg\,year^{-1}$ NO_x (as NO_2) was emitted in Europe in 1989 (Simpson, 1993). The estimates are highly uncertain for the Eastern European countries. For example, calculations of the NO_x emissions done by Pacyna et al. (1991) and Veldt (1991) lead to on the order of $7\,Tg\,year^{-1}$ NO_x for the former USSR, compared with $2.9\,Tg\,year^{-1}$ NO_x from EMEP. Source category information (Builtjes, 1992) shows that about 45% of the NO_x and VOC emissions in Europe are due to traffic, with a higher relative contribution in Western Europe (50–60%). The comprehensive National Acid Precipitation Assessment Program (NAPAP) inventory for the US and Canada (Placet et al., 1990) estimated that

TABLE 4.4   Summary of Current Work on Annual SO$_x$ and NO$_x$ Inventories

---

Default inventory: 1985, Dignon (1992)
   1985 emissions from fossil fuel combustion only
   1° × 1° resolution, (1,1) at 180° W, 90° S
   Global SO$_x$ emissions (Tg S year^{-1})
   Global NO$_x$ emissions (Tg N year^{-1})

US/Canada: NAPAP 1985, version 2
   1985 emissions from all land sources; partial emissions from shipping
   Point sources gridded individually
   Area sources gridded to 20 × 20 km NAPAP grid, aggregated to 1° × 1°
   US/Canada SO$_x$ emissions (Tg S year^{-1})
   US/Canada NO$_x$ emissions (Tg N year^{-1})
   SO$_x$ emissions in default inventory (Tg S year^{-1})
   NO$_x$ emissions in default inventory (Tg N year^{-1})

Europe: EMEP 1985 (September 1992 Version)
   1985 emissions from all land sources; ship traffic within Europe included
   150 × 150 km grid reapportioned to 1° × 1° grid
   European SO$_x$ emissions (Tg S year^{-1})
   European NO$_x$ emissions (Tg N year^{-1})
   SO$_x$ emissions in default inventory (Tg S year^{-1})
   NO$_x$ emissions in default inventory (Tg N year^{-1})

Australia: 1985, Environment Protection Authority
   Gridded to 1° resolution, plus point source information
   Some NO$_x$ emissions for point sources missing
   Australian SO$_x$ emissions (Tg S year^{-1})
   Australian NO$_x$ emissions (Tg N year^{-1})
   SO$_x$ emissions in default inventory (Tg S year^{-1})
   NO$_x$ emissions in default inventory (Tg N year^{-1})

South Africa: 1985, South African Department of National Health and Population Devleopment
   Gridded to 1° resolution; all sources included
   South African SO$_x$ emissions (Tg S year^{-1})
   South African NO$_x$ emissions (Tg N year^{-1})
   SO$_x$ emissions in default inventory (Tg S year^{-1})
   NO$_x$ emissions in default inventory (Tg N year^{-1})

Asia: 1985, Kato and Akimoto (1992)
   1985 emissions from transformation, industrial, transportant and miscellaneous
   Regional emissions from China and India; country level for all others
   Gridded to 1° × 1° grid using population file (Dignon)
   Asian SO$_x$ emissions (Tg S year^{-1})
   Asian NO$_x$ emissions (Tg N year^{-1})
   SO$_x$ emissions in default inventory (Tg S year^{-1})
   NO$_x$ emissions in default inventory (Tg N year^{-1})

---

annual NO_x emissions from anthropogenic sources (expressed as NO_2) are 20.5 Tg year^{-1} for the US and 2.1 Tg year^{-1} for Canada. Point sources account for 57% and mobile sources for 43%.

Trends in SO_x and NO_x emissions have also been developed (Hameed and Dignon, 1992). Statistical models previously introduced by the authors for estimating emission rates from rates of fuel combustion have been utilized to obtain emissions of nitrogen and sulfur gases in fossil fuel combustion for every country in the world for every year from 1970 to 1986. Changes in the global emission rates of these gases are presented together with the trends over individual continents. Global emissions of NO_x increased by nearly a third in this period, i.e., from 18 Tg N year^{-1} in 1970 to 24 Tg N year^{-1} in 1986. Emissions of SO_x increased by approximately 18%, from 57 Tg N year^{-1} in 1970 to 67 Tg N year^{-1} in 1986.

The accuracy of the anthropogenic NO_x emission data is somewhat less than for SO_x emissions but greater than for other chemicals. The greater uncertainty relative to SO_x arises because of uncertainties in the NO_x emissions associated with large stationary combustion sources and transportation factors. The greater certainty relative to other chemicals arises because NO_x anthropogenic sources are well known and measurements of NO_x from different sources are reliable.The greatest uncertainty with respect to the NO_x emissions is the possible importance of natural sources. Reports of preliminary estimates (Bouwman and van der Hoek, 1993) suggest that emissions from soils and lightning are significant in some areas. Given that detailed analysis of natural emissions is available only for North America, alternative procedures, e.g., sensitivity studies or surrogate measures such as N fertilizer use, will be needed to explore the importance of natural NO_x emissions on a global scale. Comprehensive estimates of anthropogenic NO_x sources and anthropogenic and biogenic total VOC sources worldwide are becoming available. Given that NO_x anthropogenic sources are increasing and that preliminary estimates of natural NO_x sources indicate that these source are significant globally and are probably increasing, it is most important to better characterize the contributions of soil and lightning sources to the overall x global emission budget.

4.3.2. Volatile Organic Compounds

The appropriate treatment of VOC emissions is crucial to a proper assessment of chemical reactions in the atmosphere. For example, VOCs contribute to the formation of photochemical ozone on urban and regional scales. They also play a role in acid deposition, since they contribute to the generation (or inhibition) of the radicals responsible for converting sulfur oxides to sulfuric acid and nitrogen oxides to nitric acid, and they are involved in the formation of peroxides that influence acid formation in clouds. Some VOCs also undergo phase transformation, leading to aerosols implicated in visibility degradation. Because of their central role in atmospheric chemical processes, uncertainties in VOC emissions or in the relationships between emissions and chemical processes could significantly hamper assessments of chemical impacts at all levels. Consequently, it is important to properly characterize VOC emissions and to understand the implications of emission uncertainties on assessment of the fate and impacts of these chemicals.

The characterization of VOC emissions requires not only the estimate of total VOCs, but also the characterization of individual VOC emissions by category. This is a formidable task. VOCs differ significantly in their effects on chemical processes, and studies of the impacts of these chemicals must be able to represent these differences in chemical reactivity with a reasonable degree of accuracy. This is difficult not only because significant uncertainties remain in our knowledge of the atmospheric reactions of most types of VOCs, but also because the many hundreds (if not thousands) of different types of VOCs emitted into the atmosphere can be represented realistically in chemical studies by only a limited number of VOC classes. The process of determining the most appropriate VOC classes will depend on the issue being examined. For example, studies of regional acid formation need to focus on tropospheric oxidant chemistry, while studies of stratospheric ozone depletion need to focus on halocarbon chemistry. While the different studies may require very different classes, the general procedures used to aggregate the emissions are the same. Such a procedure that allows one to intercompare source types, emission magnitudes, and rates of chemical reaction has been introduced for tropospheric studies (Middleton et al., 1990). This procedure categorizes the hundreds of individual VOCs by chemical reactivity and emission magnitudes. The importance and role of VOCs can then be investigated more accurately and efficiently.

For the United States, as shown in Table 4.5, the breakdown of emissions by human-caused source type for each category indicates that emissions from vehicles are the dominant contributors. Over 50% of the moles of carbon reacted is associated with mobile sources. These are the major components of the emissions for the aromatic and alkane classes, considered to be the most important groups in terms of overall moles of carbon reacted. These sources also dominate the emissions for classes of secondary importance. Of these classes, only the ethene class has significant contributions from other source types. Other source types are important contributors to overall emissions, as can be deduced from the methane and ethane distributions. However, in terms of moles reacted, the vehicular sources clearly dominate when anthropogenic sources are considered.

As shown in Figure 4.2, the relative importance of natural versus man-made VOCs also varies when reactivity rather than emission magnitude alone is considered. When considering the relative importance of natural versus

TABLE 4.5 Relationships Among Emission Group Reactivities, Magnitudes, and Source Contributions

RADM Class	Emission Category	% Moles Carbon Emitted[a]	% Moles Carbon Reacted[b]	% Moles Carbon Reacted by Source Type[c]							
				Area						Point	
				Fuel	Burning	Mobile 1	Mobile 2	Misc.	Waste	Comb.	Manuf.
CH4	Methane	12.88	0.02	0.01	0.00	0.01	0.00	0.00	0.01	0.00	0.00
ETH	Ethane	2.23	0.17	0.05	0.00	0.03	0.02	0.05	0.01	0.01	0.02
HC3	Propane	0.62	0.20	0.01	0.01	0.00		0.12	0.01	0.00	0.04
	Alk (0.25–0.25)	7.01	4.04	0.02	0.07	2.81	0.03	0.68	0.08	0.02	0.34
	Acetylene	2.04	0.43	0.00	0.03	0.11	0.12	0.11	0.01	0.00	0.04
	Haloalkanes	0.51	0.03					0.01			0.02
	Others (<0.25)	0.27	0.07					0.03			0.03
	Others (0.25–0.25)	2.56	1.88	1.14				0.22		0.03	0.50
HC5	Alk (0.5–1)	10.24	9.31	0.04	0.26	5.38	0.39	1.93	0.40	0.05	0.82
	Others (0.5–1)	3.14	3.24			0.19		0.97		0.00	2.08
HC8	Alk (1–2)	12.34	16.39	0.01	0.86	5.34	3.22	2.74	1.31	0.04	2.86
	Alk (>2)	2.77	4.45			1.73	1.13	1.23		0.00	0.37
	Alk/aro	1.16	1.80			0.67	0.50	0.06			0.56
	Others (>1)	0.73	1.05			0.11	0.22	0.20		0.00	0.52
OL2	Ethene	6.59	8.52	2.57	0.47	1.79	0.74	1.23	0.96	0.07	0.69
OLT	Propene	2.31	4.04	0.29	0.07	1.01	1.80	0.51	0.02	0.03	0.29
	Alkenes (prim)	1.59	2.78	0.07	0.51	0.90	0.12	0.22	0.77	0.03	0.14
	Alkenes (mix)	0.40	0.70	0.00	0.17	0.02	0.01	0.21	0.25	0.01	0.04
	Styrene	0.65	1.14				0.20	0.45		0.00	0.49

		(1)	(2)	(3)	(4)	(5)	(6)	(7)	(8)	(9)	(10)
OLI	Alkenes (int)	3.88	6.78	0.00		5.73	0.20	0.51		0.01	0.32
	Alkenes (mix)	0.40	0.70	0.00	0.17	0.02	0.01	0.21	0.25	0.01	0.04
TOL	Benzenes	3.19	1.00	0.25	0.01	0.27	0.05	0.24	0.06	0.01	0.12
	Aromatics (<2)	8.53	9.54	0.03		4.64	1.25	3.03	0.00	0.10	0.64
	Styrenes	0.33	0.36				0.07	0.14		0.00	0.15
XYL	Aromatics (>2)	10.55	18.44	0.03		9.40	1.39	5.77	0.01	0.29	1.56
	Alk/aro mix	0.12	0.19			0.07	0.05	0.01			0.06
CSL	Phenols and cresols	0.12	0.21	0.00	0.13	0.00	0.00	0.08		0.00	0.00
HCHO	Formaldehyde	0.78	1.36	0.03	0.00	0.44	0.35	0.31	0.00	0.04	0.18
ALD	Other Aldehydes	0.44	0.77			0.14	0.29	0.21			0.13
KET	Acetone	0.79	0.06	0.00	0.02		0.00	0.02	0.01	0.00	0.02
	Higher ketones	0.77	0.35					0.10			0.25
OR2	Organic acids	0.00	0.00	0.00			0.00	0.00		0.00	0.00
	Total	100	100	4.56	2.76	40.81	12.18	21.59	4.16	0.75	13.33

[a] The % moles carbon emitted are the moles carbon emitted for the emission category/total moles carbon for all the categories.

[b] The % moles carbon reacted are the moles carbon emitted × (1 − exp(k_{OH} × [OH]))/total moles reacted.

[c] The % associated with the source type is the emission breakdown within each category. Blanks indicate no emissions. Emissions <0.00 are given by 0.00. Each row sums to the % moles carbon number for each category.

anthropogenic sources along with the reactivity of the different chemicals, it is estimated that natural sources of VOCs are much more important overall for the United States. In developing global VOC inventories, VOC totals are generally considered. The level of chemical resolution discussed above has only been considered in North America and Europe. In addition, it should be noted that individual inventories exist only for a few chemicals. Global information exists for isoprene and terpenes, and regional information exists for pesticides and PAH (Graedel et al., 1993).

Even limiting the inventories to total VOCs poses problems on the global scale. Inventories for VOCs are complex not only because of the numerous source types involved, but also because the term VOC is not always given the same meaning. Sometimes chemicals of low reactivity, such as ethane, are included; sometimes they are omitted. In addition, inventories may be for anthropogenic VOCs, natural VOCs, or both. The quality of the VOC emission characterizations depends on the quality of the emission factors assumed. With respect to uncertainty, three source categories can be distinguished: technical activities that emit VOCs by combustion or evaporation, land clearing and other intentional burning, and biogenic processes—the generation of VOCs by vegetation. To date, studies of emission factors, activity rates, and chemical profiles have concentrated on processes under the first category. However, large uncertainties remain for the emission factors and speciation profiles.

VOCs are emitted into the atmosphere from natural sources in marine and terrestrial landscapes. Surface emissions of these compounds are of interest

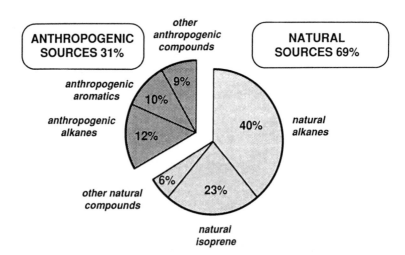

FIGURE 4.2.   Total US VOC reactivity budget. Contributions of anthropogenic and natural source chemicals.

because of their role in tropospheric chemistry and the global carbon cycle. Atmospheric concentrations of ozone, OH, CO, peroxyacetyl nitrate, organic acids, and aerosols are influenced by the emission of these compounds. Several inventories of natural VOCs on regional and global scales have been noted elsewhere (Graedel et al., 1993; Guenther, 1993). Many focus on isoprene and monoterpene emissions. These inventories suggest that annual natural VOC emissions exceed anthropogenic emissions on a global scale (Muller, 1992). On regional scales, summertime natural VOC emission rates range from several factors higher than anthropogenic emissions, e.g., a factor of 3 higher in the United States (Lamb et al., 1987) and a factor of 6 higher in Spain (Simpson and Tarrason, 1992), to several factors lower, e.g., a factor of 2 lower in Germany (Simpson and Tarrason, 1992). The most recent estimates are based on emission factors considered representative of individual surface types, light- and temperature-dependent emission algorithms, and seasonal biomass adjustments. Monthly temperatures and surface types are based on existing $1° \times 1°$ global gridded inventories. Light intensity is estimated with a model that accounts for hourly and monthly variation due to changing solar angles, monthly variation due to changes in cloudiness, and light extinction within plant canopies.

Uncertainties exist in two major areas: assignment of emission factors to individual surface types; and modeling the processes that control variations in emissions from a given surface type. VOC emission rates vary greatly among plant species. This results in large uncertainties associated with the emission factors assigned to landscapes where emission factors of the dominant plant species are unknown. Temperature, light intensity, stresses (e.g., drought and injury), phenology, and other factors play a role in regulating the emissions of some, but not all, VOCs. Many of these relationships are not well understood, and numerical algorithms exist for only a few processes (e.g., temperature and light dependence). The results of the initial inventory suggest that a majority of global natural VOC emissions are from regions with the highest uncertainties, e.g., tropical forest areas.

Anthropogenic VOC inventories currently focus on organizing estimates from small combustion sources, transport, solvent losses and the oil and gas sector. The substances usually considered are methane for some sources, non-methane VOCs, and semi- and non-volatile organic carbon (usually the higher molecular weight, more condensible VOCs). The reference year is 1985 or 1990. One of the major difficulties is the availability of information on biomass burning. Unlike statistics for traditional fuels such as coal, information on biomass fuels is sparse for much of the world where these fuels are used most extensively (Veldt, 1993).

Emission inventories for global non-methane VOCs and CO are not as comprehensive as those for NO_x (e.g., Graedel et al., 1993). Recent estimates indicate a global total for anthropogenic VOCs on the order of 142 Tg year^{-1} with

TABLE 4.6 Estimation of Anthropogenic NMVOC Emissions (1990), Including Condensable Organics

Activity	Developed Countries[a]			Developing Countries			World		
	Tg product	kg Mg^{-1}	Tg NMVOC	Tg product	kg Mg^{-1}	Tg NMVOC	Tg product	Tg NMVOC	%
Fuel production/distribution									
Petroleum	1300			1800			3100	8	6
Natural gas	1700[b]			450[b]			2150	2	1.5
Oil refining	2050	7.5[c]	3.2	900	2	1.8	2950	5	3.5
Gasoline distribution	600	3	2.0	130	3	0.4	730	2.5	2
Fuel consumption[d]									
Coal	150	6	0.9	450	6	2.7	600	3.5	2.5
Wood	250	16	4	1300	16	20.8	1550	25	18
Crop residues (incl. waste)	650	7	4.6	1400	7	9.8	2050	14.5	10
Charcoal				20	120	2.4	20	2.5	2
Dung cakes				400	7	2.8	400	3	2
Road transport								36	25
Chemical industry	300	5	1.5	30	10	0.3	330	2	1.5
Solven use	15	1000	15	5	1000	5	20	20	14
Uncontrolled waste burning								8[e]	5
Other								10[e]	7
Total								142	100

[a] OECD, CIS, Eastern Europe (24% of world population).
[b] 10^9 m^3.
[c] Eastern Europe: 2; CIS: 5; other: 0.5.
[d] Small combustion sources.
[e] Based on Watson et al. (1991).

25% due to road transport, 14% to solvent use, 13% to fuel production/ distribution, 34% to fuel consumption, and the rest due to uncontrolled burning and other minor sources (Veldt, 1993). Muller (1992) estimates that over two-thirds of the 383 Tg year^{-1} of CO emissions are from road transport, with almost half of those sources attributable to North America and Europe. The remaining CO emissions are due to biomass burning and other sources.

A summary of anthropogenic non-methane VOC (NMVOC) emissions is presented in Table 4.6. As indicated, road transport is the largest one source worldwide, with fuel consumption following as the second largest source. More detailed VOC emissions are available for the industrialized countries. For example, according to EMEP, 24.5 Tg year^{-1} VOCs (by mass) were emitted in Europe in 1989 (Simpson, 1993). The NAPAP anthropogenic VOC inventory estimates that the US emitted over 22 Tg year^{-1} of VOCs (expressed in terms of individual mass of the compounds), while Canada emitted 2.4 Tg year^{-1}. Transportation is responsible for over a third of the emissions. A recent review of the status of the US emission inventory (NRC, 1991) concluded that the uncertainties associated with emission factor determinations can be considerable and that accurate knowledge of the VOC, NO_x, and CO ratios is critical for the development of effective ozone abatement strategies in polluted areas.

The long-term trend in anthropogenic VOC emissions in the US from 1900 to 1985 (Placet et al., 1990; Gschwandtner et al., 1986) shows VOC emissions remaining relatively constant between 1900 and 1920. After 1920, VOC emissions increased as transportation, industrial processes, and solid waste disposal activity increased. Emissions rose until the early 1970s when emission controls for automobiles and industrial processes were implemented. No trends in natural VOC emissions have been estimated. The trend in the ratio of the mass of national annual anthropogenic NO_x and VOC emissions shows a fivefold increase from 1900 to 1980 (Placet et al., 1990). The uncertainty may be even higher for VOC emissions owing to gaps in the source categories considered and uncertainties in the emission factors (Baars et al., 1993). Many of these estimates may be affected by the changing economy in Eastern Europe.

4.3.3. Aerosols

Aerosols are derived from a number of human activities, mainly power plants and industrial sources, mobile sources, wood burning, re-entrained road dust, and agricultural processes. Natural sources include mainly wind-blown dust, plant pollen, wildfires, sea spray, and volcanoes. To characterize the impacts of aerosols properly, it is important to estimate the composition and size distributions of the emitted aerosols. Aerosols with diameters of 0.1–1.0 μm are most effective at extinguishing incoming solar radiation and causing haze. These aerosols are also more detrimental to human health. Soot particles can be on the order of three times more effective at extinguishing light than sulfate, nitrate, and organic carbon aerosols.

In general, the development of detailed global information on aerosol sources is in the initial stages (Graedel et al., 1993). Inventories including discrimination with respect to size and composition difference have generally been developed only for particular regions, if at all. The airborne particulate emission inventory developed for North America (Placet et al., 1990) includes emissions from anthropogenic sources (e.g., industrial facilities, power plants, motor vehicles, and furnaces) as well as from wind erosion, dust devils (intense convective wind events), and unpaved roads. Estimates of the alkalinity content and size distribution of the particulate matter are included in a preliminary fashion. These considerations of aerosol size and composition affect the reactivity of the particulates, their potential transport distance, and their lifetime in the atmosphere, as well as their potential for causing adverse impacts on human health and welfare.

Since the North American inventory was developed for the purpose of analyzing the acidic deposition phenomenon, a method has been developed to estimate alkaline particle emissions, as such particles can neutralize acids in the atmosphere. Owing to the difficulty in determining the amount of dust that remains airborne, the elemental content of the particulates, and the sporadic nature of wind action sources, these estimates of wind-borne alkaline dust are model-dependent and highly uncertain. Estimated emissions of particles in the former Soviet Union from industrial sources are available from Berlyand (1990). Industrial emissions of particles were estimated for individual cities and towns, as well as for broad economic regions and the former Soviet Union as a whole. The inventory includes estimates of emissions of specific particles, including lead, organic and inorganic dust, various metal compounds, fluorine dust, cement dust, and others. The change in emissions between 1985 and 1989 is also included. No size distributions are yet available.

For the estimation of the anthropogenic contribution to acidic or total sulfate deposition, as well as for the estimation of ambient concentrations of visibility-reducing sulfate aerosol, it is important to have data on fractions of SO_2 emissions that are converted to sulfate particulates before they enter the atmosphere. NAPAP provides emission factors for these components for a number of sources (EPA, 1989). In general, the PHOXA project (Veldt, 1991) used percentages of SO_2 emission factors. At present, these inventories must be regarded as only very rough estimates.

The first quantitative assessments of worldwide emissions of trace metals from natural and anthropogenic sources were prepared during the 1980s. However, the spatial distribution of trace element emissions is not available on a global scale (Graedel et al., 1993).

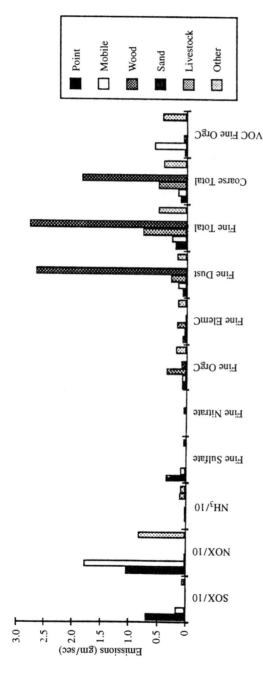

FIGURE 4.3. Metro daytime average emissions. Example of air pollutant source contributions for a metropolitan area. Daytime average Colorado Front Range, wintertime emissions. Note that the gaseous emissions (SO_x, NO_x, NH_3) are divided by 10. The VOC fine organic carbon refers to the VOC emissions counted as contributing directly to secondary organic aerosols. The others (fine sulfate, nitrate, organic carbon, and elemental carbon, and coarse total) are the primary aerosol source strengths. The source categories refer to major point sources, mobile sources (trucks and cars), wood burning (indicates residential wood burning and fireplaces), sand (indicates re-entrained road dust and street sanding), livestock (includes feedlots), and other (all remaining sources, including natural gas usage, minor point sources, and waste treatment plants).

Organic aerosols composed of organic carbon and elemental carbon (soot) particles are injected into the atmosphere mainly by fossil fuel and large-scale biomass burning. Soot particles in the atmosphere may influence the global climate directly by scattering and absorbing sunlight and indirectly by acting as cloud condensation nuclei. Elemental carbon is particularly effective in absorbing incoming solar radiation. Some attempts have been made to provide global estimates for the flux of soot from biomass burning. Penner et al. (1990) produced a model-related global emission inventory of soot particles. The inventory is divided into latitude bands. The emission levels are highest in the tropical areas where biomass burning predominates. A global biomass-burning soot flux of about $7\,Tg\,C\,year^{-1}$ has been estimated.

The anthropogenic source of soot aerosol from fossil fuel consumption has been estimated by Penner et al. (1993). This inventory, clearly a first attempt that will benefit from more extensive field measurement, suggests a global anthropogenic soot flux of about $25\,Tg\,C\,year^{-1}$, several times that from biomass burning. However, most of the anthropogenic source is in the northern hemisphere and much of the biomass burning is in the tropics or southern mid-latitudes, so the source of soot particles differs with latitude band.

On more limited scales, i.e., development of aerosol inventories for addressing regional-scale problems such as visibility degradation, it has been shown that levels of organic carbon aerosol emissions are similar to those for the soot aerosols of wood burning. For industrial and transportation sources the levels of organic carbon are slightly higher (Middleton et al., 1994). A breakdown of source contributions for various aerosol compositions and the relative impact of these sources on regional visibility in the Colorado Front Range region during a wintertime episode is given in Figure 4.3 to illustrate this point. While the organic carbon aerosols, like the nitrate and sulfate aerosols and the soil aerosol, are less effective at light extinction than soot particles, the total amount of these aerosols taken together makes them very important contributors to regional visibility degradation and probably to climate effects. Estimates of natural sources of aerosol include materials from volcanic eruptions, wildfires, sea-salt spray, and pollens. The relative importance of these sources will also need to be carefully considered in the development of global inventories. The direct emission of organic aerosols in the form of pollens in particular, as with the inventories of gaseous VOC relatives, may be an important source of aerosols in heavily vegetated areas.

4.3.4. Other Chemicals

4.3.4.1. NH₃ Emissions

The characterization of NH_3 emissions is essential for estimating the formation of acids and aerosols (see Chapters 5 and 12). As described in Graedel et al.

TABLE 4.7   Sources of Ammonia Emission and Priority for Emission Factors Given by the GEIA Working Group

Source	Emission $(Tg\ NH_3\text{-}N)^a$	Activity Level[b]	Emission Factor
Natural soils	10	Land use database	--
Wild animals		Land use database	-
Biomass burning	5	CH_4–SO_x/NO_x database	-
Coal combustion	2	SO_x/NO_x database	-
Domestic animals	32	CH_4 database	++
Fertilizer application	9	IFA/FAO	+
Fertilizer production		SO_x–NO_x database	-
Seas and oceans	13	General statistics	--
Human excretion	4	General statistics	-/+
Sewage sludge			
Automobiles	0.2	SO_x/NO_x database	-/+
Agricultural crops	No value	Land use database	--
Total on global scale	75		

[a]Base year 1990 (Schlesinger and Hartley, 1992).
[b]Priority ranges from -- (low) to ++ (high).

(1993), a number of inventories have been made by various groups. A wide variety of techniques has been used to describe emissions from waste-water treatment, refineries, ammonia synthesis, coke manufacture, mobile sources, domestic and wild animals, and fertilizer use. Emission factors vary from study to study. To use these inventories on the global scale, there is a need to compare methods and to standardize emission factors/criteria that were used. As shown in Table 4.7 (Bouwman and van der Hoek, 1993), by far the most important sources are the emissions from domestic animals. However, in highly populated metropolitan areas, industrial sources such as waste-water treatment plants and industrial sources can become dominant (Middleton et al., 1993).

4.3.4.2. Nitrous Oxide Emissions

The major problem in N_2O work is the great number and variety of minor sources in addition to the major soil emission sources. These other sources include biomass burning, land clearing, animal wastes, freshwater systems, coastal waters, industrial sources, combustion processes, and mobile sources (transportation). It is likely that other sources exist that have not been identified or for which the source strength is uncertain. As shown in Table 4.8 (Bouwman and van der Hoek, 1993), all the other sources could equal the soil sources of N_2O globally (see Chapter 3 for more discussion).

TABLE 4.8   Global Sources of N$_2$O Identified and Tentative Estimates of Annual Emission

Source	Global Emission (Tg N$_2$O-N year^{-1})
Natural soils	7.6
Cultivated soils	0.3–2
Biomass burning	0.1–2
Fossil fuel combustion	<0.1
Waste-water treatment[a]	1–2
Ocians[b]	1.4–2.6
Land use changes	0.4
Industry	0.4
Mobile sources	0.1–1.3
Aquifers	0.5–1.3
Sewage	0.2–1.9
Animal waste	0.2–0.6
Global warming	0–0.6
Atmospheric formation	?
Trash incineration	?
N deposition	?
Coastal and inland waters	?
Evapotranspiration	?
Irrigation using ground-water	?

Source: Estimates from Kalil and Rasmussen (1992), except: [a]Banin (1986); [b]Watson et al. (1992).

4.3.4.3. Anthropogenic CFCs

Many halocarbons present in the atmosphere come entirely from anthropogenic sources and are manufactured worldwide by a limited number of major industrial companies (see Chapter 7 for more details). Annual production and total sales figures covering most of the world are available from various trade organizations. As reviewed in Graedel et al. (1993), production figures by the reported companies have an expected accuracy of better than 1%. However, not all companies report the information required for inventory development. The spatial distribution of the release of halocarbons is poorly known, because data on the manufacture, shipment, and sales of the compounds are restricted by corporate confidentiality and competitive concerns. This poses a major problem for the development of a gridded global inventory for the halocarbons. To circumvent this problem, observations have been used to try to estimate the source strengths (Cunnold et al., 1986, 1994). There has been significant convergence in the release estimates for CFCl$_3$ and CF$_2$Cl$_2$ between figures derived from industrial production and those inferred from atmospheric measurements. The worldwide 1986 consumption figures given in a recently released United National Environmental Program report (UNEP, 1990) have provided new information on production in Eastern Europe and the former USSR.

4.3.4.4. Reduced Sulfur

The increased awareness of the environmental consequences of acidic precipitation and the role of sulfate aerosols in climate change has resulted in many recent measurements of reduced sulfur gases emitted to the atmosphere from the ocean and terrestrial biosphere. As reviewed by Graedel et al. (1993), the total flux of biogenic reduced sulfur to the atmosphere is calculated at $16\,\mathrm{Tg\,year^{-1}}$ and is almost entirely (over 95%) from the oceanic emission of DMA. The uncertainty in the ocean flux is approximately a factor of 2.

4.3.4.5. Carbon Dioxide

Fossil fuel combustion is the major anthropogenic source of carbon dioxide. The other major source is related to changes in land use that result in changes in the standing stock of biomass—most notably the clearing of tropical forests. There is a wide range of uncertainty for these values. In many studies, modeling efforts are used to help reconcile estimates of the natural sources of CO_2.

An example of the interactions of models with estimates of emission fluxes is provided by the work of Tans et al. (1990). These workers combined observed atmospheric concentrations of CO_2 and data on partial pressures of CO_2 in surface ocean waters to identify globally significant sources and sinks of CO_2. In their study the observed differences between the partial pressure of CO_2 in the surface waters of the northern hemisphere and in the atmosphere are too small for the ocean to be the major sink of fossil fuel CO_2. They suggest that a large amount of CO_2 is absorbed on the continents by terrestrial ecosystems.

4.3.4.6. Carbon Monoxide

On a global scale, inventories for the major sources of CO have been developed for anthropogenic sources. The most detailed, including biomass burning, was produced on $1° \times 1°$ grids (Logan and Dignon, 1992, unpublished results). Combustion of fossil fuels and industrial activity were determined using United Nations data for fossil fuel consumption together with emission factors for CO and distributing fuel use on the basis of population. CO from wood fuel was treated in a similar manner to the sources from fossil fuels. In general, this inventory and others recently reviewed (Graedel et al., 1993) indicate that in industrial countries, in highly populated areas, CO emissions from transportation are highest.

In the recent preliminary analysis of CO emissions from road transport (Samaras and Veldt, 1993), CO from this source is highest in North America, with OECD and Eastern Europe each providing similar but smaller levels. Latin America and Asia follow with less than half the levels of North America. Africa, Japan, and Oceania are each a factor of 10 lower than North America. CO emissions from road transport are highest in highly industrialized areas. On a global scale, however, biomass burning also becomes an important source.

4.3.4.7. Methane

Total emission inventories of methane are among the most complex of those for all the atmospheric gases because of the large number of methane source types. Some of the sources are anthropogenic, some are natural, and many are rather poorly defined. As reviewed by Graedel et al. (1993), emission estimates for all the methane sources must be considered to be relatively preliminary. The best-known source is perhaps the emissions from domestic animals. As indicated in Table 4.9 (Matthews and Roulet, 1993), many other sources are estimated to be equally important on a global scale. The result of research during the past 20–30 years has been a trend toward declining estimated source strengths for the major source, accompanied by an increase in the number of sources identified as potential contributors to the methane cycle. The general consensus is that methane is well on the way to being much better understood (Matthews and Roulet, 1993). Chapter 3 provides more details on the sources and trends of methane.

4.3.4.8. Hydrogen Chloride/Hydrogen Fluoride

As reviewed by Graedel et al. (1993), emissions for HCl are at present in a rather confused and incomplete state. Of many known anthropogenic sources, coal combustion and waste incineration are by far the most important. The emission

TABLE 4.9   Methane Source Overview for the GEIA Methane Working Group

	Methane (Tg year^{-1})[a]
Total sources	442–524
1. Fossil fuel	
Coal mining (no post-mining)	35–50
Natural gas (transmission loss > vented/flared loss)	25–42
2. Biomass burning (burning of forest/grass, agricultural waste)	45
3. Natural wetlands	100–115
High latitude (bogs, tundra)	30–35
Low latitude (swamps, alluvial)	70–80
4. Rice cultivation	40–70
5. Animals – mainly domestic ruminants	80
6. Termites	20
7. Oceans and coastal sediments	10
8. Landfills	20—25
9. Waste-water treatment and animal waste	30
10. Fuel combustion	28
11. Other minor (industrial, residential waste burning, peat mining, geothermal)	9

[a]These should be viewed as candidate values that satisfy the general constraints imposed by the mass balance of methane and its carbon isotopes.

factors vary and large uncertainties should be assigned to the flux estimates. For marine or coastal regions, however, any such uncertainties are dwarfed by estimates that HCl volatilization from sea salt may be as much as 100–400 Tg year^{-1}, perhaps a hundred times the anthropogenic source on a global basis. Only one emission inventory is known for HF; it treats emissions from large US and Canadian industrial sources. The inventory reports that the most significant US point source of HF is coal-fired utilities (Graedel et al., 1993). Other sources include hydrogen fluoride manufacture, the primary aluminum industry, and the phosphate fertilizer industry.

4.3.4.9. Natural Atmospheric Radionuclides

Radioactivity is a well-known health hazard. The chemicals responsible for high levels of radiation under normal conditions are attributed to natural causes. Radon 222 is a daughter of ^{226}Ra, which is a ubiquitous constituent of crustal materials. Emission of ^{222}Rn to the atmosphere follows radioactive decay of ^{226}Ra in soil and upward transport of ^{222}Rn through the soil gas to the surface. The rate of emission of ^{222}Rn from the surface in a given location is a function of factors such as the subsurface concentration of ^{226}Ra and the nature of the near-surface material, e.g., massive rock as contrasted with crushed rock, finely divided sand, or soil. Several approaches have been used to establish the radon emission rate from land. These include chamber methods, measurement of ^{210}Pb/^{226}Ra disequilibria in soil columns, and atmospheric models of varying degrees of sophistication. A review of this work (Kritz, 1993) suggests a figure of 1 atom cm^{-2} s^{-1} as the best approximation to the mean rate of radon emission from land (other than those areas covered by ice caps.)

4.4. SUMMARY AND CONCLUSIONS

Understanding of sources varies considerably from chemical to chemical. The only species for which global fluxes are regarded as well known are CFCs. The overall spatial resolution of information on sources is uniformly poor or non-existent. For specific regions, however, the spatial resolution of the emission information can be regarded as good for CO_2, NO_x and SO_x. Knowledge of the temporal resolution of sources on a global scale is almost universally fair, poor, or non-existent. The exception is CO_2 in some regions. This does not imply superior time resolution for CO_2 inventories, but rather that for most purposes annual information for CO_2 emissions is sufficient. Except for most of North America and Western Europe, detailed regional inventories are available for very few species and very few regions.

The situation should improve soon when the GEIA inventories for anthropogenic SO_x and NO_x and for natural VOCs, CO_2, and others become available.

These inventories will provide an important addition to our quantitative understanding of air pollution sources worldwide. The research reviewed in this chapter has dealt primarily or exclusively with inventories for the present time and under normal conditions. For completeness, we note here that three other types of information are often needed: (1) historical emission inventories, in which inventories for one or more epochs in the past are estimated; (2) special occurrence emission studies, in which emissions from such events as volcanic eruptions and forest fires are assessed or estimated; and (3) future emission scenarios, in which projections of such factors as population, industrial activity, and emission control technology are used to predict emissions in future epochs.

The continued development and evaluation of source characterization for air pollutants benefit considerably from interaction among the research community. Collective experience on various local, regional, and global scales shows that the use of emission information in assessment activities provides one important way of checking the databases themselves. In the development of emission inventories, especially the more complex ones, checks for errors in processing as well as in assumptions regarding the factors and other underlying information are used. However, in cases where the uncertainties or lack of information are highest, the union of emission inventories and observational data can provide a valuable consistency check on the inventories.

This discussion of sources of air pollutants attempt to provide an overview of ongoing activities aimed at continually advancing our understanding of pollution sources. Some indications of current thinking regarding levels of source strengths are provided. For more in-depth information on sources of specific chemicals, references have been provided, and the reader is encouraged to contact the literature and the researchers involved in the specific work. Much progress has been made on the characterization of sources worldwide; much more remains to be done. The continued efforts with IGAC and other world organizations to bring researchers together in many parts of the world provide a major contribution to world community understanding of chemical impacts on the environment. Through interactions at the scientific and other levels, there will continue to be more effective movements toward mutual understanding and consensus on the relationships between human activities and adverse environmental impacts.

Acknowledgments

The efforts of Sharon Blackmon and Henry Lansford in assisting in the preparation and editing of this chapter are greatly appreciated. Much of the material presented is the result of the tremendous efforts of the GEIA community. Their role in advancing our knowledge of air pollution sources is most gratefully noted.

References

Aselmann, I. and Crutzen, P.J. (1989). Global distribution of natural freshwater wetlands and rice paddies, their net primary productivity, seasonality and possible methane emissions. *J. Atmos. Chem.*, **8**, 307–358.

Baars, H.P., Builtjes, P.J.H., Pulles, M.P.J. and Veldt, C., Ed. (1993). *Proc. of the TNO/ EURASAP Workshop on the Reliability of VOC Emission Data Bases, IMW-TNO Publication P 93/040*, Delft.

Banin, A. (1986). Global budget of N_2O: the role of soils and their change. *Sci. Total Environ.*, **55**, 27–38.

Benkovitz, C. (1993). SO_x/NO_x (anthropogenic). In Bowman, A.F., Ed., *Report of the Third Workshop of the Global Emissions Inventory Activity (GEIA), RIVM Report 481507002*, pp. 1–6.

Berlyand, M.E. (1990). Emissions of harmful substances. *1989 Annual Report on the State of the Air Pollution and Anthropogenic Emissions in Cities and Industrial Centers of the Soviet Union*, State Commission on Hydrometeorology, Leningrad.

Bouwman, A.F., Ed. (1993). *Report of Third Workshop of the Global Emissions Inventory Activity (GEIA), RIVM Report 481507002*.

Bouwman, A.F. and van der Hoek, K.W. (1993). N_2O and NO_x (natural) and NH_3. In Bouwman, A.F., Ed., *Report of the Third Workshop of the Global Emissions Inventory Activity (GEIA), RIVM Report 481507002*, pp. 13–14.

Builtjes, P.J.H. (1992). The LOTOS—Long Term Ozone Simulation project summary report. *TNO Report IMW-R 92/240*.

CEC (Commission of the European Communities) (1988). Emission factors. *DG XI CORINAIR, Technical Report 88–355*.

CEC (1989). Environment and quality of life: CORINAIR working group on emissions factors for calculating 1985 emissions for road traffic, Vol. 1: Methodology and emission factors. Final Report EUR 12260 EN.

CEC (1991). *Default Emission Factors Handbook*, 2nd Edn., CORINAIR Inventory Project, CITEPA, Paris.

Couch, G. R. (1988). Lignite resources and characteristics. *Coal Research Report IEACR/ 13*, International Energy Agency, London.

Cunnold, D.M., Fraser, P.J., Prinn, R.G., Simmonds, P.G., Alyea, F.N., Crawford, A.J. and Rasmussen, R.A. (1992). Global trends and annual releases of $CFCl_3$ and CF_2Cl_2 estimated from ALE/GAGE measurements from July 1978 to June 1991., *J. Geophys. Res.*, **99**, 1107–1126.

Cunnold, D.M., Prinn, R.G., Rasmussen, R.A., Simmonds, P.G., Alyea, F.N., Cardelino, C.A., Crawford, A.J., Fraser, P.J. and Rosen, R.D (1986). Atmospheric lifetimes and annual release estimates for $CFCl_3$ and CF_2Cl_2 from 5 years of data. *J. Geophys. Res.*, **91**, 10,797–10,817.

Dignon, J. (1992). Private communication, Lawrence Livermore Lab., CA.

Eischeid, J.K., Diaz, H.F., Bradley, R.S. and Jones, P.D. (1991). A comprehensive precipitation data set for global land areas. *Report DOE/ER-69017T-H1*, US Department of Energy, Washington, DC.

EPA (US Environmental Protection Agency) (1985). Compilation of Air Pollutant Emission Factors, *Rep. AP-42*, Office of Air Quality Planning and Standards, Research

Triangle Park, NC.

EPA (1986). VOC emission factors for NAPAP emission inventory. *Report EPA-60017–86–052*, US Environmental Protection Agency, Washington DC.

EPA (1987). Criteria pollutant emission factors for the 1985 NAPAP emission inventory. *Report EPA-60017–87–015*, US Environmental Protection Agency, Washington, DC.

EPA (1989). The 1985 NAPAP emissions inventory (version 2): development of the annual data and modelers' tapes. *Report EPA-60017–89–012a*, Applied Technology Corporation, Chapel Hill, NC.

EPA (1990). Development and selection of ammonia emission factors for the 1985 NAPAP emission inventory. *Report EPA-60017–90–014*, US Environmental Protection Agency, Washington, DC.

EPA (1991a). Crosswalk/air toxic emission factor data base management system user's manual, version 1.2. *Report EPA-450/2–91–028*, Radian Corp., Research Triangle Park, NC.

EPA (1991b). Volatile organic compound (VOC)/particulate matter (PM) speciation data system user's manual, version 1.4. *Report EPA-450/4–91–027*, Radian Corp., Research Triangle Park, NC.

EPA (1991c). User's guide to the personal computer version of the biogenic emissions inventory system (PC-BEIS). *Report EPA-45014–91–017*, Office Air Quality Planning Standards, US Environmental Protection Agency, Washington, DC.

Feldman, G.C., Kuring, N., Ng, C., Esaisas, W.E., McClain, C.R., Elrod, J.A., Maynard, N., Endres, D., Evans, R., Brown, J., Walsh, S., Carle, M. and Podestra, G (1989). Availability of the global data set. *EoS Trans. AGU*, **70**, 634–641.

Gates, W.L. and Nelson, A.B. (1975). A new (revised) tabulation of the Scripps topography on a 1° global grid. Part I: Terrain heights. *Report R-1276-1-ARPA*, Rand Corp., Santa Monica, CA.

Graedel, T.E., Bates, T.S., Bouwman, A.F., Cunnold, D., Dignon, J., Fung, I., Jacob, D.J., Lamb, B.K., Logan, J.A., Marland, G., Middleton, P., Pacyna, J.M., Placet, M. and Veldt, C. (1993). A compilation of inventories of emissions to the atmosphere. *Global Biogeochem. Cycles*, **7**, 1–26.

Gschwandtner, G., Gschwandtner, K., Eldridge, K., Mann, C. and Mobley, D. (1986). Historic emissions of sulfur and nitrogen oxides in the US from 1980 to 1990. *J. Air Pollut. Control Assoc.*. **36**, 139.

Guenther, A. (1993). VOC (Natural). In Bouwman, A.F., Ed., *Report of the Third Workshop of the Global Emissions Inventory Activity (GEIA)*, RIVM Report 481507002 p. 7.

Hahn, C., Warren, S.G., London, J., Jenne, R.L. and Chervin, R.M. (1988). Climatological data for clouds over the globe from surface observations, *Reo, NDP026*, Carbon Dioxide Inf. Anal. Cent., Oak Ridge Nat. Lab., Oak Ridge, TN.

Hameed, S. and Dignon, J. (1992). Global emission of nitrogen and sulfur oxides in fossil fuel combustion 1970–1986. *J. Air Waste Manag. Assoc.*, **42**, 159–163.

Hansen, J. and Lebedeff, S. (1988). Global surface air temperatures: updates through 1987. *Geophys. Res. Lett.*, **15**, 323–326.

Jones, P.D., Raper, S.C.B., Bradley, R.S., Diaz, H.F., Kelly, P.M. and Wigley, T.M.L. (1986a). Northern Hemisphere surface air temperature variations, 1981–1984. *J. Climatol. Appl. Meteorol.*, **25**, 161–179.

Jones, P.D., Raper, S.C.B. and Wigley, T.M.L. (1986b). Southern Hemisphere surface air temperature variations, 1851–1984. *J. Climatol. Appl. Meteorol.*, **25**, 1213–1230.

Jones, P.D., Wigley, T.M.L. and Wright, P.B. (1986c). Global temperature variations between 1861 and 1984. *Nature*, **322**, 430–434.

Kato, N. and Akimoto, H. (1992). Atmospheric emissions of SO_2 and NO_x in Asia: emission inventories. *Atmos. Environ.*, **26A**, 2997–3017.

Khalil, M.A.K. and Rasmussen, R.A. (1992). The global sources of nitrous oxide. *J. Geophys. Res.*, **97**, 14651–14660.

Kritz, M. (1993). Radionuclides. In Bouwman, A.F., Ed., *Report of the Third Workshop of the Global Emissions Inventory Activity (GEIA), RIVM Report 481507002.*

Lamb, B., Guenther, A., Gay, D. and Westberg, H. (1987). A national inventory of biogenic hydrocarbon emissions. *Atmos. Environ.*, **21**, 1695–1705.

Lerner, J., Mathews, E. and Fung, I. (1988). Methane emissions from animals: a global high-resolution data base. *Global Biogeochem. Cycles*, **2**, 139–156.

Levitus, S. (1982). *Climatological atlas of the world. NOAA Professional Paper No. 13*, US Government Printing Office, Washington, DC.

Lubkert, B. and Zierock, K.H. (1989). European emission inventories–a proposal of international worksharing, *Atmos. Environ.*, **23**, 37–48.

Mannini, A., Daniel, M., Kirchner, A. and Soud, H. (1990). *World coal-fired power stations. Coal Research Report IEACR/28*, International Energy Agency, London.

Matthews, E. (1983). Global vegetation and land use: new high resolution data bases for climate studies. *J. Climatol. Appl. Meteorol.*, **22**, 474–487.

Matthews, E. and Fung, I. (1987). Methane emission from natural wetlands: global distribution, area, and environmental characteristics of sources. *Global Biogeochem. Cycles*, **1**, 61–87.

Matthews, E. and Roulet, N. (1993). Methane. In Bouwman, A.F., ED., *Report of the Third Workshop of the Global Emissions Inventory Activity (GEIA), RIVM Report 481507002.*

Matthews, E., Fung, J. and Lerner, J. (1991). Methanic emission from rice cultivation: Geographic and seasonal distribution of cultivated areas and emissions, *Global Biogeochem. Cycles*, **5**, 3–24.

MHPPE (1980). *Handbook of Emission Factors, Part 1. Nonindustrial Sources.* Ministry of Housing, Physical Planning and Environment, The Hague.

MHPPE (1983). *Handbook of Emission Factors, Part 2. Industrial Sources.* Ministry of Housing, Physical Planning and Environment, The Hague.

MHPPE (1988). *Handbook of Emission Factors, Part 3. Stationary Combustion Sources,* Ministry of Housing, Physical Planning and Environment, The Hague.

Middleton, P. and members of the technical team (1993). *Brown Cloud II: The Denver Air Quality Modeling Study Final Technical Report,* pp. 1–123, (available from P. Middleton, ACT, c/o UCAR, Box 30000, Boulder, CO).

Middleton, P. and Lansford, H. Eds (1994). *Fourth Global Emissions Inventory Activity Workshop Summary* available from P. Middleton, ACT, c/o UCAR, Box 30000, Boulder, CO.

Middleton P. Stockwell,, W. R. and Carter, W.P.L. (1990). Aggregation and analysis of volatile organic compound emissions of regional modeling. *Atmos. Environ.*, **24A**, 1107–1133.

Misenheimer, D., Battye, R., Clowers, M.R. and Werner, A.S. (1985). Hydrogen chloride and hydrogen fluoride emission factors for the NAPAP emission inventory. *Report EPA-60017-85-041*, US Environmental Protection Agency, Washington, DC.

Muller J.-F. (1992). Geographical distribution and seasonal variation of surface emissions and deposition velocities of atmospheric trace gases. *J. Geophys. Res.*, **97**, 3787–3804.

NGDC (National Geophysical Data Center) (1988). *Digital relief of the surface of the earth. NGDC Data Announcement 88-MGG-02*, National Geophysical Data Center, NOAA E/GC3, Boulder, CO.

NRC (National Research Council) (1991). *Rethinking the Ozone Problem in Urban and Regional Air Pollution*, National Academy Press, Washington, DC.

Olson, J. and Watts, J. (1982). *Map of Major World Ecosystem Complexes*, Environmental Sciences Division, Oak Ridge, National Laboratory, Oak Ridge, TN.

Oort, A.H. (1983). Global atmospheric circulation statistics, 1958–1973. *NOAA Professional Paper*, **14**, US Department of Commerce, Rockville, MD.

Pacyna, J.M. (1986). Emission factors of atmospheric elements. In Nriagu, J.O. and Davidson, C.I., Eds., *Toxic Metals in the Atmosphere*. Wiley, New York, NY, pp. 1–32.

Pacyna, J.M., Larssen, S. and Semb, A. (1991). European survey for NO_x emissions with emphasis on Eastern Europe, *Atmos. Environ.*, **25A**, 425–439.

Penner, J.E., Atherton, C.S. Dignon, J. Ghan, S.J. Walton J.J. and Hameed, S. (1991), Tropospheric nitrogen: a three-dimensional study of sources, distributions, and deposition, *J. Geophys. Res.*, **96**, 959–990.

Penner, J.E. Eddleman, H. and Novakov, T. (1992). Towards the development of a global inventory for black carbon emissions. *Atmos. Environ.*, **26A**, 1277–1295.

Penner, J.E., Ghan, S.J., Walton, J.J. and Dignon, J. (1990). The global budget and cycle of carbonaceous soot aerosols. *7th Int. Symp. of Comm. on Atmospheric Chemistry and Global Pollution*, Chamrousse, France.

Placet, M., Battye, R.E., Fehsenfeld, F.C. and Bassett, G.W. (1989). *Emissions involved in acidic deposition processes. State-of-Science/Technology Report 1*, National Acid Precipitation Assessment Program (NAPAP), Washington, DC.

Reynolds, R.W., (1982). A monthly averaged climatology of sea surface temperature. *NOAA Technical Report NWS 31*, National Oceanic and Atmospheric Administration, Silver Spring, MD.

Reynolds, R.W. (1988). A real-time global sea surface temperature analysis, *J. Climate*, **1**, 75–86.

Rossow, W.B. and Schiffer, R.A. (1991). ISCCP cloud data products. *Bull. Am. Meteorol. Soc.*, **72**, 2–20.

Russell, G.L. and Miller, J.R. (1990). Global river runoff calculated from a global atmosphere general circulation model. *J. Hydrol.*, **117**, 241–254.

Samaras, Z. and Velt, C. (1993). Global emissions from road transport. In Bouwman, A.F., Ed., *Report of the Third Workshop of the Global Emissions Inventory Activity (GEIA), RIVM Report 481507002*.

Schlesinger, W.H. and Hartley, A.E. (1992). A global budget for NH_3. *Biogeochemistry*, **15**, 191–211.

Seydlitz Weltatlas (1984). Geographische Verlagsgesellschaft, Cornelsen and Schroedel, Berlin.

Shea, D. (1986). Climatological atlas: 1950–1979. Surface air temperature, precipitation, sea-level pressure, and sea-surface temperature (45S–90N). *Tech. Note NCAR/TN-269 + STR*, National. Center for Atmospheric Research, Boulder, CO.

Simpson, D. (1993). Photochemical model calculations over Europe for two extended summer periods: 1985 and 1989. Model results and comparison with observations, *Atmos. Environ.* **27A**, 921–943.

Simpson, D. and Tarrason L. (1992). Natural hydrocarbons in EMEP MSC-W oxidant model. *EMEP MSC-W Note 1/92*.

Slutz, R., Lubker, S., Hiscox, J., Woodruff, S., Jenne, R., Joseph, D., Steurer, P. and Elms, J. (1985). *Comprehensive Ocean–Atmosphere Data Set (COADS) Release 1*, National Research Laboratory Climate Research Program, Boulder, CO.

Tans, P.P., Fung, I.Y. and Takashashi, T. (1990). Observational constraints on the global atmospheric CO_2 budget. *Science*, **247**, 1431–1438.

Tucker, C.J., Fung, I.Y., Keeling, C.D. and Gammon, R.H. (1986). Relationship between atmospheric CO_2 variations and a satellite-derived vegetation index. *Nature*, **319**, 195–199.

UNEP (1990). *Report of the executive director of the United National environment program*, secretariat of the Montreal Protocol. Addendum: revised report on data on production, imports, exports and consumption of substances listed in Annex A of the Montreal Protocol. *UNEP/OZl.Pro.2/2/Add.4/Rev.1*.

Veldt, C. (1993). VOC (anthropogenic). In Bouwman A.F., Ed., *Report of the Third Workshop of the Global Emissions Inventory Activity (GEIA), RIVM Report 481507002*.

Veldt, D. (1991). Emissions of SO_x NO_x, VOC, and CO from East European countries. *Atmos. Environ.*, **25A**, 2683–2700.

Watson, J.J., et al. (1991). *EPA–600/8–91–002*.

Watson, R.T., Miera, L.G., Sanhueza, E. and Janetos, A (1992). The supplementary report to the IPCC scientific assessment. In Houghton, J.T., Callander, B.A. and Varney, S.K., Eds., *Climate Change 1992*. Cambridge University Press, New York, NY. pp. 25–46.

Williams, E.J., Guenther, A. and Fehsenfeld, F. C.. (1992a). An inventory of nitric oxide emissions from soils in the United States. *J. Geophys. Res.*, in press.

Wilson, M.F. and Henderson-Sellers, A. (1985). A global archive of land cover and soils data for use in general circulation climate models. *J. Climatol.*, **5**, 119–143.

Woodruff, S.D., Slutz, R.J., Jenne, R.L. and Steurer, P.M. (1987). A comprehensive ocean-atmosphere data set. *Bull. Am. Meteorol. Soc.*, **68**, 1239–1250.

Zobler, L. (1986). A world soil data file for global climate modeling. *NASA Technical Memorandum 87802*, National Aeronautics and Space Administration, Washington, DC.

5

Atmospheric aerosols

Rudolf F. Pueschel

5.1. INTRODUCTION AND OVERVIEW

5.1.1. Definition

Aerosols are defined as particles and/or droplets suspended in air. Some have anthropogenic origin and some are naturally produced. They are generated by the mechanical disintegration of soil and sea-spray yielding primary particles of sizes generally larger than $1\,\mu m$, and by gas-to-particle conversions that result in submicron secondary aerosols. Although highly variable in space and time, they are always present in the atmosphere.

5.1.2. Climate Forcing

Aerosol and cloud particles exert a variety of important influences on the Earth's climate. Aerosols are part of the Earth–atmosphere climate system because they interact with both incoming solar and outgoing terrestrial radiation. They do this directly through scattering and absorption and indirectly through effects on clouds. The effect of aerosols on long-wave terrestrial (infrared) radiation is much smaller than their interaction with solar energy, because the transparency of aerosols increases at longer wavelengths and because they are most concentrated in the lower troposphere where the air temperature, which governs emissions, is similar to the surface temperature.

Submicrometer aerosols usually predominate in terms of number of particles per unit volume of air. Particles of such a size scatter radiation from the Sun very effectively because they have dimensions close to the wavelengths of visible light. Light absorption is dominated by particles containing elemental carbon (soot) produced by incomplete combustion of fossil fuels and by biomass burning. Light scattering dominates globally, although absorption can be significant at high latitudes. Light absorption lowers the aerosol single-scatter albedo ω_0, a measure of the fraction of light attenuation (absorption plus scattering) that is due to scattering alone. Whether the aerosol warms or cools the Earth–atmosphere system depends on the magnitude of its single scatter albedo in relation to the surface albedo.

Major volcanic eruptions produce stratospheric aerosols that heat the stratosphere by several degrees Kelvin and are thought to significantly cool the troposphere (e.g., Lacis et al., 1992). The 1815 eruption of Indonesia's Tambora, the largest on record, is widely believed to have caused June snowstorms and severe crop failures at mid-latitudes during the following "year without summer". The substantially larger eruptions of prehistory call attention to volcanism as an important natural agent of climate change (Sigurdsson and Laj, 1992).

Stratospheric heating and increased aerosol surface area can lead to important dynamical and chemical effects, such as lifting the stratospheric ozone layer via convective motions (Kinne et al., 1992) and causing important changes in ozone destruction rates (Prather, 1992). In the Arctic and Antarctic stratospheres the coldest wintertime temperatures cause nitric acid and water vapor to condense (Toon et al., 1986), forming polar stratospheric clouds (PSCs) that facilitate ozone depletion both by removing nitrogen from the stratosphere (hence curtailing the reaction path that protects ozone from chlorine-catalyzed destruction) and by providing reaction surfaces (which convert chlorine from benign to reactive forms).

In the troposphere, man-made sulfate particles, once thought to have significance only near their urban sources, are now recognized to have global radiative effects (Charlson et al., 1992) that may have significantly offset the warming induced by the greenhouse gases released in industrial times (IPCC, 1990). Most sulfate particles originate from sulfur dioxide (SO_2). The emission of man-made SO_2 had increased to about $150\,Tg\,year^{-1}$ in 1980 from about $10\,Tg\,year^{-1}$ in pre-industrial times (World Resources, 1989). Smoke particles from biomass burning have been postulated to produce a similar cooling (Penner et al., 1992), while sulfate particles produced from biogenic dimethylsulfide have been hypothesized to influence albedo, with possible feedback effects on their production (Charlson et al., 1987).

5.1.3. Effects on Clouds

The effects of aerosols on clouds are caused by a small but important subset of aerosol particles called cloud condensation nuclei (CCNs). These are particles with an affinity to water vapor, which condenses upon them preferably, causing clouds to form at supersaturations of a fraction of 1% (Köhler, 1926). Without CCNs, supersaturation of several hundred per cent would occur in the atmosphere before clouds could form. Both the radiative fluxes and the global water cycle under such conditions would be critically different from what they actually are.

The concentration, size, and water solubility of CCNs have an immediate influence on the concentration and size of cloud droplets, which in turn determine the radiative properties of clouds (Twomey, 1977). Physical (coagulation) and chemical (surface adsorption of gases) interactions between natural and man-made substances can influence the cloud-forming ability of aerosols, thereby affecting the frequency of formation, lifetime, and radiative properties of clouds. At constant water vapor mixing ratios and temperature, increased concentrations of CCNs result in an increased number of cloud droplets of smaller size, thereby enhancing the short-wave albedo of clouds. Anthropogenic sulfate aerosols and smokes from biomass burning dominate man-made albedo increases of clouds. However, soot particles incorporated in sufficient amounts into cloud droplets can reduce cloud albedo (Twohy et al., 1989). In contrast, the perturbation in long-wave absorption by cloud modification is negligible, because tropospheric clouds are already optically thick at infrared wavelengths (Paltridge and Platt, 1976).

An increase in cloud droplet concentration due to a decrease in mean droplet size can inhibit growth to precipitation particle sizes and thus extend cloud lifetimes (e.g., Twomey, 1991). The resultant increase in fractional cloud cover would affect radiation transfer through the atmosphere. Inhibited precipitation development could further alter the amount and distribution of water vapor and heat in the atmosphere and thereby modify the Earth's hydrological cycle. Changes in global weather patterns as well as in the concentration of water vapor, the dominant greenhouse gas, could be the consequence.

5.1.4. Air Pollution

Aerosols play a role in air pollution, defined as the presence in the atmosphere of a substance in such amounts as to adversely affect humans, animals, vegetation, or materials (Williamson, 1973). Chronic exposure to high concentrations of particulate pollutants may be injurious to the lung. In addition, aerosols can play a synergistic role in intensifying the toxic effects of gases such as SO_2 and NO_x, in catalyzing the oxidation of SO_2 to H_2SO_4, and in increasing atmospheric turbidity and reducing visibility. There are instances of increases in both mortality and morbidity due to particulate air pollutants. Examples include an episode in Belgium's Meuse Valley in 1930 when stagnation of the overlying air caused a build-up of pollutant concentrations during a week-long period. Many people became ill and 60 died. Other episodes are an incident in Donora, Pennsylvania in 1948 when half the population of 14,000 became ill and 20 died, and a "killer fog" in London in 1952 that resulted in 4000 deaths.

The adverse effects of air pollutants are often so insidious that their exact toll is difficult to assess. It has been estimated, however, that the annual loss from damage to crops, plants, trees, and materials in the United States alone may amount to as much as $5 billion, and that health expenses and lost income due to disease range between $2 billion and $6 billion (see Chapter 1).

5.1.5. Relevant Aerosol Characteristics

Aerosol characteristics that determine environmental effects are physical (size, shape, number), chemical (composition), and optical (refractive index) properties. These properties are difficult to assess for a variety of reasons. One of these is the strong variability of aerosol concentrations in space and time. This is due to the fact that atmospheric residence times of particles are shorter than mixing times within each hemisphere, and these are still shorter than mixing times between hemispheres. As a consequence, measurements made at any given time and place are not necessarily typical for other times and locations. Satellites that could provide adequate global and temporal coverages currently have too limited a capability for delineating relevant aerosol characteristics, and aircraft and ground measurements are too spotty in space and time to adequately determine the global distribution of aerosols. Moreover, aerosols cover several orders of magnitude in concentration (from less than one to several million per cubic centimeter) and size (from units of Ångströms for molecular clusters to millimeters for precipitation particles). No single instrument or technique has the capability to adequately assess atmospheric aerosol properties over this large a size range. Also, in contrast with trace gases, the concentration of aerosol particles is so sparse that measurements have significant experimental and statistical uncertainties.

What follows is an overview of (1) formation, evolution, transformation, and fate, (2) effects on climate, and (3) role in heterogeneous chemistry, including polar stratospheric clouds of man-made particles and the natural atmospheric background aerosol.

5.2. FORMATION, TRANSFORMATION, REMOVAL

5.2.1. Primary and Secondary Aerosols

Aerosols are formed by two general mechanisms. These are the break-up of material and the agglomeration of molecules. Accordingly, atmospheric aerosols may be either primary or secondary. If primary they are due to direct emission of particulate material into the atmosphere from both anthropogenic (e.g., urban/industrial processes, land use practices) and natural (e.g., volcanism, wind-blown dust, sea-spray) activities. Secondary aerosols result from particle formation by gas reactions.

5.2.2. Meteorological Effects

The duration and rate of generation at the Earth's surface and the subsequent atmospheric dispersion and residence time of aerosols are critically dependent on types of emissions, on terrain, and on the prevailing meteorology. In still air, aerosol fluxes from the surface into the boundary layer occur as long as a positive radiation balance exists. Exchange stops by the time a surface temperature inversion has formed, and particles settle back to the Earth's surface from the boundary layer when the radiation balance is negative, e.g., at dusk. This is one reason for strong diurnal and seasonal variabilities of tropospheric concentrations. Temperature inversions aloft frequently cause aerosol layers (gradients of aerosol abundances) to form. Temperature inversions near the surface, initiated by low ground temperatures due to, e.g., ocean currents from Antarctica, greatly enhance smog episodes in cities along the Pacific coast from Santiago de Chile to Los Angeles. Wind and its causes (atmospheric temperature and pressure gradients) are important factors in the generation of primary aerosols.

5.2.3. Emission Rates

5.2.3.1. Man-Made Aerosols

Man's utilization of natural resources requires changing the physical form of these resources. Common examples are pulverization of coal, rock crushing, and cement manufacturing. The resulting emissions of primary particles can cause a severe local nuisance unless they are properly controlled.

Specific anthropogenic aerosol emissions result from fuel consumption in stationary sources (steam–electrical, industrial, commercial, institutional, and residential), industrial process losses (iron and steel mills, copper, lead, and zinc smelters, acid-manufacturing plants, petrochemical works, pulp mills, grain mills, etc.), solid waste disposals (municipal and on-site incineration and open and conical burning), transportation (gasoline and diesel vehicles, railroads, shipping vessels, aircraft, and non-highway machinery), agricultural activities including burning, and miscellaneous (forest and structural fires, coal refuse burning, organic solvent evaporation, and gasoline marketing).

Two approaches have been used to estimate global pollution emissions. Either the US emission data are extrapolated to the world, or global emission assessments are based on energy consumption and/or fossil fuel combustion rates by various nations. Both approaches yield estimates that 93–95% of the industrial aerosol precursor gases—hydrocarbons (HCs), nitrogen dioxide (NO_2) and sulfur dioxide (SO_2)—originate in the northern hemisphere, of which SO_2 emissions are of prime interest to man-made aerosol formation (Bates et al., 1992). In comparison with SO_2, the H_2S emissions from chemical processing and sewage treatment are rather small. Extrapolating US emission data of direct particle production by source category and particle size worldwide, estimates of global emissions of total anthropogenic aerosols (primary and secondary) are 323–690 Tg year^{-1}, the majority of which are contributed by submicrometer particles of diameter less than 1 μm (SCEP, 1970; see Table 5.1).

Emissions from agricultural burning are the major source of gaseous and particulate pollutants in the southern hemisphere. A combination of field measurements and meteorological diffusion modelling yields emission factors of 8×10^{-3} (8 kg of particulates per 1000 kg of material burned). With estimates of $2-5$ Mg ha^{-1} year^{-1} of burnable material and a burnable area of up to 18×10^8 ha in 44 countries in tropical Africa, 30 countries in Central America, 10 countries in South America, 19 countries in tropical Asia, and 16 countries and islands in Oceania, the total emission of particulate material from agricultural burning is $29-72$ Tg year^{-1} (Bach, 1976).

The principal *aerosols of sulfur* in the atmosphere, mainly oxidation products of SO_2, are acidic (H_2SO_4; NH_4HSO_4) and neutral (($NH_4)_2SO_4$) sulfates. The annual rate of SO_2 emissions has increased from 10 Tg year^{-1} in 1860 to 150 Tg year^{-1} in 1980 (World Resources, 1989). The sources of man-made SO_2 emissions are coal combustion (68%), petroleum combustion (16%), smelting (11%), and petroleum refining (5%) operations. The increase in these activities in industrial times led to an increase in tropospheric non-sea-salt (NSS) sulphate from 0.19 to 0.53 Tg in the northern hemisphere and from 0.16 to 0.23 Tg in the southern hemisphere (Langner et al., 1992). The 1990 Clean Air Act Amendments in the United States call for a reduction by 50% of sulfur emissions and considerable reductions in nitrogen dioxide emissions. European countries are also moving rapidly toward emission reductions (e.g., Iversen et al., 1991). To the extent that these reductions are accomplished through reduced fuel use, they will have the double benefit of less sulfate particle generation and slower build-up of carbon dioxide in the air. Opposing a reduction of SO_2 emissions from an effective control strategy in Europe and America, however, is the largely uncontrolled increase in emissions of sulfur due to industrialization in developing countries, notably China.

Nitrate aerosol originates primarily from NO and NO_2. From the sulfate/nitrate ratio of about 5:1, determined from 24 h high-volume samples from 217 urban stations across the US, and similar atmospheric residence times of both ions, the man-made nitrate emissions are $14-40$ Tg year^{-1} (Ludwig et al., 1970).

Extrapolating US emission values, global man-made *gaseous hydrocarbon* (HC) emissions are on the order of $80-90$ Tg year^{-1}, 31% of which have been classified as reactive, i.e., they can participate in photochemical reactions in the atmosphere to form aerosols. The major contribution to this total amount is from petroleum combustion processes, including petroleum evaporation and transfer losses (55%), incineration (30%), solvent usage (10%), and others (5%). Bach (1976) assumed that about 75% of the 1968 US emission rate of 29 Tg year^{-1} consisted of non-methane compounds. If the world HC emission rate were that of the US, and if one-third of that amount were converted to HC aerosols, then a global production rate of about 15 Tg year^{-1} of man-made HC aerosols of submicron size could be expected.

Another class of anthropogenic aerosol with significance to both climate and heterogeneous chemistry is that of *black carbon or soot*. Black carbon aerosol (BCA) is produced during the incomplete combustion of fuels. The two most important sources are fossil fuel combustion and biomass burning. One estimate (Penner et al., 1993) puts the global emissions of BCA at $24\,Tg\,year^{-1}$. Measurements show that atmospheric concentrations of BCA vary between 200 and $800\,ng\,m^{-3}$ in rural regions of the northern hemisphere but decrease to between 5 and $20\,ng\,m^{-3}$ over ocean areas. Only a few $ng\,m^{-3}$ of BCA occur at the South Pole and in the upper troposphere, while only a fraction of an $ng\,m^{-3}$ of BCA is present in the stratosphere. Assuming an average tropospheric BCA concentration of $10\,ng\,m^{-3}$ and an average tropopause height of $10\,km$, the average lifetime of BCA in the troposphere is 0.42 years or 152 days. Stratospheric concentrations are commensurate with polar route emissions from the current subsonic commercial aircraft fleet of $1.9 \times 10^9\,g\,year^{-1}$ of BCA (Turco, 1992), based on a fuel consumption of $1.5 \times 10^{14}\,g\,year^{-1}$ and an emission factor of 10^{-4}.

5.2.3.2. Natural Aerosols

However important anthropogenic emissions may be close to their source, they are nevertheless unmatched on a global basis with the break-up and emission of particles by natural means. Eolian forces generating particles at the Earth's surface and transporting them either over land (wind-blown dust) or from sea to land (sea-salt spray), and volcanic forces ejecting particles from the Earth into the atmosphere, are responsible for the bulk of the natural primary aerosol. Other aerosol substances that may be locally important are those from wildfires and wind-blown dust. Only negligible contributions result from meteoritic dust originating in space.

Empirical relationships between wind speed u $(m\,s^{-1})$ and mass M $(\mu g\,m^{-3})$ of suspended particles (Jaenicke, 1986) are

$$M = 4.3 \exp (0.16u) \tag{5.1}$$

with u between 1 and $21\,m\,s^{-1}$ over land, and

$$M = 52.8 \exp (0.30\,u) \tag{5.2}$$

with u between 0.5 and $18\,m\,s^{-1}$ over water. Thus it follows that *wind-blown dust* is nature's major contribution to the primary continental aerosol. Soil material transported from arid regions by wind is responsible for the distribution of certain clay materials in oceanic sediments, and Saharan dust has a significant impact on the aerosol chemistry over the tropical North Atlantic. Estimates of global input from wind-blown dust range from 100 to $500\,Tg\,year^{-1}$. These are occasionally influenced by man. For example, dust storms over America's Great Plains are related to agricultural activities, and Africa's Sahel has been rendered vulnerable to wind erosion by cattle overgrazing. The Great Plains are subject annually to occurrences of 45 h of dust stemming from dust storms of an average duration of 6 h, yielding dust concentration up to $5000 \, \mu g \, m^{-3}$. Dust loadings of $300-800 \, \mu g \, m^{-3}$ are found regularly in the lowest 5–10 km of the atmosphere over the Rajputana desert between India and Pakistan. Storms over the deserts and loess terrains of China and Mongolia produce dust that can be traced by satellites over distances of several thousand kilometers. Residues of Saharan dust amounting to average concentrations of $61 \, \mu g \, m^{-3}$ at 1.5–3.0 km altitude have been measured thousands of kilometers from their source over Barbados (Prospero and Carlson, 1972).

Primary *marine aerosols* result from the injection of sea-salt into the atmosphere. Estimates of the mass of airborne *sea-salt particles* at sea vary between 10 and $20 \, \mu g \, m^{-3}$ (Prospero and Carlson, 1972). Assuming that the majority of sea-salt is dispersed within the planetary boundary layer 1000 m deep, the global concentration amounts to about 11 Tg. With a production rate of $1000 \, Tg \, year^{-1}$ (Erickson, 1959), the resulting average residence time is 1.1×10^{-2} years or 4 days. Thus about 90% of marine aerosols settle out over the oceans. The remainder of $100 \, Tg \, year^{-1}$ is carried across the ocean shores into continental air. Once over land, the concentration of sea-salt particles decreases rapidly to about 15% at 4 km and to about 8% at 30 km inland.

The frequency and intensity of *volcanic eruptions* (Lang, 1974) and estimations of the height of the eruption clouds and amounts of ejected material yield estimates of aerosol emissions due to volcanism between 25 and $550 \, Tg \, year^{-1}$. Most of the volcanic material is injected into the stratosphere. In contrast, the estimates of terrestrial accretion of *particles of extraterrestrial origin* fluctuate between 0.02 and $10 \, Tg \, year^{-1}$. Forest fires have been estimated to contribute between 3 and $150 \, Tg \, year^{-1}$ of particulate material into the atmosphere.

The majority of secondary natural aerosols (those formed from gases) result from oxidation of sulfur gas emissions from land and sea. These source gases include H_2S, DMS, methanethiol, carbon disulfide (CS_2), carbonyl sulfide (COS), methyl mercaptans, and others. Reactions in the atmosphere of hydroxyl radicals (OH) with these sulfur compounds form sulfur dioxide (SO_2), which is a precursor gas that can be further oxidized to form sulfate aerosol particles. From tropospheric measurements of CS_2 and the rate constant for its conversion to COS, Turco et al. (1980) estimate that 5 Tg of OCS per year could be generated from CS_2 in the atmosphere. In addition, there are direct sources of OCS which include the refining and combustion of fossil fuels ($1 \, Tg \, year^{-1}$), natural and agricultural fires ($0.2-0.3 \, Tg \, year^{-1}$), and soils ($0.5 \, Tg \, year^{-1}$), yielding a total influx of from 1 to $10 \, Tg \, year^{-1}$, up to 50% of which may be anthropogenic.

The principal sources for natural H_2S are sulfate respiration and the decomposition of amino acids containing various thiol groups. Most natural sulfur gas on land ($62\,Tg\,year^{-1}$) is produced from decaying vegetation and animals. In the marine environment most volatile sulfur is emitted in the form of dimethylsulfide (DMS) which is excreted by living planktonic algae (see Chapter 8 for more details). The only significant non-biological natural S flux is the emission of SO_2 and H_2S by volcanoes and fumaroles. This process releases on the order of $26\,Tg\,year^{-1}$, or about 10–20% of the total natural flux of gaseous sulfur to the atmosphere, of which $4\,Tg\,year^{-1}$ are from volcanic eruptions.

Naturally produced secondary nitrate aerosols result mostly from NO and amount to $75\,Tg\,year^{-1}$, 80% of which are submicron size. Global emissions of natural hydrocarbons from vegetation (mostly terpene compounds) and from the soil amount to about $150\,Tg\,year^{-1}$, of which approximately half are converted to aerosol, resulting in an annual emission rate of about $75\,Tg\,year^{-1}$.

5.2.3.3. A Comparison

Table 5.1 summarizes existing estimates of natural and man-made aerosol emissions and/or aerosol precursor gases. The wide range of emission estimates indicates a great difficulty in assembling global or regional budgets for atmospheric aerosol. This is due to a small magnitude of aerosol mass concentrations in air compared with trace gas concentrations, and the amounts processed through the atmospheric portion of the elemental cycles that are involved. The resulting chemical heterogeneity, in addition to large temporal and spatial variabilities, adds to the difficulty in assembling atmospheric aerosol budgets.

In spite of a wide range of emission estimates, Table 5.1 suggests that global aerosol production seems to be dominated by natural sources. However, because two-thirds of the Earth's surface is water, another major percentage is polar, which leaves only about 2.5% of the Earth's surface, mainly in the eastern US, Europe, Japan and parts of Australia and Asia, where anthropogenic aerosol production occurs. The natural production over this same area amounts to only a fraction of the man-made production. Anthropogenic SO_2 emissions to the atmosphere (primarily from fossil fuel combustion and metal smelting) have increased over the industrial period and now exceed natural emissions of sulfur-containing gases on a global basis. They completely blanket natural emissions in industrialized regions. Hence it is safe to conclude, on the basis of Table 5.1, that urban/industrial aerosols are dominated by anthropogenic material. Plumes of light-scattering aerosol extending from industrial regions to the marine atmosphere have, like wind-blown desert dust, been discerned in satellite measurements. Examples of man-induced deteriorating air quality are the permanently poor visibility of generally less than $10\,km$ in almost any European country, and smog episodes in cities along the west coast of the Americas, with Los Angeles probably the best-known example.

TABLE 5.1 Estimates of Global Emissions of Aerosols

Source	Strength (Tg year^{-1})	
	Natural	Anthropogenic
Primary particle production		
Transportation		2
Stationary fuel sources		43
Fly ash from coal		36
Non-fossil fuels		8
Petroleum combustion		2
Industrial processes		56
Iron and steel industry		9
Incineration		4
Cement production		7
Solid waste disposal		2
Miscellaneous		16–29
Soot		24
Agricultural burning		29–72
Sea salt	300–2000	
Soil dust	100–500	
Volcanic particles	25–300	
Meteoritic debris	0–10	
Forest fire smoke	3–150	
Subtotal	428–2810	215–260
Secondary particle production		
Sulfates from SO$_2$		70–220
Sulfates from H$_2$S	105–420	
Sulfates from DMS	16–32	
Sulfates from volcanoes	9	
Biomass burning		3
Nitrate from NO$_x$	75–700	23–40
Ammonium from NH$_3$	269	
Carbonates from hydrocarbons	15–200	15–90
Subtotal	195–1220	108–350
Total	623–4030	323–610

Both the optically effective and cloud-nucleating fractions of the atmospheric aerosol in urban/industrial regions are frequently dominated by sulfate particles. Residence times of SO$_2$ and SO$_4^{2-}$ in the troposphere are short (several days) compared with mixing times in the troposphere (several months within hemispheres; about a year between hemispheres). Therefore anthropogenic aerosols in general and man-made sulfates in particular are concentrated mainly in industrial regions and areas downwind of them, largely in mid-latitudes. Concentrations of aerosol sulfate and cloud condensation nuclei within 1000 km or more downwind of regions of industrial emissions are an order of magnitude or more greater than those in remote regions (Schwartz, 1989). Because more than 90% of industrial SO_2 is emitted in the northern hemisphere (Bates et al., 1992), anthropogenic sulfate is confined largely to that hemisphere. Widespread distribution of anthropogenic sulfate in the northern hemisphere is evidenced by elevated concentrations of non-sea-salt sulfate in the aerosol, in higher acidic precipitation at remote sites in the northern hemisphere relative to those in the southern hemisphere, and by increases in concentrations of sulfate in northern hemisphere glacial ice over the past 100 years (Mayeewski et al., 1990) that closely match the record of anthropogenic emissions. Mass concentrations of non-sea-salt sulfate appear to be enhanced by 30% or more over the natural background in much of the marine northern hemisphere (Schwartz, 1989).

Considering an energy demand proportional to the increase in the Earth's population, for which there is no limit in sight, it is clear that the man-made component of the atmospheric aerosol is going to increase, and inadvertent climate modification on scales large enough to affect habitability will have to be reckoned with (Kondratyev, 1986).

5.2.4. Transformation

While airborne, transformation of aerosol takes place owing to physical (coagulation between particles) and chemical (condensation of vapors and surface reactions) modifications. We have seen that agglomeration of gaseous molecules to form solid or liquid aerosols is an important mechanism of their formation. Further deposition of gases on airborne particles is a significant phenomenon of aerosol transformation.

5.2.4.1. Condensation

Interaction of water vapor, the most abundant gas in the atmosphere, with a fraction of the aerosol results in the development of clouds and fogs and is probably the most important process of particle transformation by agglomeration. Water vapor condenses preferentially on hygroscopic salt particles or on hygroscopic molecules such as SO_3. As agglomeration proceeds, the particles grow, and non-spherical crystals become more spherical as water condenses on them. With the exception of a small class of hydrophobic aerosols, there is a continuous increase in the geometric mean diameter of sulfates and other salts at increasing ambient relative humidity (Blanchet and List, 1983). The change in size and/or shape and/or refractive index with increasing relative humidity results in an increase in the aerosol optical thickness in the visible and near-IR regions as long as the atmospheric water vapor partial pressure is greater than approximately 5 mbar. There is little or no correlation between the spectral aerosol optical thickness and relative humidity when the relative humidity is in the low (0.45–0.75) range. A phase change from solid to liquid, or vice versa, at the point of deliquescence results in an abrupt change in size (and possibly shape) at a relative humidity that is typical for a given salt. It is thus possible to use the deliquescence temperature to classify a hygroscopic aerosol type, e.g., ammonium sulfate, and distinguish it from acidic sulfuric acid and ammonium bisulfate (Butcher and Charlson, 1972).

Particles with the greatest affinity for water act as CCNs, i.e., centers around which cloud drops can develop when the air is near saturation with respect to water (Köhler, 1926). The chemical composition of condensation nuclei is important for the process of cloud droplet formation (Pruppacher and Klett, 1978). The effect of different soluble compounds, e.g., NaCl and $(NH_4)_2SO_4$, on the hygroscopic growth and nucleation process is less important than is the difference between soluble and non-soluble material. Thus nucleation is a scavenging process that assists in preferential precipitation of the more hygroscopic species of aerosols. Less hygroscopic or even hydrophobic species remain in the cloud and may be transported farther in the atmosphere after the cloud has dissipated. The more hygroscopic a CCN, the earlier it will become a droplet as water saturation is approached. At a given composition, larger CCNs will acquire water first. Consequently, the larger droplets have nucleated on the larger aerosol particles (Twohy et al., 1989). If the saturation increases further, smaller droplets become activated and dilute more quickly than the larger droplets because of their larger ratio of surface area to volume. Because sulfates are generally smaller than sea-salt particles, the size-selective nucleation process will result in the sulfate-dominated particles growing into smaller cloud droplets and the supermicrometer, e.g., sea-salt and soil dust, particles becoming the larger droplets. Soil dust and soot particles, although initially hydrophobic, can be made hygroscopic while airborne through the acquisition of SO_4^{2-} and other salts by coagulation.

The process of cloud formation was first explained on thermodynamical grounds by Köhler (1926). Figure 5.1 shows the calculated saturation ratio (expressed as relative humidity) versus particle size for one particular salt. At very low humidities the hygroscopic aerosol (salt) particle is dry and crystalline. Increasing the relative humidity initially causes some water to be adsorbed on the particle surface, but the amount is insufficient to dissolve the salt and the particle remains largely crystalline. Hence the growth in particle size is small. At a certain critical relative humidity, which is typical for the type of salt particle, the crystal takes up enough water to form a saturated solution. At this point the size of the particle increases abruptly by a factor of 2–3. At higher relative humidities the increase in size follows a growth equation based on a combination of Raoult's law and the Kelvin relation for the dependence of the vapor pressure of a particle

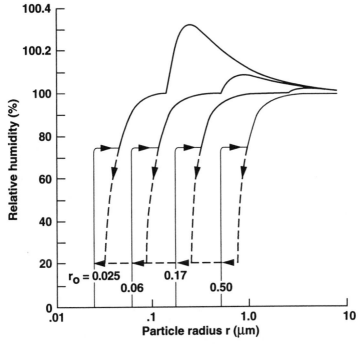

FIGURE 5.1. Köhler diagram: variation in particle size with relative humidity p/p_s, for a salt deliquescing at $p/p_s = 0.75$ (typical for NaCl) for CCNs of various sizes. (After Butcher and Charlson, 1972.)

on its radius. For a particle in equilibrium with its environment, the ratio of the actual vapor pressure of the solution droplet, p, to the saturated vapor pressure of water, p_s, is

$$\frac{p}{p_s} = \exp\left(\frac{2\sigma V}{RTr}\right)\left(1 + \frac{imM_w}{M_s(\frac{4}{3}\pi r^3 \rho - m)}\right)^{-1} \tag{5.3}$$

where σ is the surface tension of a solution droplet, V is the molar volume of the liquid phase, T is the temperature, r is the droplet radius, i is the van't Hoff factor, i.e., the average number of moles of dissolved species produced per mole of solute, m is the mass of solute in the droplet, M_s is the molecular weight of solute, ρ is the density of a droplet, and M_w is the molecular weight of water. The validity of particle growth as a function of relative humidity has been tested and confirmed experimentally (Tang et al., 1977).

An important feature displayed by each growth curve in Figure 5.1 is the maximum in the domain of supersaturation above 100% relative humidity. It indicates the degree of supersaturation required for a particle to form a stable cloud droplet. Once a particle has grown just beyond the maximum, it enters a region of instability and must grow further, provided that the water supply in the ambient air suffices.

Upon decreasing the relative humidity to values below the deliquescence point, the solution becomes supersaturated. Recrystallization does not occur as spontaneously as deliquescence, so that the particle size moves along the extended Köhler curve for a while before it shrinks to a value near the dry radius at low humidities (< 20%). This hysteresis indicates an important mediation of aerosol size by clouds, with consequences for atmospheric optics and chemistry. Many cycles of cloud particle condensation and evaporation may take place long prior to the formation of precipitable particles, affecting the size and elemental composition of aerosols. Thus a cloud upon evaporation leaves behind an "activated" aerosol particle of a size that is larger than was the CCN upon which the droplet had initially formed. The larger surface area renders the particle more active both optically and chemically.

5.2.4.2. Oxidation and Neutralization

Next to condensation, important processes of particle transformation are oxidation and neutralization reactions on existing particles. In the cloudless atmosphere, gas phase oxidation of DMS, OCS, or CS_2 to SO_2 is proportional to the concentration of hydroxyl (OH), which varies substantially in space and time. It is most abundant in the boundary layer and in the upper troposphere during summer. Aqueous phase oxidation of SO_2 is proportional to cloudiness and, in winter, dominates in the lower and middle troposphere where clouds are abundant. Studies of the oxidation of SO_2 to SO_4^{2-} in the atmosphere indicate that condensed, aqueous phase reactions are important and can even be the main mechanism under atmospheric conditions when the aqueous phase is present in clouds, fogs, and high-humidity hazes. Thus, in a mixture of hygroscopic and hydrophobic particles, the oxidation of SO_2 will be favored in or on the former at high humidity.

Further reaction between OH and SO_2 results in the oxidation of SO_2 to SO_4^{2-} to form an H_2SO_4 aerosol. Acidic aerosols (H_2SO_4 and NH_4HSO_4) can be neutralized to $(NH_4)_2SO_4$ in the presence of ammonia vapor. Chemical reactivity of these water-soluble fractions is important with respect to various atmospheric and biogeochemical processes. Faster photochemical production of SO_4^{2-} from SO_2 gas-to-particle conversion takes place in summer, since the SO_2 and OH concentrations are higher in summer than in winter. The result is seasonal and diurnal variability in the physical and chemical properties of aerosol. Both aerosol number and volume concentrations and the SO_4^{2-} concentration are highest in the summer troposphere over the rural south–central United States.

5.2.4.3. Coagulation

Through Brownian motion, particles smaller than $1\,\mu m$ diameter continually migrate through the air and occasionally collide with other particles. A certain fraction of collisions will make the particles stick to one another, leaving them united to form a larger particle. This process causes a continual removal of small particles and an enhancement of larger ones. The rate at which this occurs depends upon the composition of the particles, the pressure-dependent velocity of their movements, the cross-sectional area of each particle and, most importantly, the particle concentration.

5.2.5. Removal

5.2.5.1. Dry Deposition

Diffusion and sedimentation are processes by which particles are brought into contact with the Earth's surface. Aerosol particles may migrate in and out of a control volume either by their own thermal agitation (Brownian motion) or by turbulent eddying of air. Brownian diffusivity in the atmosphere is exceedingly small ($10^{-8} < D < 10^{-4}\,cm^2\,s^{-1}$) for realistic particles sizes ($0.01 < r < 10\,\mu m$) compared with turbulent eddy diffusivity D_T for mass transport ($10^2 < D_T < 10^5\,cm^2\,s^{-1}$). The magnitude of diffusivity depends on the proximity to the ground and on conditions of local hydrostatic stability of the air. In general, turbulent diffusion will exceed the effect of Brownian motion for atmospheric transport, except in very thin layers near surfaces such as leaves or stones, where the air motion will be slow and laminar in nature.

The transfer of gases or aerosol particles from turbulent air to an underlying boundary depends on the properties of the flow near the surface as well as on the nature of the surface itself. In a simplified picture, transfer takes place through a turbulent boundary layer into a viscous or laminar sublayer adjacent to the boundary. The boundary layers are identified with a velocity gradient and with a gradient in concentration of diffusing aerosols. In the turbulent layer these gradients will be essentially similar in thickness because of the dominance of the eddying motion of the turbulence. In contrast, in the viscous sublayer, defined by gradients of properties, the nature of the transfer mechanism will depend primarily on molecular properties of the air. Its thickness will then vary with the molecular diffusivity of the species being transferred. Since molecular transfer is quite slow compared with that in the turbulent layer, the mechnism of transport in the sublayer is the rate-limiting mechanism of the process, and deposition of aerosols on surfaces will vary with the molecular (or Brownian) diffusivity as indicated in Figure 5.2, which shows the deposition velocity as a function of particle diffusivity and particle size for flow over smooth boundaries.

In the absence of horizontal concentration gradients the loss rate per unit volume for diffusional transport is given as

$$L_D = -D_T \, \delta^2 n/\delta Z^2 \qquad (5.4)$$

As particles increase in size, first through their formation by the agglomeration of molecules in condensing vapors and later by the coagulation of pairs of particles, the effect of gravity may become more important than that of Brownian motion. Sedimentation then becomes a factor which removes aerosols from the air. In a dry, still atmosphere the upper limit for aerosol sizes is determined by sedimentation. The gravitational force experienced by a particle is $f_g = mg$, where m is the mass of the particle and g is the acceleration due to gravity. Once the particle has acquired a downward velocity, it will experience an opposing force $f_d = 6\pi\eta r v$ due to the friction with air, where η is the viscosity of air, r is the particle radius, and v is its velocity. Equating these two forces results in a settling velocity of particles

$$v_s = 2r^2 \rho g/9\eta \qquad (5.5)$$

with which a particle of a given size will approach the ground. The deposition velocity for particles settling out at the surface is identical with the sedimentation velocity. Thus the rate of sedimentation increases rapidly with particle size. Other factors influencing settling rates, albeit less importantly than does particle size, are particle shape and particle density. The deposition flux F_d can be estimated by

$$F_d = v_s \delta n/\delta Z \qquad (5.6)$$

where $\delta n/\delta Z$ is the concentration gradient of particles with altitude.

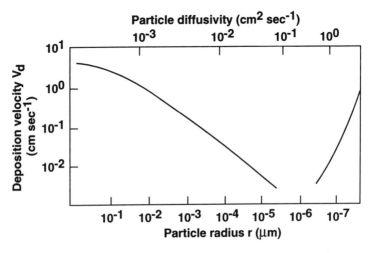

FIGURE 5.2. Deposition velocity as function of particle size. (After Hidy, 1973.)

Dry deposition of aerosol is thus proportional to its surface concentration. Its removal rate is determined by the deposition velocity, the vertical turbulent diffusion coefficient of the boundary layer, and the vertical distance between the surface and the height at which the aerosol occurs.

5.2.5.2. Wet Removal

In addition to removal by dry deposition, removal of aerosols near the surface takes place by precipitation, preceded by in-cloud and subcloud scavenging. The estimation of the loss of particles by rain-out involves complicated and interacting factors that may contribute to scavenging of aerosols by cloud droplets and ice crystals. One mechanism for trapping particles is nucleation of the hydrometeor (see Section 5.2.4). Others involve migration of the particles to the cloud drop by (a) thermophoretic forces during condensation or evaporation, (b) electrical interactions, or (c) thermal agitation. The amount of aerosol removed from air by cloud formation can be approximated by

$$\Delta n = E_n\, n_a \tag{5.7}$$

where E_n is a cloud scavenging efficiency that involves many factors (solubility, particle size, etc.) and n_a is the particle concentration in air. Δn must be proportional to the amount of water vapor ingested by the cloud during formation. The amount of water vapor removed from the air during cloud formation can be approximated as

$$\Delta q = q_a - q_s(T_c) \tag{5.8}$$

where q_a is the water vapor content of air and $q_s(T_c)$ is the corresponding saturated humidity of air at cloud temperature T_c. The scavenging ratio $R = n_c/n_a$, where n_c is the aerosol concentration in cloud droplets, can be expressed as

$$R = \frac{\Delta n/\Delta q}{n_a} = \frac{E_n}{q_a - q_s(T_c)} \tag{5.9}$$

In-cloud scavenging of particles (Junge and Gustafson, 1957) is dependent on the average rate of formation of precipitation and the liquid water content of the rain cloud. Empirical relationships show an efficiency of wet removal that is approximately 10% of the liquid water content. Typical values of liquid water content are $1-2\,\text{g m}^{-3}$ for precipitating clouds and $0.05-0.25\,\text{g m}^{-3}$ for convective and stratiform clouds (Mason, 1971). Wet removal efficiencies E_n will therefore lie in the range $0.5\% < E_n < 20\%$.

Scavenging will also occur as the larger cloud particles fall downward. If the falling hydrometeors are evaporating, a temperature gradient could develop between them and the surrounding air, such that thermophoresis may have to be considered. Scavenging of particles by hydrometeors falling from clouds located above the aerosol and isolated from it by stratification will eventually cleanse the entire underlying air layer and dissolve its contents in the total quantity of deposited water if the precipitation event lasts long enough. Hence in the limit an inverse relationship between concentration in precipitation and amount of precipitation would result. The process involved is analog to that of radioactive decay, where it is commonly assumed that the rate of removal at any instant is proportional to the quantity of radioactive material. Therefore at any instant of a precipitation event the rate of change with time in particle concentration can be written as

$$\Delta n_a / \Delta t = -\lambda n_a \tag{5.10}$$

where λ is a scavenging rate. Empirical data show $0.5 \times 10^{-5} < \lambda < 15 \times 10^{-5}$ for stratiform systems and $5 \times 10^{-5} < \lambda < 3 \times 10^{-3}$ for convective systems (Hicks, 1986). Integration leads to

$$n_a(t) = n_{a0} \exp(-\lambda t) \tag{5.11}$$

where $n_a(t)$ is the concentration as function of time and n_{a0} is the concentration at the onset of precipitation at $t = 0$.

If uniformly mixed throughout a layer of the atmosphere of depth h, the total quantity removed in an event of duration t is obtained by integration as

$$h(n_{a0} - n_{at}) = h n_{a0} [1 - \exp(-\lambda t)] \tag{5.12}$$

If rainfall rate is I and p ($= It$) is total precipitation, then the concentration in the precipitation falling in the course of an event is

$$n_p = \frac{h n_{a0} [1 - \exp(-\lambda t)]}{It}$$

$$n_p = \frac{h n_{a0} [1 - \exp(-\lambda p / I)]}{p}$$

$$R = n_p / n_{at} = \frac{h \exp(1 - \lambda_p / I)}{p} \tag{5.13}$$

In this derivation of R it is assumed that λ is constant, which is a simplification. In reality, λ is expected to vary with particle size, particle solubility, precipitation rate, etc.

The best quantitative information on the deposition (dry and wet) fluxes of NSS sulfate comes from ice core records obtained in Greenland, showing an enhancement by a factor of 2–4 over the last hundred years (Neftel et al., 1985). This agrees well with calculated changes of 2–5 in total NSS sulfate by Langner et al. (1992) over 58% of the northern and over 16% of the southern hemisphere surface area.

5.2.6. Residence Times

The rates of formation, agglomeration, coagulation, and removal, in combination with a size-dependent water solubility of particles, determines the atmospheric residence times of aerosols. They can be described by an empirical equation (Jaenicke, 1986)

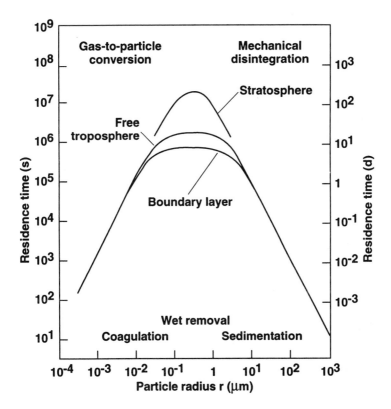

FIGURE 5.3. Atmospheric residence times as function of particle size. (After Jaenicke, 1986.)

$$\frac{1}{\tau} = \frac{r^2}{FR^2} + \frac{R^2}{Kr^2} + \frac{1}{\tau_{wet}} \tag{5.14}$$

where τ is the residence time, τ_{wet} is the wet removal residence time, K and F are constants, and $r = 0.3\,\mu m$ is the radius of particles whose residence time is the longest.

Figure 5.3 is a graphical depiction of residence time as function of particle size. It shows that the residence for small and large particles is relatively short. Small particles are subject to Brownian motion, resulting in a high rate of coagulation. The large particles have settling velocities of several $cm\,s^{-1}$, resulting in a fairly high rate of sedimentation. Residence time is largest in the stratosphere. This is due to the fact that volcanic eruptions, the major source of stratospheric particles and their precursors, are rather sparse. In addition, the absence of water vapor in significant amounts in the stratosphere prevents cloud formation which renders wet removal ineffective. Such source and sink limitations let coagulation and sedimentation proceed very effectively to produce the surface area distribution shown as a dashed curve in Figure 5.6 (see Section 5.3).

5.2.7. Size Distributions

With the process of coagulation determining the lower end of the size spectrum, and that of removal the upper end, the question arises whether a unique size distribution of particles in the atmosphere can be established which, according to Friedlander (1970), should be called the "self-preserving" distribution.

Studies of atmospheric aerosol have shown that there are generally three peaks within the particle size range between 0.01 and $10\,\mu m$, dubbed by Whitby (1978) the nucleation, the accumulation, and the coarse particle modes. Figure 5.4 summarizes the formation, transformation, and removal processes (after Whitby, 1978). It is a schematic of the distribution of aerosol surface area with particle size, showing the three major modes, the main source of mass for each mode, the principle processes involved in inserting mass into each mode, and the principle removal mechanisms. The nucleation and coarse particle modes are prevalent near the sources of particles. As an aerosol evolves with time away from sources, the dominant mode is the accumulation mode around $0.1\,\mu m$ diameter. Figure 5.5 gives an example of a particle size distribution that is frequently observed in the free troposphere.

A striking feature of atmospheric aerosols is a steady decrease in particle number per unit volume of particles above $0.1\,\mu m$ size, which prompted Junge (1961) to point out first that the distribution of particles larger than $0.1\,\mu m$ can often be approximated by a power law

$$\frac{\Delta n(r)}{\Delta r} = Cr^{-(\beta + 1)} \tag{5.15}$$

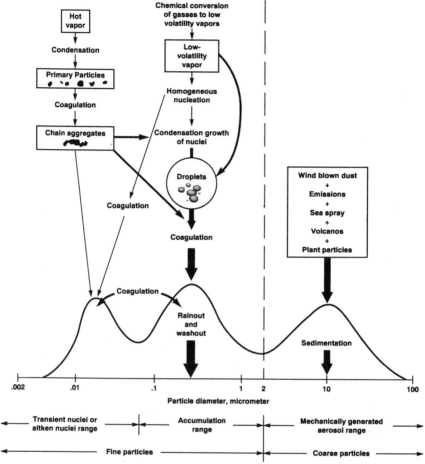

FIGURE 5.4. Schematic of an atmospheric aerosol surface area distribution showing the three modes, the major sources for and principal processes involved in inserting mass into each mode, and the principal removal mechanisms. (After Whitby, 1978.)

where Δn is the concentration of particles whose radius r lies between r and $r + \Delta r$, and C is a constant. The value of β usually averages 3, although variations between 2.5 and 4 are not uncommon, depending upon the type of aerosol and its atmospheric residence time. The power law distribution is a special case of a log-normal distribution

$$\frac{\Delta n(r)}{\Delta r} = \frac{n_0}{r \ln \sigma_g \sqrt{2\pi}} \exp\left[-\frac{1}{2} \left(\frac{\ln r - \ln r_g}{\ln \sigma_g} \right)^2 \right] \tag{5.16}$$

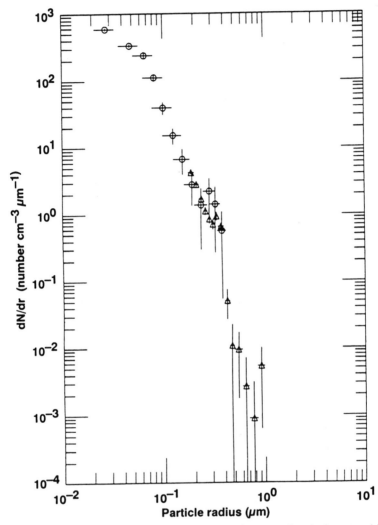

FIGURE 5.5. Aerosol size distribution measured in the free troposphere by impactors (circles) and optical particle counters (triangles). (After Pueschel et al., 1994a.)

which more universally describes each particle mode by the concentration per unit volume, n_0, the geometric mean radius r_g, and the geometric standard deviation σ_g. The great range of particle sizes, with the ratio between the extreme sizes often on the order 100:1 or greater, renders the log-normal distribution a suitable tool for describing the distribution of particle sizes. According to the log-normal distribution law, one of the most important in nature, it is the logarithm of particle size that is normally distributed.

5.2.8. Composition

The composition of the aerosol depends upon the sources and subsequent transformation while airborne. In a first approximation it is thus possible to distinguish continental from maritime aerosols, although mixing between the two types is often the case. The composition of continental aerosols, furthermore, varies from urban to rural areas, determined by the local types of natural and anthropogenic sources and the distance from these sources. In general, the aerosol represents a mixture of substances from various sources, and any specific component may have more than one origin.

Sulfate is the dominant inorganic constituent of aerosols, except for the marine aerosol which is dominated by sodium chloride (NaCl). SO_4^{2-} mass fractions range from 22% for continental aerosols to 75% in the Arctic and Antarctic regions. The origin of sulfates over the continent is SO_2 gas-to-particle conversion, because (1) the sulfur content of the Earth's crust is too low to provide a significant source of primary sulfates and (2) sulfate is concentrated in particles of submicron size.

Near sources of gaseous ammonia, NH_4^+ is the principal cation associated with sulfate in the continental aerosol. The degree of neutralization of sulfuric acid in parts of the troposphere depends on the supply of NH_3 relative to the rate of formation of H_2SO_4. Molar NH_4^+/SO_4^{2-} ratios range from 1 to 2, suggesting a composition intermediate between NH_4HSO_4 and $(NH_4)_2SO_4$. Over much of the oceans, in the continental remote troposphere, and in the stratosphere the major aerosol component is H_2SO_4. As was pointed out earlier, sulfate in the aerosol shows seasonal variability. The mid-latitude sulfate aerosol concentration is highest in summer owing to both higher SO_2 emission rates and greater OH concentrations. The Arctic aerosol, in contrast, has a high SO_4^{2-} content in winter.

Long-range transport of anthropogenic aerosol results in modifications of the atmosphere long distances away from the particle sources. Polar regions are sinks for most atmospheric trace constituents, including aerosols. Once introduced into the Arctic atmosphere, a combination of different source regions, thermal atmospheric stability, reduced photochemistry, low scavenging rates, and slow meridional transport produces a complex, vertically layered atmosphere in which very clean and quite polluted air can alternate and can potentially last for a long time. Early airborne observations made it clear that the particulate portion of this pollution contributed to a significant degradation of visibilities to as low as 3–8 km. These episodes of anthropogenic pollution (haze events) are also common at surface locations around the Arctic rim in winter. The haze contains secondary aerosol (mostly acidic sulfates) and primary aerosol of industrial origins, mixed with natural aerosols made up of soil and volcanic material. Concentrations in winter are typically one magnitude higher than in summer throughout most of the Arctic troposphere. The covariation in SO_4^{2-} with vanadium, which is partly a product of fossil fuel combustion, and with the concentrations of radon and its decay product ^{210}Pb, as well as air mass trajectory analyses, suggests that Arctic air pollutants originate at mid-latitudes, mainly in Europe, and are transported to the Arctic via Russia.

In contrast, a natural source of the SO_4^{2-} content, similar in abundance to that in the Arctic, must be invoked for the Antarctic aerosol because of the absence of anthropogenic pollution in this region. The fine and coarse aerosol particle composition in the Antarctic atmosphere shows sulfur preferentially as $CaSO_4$ or NH_4HSO_4 for particles in the size range 0.1–2.0 µm, with a higher $CaSO_4$ particle concentration in summer. Other types of particles are NaCl and $MgCl_2$ from sea-salt. A large number of these show small (1%) amounts of sulfur, indicating reactions of these particles with gaseous sulfur.

Sulfate in the marine aerosol is present in both the accumulation and coarse particle modes, indicating that its ionic composition is quantitatively different from that of the continental aerosol, where SO_4^{2-} is found mainly in the accumulation mode. The coarse particle fraction is associated with sea-spray, whereas the submicrometer-size fraction again results from SO_2 gas-to-particle conversion. The major gaseous precursor of sulfate in the unperturbed marine atmosphere is dimethyl sulfide, a biogenic compound emanating from the sea surface. Oxidation pathways for dimethyl sulfide are discussed in Chapter 8. The main anions of the marine aerosol are Na, Cl, Mg, K, and Ca, which make up the coarse particle mode.

Particulate nitrate in the marine aerosol is associated mainly with coarse particles. Its source must be gaseous nitric acid, because sea-water contains insignificant amounts of nitrate. That gaseous nitrate condenses preferentially onto coarse particles, in contrast with sulfate, has to do with its volatility, which is much higher than that of sulfuric acid, thus preventing simultaneous condensation in the same size range. The association of nitrate with the bulk of sea-salt may also result from the dissolution of nitric acid in the sea-water droplets at relative humidities in excess of 75%. In the continental aerosol, nitrate is distributed over the whole 0.1–10 µm size range. In coarse particles it is mainly balanced by sodium, and in the submicrometer particles by ammonium.

Some trace components of the aerosol are considerably enriched compared with their abundances in the Earth's crust and in the oceans. The enrichment factor EF is defined as

$$EF(X) = \frac{F((X)/(Ref)_{aerosol}}{(X)/(Ref)_{source}} \tag{5.17}$$

where X is the element under consideration and Ref is an appropriate reference element. Useful reference elements are Al for crustal materials and Na for seawater. There is sometimes a marked difference in the composition of aerosols in cities dominated by sulfurous smog and those with photochemical smog. In the latter there is a considerably higher concentration of organic aerosols and nitrates, and very often a lower concentration of sulfates. In a city such as Los Angeles, probably 20–40% of the aerosols are primary pollutants and 30% are secondary. The remainder appears to be representative of the natural background concentration. In cities with sulfurous smog, much of the aerosol matter is ash or waste from chemical processes. Table 5.2 summarizes what is known about the chemical composition of the three major atmospheric aerosol types.

TABLE 5.2   Mass Concentrations ($\mu g\,m^{-3}$) of Components in Three Aerosol Types

Element or Compound	Urban Aerosol		Continental	Marine
	Sulfurous Smog	Photochemical Smog		
SO_4^{2-}	14.0	16.5	0.5–5	2.6
NO_3^-	3	10	0.4–1.4	0.05
Cl^-	3.2	0.7	0.08–0.14	4.6
Br^-	0.1	0.5	–	0.02
NH_4^+	4.8	6.90	0.4–2.0	0.16
Na^+	1.2	3.1	0.02–0.08	2.9
K^+	0.4	0.9	0.03–0.1	0.1
Ca^{2+}	1.6	1.9	0.04–0.3	0.2
Mg^{2+}	0.6	1.4	–	0.4
Al_2O_3	3.6	6.4	0.08–0.4	–
SiO_2	5.9	21.1	0.2–1.3	–
Fe_2O_3	5.3	3.8	0.04–0.4	0.07
CaO	–	–	0.06–0.18	–
Organics	27.1	30.4	1.1	0.9
Sum	43.5	75.7	1.8–11.0	11.2

Selected mass fractions, molar and elemental ratios

SO_4^{2-} (%)	29.5	21.8	30.2–45.7	22.6
NO_3^- (%)	6.3	12.8	13.3–22.7	0.44
NH_4^+/SO_4^{2-}	1.9	2.2	2.1–3.4	0.47
Si/Al	1.4	2.9	1.9–2.8	–
Fe/Al	1.9	0.8	0.6–1.2	–
Na/Al	0.6	0.9	0.4–0.6	–
K/Al	0.2	0.3	0.5–0.6	–
Ca/Al	0.8	0.6	1.9–2.1	–

5.2.9. Spatial Distributions

In the boundary layer, size distributions of the optically important 0.1–23.5 μm diameter range are typically bimodal with modes around 0.1 μm and several microns diameter, which are typical modal radii for the accumulation and coarse particle modes, respectively. The coarse mode decreases rapidly with increasing altitude owing to preferential settling and interactions with clouds. The relatively long-lived accumulation mode, of greater importance to long-range transport of sulfur pollutants, shows both horizontal and vertical variabilities. While horizontal homogeneity of aerosol concentrations over tens to hundreds of kilometers has been observed, vertically the accumulation mode forms layers. Typically there is a boundary layer aerosol mass concentration of 10–20 μg m^{-3} below 900 mbar, then a band between 700 and 900 mbar with 2–5 μg m^{-3}, and a layer above 700 mbar that contains less than 2 μg m^{-3} of particles. Finer vertical structures within each of these layers are frequently observed. The accumulation mode concentration off the Virginia coast can be twice that of Bermuda. At Bermuda the particulate burden reaching the islands from the west is enhanced by about 50% over the fluxes of background aerosol from the south and/or east.

Size distribution and composition of lower tropospheric aerosols are strongly influenced by prevailing meteorology. For example, sulfur-rich stratified aerosol layers are a common feature from 0 to 3000 m off the northeast American coastline. These occur under clear air conditions with airflow from west to east, but change dramatically during the encroachment of warm frontal systems. This causes the sulfate content to decrease several-fold, and chloride to become the most common water-soluble anion in the lower 3000 m, probably owing to increased vertical mixing and dilution of pollutant aerosols. Thus the structure and stability of stratified atmospheric layers are important in studies of the ocean-atmosphere chemistry problem.

5.3. STRATOSPHERIC AEROSOLS

A negative lapse rate above the tropopause, caused by absorption of solar radiation by ozone, results in a fairly dense aerosol layer, dubbed the Junge layer after its discoverer (Junge, 1961). Because of the high degree of stability of the stratosphere, there is little vertical convection. This and the absence of clouds due to a lack of moisture result in aerosol residence times of a year or more (compared with days in the troposphere). This distinction between stratospheric and tropospheric aerosols justifies a separate discussion of stratospheric aerosol.

During large volcanic eruptions, large amounts of gases in the form of sulfur dioxide and hydrogen sulfide, and large numbers of particles such as ash and sulfates, are injected into the stratosphere. Relatively small but sulfur-rich volcanic eruptions can have atmospheric effects equal to or even greater than much larger sulfur-poor eruptions. Small eruptions are probably the most frequent source of stratospheric aerosols. Figure 5.6 shows an example of a stratospheric background (dashed curve) and a volcanically enhanced (solid curve) stratospheric aerosol size distribution.

During volcanic quiescence, the origin of the stratospheric aerosol is not exactly known. We know that there is a common source of sulfur compounds for the stratosphere in both the northern and southern hemispheres. The precursor gas of the stratospheric aerosol is SO_2 of mostly volcanic, but also possibly of biogenic or anthropogenic, origin. Carbonyl sulfide (OCS), an oxidation product of carbonyl disulfide (CS_2) with an atmospheric lifetime of about 1 year, is a candidate for stratospheric sulfur aerosols.

The sulfuric acid aerosol is formed by photochemical reaction of sulfur gases with water vapor in the stratosphere. Particle formation is strongly temperature-dependent. Extremely explosive eruptions may also inject amounts of chlorine into the stratosphere in the form of hydrogen chloride to induce chemical reactions that lead to ozone losses similar to those caused by chlorofluoromethanes (see Section 5.4.2). Particles are formed in the stratosphere from sulfur gases either by some sort of *in situ* homogeneous nucleation mechanism, including binary and ternary nucleation, or by binary heterogeneous nucleation onto particulates (i.e., condensation nuclei). These could be injected into the stratosphere from the troposphere or from outer space (such as with meteoritic debris or ion clusters). If homogeneous nucleation occurs, the formation of new particles involving H_2SO_4–HNO_3–H_2O ternary reactions may be the most favorable mechanism for the formation of stratospheric aerosols (Kiang et al., 1975). This is because vapor pressures for the ternary system H_2SO_4–HNO_3–H_2O (with weight composition around 70–80% H_2SO_4, 10–29% HNO_3, 10–20% H_2O) are at $-50\,°C$ low enough to prevent evaporation. If this mechanism operates, acids of nitrogen should be present in stratospheric aerosols in addition to H_2O and H_2SO_4. Indeed, Farlow et al. (1978) tentatively identified two forms of nitrosyl sulfuric acid ($NOHSO_4$ and $NOHS_2O_7$) in stratospheric aerosols. The first of these can be formed either directly from gas reactions of NO_2 with SO_2, or by gas–particle interactions between NO_2 and H_2SO_4. The second product may form when SO_3 is involved. Estimates based on these reactions suggest that the maximum quantity of NO that might be absorbed in stratospheric aerosols could vary from one-third to twice the amount of NO in the surrounding air. If these reactions occur in the stratosphere, then a mechanism exists for removing nitrogen oxides from that region by aerosol particle fall-out. This process may typify a natural means that helps cleanse the lower stratosphere of excessive pollutants and results in ozone depletion. Burley and Johnston (1992) revived

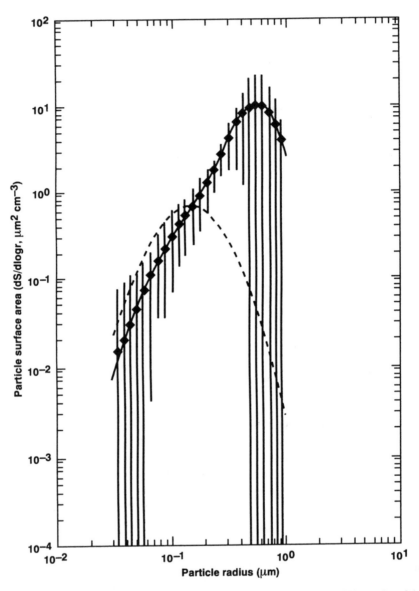

FIGURE 5.6. Mean and standard deviation of the surface area distribution of 28 samples of the Pinatubo volcanic aerosol collected between August 1991 and March 1992. The dashed curve represents the surface area distribution of a "background" sample collected over Antarctica in August 1987. (After Pueschel et al., 1994b.)

the nitrosyl sulfuric acid issue in connection with ozone depletion by heterogeneous reactions of NO_x with stratospheric aerosols (see Section 4.2).

Because of a strong temperature dependence of homogeneous nucleation rates (Yue and Deepak, 1982), the particles appear to be formed in polar regions during the cold winter and advected to lower latitudes. Since most of them do not seem to contain inclusions, it is believed that they are the product of homogeneous nucleation of sulfuric acid and water. The larger sulfuric acid aerosol particles, however, do sometimes contain some sort of insoluble inclusion (or at least some material remnant which survives evaporation). It is not certain whether this is an indication of a heterogeneous nucleation process or just the fact that these are aged aerosol particles which have undergone coagulation with available material.

In general, there is a tropical source for the reacting gases, a slow average transport to higher latitudes, and a formation of aerosol particles by growth onto the condensation nuclei (CNs) produced in the winter polar vortices. The sulfuric acid vapor which gives rise to the new particle formation in the winter polar vortices can also originate at mid-latitudes by evaporation of aerosols at high altitudes (above about 30 km the temperature is too high to support the aerosol and the particles evaporate) which leads to a supply of gas phase sulfuric acid which will migrate to the poles. During the winter descent of air, this sulfuric acid is carried to lower levels. Since the temperature is low enough, nucleation will occur. Since this air is relatively particle free, coagulation will not reduce the number density too quickly. With the breakdown of the polar vortex, these particles are advected to the mid-latitudes and serve to maintain the sulfate layer by supplying new particles.

With regard to the fate of stratospheric aerosols, it is believed that the particles near the bottom of the aerosol layer are removed by being mixed out into tropospheric air. The particles are probably mainly removed from the stratosphere at tropopause folds, which are a major sink for stratospheric air. Subsidence of air near the poles in winter may also account for a significant loss of stratospheric aerosol particles. Polar regions in the winter appear to be one of the most important sinks for stratospheric aerosols. Thus few if any of the particles injected into the stratosphere by the 1985 eruption of Redoubt volcano (60.5° N, 152.0° W) were observed in the tropics afterwards.

5.4. EFFECTS

5.4.1. Climatic Changes

Aerosol and cloud particles exert a variety of important influences on the Earth's climate and on the atmosphere's chemical composition and dynamics. An interplay between atmospheric, geophysical, solar, and geographic factors influences the climatic effects of aerosols (Kondratyev, 1986). While population growth and growing industrialization may lead to unexpected (inadvertent) climatic changes, natural factors should predominate until the end of the century.

5.4.1.1. Direct Radiative Forcing

Natural aerosol constituents contribute to light scattering in the unperturbed atmosphere. These are submicron particles of sulfates and organic carbon (such as terpenes) produced by partial atmospheric oxidation of, respectively, gaseous biogenic sulfur and organic compounds. Sea salt and wind-blown dust can contribute locally, as can smoke and ash from volcanoes and wildfires. These sporadic events are globally less significant, however, because the particles are either generated too infrequently or they are large and short-lived, and thus transported only short distances.

The mid-visible optical thickness, a measure of the attenuation of sunlight in the free troposphere due to terpenes and sulfates, is $\tau \approx 0.01 \pm 0.005$ (Russell et al., 1993). With an average height of the tropopause of 15 km, the resulting extinction due to background aerosol scattering is about 7×10^{-4} km^{-1}. This extinction value compares favorably with a computed light extinction based on *in situ* measured size distributions in the Pacific Basin free troposphere of about 2×10^{-4} km^{-1} (Pueschel et al., 1994a).

Critical for the direct climatic effects of anthropogenic sulfates is the amount of secondary aerosol formed from SO_2 emitted by fossil fuel combustion. About half the SO_2 is dry-deposited on the Earth's surface before oxidation to sulfates, and most of the remainder is oxidized in cloud droplets, such that only 6% of the sulfur emitted per year from anthropogenic activities is available for the gas phase production of new particles (Langner et al., 1992). However, because a cloud that may wet-oxidize SO_2 to produce H_2SO_4 goes through many cycles of condensation and evaporation before the droplets become large enough to precipitate out, cloud-oxidized sulfate particles can be released from an evaporating cloud and also participate in a direct forcing of climate. Therefore it appears that approximately 40–50% of the SO_2 emissions from the burning of fossil fuel are converted to new particles (Charlson et al., 1992).

The direct (non-cloud) radiative forcing by man-made sulfates has been estimated from empirical relations between light extinction and sulfate concentrations. Assume an annual emission of anthropogenic SO_2 of 90 Tg year^{-1}, 40% of which reacts to produce SO_4^{2-} particles covering 5×10^{14} m^2 of the Earth's surface with a residence time of 0.02 years. Under these conditions the anthropogenic sulfate burden is 4.6×10^{-3} g m^{-2} (Charlson et al., 1992). With a mid-visible sulfate scattering cross-section of 5 m^2 g^{-1} at a low (30%) relative humidity, the global optical depth attributable to dry anthropogenic sulfates is $\tau_{SO_4^{2-}} \approx 0.02$. This is approximately seven times the dry background aerosol optical depth that was calculated by Pueschel et al. (1994a) from measured size distributions in the Pacific Basin free troposphere. Allowing for a relative increase in scattering due to larger particle sizes associated with deliquescent or hygroscopic accretion of water as the relative humidity increases, the globally averaged optical depth is $\tau \approx 0.04$ (Charlson et al., 1992). This value is four times the optical depth typically measured in the free troposphere (e.g., Russell et al., 1993). The corresponding radiative transfer forcing is $\Delta F_r = -1.3\,W\,m^{-2}$, assuming a global top-of-the-atmosphere radiative flux of $1370\,W\,m^{-2}$, of which 76% reaches the surface given a global mean albedo of 15% through 60% cloud cover (Charlson et al., 1991).

Kiehl and Briegleb (1993) argue that a specific extinction of $5\,m^2\,g^{-1}$ is valid only for mid-visible wavelengths. It is larger for wavelengths less than $0.55\,\mu m$ and decreases rapidly for wavelengths greater than $0.55\,\mu m$, thereby reducing the aerosol climate forcing. They also assume an asymmetry parameter that is larger than the one used by Charlson et al. (1992). As a consequence, their radiative transfer forcing, at similar anthropogenic aerosol optical thickness, is reduced by about a factor of 2 compared with the value of Charlson et al. (1992).

Several attempts (e.g., Enghardt and Rodhe, 1993) have been made to validate climate forcing by sulfate aerosols. These have involved searching for differences in the evolution of the hemispheric annual mean temperatures that should be associated with the difference in hemispheric sulfate loadings. This comparison is made difficult because of the large difference in proportions of land and sea between the two hemispheres. Locations of polluted regions mainly over the continents in the northern hemisphere (NH) and of clean regions over the oceans in the southern hemisphere (SH) add further problems. These differences are probably also the reason why model estimates of the greenhouse effect generally indicate a more rapid warming in the NH than in the SH (e.g., Gates et al., 1992). An additional complicating factor may be the increased cloudiness observed at several continental sites in the NH (Henderson-Sellers, 1986), which is likely to be connected to a decrease in the daily annual temperature range (Karl et al., 1991). Although some of the studies indicate a smaller warming of the NH than the SH, the results so far have been inconclusive.

Opposing the cooling forcing by SO_4^{2-} and enhancing the warming potential of greenhouse gases are radiation-absorbing aerosols. Light absorption is highest for particles containing elemental black carbon (soot) produced by the incomplete combustion of carbonaceous fuel. During dust storms, soil particles (primary, natural), second only to black carbon (soot; secondary, anthropogenic) aerosol (BCA), can also make a contribution to the absorption of solar radiation in the lower atmosphere. However, it is BCA that has the highest absorption cross-section, on the order of $10\,m^2\,g^{-1}$ (Faxvog and Roessler, 1978). It dominates the absorption of light in most environments (Clarke and Charlson,

1985) and can play a role in radiative transfer and in the effects of aerosols on climate (Ackerman and Toon, 1981). In particular, it can offset the cooling effect attributed to anthropogenic sulfates (Charlson et al., 1992; Kiehl and Briegleb, 1993) and accelerate atmospheric warming by the greenhouse effect (IPCC, 1990). Charlock and Sellers (1980) estimate that if BCA decreases the albedo of aerosols, with an assumed global average of optical depth of 0.125, from 0.95 to 0.75, the radiative effect of the aerosols changes from a net cooling of $-1.2\,°C$ to a net warming of $0.5\,°C$. Pollack et al. (1976) postulate that $\varphi_0 = 0.98$ in the stratosphere changes cooling to warming.

Absorption of radiation from the Sun by tropospheric aerosols is one of the major sources of diabatic heating of as much as $5\,°C\,day^{-1}$ in dense haze layers in the lower troposphere, thereby affecting the daytime boundary layer structure (Asano and Shiobara, 1989). Calculations performed on a radiative model of the atmosphere that incorporates dust as an absorber and scatterer of infrared radiation show an increased greenhouse effect for a reference wavelength of $0.55\,\mu m$, where the net upward flux at the surface is reduced by 10% owing to the strongly enhanced downward emission. There is a substantial increase in the cooling rate near the surface, but the mean cooling rate throughout the lower troposphere is only 10% (Harshvardhan and Cess, 1978). Total heating rates of 0.175 and $0.235\,K\,h^{-1}$ have been deduced for hazy and foggy atmospheres, respectively, with a new airborne radiometric system with a time resolution as high as 60 ms. The aerosol contributions to these heating rates have been found to be 0.065 and $0.235\,K\,h^{-1}$, respectively (Ackerman and Valero, 1984). These results indicate a possibility of aerosol absorption inhibiting local precipitation.

The $10\,m^2\,g^{-1}$ absorption cross-section for soot can be increased if soot is mixed in certain ways with other types of aerosol. Any situation from pure soot particles existing separate from transparent, e.g., $(NH_4)_2SO_4$, particles (external mixtures) through to sulfate particles incorporating some mass fraction of the soot (internal mixtures) is possible. Owing to focusing inside the sulfate, the mixed particles can absorb 2.4 times as much light as an external mixture with the same bulk chemical composition (Heintzenberg, 1978). Incorporated in sufficient amounts into cloud drops, soot aerosol can affect the cloud albedo (Twohy et al., 1989). If the BCA is located in the interior of a drop, the specific absorption coefficient for the BCA is increased from about 10 to $20\,m^2\,g^{-1}$. If the BCA particles are concentrated near the surface of the drop, the absorption coefficient may be enhanced by two to three orders of magnitude (Bhandari, 1986). Heating rates twice as high for mixed soot/transparent particles than for a population of independent soot and transparent particles with the same bulk composition have been calculated for Arctic haze (Wendling et al., 1985).

Estimates of mass ratios of SO_4^{2-} to BCA in aerosol and rain in near-source regions vary between 2 and 4 (Penner et al., 1993). Because the mid-visible scattering cross-section of SO_4^{2-} and the absorption cross-section of BCA are nearly equal $(10\,m^2\,g^{-1})$ at representative relative humidities in the planetary boundary layer, these concentration ratios imply single-scattering albedo values $0.9 < \varphi < 0.95$. Additional oxidation of SO_2 downwind from sources increases the concentration of SO_4^{2-}, but not of BCA, thereby increasing the ratio of scattering to absorption. Measurements in the free troposphere yield BCA concentrations that barely reach 1% of the total aerosol. Over surfaces with low albedo ($\varphi < 0.1$ is characteristic of most of the Earth), such aerosols cool rather than warm the Earth-atmosphere system (Coakley et al., 1983). Thus the scattering by SO_4^{2-} should dominate absorption by BCA at most latitudes, but absorption can dominate at high latitudes, especially over highly reflecting snow- or ice-covered surfaces (Blanchet, 1989).

5.4.1.2. Indirect Radiative Forcing

Increased concentrations of CCNs result in increased concentrations of cloud droplets, resulting in enhanced short-wave albedo of clouds (Twomey, 1977; Coakley et al., 1983). Sulfate aerosols appear to act as major anthropogenic CCNs (Charlson et al., 1992), but particles of smoke from biomass burning may be important as CCNs in some circumstances (Penner et al., 1992). About half of the anthropogenic SO_4^{2-} particles are active as CCNs. Estimates place the CCN numbers in the northern hemisphere 30–50% above those in the southern hemisphere. A 30% increase in the number of CCNs could imply a negative climate forcing at least as large as the direct sulfate forcing (Schwartz, 1988), resulting in a total negative sulphate aerosol forcing of the same magnitude as the positive forcing due to the increases in CO_2 $(1.5\,W\,m^{-2})$ and other greenhouse gases (IPCC, 1990). Such a large total sulfate aerosol forcing is incompatible with a lack of differences in hemispheric temperatures.

Radiative properties of cirrus clouds and latent heat evolution rates of convective clouds are both highly sensitive to microphysical constraints provided by nuclei from the stratosphere or the Earth's surface. Such nuclei give rise to ice particles at temperatures below $0\,°C$ and to large cloud droplets $(r > 40\,\mu m)$ at temperatures both below and above $0\,°C$ at mid-levels in the atmosphere. In the troposphere at high altitudes near $-40\,°C$, cirrus crystals form on soluble H_2SO_4 which dilute to cloud droplets and freeze by homogeneous nucleation; alternatively, mineral particles advected from the surface may form crystals at somewhat higher temperatures. Incorporation of high concentrations of nuclei into regions of cirrus formation would be expected to lead to greater optical depth (more smaller crystals and greater surface area for a given mass of ice).

Because cloud droplets form by condensation of water on existing CCN particles, the concentration, size, and water solubility of CCNs have an immediate influence on the concentration, size, and chemical make-up of the cloud droplets. The short-wave radiative properties of clouds change even if the macroscopic and thermodynamic properties of these clouds are not affected by the aerosol. In contrast, the perturbation in long-wave absorption by tropospheric clouds arising from an increase in cloud droplet concentration is negligible, because tropospheric clouds are already optically thick at infrared wavelengths (Paltridge and Platt, 1976).

A decrease in mean droplet size associated with an increase in cloud droplet concentration is expected to also inhibit precipitation development and to increase cloud lifetimes (Albrecht, 1989). Such an enhancement of cloud lifetime and the resultant increase in fractional cloud cover would increase both the short- and long-wave radiative influence of clouds. Because this effect would predominantly influence low clouds for which the short-wave influence dominates (Paltridge and Platt, 1976), the net effect would be one of further cooling. Inhibited precipitation development might further alter the amount and vertical distribution of water and heat in the atmosphere and thereby modify the Earth's hydrological cycle. Although these effects cannot yet be quantified, they have the potential of inducing major changes in global weather patterns as well as in the concentration of water vapor, the dominant greenhouse gas (Twomey, 1991).

5.4.1.3. Stratospheric Aerosol Radiative Forcing

Particulate matter normally found in the stratosphere above the tropopause may also influence the terrestrial radiation balance, catalyze heterogeneous chemical reactions, and serve as a tracer of atmospheric motion. Unaffected by volcanic eruptions, the stratospheric aerosol optical depth is too small to significantly affect the Earth's climate (Pollack et al., 1981). Enhanced stratospheric haze layers resulting from the injection of aerosol-forming material into the stratosphere by volcanic eruptions, however, can alter that pattern.

Major volcanic eruptions typically inject huge quantities of terrestrial material into the stratosphere (Lang, 1974), thereby producing primary (ash) and secondary (sulfuric acid from SO_2) aerosols. Such volcanic aerosols have long been suspected to cause short-term climate changes (Pollack et al., 1976). Because of their small (submicron) size, these aerosols are more effective at attenuation by scattering of incoming solar radiation (the albedo effect) than they are at absorbing the long-wave terrestrial radiation (the greenhouse effect). Thus the particles change the Earth's radiation balance by reflecting more of the Sun's energy back to space, at the same time permitting the planet to cool radiatively at about the same rate as before an eruption. The result is a net loss of energy for the Earth-atmosphere system or a cooling of the below-aerosol atmosphere and the surface. Because volcanic dust depletes some of the energy available to the Earth's climate system, thereby forcing the system to adapt to a new equilibrium state, the change in the Earth's radiation budget that is initiated by volcanic eruptions is termed volcanic aerosol forcing. This was observed after the 1984 eruption of the Mexican volcano El Chichon, when the solar radiation received at the surface at several locations in the US and in Hawaii was depleted by 25%, diffuse sky radiation was enhanced by up to a factor of 3, and global radiation was lowered by up to 5% (Rao and Takashima, 1986).

Volcanic aerosol-enhanced clear-sky albedos are most effective at the lowest surface albedos and decrease monotonically with increasing background albedo. For example, Valero and Pilewski (1992) show that surface albedos of 0.05 (typical for oceans) and 0.30 (characteristic for land areas) increase at $0.55 \, \mu m$ wavelength by 0.052 and 0.032, respectively, when a volcanic aerosol layer is inserted above the background concentration. Thus a small rise in aerosol optical depth substantially increases the Earth–atmosphere albedo over the relatively dark surfaces of the cloud-free oceans, but has a smaller or no impact on the total albedo over brighter surfaces such as clouds and light sand or alkali deserts. This means that aerosol climate forcing varies regionally and with latitude and season. However, volcanic aerosols from large eruptions generally spread over most of a hemisphere and remain airborne for about a year or more. Therefore the global (or hemispheric) mean is an appropriate case for studying the extent of climate forcing by individual volcanic eruptions.

The most common method of estimating volcanic aerosol forcing is to calculate the radiative transfer on the basis of assumptions or measurements of aerosols and their properties. For example, Minnis et al. (1993) analyzed direct Earth Radiation Budget Experiment (ERBE) satellite (ERBS) measurements of the radiation budget, as well as optical properties of the aerosol, to quantify the volcanic forcing by the 1991 eruption in the Philippines of Mount Pinatubo. They compared post-Pinatubo monthly means of incoming total, reflected short-wave and outgoing long-wave radiation with the average ERBS monthly means for 1985–1989. The difference between a given post-Pinatubo monthly mean and its corresponding pre-Pinatubo 5 year monthly average is defined as an anomaly. In this way it was determined that the globally averaged anomalous radiative cooling caused by Mount Pinatubo between -40 and $+40$ degrees latitude during August and September 1991 was $-2.7 \pm 1.0 \, W \, m^{-2}$. This estimate compares favorably with results from a theoretical model that predicts a net forcing of $-2.8 \, W \, m^{-2}$ (Lacis et al., 1992) for a global mean of a volcanic aerosol optical depth of 0.15. Such an optical depth was actually observed after the 1991 eruption of Mount Pinatubo (Russell et al., 1993).

The influence of existing tropospheric clouds on reduction of the short-wave albedo (increased background albedo) and a decreased temperature difference between aerosol layer and emitting troposphere by stratospheric aerosols has been investigated by Kinne (1993). His calculations agree with the infrared forcing of Lacis et al. (1992), but the solar forcing is lower by about 50% owing to the presence of tropospheric clouds. Thus the net global Pinatubo volcanic aerosol forcing is about $-2 \, W \, m^{-2}$ for the troposphere and about $-1 \, W \, m^{-2}$ for the

Earth–atmosphere system. Figure 5.7 summarizes various aerosol climate forcings in relation to those from greenhouse gases.

The estimated global net flux changes between -1 and $-3\,\mathrm{W\,m^{-2}}$ are characteristic for the time period of several months following the June 1991 eruption of Pinatubo. These may be compared with the value of $+1.25\,\mathrm{W\,m^{-2}}$ for the radiative forcing due to the increase in carbon dioxide measured since the industrial revolution (IPCC, 1990), and with the estimate of $-1\,\mathrm{W\,m^{-2}}$ due to a direct (non-cloud) effect of anthropogenic sulfates (Charlson et al., 1992). Thus it is clear that the Pinatubo climate forcing, similar to many other volcanic forcings, had the potential to initially offset and later delay CO_2-induced greenhouse warming. This would be the case as long as the mid-visible optical depth exceeded about one tenth, approximately 10 times the background aerosol optical depth.

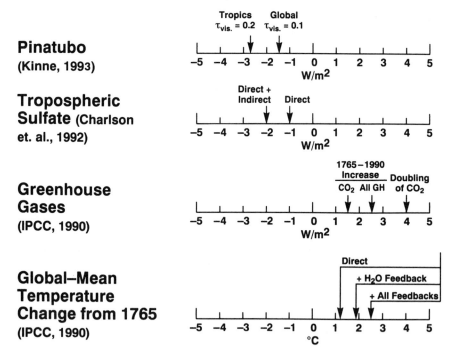

FIGURE 5.7. Pinatubo aerosol radiative forcing (Kinne, 1993), shown in relation to radiative forcing by anthropogenic sulfates (Charlson et al., 1992) and greenhouse gases (IPCC, 1990).

An influence of tropospheric clouds on volcanic aerosol forcing raises an important question of possible feedback of volcanic aerosol on these clouds. After an eruption, volcanic debris in the stratosphere enters the troposphere, primarily as a result of tall convective storms and tropopause folds. Significantly enhanced tropospheric aerosol loading attributed to Mount Pinatubo was observed by lidars as far north as 40° N during August 1991 (Post et al., 1992). These aerosols were available for incorporation into clouds, especially at the upper levels. Sulfate aerosols function as efficient cloud condensation nuclei, and volcanic ash can act as an ice nucleus at temperatures below $-16\,°C$ (Mason and Maybank, 1958). Greater concentrations of cloud and ice nuclei tend to increase the number and reduce the effective radius of the hydrometeors in the cloud. At constant liquid and ice-water contents, cloud albedo increases as effective particle radius decreases. Such an indirect effect could explain the observation by Minnis et al. (1993) of increased deep convective cloud albedo for short-wave fluxes with no effect on the long-wave flux, because optically thick (high albedo) clouds are opaque to long-wave radiation. Although the indirect effects have been reported only for deep clouds, they may also occur for other cloud types. For example, diffusion of volcanic aerosols across the tropopause may alter the optical properties of high, thin cirrus clouds or enhance the generation of clouds (Sassen, 1992). Thus it appears that volcanic radiative effects are more complex than simple models that depict direct forcing by a single aerosol layer distributed uniformly over the background. Indirect effects are not now included in most climate models.

As has been observed after the 1982 El Chichon eruption (Post, 1986), sulfuric acid droplets are indeed transported from the stratosphere into the upper troposphere. This can take place by gravitational settling (if particles are larger than $0.7\,\mu m$ diameter) or dynamically by stratospheric–tropospheric exchange processes such as tropopause folding. The result is an elevated tropospheric concentration of supercooled sulfuric acid droplets (Sassen 1992). These grow by the accretion of water vapor, which is more abundant in the troposphere than in the stratosphere. The droplets can nucleate ice to increase the cirrus particle population by as much as a factor of 5, resulting in a net radiative forcing (surface warming) by as much as $8\,W\,m^{-2}$ (Jensen and Toon, 1992).

Potential anthropogenic perturbations of the stratospheric aerosol are possible because of aircraft and space shuttle operations. The potential effects of emissions of sulfur dioxide gas and soot granules by supersonic transport and of aluminum oxide particulates from space shuttle rocket launches on stratospheric aerosols were estimated by Turco et al. (1980). They used an interactive particle–gas model of the stratospheric aerosol layer to calculate changes due to exhaust emissions, and a radiation transport model to compute the effect of aerosol changes on the Earth's average surface temperature. They conclude that a release of small particles (soot or aluminum oxide) into the stratosphere should not lead to a corresponding significant increase in the concentration of optically active aerosols, because increase in large particles is severely limited by rapid loss of small seed particles via coagulation. This conclusion suggests that a fleet of several hundred advanced supersonic aircraft operating daily at 20 km altitude or one space shuttle launch per week could produce a roughly 20% increase in the large-particle concentration of the stratosphere. Aerosol increases of this magnitude would reduce global surface temperature by less than 0.01 K. This conclusion is in agreement with the findings of Pollack et al. (1976), who concluded from terrestrial and radiative transfer calculations that the climate is unlikely to be affected by supersonic transport and space shuttle operations during the next several decades.

Sampling of soot particles in the stratosphere by impactors documents a current stratospheric soot mass loading of $0.6 \, ng \, m^{-3}$, which is equivalent to 0.01% of the total aerosol mass after the eruption of Mount Pinatubo in 1991. This low concentration is commensurate with current air traffic, with realistic assumptions of soot emission factors by jet engines, polar route mileage flown in the stratosphere, and particle residence times in the stratosphere. Independent studies of the absorption coefficients (Pueschel et al., 1992), in combination with total extinction as, for example, measured by a sun photometer aboard the ERBE satellite in the Stratospheric Aerosol and Gas Experiment (SAGE II; Oberbeck et al., 1989), yield a current single-scatter albedo of the stratospheric aerosol of $\varphi_o \approx 0.99 \pm 0.01$. This value is an upper limit for two reasons. First, the 1992 post-Pinatubo high sulfuric acid aerosol concentration will decay with an e-folding time of approximately 1 year, as found in previous eruptions (Post, 1986). Second, the present soot concentration is expected to double owing to planned supersonic commercial air traffic, based on expected fuel consumption and emission factors. Such an increase in aerosol black carbon is expected to decrease the single scatter albedo by 1%, which could change the sign of climate forcing from cooling to warming (Pollack et al., 1976) and increase stratospheric temperatures by several degrees (Lacis et al., 1992). Thus the 1% decrease in stratospheric aerosol single-scatter albedo due to planned supersonic commercial transport could affect tropospheric climate and significantly affect heterogeneous stratospheric chemistry (Chapter 11).

The approaches chosen to show the effect of volcanic aerosol on radiative transfer are scientifically solid, whether they are based on modelling only (e.g., Lacis et al., 1992) or on a combination of models with satellite (e.g., Minnis et al., 1993) or in situ observations (e.g., Kinne, 1993). Nevertheless, numerous attempts to prove an effect on air or sea surface temperature, the ultimate climate response parameter, have produced only mixed results (e.g., Angell and Korshover, 1983). The reasons are similar to those complicating the observation of an effect of tropospheric aerosols on temperature, as was discussed earlier. Volcanic forcing, in addition, is too short-lived to overcome the thermal inertia of the the Earth's surface, particularly the oceans. Radiative transfer forcing during the immediate post-eruption period is temporary and will change with time owing to changes in the volcanic aerosol characteristics. Gravitational settling will remove the relatively large primary ash aerosols preferentially, while smaller particles stay behind and will be replenished by the generation of submicron particles due to SO_2 gas–H_2SO_4 droplet conversion for many more months after an eruption. For example, size distribution measurements after the 1991 Pinatubo eruption showed that particle size initially increased from a pre-eruption effective radius (defined as the ratio of the third and second moments of the size distribution) of approximately 0.25 μm to 0.8 μm (observed 18 months after the eruption). From then on it decreased exponentially back to < 0.30 μm after 3.0 years following the eruption. Theoretical calculations show infrared forcing (greenhouse warming) to be a strong function of particle size, while the opposing short-wave forcing is relatively insensitive to particle size if the effective radius is larger than the effective solar wavelength (Lacis et al., 1992). Greenhouse warming relative to albedo cooling will therefore be greatest shortly after an eruption.

Only recently was it possible to pick out an overall global cooling effect of 0.1–0.5 °C by volcanoes from the randomly fluctuating natural temperature patterns from one year to the next (Kerr, 1989). This required numerous case studies before making a link between volcanic eruptions and a particular weather pattern. The number of large eruptions in historic times is almost too sparse to build a statistically significant case, and a cooling effect of the few eruptions on record is masked by the warming due to El Niño and greenhouse gases (Mass and Portman, 1989).

The complexity of volcanic aerosol climatic forcing is illustrated by the fact that 1991–1992 was marked by an unusually mild winter throughout North America and much of northern Eurasia, despite expected post-Pinatubo cooling (Kerr, 1993). This counter-intuitive effect of raising winter temperature in one part of the globe by altering weather patterns in another has been explained by Robock and Mao (1992). They surveyed winter temperatures at sites throughout the northern hemisphere immediately following what they estimated were the 12 largest volcanic eruptions since hemispheric records have been kept. After correcting for the climate effects of El Niño's warm Pacific waters, it was found that the winter after each eruption was unusually mild across northern Eurasia. The phenomenon is explained by surface weather patterns that are altered by the stratospheric aerosol through a chain of physical processes. First the aerosol absorbs solar radiation and warms the tropical stratosphere where sunshine is abundant even in winter. This then intensifies the stratospheric temperature difference between the tropic and polar regions, strengthening the potential vorticity of the polar vortex. This in turn redirects the jet streams below the stratosphere that guide storms. The result is a surge of warm air north into North America and northern Eurasia and of cold air down over Greenland.

5.4.2. Heterogeneous Chemistry

5.4.2.1. Stratospheric Ozone Depletion

Aerosol surfaces in the atmosphere permit chemical reactions to occur which would otherwise be kinetically unfavorable. Heterogeneous reactions initiated and/or catalyzed by atmospheric particles can lead to major shifts in gas phase photochemistry.

In the stratosphere, catalysis of nitrate transformations on or in polar stratospheric cloud (PSC) particles is now proven to be capable of inverting the standard relationships between chlorine reservoirs and ClO_x radicals in the polar regions, and a direct consequence has been massive Antarctic ozone depletions. In the cold Antarctic stratosphere, particles of nitric acid n-hydrate (type I polar stratospheric clouds) first condense at about 193 K; water ice particles (type II clouds) form when the temperature falls to about 187 K (e.g., Toon et al., 1986). These polar stratospheric clouds provide surfaces for reactions that are the key to the Antarctic ozone hole (see Chapter 11). The heterogeneous reactions convert inert hydrogen chloride (HCl) and chlorine nitrate ($ClONO_2$) to reactive molecular chlorine (Cl_2) and hypochlorous acid (HOCl):

$$HCl + ClONO_2 \rightarrow Cl_2 + HNO_3$$

$$H_2O + ClONO_2 \rightarrow HOCl + HNO_3 \tag{5.18}$$

The molecular chlorine and hypochlorous acid are gases that can be photolyzed by solar radiation to form chlorine radicals which can catalyze ozone destruction. In contrast, the HNO_3 produced by those heterogeneous reactions remains condensed on the cloud particles. This condensation ties up the nitrogen compounds that, were they in the gaseous phase, would react with active chlorine to re-form inert $ClONO_2$. During the winter, some of the relatively large type II particles have settling velocities large enough to transport the HNO_3 to lower altitudes. This process of denitrification physically separates HNO_3 from ozone-reactive Cl radicals, thereby extending the lifetime of active chlorine.

In polar spring, when the Sun reappears, photons dissociate Cl_2 into chlorine atoms (Cl) that attack ozone, yielding molecular oxygen and chlorine monoxide (ClO), which forms a dimer that also is photolyzed to Cl to destroy more ozone:

$$2(Cl + O_3 \rightarrow ClO + O_2)$$

$$ClO + ClO \rightarrow Cl_2O_2$$

$$Cl_2O_2 + h\nu \rightarrow Cl + ClOO$$

$$ClOO \rightarrow Cl + O_2 \tag{5.19}$$

The validity of this chemical ozone depletion scheme has been verified by many studies, such as satellite observations (McCormick et al., 1989), lidar measurements (Gobbi et al., 1991), balloon observations (Deshler et al., 1992), and nitric acid-specific aerosol formation (Pueschel et al., 1989 and 1990). These workers all found nitric acid trihydrate formation at the appropriate temperatures, corresponding to concentrations of NO_x and H_2O vapors that were measured independently.

The major sources of the HCl in the stratosphere are industrial chlorine compounds such as chlorofluorocarbons (CFCs) used by man as refrigerants and aerosol propellants. CFC production has been increasing steadily since the 1950s. The 1990 Montreal Protocol forced severe restrictions on the production of chlorofluorocarbons, resulting in a leveling of atmospheric concentration (see Chapter 7 for details).

Aerosol added to the stratosphere can affect ozone concentrations in several ways. First, sunlight absorbed by the aerosol particles can heat the stratosphere, altering its usual circulation patterns (Kinne et al., 1992). Secondly, the aerosol could increase chlorine- and bromine-catalyzed ozone depletion by providing surfaces for heterogeneous reactions (Prather, 1992). Thus aerosols from Mount Pinatubo could be responsible for the increase in size of the Antarctic ozone hole this past year (Solomon et al., 1993), which was about 25% bigger than average and the largest on record. Mount Pinatubo injected 15–30 Tg of SO_2 directly into the stratosphere (Brasseur and Granier, 1992). Within several months of the eruption, stratospheric H_2SO_4 droplets had spread from the tropics to high latitudes, increasing the aerosol layer some 20–50 times over its background levels (Brock et al., 1993). Because of an exponential decay (Post, 1986), it will be several years before aerosols approach pre-Pinatubo levels.

Enhanced aerosols from volcanoes such as Mount Pinatubo are accelerating ozone depletion at high latitudes primarily through the reaction of $ClONO_2$ with water. Laboratory studies have shown that this reaction proceeds on dilute sulfate aerosols (Luo et al., 1993) and that the reaction rate is strongly dependent on the fraction of sulfuric acid in solution. The colder the temperature, the greater are the water content and holding capacity of HCl and the faster the reaction proceeds. Thus the $ClONO_2$ and H_2O are most likely to trigger extra ozone depletion in the coldest regions of the stratosphere near the poles and in the tropics.

A second temperature-independent mechanism is the reaction of dinitrogen pentoxide (N_2O_5) with H_2O to form nitric acid (Rodriguez et al., 1991):

$$NO_2 + NO_3 \rightarrow N_2O_5$$

$$N_2O_5 + H_2O \rightarrow 2HNO_3 \tag{5.20}$$

However, before this former reaction, chlorine could have been bound as follows:

$$NO_2 + ClO \rightarrow ClONO_2 \qquad (5.21)$$

The H_2SO_4/H_2O aerosol removes NO_2, allowing ClO and BrO concentrations to rise. The active halogen species can then catalytically destroy ozone:

$$ClO + BrO \rightarrow Cl + Br + O_2$$

$$Cl + O_3 \rightarrow ClO + O_2$$

$$Br + O_3 \rightarrow BrO + O_2$$

$$\overline{\text{net: } 2O_3 \rightarrow 3O_2} \qquad (5.22)$$

Prather (1992) argues that an increased effective surface area of sulfuric acid after volcanic eruptions can induce heterogeneous reactions involving $ClONO_2$, and secondarily N_2O_5, to suppress NO_x abundances by more than a factor of 10 relative to gas phase chemistry. When NO_x levels fall below a threshold, e.g., 0.6 ppb at 24 km in mid-latitudes, the chlorine-catalyzed loss of O_3 proceeds at rates comparable with those during the formation of the Antarctic ozone hole, more than 50 ppb day^{-1}. Such losses might have occurred in the most volcanically perturbed regions over the tropics and mid-latitudes following the eruption of Mount Pinatubo. Various investigators (e.g., Brock et al., 1993) measured up to 20-fold increases in the stratospheric particle surface area due to the 1991 Pinatubo volcanic eruption. With a background surface area of 0.4 $\mu m^2 cm^{-3}$ (Pueschel et al., 1989), the Pinatubo-enhanced volcanic aerosol surface area is close to the threshold of 10 $\mu m^2 cm^{-3}$ that Prather (1992) defines as the transition region from negligible aerosol-induced ozone loss rates to those in excess of 10 ppb day^{-1}.

Mount Pinatubo was more than twice as effective as El Chichon in causing rapid O_3 loss, because of the increase in stratospheric chlorine (from CFCs) since 1982 (Rodriguez et al., 1991) and because overall global losses associated with a volcanic eruption are approximately linearly proportional to the amount of sulfate surface area. The build-up of atmospheric chlorine, the intensity of volcanic eruptions (Brasseur et al., 1990), and the intensity of projected aircraft emissions all play a role in the forecast of future sulfate ozone depletion.

Dustsonde-measured baseline concentrations during periods of volcanic quiescence indicate that the stratospheric background aerosol may be increasing by as much as 5–10% year^{-1} (Hofmann, 1990). Whether this increase is of natural or anthropogenic causes could not be determined because of inadequate information on sources. Carbonyl sulfide (OCS), thought to be the dominant non-volcanic and possibly anthropogenic source of stratospheric H_2SO_4 vapor (Crutzen, 1976), is especially problematic. Nevertheless, studies of catalytic destruction of ozone by anthropogenic sulfates seem to be in order. The revival of plans for supersonic commercial aircraft promises to further enhance interest in anthropogenic (soot and sulfuric acid) aerosol-catalyzed reactions.

5.4.2.2. Tropospheric Heterogeneous Chemistry

Aerosol particles also play a crucial role in the depletion of ozone in surface air, as is often observed in the Arctic spring. In contrast with stratospheric chemistry, Cl plays no known role in this process. Solar radiation reaching the Earth's surface is not energetic enough to photodissociate chloroform and other chlorinated hydrocarbons (CFCs). Brominated and iodinated hydrocarbons, however, largely of oceanic origin, do absorb solar radiation at the surface sufficiently to create atomic bromine and ozone loss by the following cycle (Barrie et al., 1988):

$$CHBr_3 + h\nu \rightarrow Br + CHBr_2 \rightarrow nBrO_x$$

$$Br + O_3 \rightarrow BrO + O_2$$

$$BrO + BrO \rightarrow 2Br + O_2(+ h\nu) \rightarrow 2Br + O_2$$

$$BrO + h\nu \rightarrow Br + O \tag{5.23}$$

This cycle of Br reacting with ozone, followed by the self-reaction of the BrO produced, represents a catalytic loss mechanism for O_3 as Br is regenerated. However, the radicals Br and BrO are rapidly converted to the non-radical (inert) species hydrobromic acid (HBr), hydrobromous acid (HOBr), and bromine nitrate ($BrNO_3$). McConnell et al. (1992) proposed that cycling of inorganic bromine between aerosols and the gas phase could maintain sufficiently high levels of Br and BrO to destroy ozone, but they did not specify a mechanism for aerosol phase production of active bromine species. Fan and Jacob (1992) proposed such a mechanism, based on known aqueous phase chemistry. It rapidly converts HBr, HOBr and $BrNO_3$ back to Br and BrO radicals. For example, $BrNO_3$ should be rapidly scavenged by the aerosol and hydrolyzed to HOBr(aq) in solution, followed by subsequent reaction of HOBr(aq) and Br$^-$ to produce Br_2(aq):

$$BrNO_{3(g)} + H_2O \rightarrow HOBr_{(aq)} + HNO_{3(aq)}$$

$$HOBr_{(aq)} + Br^- + H^+ \rightarrow Br_{2(aq)} + H_2O \tag{5.24}$$

As Br_2(aq) is produced, it volatilizes to the gas phase were it is photodissociated to Br. Similarly to the situation in the stratosphere, this mechanism should be particularly efficient in the presence of high concentrations of sulfuric acid aerosols that are frequently observed during boundary layer ozone depletion events.

5.5. SUMMARY AND FUTURE NEEDS

Atmospheric sulfur gases may arise either from industrial activity or from natural sources such as volcanoes and the oceans. These gases are transformed in the atmosphere to sulfate particles and/or sulfuric acid droplets. Particles reflect sunlight back to space, thus decreasing the heat input to the Earth. Potentially, the temperature of the Earth–atmosphere system is thereby lowered. The amount of sulfur emitted globally by human activities has increased from about $10 \, Tg \, year^{-1}$ in 1880 to about $150 \, Tg \, year^{-1}$ in 1990. Therefore reflection of sunlight has possibly increased, perhaps enough over time to reduce the amount of heating induced by trapping of surface infrared radiation.

Several large volcanic eruptions in the last three decades, for which the change in sunlight reaching the Earth's surface has been measured, provide proof of radiative forcing by volcanic aerosols. Although the intensity and frequency of eruptions are unpredictable, powerful eruptions generate large quantities of stratospheric sulfuric acid with e-folding times of about 1 year. As a consequence, particles spread wide enough over one or both hemispheres to globally average their effects.

In the visible it appears that an extinction exceeding $3 \times 10^{-4} \, km^{-1}$ is outside $1 \, \sigma$ error bars of measurements. Assuming an average tropopause height of $10 \, km$, this extinction implies an optical depth of $\tau \approx 0.003$. This is close to the "background" optical depths measured with the NASA Ames autotracking airborne sun photometer at Mauna Loa, Hawaii and aboard Ames aircraft prior to the 1991 eruption of Pinatubo volcano. Therefore an anthropogenic SO_4^{2-} optical depth of $\tau \approx 0.04$, shown by Charlson et al. (1992) to yield a direct negative climate forcing of about $1 \, W \, m^{-2}$, in the free troposphere would be more than 10 times the "background" optical depth of the Pacific Basin free troposphere. Such an increase in optical depth would easily be detectable with existing aerosol instruments, especially on airborne platforms where they can separate boundary layer from free tropospheric and volcanically influenced stratospheric contributions. It might be difficult, however, to delineate an anthropogenic sulfate effect within the boundary layer, where the optical depth generally exceeds 0.04.

The man-made sulfur component is SO_2 from the combustion of fossil fuel. It is injected into the boundary layer. The effects are therefore largely confined to the lower troposphere. Assessment of global effects is a difficult undertaking for several reasons. Sources of anthropogenic aerosols are not uniformly distributed. Anthropogenic particles are short-lived in the atmosphere. As a consequence, their spatial concentration is highly variable. Rates of secondary particle formation and the time evolution of the size distribution, both of which determine the aerosol optical and cloud nucleating capabilities, are dependent on the production of condensable material and the concentration and size distribution of aerosol particles already present. Aerosol formation and removal processes are covariant with diurnally, synoptically, and seasonally varying features of the Earth–atmosphere system. A task that appears in order is screening of historic optical depth data for trends that would be proportional to the trend in atmospheric sulfur emissions.

Large gaps exist in the knowledge of anthropogenic aerosols, preventing quantification of their influence for use in climate models. What is needed is a coupling of (1) the physical–chemical processes that produce them and the meteorological processes that distribute and remove them with (2) the physical and optical characteristics that determine radiative transfer and cloud micro-physical effects.

Globally representative measurements of aerosol and cloud properties may conceivably be obtained by satellites equipped with photopolarimeters to measure the radiance and polarization of reflected sunlight in the spectral range from the near ultraviolet to the near infrared (Hansen et al., 1990). Specific cloud properties such as height, optical depth, particle phase, and effective size may also be derived from satellites. However, validation and improvement of inferences drawn from satellite measurements require that such satellite monitoring be tied to concurrent ground-based and *in situ* aircraft measurements of optical and microphysical cloud and aerosol parameters. This is particularly true for the vertical inhomogeneities, because satellites (like ground-based measurements) provide only height-integrated values of the aerosol parameters. Nevertheless, vertical distribution information is required, because effects of light scattering and absorption are altitude-dependent, as are cloud properties. The climatic effect (heating versus cooling) depends on the ambient temperature of an aerosol of given scattering/absorption characteristics.

During periods of volcanic quiescence, stratospheric aerosol formation processes depend on transport. Unfortunately, there is much uncertainty regarding the meridional transport of the stratospheric gases and particles. It is possible that the aerosols themselves could be used as tracers. In the winter of 1982–1983 the northward transport rate of El Chichon aerosols was observed by six lidars stationed between 19° and 55° latitude. The transport rate of the cloud maximum was estimated to be approximately 0.3° day^{-1}, and on January 1, 1983 it was located near 30° N. Within the uncertainty of the measurements, the cloud maximum optical thickness appeared to remain constant with latitude. On the other hand, the stratospheric aerosol concentration at 40° N began to rise a few weeks after the eruption. Nearly identical transport behavior was observed from sun photometer data following the eruption of Agung.

Thus a lidar network consisting of perhaps 20 lidars over a region perhaps as large as the United States could determine the latitudinal transport of stratospheric aerosols after volcanic eruptions. If the system were operated in some sort of unmanned mode with identical lasers and data reduction schemes, one should be able to obtain an excellent picture of stratospheric motions. In particular, tropopause fold events could be studied, with a view toward measuring the fluxes of water vapor, sulfur compounds, and particulates into the stratosphere, as well as the downward flux of stratospheric aerosols as a potential source of nuclei for upper tropospheric cirrus cloud particles.

The mean circulation of the quasi-biennial oscillation strongly affects the distribution of aerosols. The associated presence or absence of a subtropical barrier in potential vorticity will discourage or allow transport of aerosols out of the tropics by extratropical Rossby waves. The time evolution of volcanic veils may be used to infer the morphology of the potential vorticity barrier.

Aerosol composition is a strong function of air parcel history. Chemical general circulation models (GCMs) may soon be capable of representing the overlap of tracer tongues, but at present this is limited by computational capabilities. To form aerosols containing constituents with different geographical sources probably requires differential air mass advection in the vertical, coupled with small scale mixing due to gravity waves. GCMs are still a long way from being able to represent these small scale mixing processes adequately.

In the discipline of atmospheric chemistry, mechanisms of the heterogeneous reactions are only beginning to draw attention but are at the heart of several critical environmental problems. Chlorine activation is known to take place on stratospheric particle surfaces at the poles and potentially in other localities. Bromine activation has been shown to occur in surface air in the Arctic. Aspects of aerosol physical chemistry underpin the heterogeneous mechanistics. Compositions, vapor pressures, and (in the PSC instance) adsorption equilibria all pertain to aerosol reactivities. They are all influenced by particle surface characteristics. Fundamental physical properties, including morphologies and size distributions, also enter into the determination of heterogeneous reaction rates. Optical phenomena are central to the measurement of these properties. The process of nucleation, through which the aerosols under consideration are formed originally, remains obscure at all stages for polar stratospheric clouds.

The following paragraphs offer a list of topics for which major questions are outstanding regarding the influence of atmospheric particles on chemistry, along with a set of suggestions for relevant experiments, including laboratory, *in-situ*, and numerical approaches. The major thrust is toward the PSCs because they are of immediate and critical concern, but many of the overall concepts apply to sulfate as well. The problems presented are subdivided into sections focusing on the physical chemistry of heterogeneous reactions, aerosol properties bearing on them, and fundamental microphysical issues such as nucleation. Some broader-scale objectives for future polar field campaigns are also discussed.

Composition of polar stratospheric clouds is perhaps the most basic facet of their physical chemistry. It determines nitrate reactivity but remains poorly understood on several key levels. For example, nitric acid trihydrate (NAT), which probably comprises the bulk of type I PSC material, can exist under a wide range of water concentrations. The water mole fraction in turn specifies HCl holding capacities, and efficiencies for the reaction between nitrate and HCl. Actual water concentrations in the NAT particle, however, remain unknown. *In situ* mass spectroscopy of the type I PSC contents is a possibility, and it may be possible to obtain rough measurements of solid phase H_2O/HNO_3 ratios. *In situ* infrared spectroscopy of the type I clouds could also provide clues to their composition but would have to be preceded by laboratory documentation of NAT absorption bands. A third possibility is to capture PSC crystals and analyze them wet chemically for the same information, but their ease of evaporation makes this a difficult task.

Solid vapor pressures are linked closely to composition, but our understanding of PSC vapor pressures is again incomplete. Effects of impurities, coatings and co-condensation of water and NAT have not been investigated. While partial pressures of H_2O and HNO_3 are available for NAT alone, they have not been correlated in detail with the water concentration of the lattice. Adsorptive thermodynamics is a theme related to PSC vapor pressures. An adsorption equilibrium constant defines the tendency for a molecule, particularly a heterogeneously active nitrate or acid, to cover a cloud particle surface. The equilibrium can in turn be converted into a surface-binding enthalpy, which enters into calculation of rates for several elementary steps occurring early in the surface reaction process. These include desorption and two-dimensional diffusion. It could be profitable to undertake determination of adsorption isotherms for the PSC heterogeneous reactants on NAT and ice. Preliminary values have already been reported for HCl (Elliot et al., 1990) and reflect formation of two hydrogen bonds with the crystal. Nitric acid adsorption might also be measurable, but values for nitrates would be complicated by competition from surface reactions.

Although a growing body of circumstantial evidence involving Lagrangian photochemical models points to type I particles as the chief chlorine activators, the PSC/ClO$_x$ connection merits strengthening. In the laboratory, continued effort is needed in measuring collisional efficiencies for nitrate reactions on simulated PSC materials. Early experiments were plagued by improper characterization of solids. Studies of $ClONO_2$ + HCl on water ice, for example, were conducted at pressures orders of magnitude above the stability threshold for hydrochloric acid hydrates. NAT efficiencies may be more realistic at the moment, but the dependence of reactivity on HCl solubilities is not well established and, as mentioned above, stratospheric HCl mole fractions cannot yet be predicted with confidence. Aerosol chlorine activation could perhaps be verified directly by monitoring air upstream and downstream from a single cloud.

The possibility exists that certain subtleties of heterogeneous chemistry are being overlooked because attention has centered on the net chlorine-activating reactions of $ClONO_2$ and N_2O_5. ClO dimerization, for example, may be augmented on PSC surfaces, and as the rate controlling step in an odd-oxygen removal cycle, this recombination is clearly distinct from the nitrate transformations. HO_x species and their reservoirs are decoupled to some extent from the chlorine and nitrogen balances which limit ozone lifetimes, and so heterogeneous interactions of the hydrogen families remain unexplored. It could be instructive to systematize the thermodynamic and kinetic features required for significant heterogeneous catalysis of a gas phase reaction. The ramifications of candidate processes thus highlighted could be tested by photochemical simulation. Another logical priority is detailed bridging of polar photochemical and microphysical calculations through a conceptual model of PSC surface processes.

Of the particle characteristics that bear on heterogeneous chemistry, shape is an obvious physical property which contributes to the reactivity of nitrates or acids on stratospheric aerosols. An irregular surface exhibiting hopper characteristics or graininess will add to the available heterogeneous transformation area. Ice replicator samples have provided evidence that type II polar water ice particles are hexagonal and columnar (Goodman et al., 1989). Those type II particles had a surface area that was only 10% of the total aerosol surface area. They were, however, unexpectedly large and might effectively dehydrate the stratosphere by sedimentation. Departures from the familiar hexagonal–columnar habit are possible at different temperatures but have not been catalogued. Type I clouds have not been intensely scrutinized. Lidar depolarization data have, however, recently permitted discrimination of two new type I particle categories (Browell et al., 1990), perhaps corresponding to diverging morphologies.

Shape is in addition a function of the thermophysical aerosol state. Although polar air lies well below fusion temperatures for liquid water or aqueous nitric acid, there is the possibility of freezing point depression by unidentified impurities. It might be advisable to verify that the PSCs actually exist as solids. Vapor samples near the condensation point could be cooled under controlled conditions and the particle formation process observed, either in the laboratory or perhaps even *in situ*. Particle size is a related matter and further monitoring of the sizes of mid-latitude polar stratospheric aerosols is essential.

Particle optical effects are critical in calibration and operation of optical particle counters used to study aerosols, and consequently there is considerable interest in the accuracy with which they can determine particle characteristics, which in turn strongly depends on shape and refractive index of the particles. Dependence of refractive index on composition has not been delineated for PSC materials, and theoretical calculations would be desirable for the optical properties of non-spherical particles. Infrared spectroscopy of the type I clouds could also be classified under this heading.

Major uncertainties permeate calculations of atmospheric aerosol production, especially with regard to energy barriers against nucleation. It is currently thought that background sulfate aerosols entering the polar stratosphere act as nuclei for type I PSCs, and type I in turn for type II. Laboratory measurements of the saturation ratio necessary to achieve particle formation could improve the modeling situation for each step in this sequence. The initial background sulfate nuclei are present in the mid-latitude stratosphere as liquid, and it has been presumed that the strong supercooling to which they are subjected in the polar vortices results in freezing. This is only conjecture, however, and laboratory or *in situ* verification could be enlightening.

Acknowledgments

This article started with a literature search that produced close to 1000 references. I scanned most of them and used many. I would like to thank all these authors for their contributions. My particular thanks go to Patrick Hamill and Scott Elliott, who contributed sections on stratospheric aerosols and on heterogeneous chemistry of the stratosphere, respectively, to a workshop report that I edited and from which I drew material for this article. A second local source is a series of proposals that I wrote during the past 8 years, notably with Phil Russell as co-proposer, to whom, therefore, I owe much of the material presented. Last but not least, I want to thank Kenneth Snetsinger for editorial comments and Hanwant Singh for his encouragement in writing this chapter.

References

Ackerman, T.P. and Toon, O.B. (1981). Absorption of visible radiation in atmospheres containing mixtures of absorbing and nonabsorbing particles. *Appl. Opt.* **20**, 3661–3668.

Ackerman, T.P. and Valero, F.P.J. (1984). The vertical structure of Arctic haze as determined from airborne net-flux radiometer measurements. *Geophys. Res. Lett.*, **11**, 469–472.

Albrecht, B.A. (1989). Aerosols, cloud microphysics, and fractional cloudiness. *Science*, **245**, 1227–1230.

Angell, J.K. and Korshover, J. (1983). Global temperature variations in the troposphere and stratosphere, 1958–1982. *Mon. Weather Rev.*, **111**, 901–921.

Asano, S. and Shiobara, M. (1989). Aircraft measurements of the radiative effects of tropospheric aerosols. I. Observational results of the radiation budget. *J. Meteorol. Soc. Jpn.*, **67**, 847–861.

Bach, W. (1976). Global air pollution and climatic change. *Rev. Geophys.*, **14**, 429–474.

Barrie, L.A., Bottenheim, J.W., Schnell, R.C., Crutzen, P.J. and Rasmussen, R.A. (1988). Ozone destruction and photochemical reactions at polar sunrise in the lower Arctic atmosphere. *Nature*, **334**, 138–141.

Bates, T.S., Lamb, B.K., Guenther, A., Dignon, J. and Stoiber, R.E. (1992). Sulfur emissions to the atmosphere from natural sources. *J. Atmos. Chem.*, **14**, 315–338.

Bhandari, R.A. (1986). Specific absorption of a tiny absorbing particle embedded within a nonabsorbing particle. *Appl. Opt.*, **25**, 3331–3333.

Blanchet, J.P. (1989). Toward estimation of climatic effects due to Arctic aerosols. *Atmos. Environ.*, **32**, 2609–2625.

Blanchet, J.P. and List, R. (1983). Estimation of optical properties of Arctic haze using a numerical model. *Atmos.-Ocean*, **21**, 444–464.

Brasseur, G. and Granier, C. (1992). Mount Pinatubo aerosols, chlorofluorocarbons, and ozone depletion. *Science*, **257**, 1239–1242.

Brasseur, G.P., Granier, C. and Walters, S. (1990). Future changes in stratospheric ozone and the role of heterogeneous chemistry. *Nature*, **348**, 626–628.

Brock, C.A., Jonsson, H.H., Wilson, J.C., Dye, J.E., Baumgardner, D., Borrman, S., Pitts, M.C., Osborne, M., DeCoursey, R.J. and Woods, D.C. (1993). Relationships between optical extinction, backscatter and aerosol surface and volume in the stratosphere following the eruption of Mt. Pinatubo. *Geophys. Res. Lett.*, in press.

Browell, E.K., Butler, C.F., Ismail, S., Robinette, P.A., Carter, A.F., Higdon, N.S., Toon, O.B., Schoeberl, M.R., and Tuck, A.F. (1990). Airborne lidar observations in the wintertime arctic stratosphere: polar stratospheric clouds. *Geophys. Res. Lett.*, **17**, 385–388.

Burley, J.D. and Johnston, H.S. (1992). Nitrosyl sulfuric acid and stratospheric aerosols. *Geophys. Res. Lett.*, **19**, 1363–1366.

Butcher, S.S. and Charlson, R.J. (1972). *An Introduction to Air Chemistry*, Academic Press, New York, NY.

Charlock, T.P. and Sellers, W.D. (1980). Aerosol effects on climate: calculations with time-dependent and steady state radiative–convective models. *J. Atmos. Sci.*, **37**, 1327–1341.

Charlson, R.J., Langner, J., Rodhe, H., Leovy, C.B. and Warren, S.G. (1991). Perturbation of the northern hemisphere radiative balance by backscattering from anthropogenic sulfates. *Tellus*, **43AB**, 152–163

Charlson, R.J., Lovelock, J.E. Andreae, M.O. and Warren, S.G. (1987). Oceanic phytoplankton, atmospheric sulfur, cloud albedo and climate. *Nature*, **326**, 655–661.

Charlson, R.J., Schwartz, S.E., Hales, J.M., Cess, R.D., Coakley, J.A., Jr., Hansen, J.E. and Hofmann, D.J. (1992). Climate forcing by anthropogenic aerosols. *Science*, **255**, 423–430.

Clarke, A.D. and Charlson, R.J. (1985). Radiative properties of the background aerosol-absorption component of extinction. *Science*, **229**, 263–265.

Coakley, J.A. Jr., Cess, R.D. and Yurevich, F.B. (1983). The effect of tropospheric aerosols on the earth's radiation budget: a parameterization for climate models. *J. Atmos. Sci.*, **40**, 161–171.

Crutzen, P.J. (1976). The possible importance of CSO for the sulfate layer of the stratosphere. *Geophys. Res. Lett.*, **3**, 73–76.

Deshler, T., Adriani, A., Gobbi, G.P., Hofmann, D.J., DiDonfrancesco, G. and Johnson, B.J. (1992). Polar stratospheric clouds over McMurdo, Antarctica, during the 1991 spring: lidar and particle counter measurements. *Geophys. Res. Lett.*, **19**, 1819–1822.

Elliott, S., Turco, R.P., Toon, O.B. and Hamill, P. (1990). Incorporation of stratospheric acids into water ice. *Geophys. Res. Lett.*, **17**, 425–428.

Enghardt, M. and Rodhe, H. (1993). A comparison between patterns of temperature trends and sulfate aerosol pollution, *Geophys. Res. Lett.*, **20**, 117–120.

Erickson, E. (1959). The yearly circulation of chloride and sulfur in nature: meteorological, geochemical and petrological implications. *Tellus*, **11**, 375–403.

Fan, S.M. and Jacob, D.J. (1992). Surface ozone depletion in Arctic spring sustained by bromine reactions on aerosols. *Nature*, **359**, 522–524.

Farlow, N.H., Snetsinger, K.G., Hayes, D.M., Lem, H.Y. and Tooer, B.M. (1978). Nitrogen–sulfur compounds in stratospheric aerosols. *J. Geophys. Res.*, **83**, 6207–6211.

Faxvog, F.R. and Roessler, D.M. (1978). Carbon aerosol visibility versus particle size distributions. *Appl. Opt.*, **17**, 3859–3862.

Friedlander, S.K. (1970). The self-preserving particle size distribution for coagulation by Brownian motion—Smoluchowski coagulation and simultaneous Maxwellian condensation. *J. Aerosol Sci.*, **1**, 115–123.

Gates, W.L., Mitchell, J.F.B., Boer, G.J., Cubasch, U. and Meleshko, V.P. (1992). Climate modelling, climate prediction and model validation. In Houghton, J.T., Callander, B.A. and Varney, S.K., Eds., *IPCC, The supplementary Report*, Cambridge University Press, Cambridge, pp. 97–134.

Gobbi, G.P., Adriani, A., Deshler, T. and Hoffman, D.J. (1991). Evidence for denitrification in the 1990 Antarctic spring stratosphere. I—Lidar and temperature measurements. II—Lidar and aerosol measurements. *Geophys. Res. Lett.*, **18**, 1995–2002.

Goodman, J., Toon, O.B., Pueschel, R.F., Snetsinger, K.G. and Verma, S. (1989). Antarctic stratospheric ice crystals. *J. Geophys. Res.*, **94**, 16449–16458.

Hansen, J., Rossow, W. and Fung, I. (1990). The missing data on global climate change. *Iss. Sci. Technol.*,**7**, 62–69.

Harshvardhan, M.R. and Cess, R.D., (1978). Effect of tropospheric aerosols upon atmospheric infrared cooling rates. *J. Quantum Spectrosc. Radiat. Transfer*, **19**, 621–632.

Heintzenberg, J. (1978). Light scattering parameters of internal and external mixtures of soot and nonabsorbing material in the atmospheric aerosol. *Proc. Conf. on Carbonaceous Particles in the Atmosphere, CONF-7803101*, Berkeley, CA, pp. 278–281.

Henderson-Sellers, A. (1986). Increasing cloud in a warming world. *Climatic Change*, **9**, 267–309.

Hicks, B.B. (1986). Differences in wet and dry particle deposition parameters between north America and Europe. In Lee, S.D., Schneider, T., Grant, L.D. and Verkerk, P.J., Eds., *Aerosols*, Lewis Publ., Chelsea, MI, pp. 973–982.

Hidy, G.M. (1973). Removal processes of gaseous and particulate pollutants. In Rasool, S.I., Ed., *Chemistry of the Lower Atmosphere*. Plenum, New York, NY, pp. 121–176.

Hofmann, D.J. (1990). Increase in the stratospheric background sulfuric acid aerosol mass in the past 10 years. *Science*, **248**, 996–1000.

IPCC (Intergovernmental Panel on Climate Change) (1990). *The Scientific Assessment*, Houghton, J.T., Jenkins, G.T. and Ephraums, J.J., Eds., Cambridge University Press, Cambridge.

Iversen, T.B.N., Halvorsen, N.E., Mylona, S. and Sandnes, H. (1991). Calculating budgets for airborne acidifying components in Europe, 1985, 1987, 1989, and 1990. *EMEP/MSC-W Report 1/91*, Norwegian Meteorological Institute, Oslo.

Jaenicke, R. (1986). Physical characterization of aerosols. In Lee, S.D., Schneider, T., Grant, L.D. and Verkerk, P.J., Eds., *Aerosols*, Lewis Publ., Chelsea, MI, pp. 97–106.

Jensen, E.J. and Toon, O.B. (1992). The potential effects of volcanic aerosols on cirrus cloud microphysics. *Geophys. Res. Lett.*, **19**, 1759–1762.

Junge, C.E. (1961). Vertical profiles of condensation nuclei in the stratosphere. *J. Meteorol.*, **18**, 501–509.

Junge, C.E. and Gustafson, P.E. (1957). On the distribution of seasalt over the United States and its removal by precipitation. *Tellus*, **9**, 164–173.

Karl, T.R., Kukla, G., Razuvayev, V.N., Changery, M.J., Quayle, G., Heim, R.R. Jr., Easterling, D.R. and Fu, C.B. (1991). Global warming: Evidence for asymmetric diurnal temperature change. *Geophys. Res. Lett.*, **18**, 2253–2256.

Kerr, R.A. (1989). Volcanoes can muddle the greenhouse. *Science*, **245**, 127–128.

Kerr, R.A. (1993). Volcanoes may warm locally while cooling globally. *Science*, **260**, 1232.

Kiang, C.S., Cadle, R.D. and Yue, G.K. (1975). H_2SO_4-HNO_3-H_2O ternary aerosol formation mechanism in the stratosphere. *Geophys. Res. Lett.*, **2**, 41–44.

Kiehl, J.T. and Briegleb, B.P. (1993). The relative roles of sulfate aerosols and greenhouse gases in climate forcing. *Science*, **260**, 311–314.

Kinne, S.A. (1993). Personal communication.

Kinne, S.A., Toon, O.B. and Prather, M.J. (1992). Buffering of stratospheric circulation by tropical ozone: a Pinatubo case study. *Geophys. Res. Lett.*, **19**, 1927–1930.

Köhler, H. (1926). Zur Kondensation an hygroskopischen Kernen und Bemerkungen über das Zusammenfliessen der Tropfen. *Meddn. St. met.-hydrogr. Anst.*, **3**, 24–30.

Kondratyev, K. (1986). *Changes in global climate*, Balkema, Rotterdam.

Lacis, A., Hansen, J. and Sato, M, (1992). Climate forcing by stratospheric aerosols. *Geophys. Res. Lett.*, **19**, 1607–1610.

Lang, L.J., Ed. (1974). *The World Almanac*, Newspaper Enterprise Assoc., Cincinnati, OH, pp.428–429.

Langner, J., Rodhe H., Crutzen, P.J. and Zimmerman, P. (1992). Anthropogenic influence on the distribution of tropospheric sulfate aerosol. *Nature*, **359**, 712–716.

Ludwig, J.H., Morgan, G.B. and McMullen, T.B. (1970). Trends in urban air quality. *EOS Trans. AGU*, **51**, 468–475.

Luo, B.P., Clegg, S.L., Peter, Th., Mueller, R. and Crutzen, P.J. (1993). HCl solubility and liquid diffusion in aqueous sulfuric acid under stratospheric conditions. *Geophys. Res. Lett.*, in press.

McConnell. J.C., Henderson, G.S., Barrie, L., Bottenheim, J., Niki, H., Langford, C.H. and Templeton, E.M.J. (1992). Photochemical bromine production implicated in Arctic boundary-layer ozone depletion. *Nature*, **355**, 150–152.

McCormick, M.P., Trepte, C.R. and Pitts, M.C. (1989). Persistence of polar stratospheric clouds in the southern polar region. *J. Geophys. Res.*, **D94**, 11,241–11,251.

Mason, B.J., (1971). *The Physics of Clouds*, Clarendon Press, Oxford.

Mason, B.J. and Maybank, J. (1958). Ice nucleating properties of some natural mineral dusts. *Q. J. Meteorol. Soc.*, **86**, 235–241.

Mass, C.F. and Portman, D.A. (1989). Major volcanic eruptions and climate: a critical evaluation. *J. Climatol.*, **2**, 566.

Mayewski, P.A., Lyons, W.B., Spencer, M.J., Twickler, M.S., Buck, C.F. and Whitlow, S.

(1990). An ice core record of atmospheric response to anthropogenic sulphate and nitrate. *Nature*, **346**, 554–556.

Minnis, P., Harrison, E.F., Stowe, L.L., Gibson, G.G., Denn, F.M., Doelling, D.R. and Smith, W.L. Jr. (1993). Radiative climate forcing by the Mount Pinatubo eruption. *Science*, **259**, 1411–1415.

Neftel, A., Beer, J., Oeschger, H., Zuercher,F. and Finkel, R.J. (1985). Sulphate and nitrate concentrations in snow. *Nature*. **313**, 611–613.

Oberbeck, V.E., Livingston, J.M., Russell, P.B., Pueschel, R.F., Rosen, J.N., Osborne, M.T., Kritz, M.A., Snetsinger, K.G. and Ferry, G.V. (1989). SAGE II aerosol validation: selected altitude measurements, including particle micromeasurements. *J. Geophys. Res.*, **94**, 8367–8380.

Paltridge, G.W. and Platt, M.R. (1976). *Radiative Processes in Meteorology and Climatology*, Elsevier, Amsterdam.

Penner, J.E., Dickinson, R.E. and O'Neill, C.A. (1992). Effects of aerosol from biomass burning on the global radiation budget. *Science*, **256**, 1432.

Penner, J.E., Eddleman, H. and Novakov, T. (1993). Towards the development of a global inventory for black carbon emissions. *Atmos. Environ.*, **27A**, 1277–1295.

Pollack, J.B., Toon, O.B., Sagan, C., Summers, A., Baldwin, B. and Van Camp, W. (1976). Volcanic explosions and climatic change: a theoretical assessment. *J. Geophys. Res.*, **81**, 1071–1083.

Pollack, J.R., Toon, O.B. and Wiedman, D. (1981). Radiative properties of the background stratospheric aerosols and implications for perturbed conditions. *Geophys. Res. Lett.*, **8**, 26–28.

Post, M.J. (1986). Atmospheric purging of ElChichon debris. *J. Geophys. Res.*, **91**, 5222–5228.

Post, M.J., Grund, C.J., Langford, A.O. and Proffitt, M.H. (1992). Observations of Pinatubo ejecta over Boulder, Colorado by lidards of three different wavelengths. *Geophys. Res. Lett.*, **19**, 195–198.

Prather, M. (1992). Catastrophic loss of stratospheric ozone in dense volcanic clouds. *J. Geophys. Res.*, **97**, 10187–10191.

Prospero, J.M. and Carlson, T.N. (1972). Vertical and areal distribution of Saharan dust over the western equatorial North Atlantic Ocean. *J. Geophys. Res.*, **77**, 5255–5265.

Pruppacher H.R. and Klett, J.D. (1978). *Microphysics of Clouds and Precipitation*. Reidel, Dordrecht.

Pueschel, R.F., Blake, D.F., Snetsinger, K.G., Hansen, A.D.A., Verma, S. and Kato, K. (1992). Black carbon (soot) aerosol in the lower stratosphere and upper troposphere. *Geophys. Res. Lett.*, **19**, 1659–1662.

Pueschel, R.F., Snetsinger, K.G., Goodman, J.K., Toon, O.B., Ferry, G.V., Oberbeck, V.R., Livingston, J.M., Verma, S., Fong, W., Starr, W.L. and Chan, K.R. (1989). Condensed nitrate, sulfate and chloride in Antarctic stratospheric aerosols. *J. Geophys. Res.*, **94**, 11271–11284.

Pueschel, R.F., Snetsinger, K.G., Hamill, P., Goodman, J.K. and McCormick, M.P. (1990). Nitric acid in polar stratospheric clouds—similar temperature of nitric acid condensation and cloud formation. *Geophys. Res. Lett.*, **17**, 429–432.

Pueschel, R.F., Livingston, J.M., Ferry, G.V. and DeFelice, T.E. (1994a). Aerosol abundances and optical characteristics in the Pacific Basin free troposphere. *Atmos.*

Environ., **28**, 951–960.

Pueschel, R.F., Russell, P.B., Allen, D.A., Ferry, G.V., Snetsinger, K.G., Livingston, J.M. and Verma, S. (1994b). Physical and optical properties of the Pinatubo volcanic aerosol: aircraft observations with impactors and a Sun-tracking photometer. *J. Geophys. Res.*, **99**, 12915–12922.

Rao, C.R.N. and Takashima, T., (1986). Solar radiation anomalies caused by the El Chichon volcanic cloud. *Q. J. R. Meteorol. Soc.*, **112**, 1111–1126.

Robock, A. and Mao, J. (1992). Winter warming from large volcanic eruptions. *Geophys. Res. Lett.*, **19**, 2505–2509.

Rodriguez, J.M., Ko, M.K. and Sze, N.D. (1991). Role of heterogeneous conversion of N_2O_5 on sulphate aerosols in global ozone losses. *Nature*, **353**, 134–137.

Russell, P.B., Livingston, J.M., Dutton, E.G., Pueschel, R.F., Reagan, J.A., DeFoor, T.E., Box, M.A., Allen, D., Pilewski, P., Herman, B.M., Kinne, S.A. and Hofman, D.J. (1993). Pinatubo and pre-Pinatubo optical depth spectra: Mauna Loa measurements, comparisons, inferred particle size distributions, radiative effects, and relationship to lidar data. *J. Geophys. Res.*, **D98**, 22969–22985.

Sassen, K. (1992). Evidence of liquid-phase cirrus cloud formation from volcanic aerosols: climate implications. *Science*, **257**, 516–519.

SCEP (1970). Report of the study of critical environmental problems. In *Man's Impact on the Global Environment*, MIT Press, Cambridge, MA.

Schwartz, S.E. (1988). Are global cloud albedo and climate controlled by marine phytoplankton? *Nature*, **336**, 441–445.

Schwartz, S.E. (1989). Acid deposition: unraveling a regional phenomenon. *Science*, **243**, 753–763.

Sigurdsson, H. and Laj, P. (1992). In Nierenberg, W.A., Ed., *Encyclopedia of Earth Systems Science*, Vol. 1, Academic Press, San Diego, CA, p. 183.

Solomon, S., Sanders, R.W., Garcia, R.R. and Keys, J.G. (1993). Increased chlorine dioxide over Antarctica caused by volcanic aerosol from Mount Pinatubo. *Nature*, **363**, 245–248.

Tang, I.N., Munkelwitz, H.R. and Davis, J.G. (1977). Aerosol growth studies. *J. Aerosol Sci.*, **8**, 149–159.

Toon, O.B., Hamill, P., Turco, R.P. and Pinto, J. (1986). Condensation of HNO_3 and HCl in the winter polar stratosphere. *Geophys. Res. Lett.*, **13**, 1284–1287.

Turco, R.P. (1992). Upper atmosphere aerosols: Properties and natural cycles. In *The Atmospheric Effects of Stratospheric Aircraft: A First Program Report, NASA Ref. Pub. 1272*, pp. 63–82.

Turco, R.P., Whitten, R.C., Toon, O.B., Pollack, J.B. and Hamill, P. (1980). OCS, stratospheric aerosols and climate. *Nature*, **283**, 283–286.

Twohy, C.H., Clarke, A.D., Warren, S.G., Radke, L.F. and Charlson, R.J. (1989). Light-absorbing material extracted from cloud droplets and its effect on cloud albedo. *J. Geophys. Res.*, **94**, 8623–8631.

Twomey, S. (1977). *Atmospheric Aerosols*, Elsevier, New York.

Twomey, S. (1991). Aerosols, clouds and radiation. *Atmos. Environ.*, **25A**, 2435–2442.

Valero, F.P.J. and Pilewski, P. (1992). Latitudinal survey of spectral optical depths of the Pinatubo volcanic cloud: derived particle sizes, columnar mass loadings, and effects on planetary albedo. *Geophys. Res. Lett.*, **19**, 163–166.

Wendling, P., Wendling, R., Renger, W., Covert, D., Heintzenberg, J. and Moerl, P. (1985). Calculated radiative effects of Arctic haze during a pollution episode in spring 1983 based on ground-based and airborne measurements. *Atmos. Envir.*, **19**, 2181–2193.

Whitby, K.T. (1978). The physical characteristics of sulfur aerosols. *Atmos. Environ.*, **12**, 135–159.

Williamson, S.J. (1973). *Fundamentals of Air Pollution*, Addison-Wesley, Reading, MA.

World Resources Institute and the International Institute for Environment and Development (1988). *World Resources 1988–89*, Basic Books, New York, NY.

Yue, G.K. and Deepak, A. (1982). Temperature dependence of the formation of sulfate aerosols in the stratosphere. *J. Geophys. Res.*, **87**, 3128–3134.

6

Reactive odd-nitrogen (NO_y) in the atmosphere

James M. Roberts

6.1. INTRODUCTION

The chemistry of the atmosphere is dominated by acidic and oxidizing conditions. In this environment a complex set of reactions takes place that profoundly affects human health, the viability of ecosystems, and ultimately the

176

balance of the global climate. Reactive nitrogen plays a central role in the chemistry of the lower atmosphere (Leighton, 1961; Crutzen, 1979; Singh, 1987; Finlayson-Pitts and Pitts, 1986). It is involved in the production of atmospheric acidity and ozone on urban, regional, and global scales. In addition, certain reactive nitrogen compounds have phytotoxic and mutagenic properties. A detailed knowledge of the sources to, chemical processing of, and removal of reactive nitrogen from the atmosphere is crucial to the understanding of the chemistry of the atmosphere and how human activities impact the atmosphere.

Reactive nitrogen, abbreviated NO$_y$, has traditionally been defined as the sum of the simple oxides of nitrogen (NO$_x$), i.e., nitric oxide (NO) and nitrogen dioxide (NO$_2$), and all compounds that are atmospheric products of NO$_x$. This includes the compounds nitric acid (HNO$_3$), nitrous acid (HNO$_2$), nitrate radical (NO$_3$), dinitrogen pentoxide (N$_2$O$_5$), peroxynitric acid (HNO$_4$), peroxyacetic nitric anhydride (PAN) and its homologues, alkyl nitrates (RONO$_2$), peroxyalkyl nitrates (ROONO$_2$), and nitro-aromatic compounds (RNO$_2$). The compounds nitrous oxide (N$_2$O), and ammonia (NH$_3$), while sources of NO$_x$ in some regions of the atmosphere, are not considered reactive nitrogen compounds. Part of this distinction arises from the chemical methods used to measure total NO$_y$, as will be discussed below. The nomenclature of the above NO$_y$ compounds can be gleaned from tradition chemical sources and several special discussions of RC(O)OONO$_2$ and RONO$_2$ compounds (Martinez, 1980; Roberts 1990).

This chapter will be organized into sections dealing with sources of reactive nitrogen to the atmosphere, chemistry of reactive nitrogen compounds relevant to the atmosphere, and ambient measurements of various compounds and their theoretical interpretation. Each section will be further subdivided by compound or compound class.

6.2. SOURCES OF NO$_y$ TO THE TROPOSPHERE

6.2.1 NO$_x$

The emission of reactive nitrogen to the troposphere occurs primarily as NO and NO$_2$. The important sources of NO$_x$ to the atmosphere are combustion, lightning, and biological activity. Several studies have evaluated these sources and have attempted to arrive at global estimates (Logan, 1983; Stedman and Shetter, 1983; Dignon and Hameed, 1989; Lobert et al., 1990). Table 6.1 summarizes the results of these studies. There is considerable uncertainty in some of these estimates, as indicated by the range given in parentheses.

The emission of NO$_x$ from combustion results from both fossil fuel usage and biomass burning. Fossil fuel usage is closely related to industrialization and population and is therefore heavily weighted toward the northern hemisphere.

TABLE 6.1   Emissions of NO_x to the Atmosphere

Source	Rate $(Tg\,N\,year^{-1})$	Reference
Fossil fuel combustion	24 $(17-31)$[a]	Hameed and Dignon (1992)
Biomass burning	5.6 $(2.5-8.7)$	Lobert et al. (1990)
Lightning	8 $(2-20)$	Logan (1983)
Soils	8 $(4-16)$	Logan (1983)
	10	Stedman and Shetter (1983)
	20	Davidson (1991)
Oxidation of ammonia	5 $(1-10)$	Logan (1983)
Oceanic[b]	< 1	Logan (1983)
Stratosphere	≈ 0.5[c]	Logan (1983)
	1	Stedman and Shetter (1983)
Total	53[d] $(29-88)$	

[a]Includes the fractional range of uncertainty estimated by Logan (1983).
[b]From photolytic or biotic degradation.
[c]Approximately 75% of this nitrogen is expected to be present as HNO_3.
[d]Includes the most conservative estimate of soil emissions.

The estimates given in Table 6.1 are based on NO_x-fuel consumption empirical relationships. The total emission from fossil fuel combustion was then obtained from fuel usage data. Trends in NO_x emissions by continent have been estimated by Dignon and Hameed (1989) from 1860 to 1980 and show that emission from Asia is now the largest and fastest-growing contributor. The emission of NO_x is a direct result of industrial activity and is therefore tied to economic status and growth. There is tremendous economic pressure driving the economic development of Asia, so growth in NO_x emissions is likely to continue. The second and third largest contributors (by continent) to fossil-fuel-derived NO_x are North America and Europe. Many areas of these continents are quite developed and are actually declining in emission owing to conservation and emission controls. For example, it has been estimated that NO_x emissions in the southern California area have actually declined in the last 20 years (Kuntasal and Chang, 1987). It is important to note that the future scenario of NO_x emission from fossil fuel combustion is intimately connected to that of CO_2, so that limitations on the rate (or rate of growth) of CO_2 emission will directly affect that of NO_x.

Biomass burning is most prevalent in the tropics and is somewhat seasonal, since it is linked to agricultural practices. The magnitude of NO_x emissions from biomass burning has recently been investigated by Lobert et al. (1990). These studies were done by controlled burning of representative samples of biomass and analysis of the products. Biomass burning typically takes place at temperatures low enough such that NO_x is not formed from air, only from oxidation of nitrogen originally in the fuel. NO_x was the most abundant odd-nitrogen compound identified in the work of Lobert et al. (1990) (12% of biomass N), with smaller amounts of NH_3, HCN and CH_3CN. Almost 70% of the nitrogen was not identified, but other studies indicated that most of this (50% overall) was N_2. Lobert et al. (1990) used estimates of global biomass carbon burning to derive a global estimate for the NO_x from this source. Future scenarios for biomass burning are more difficult to project than for fossil fuel combustion, since a great deal of concern is being focused on tropical deforestation and these practices are likely to change in the future.

Lightning is a major global source of NO_x. The total magnitude of this source is the subject of considerable uncertainty. A number of studies have attempted to combine theoretical models or atmospheric observations of NO_x produced in individual events with global lightning frequency (see Logan, 1983; Goldenbaum and Dickerson, 1993). The value given in Table 6.1 is towards the low end of the range of source estimates; however, it is consistent with an estimate based on nitrate deposition at the poles (Albritton et al., 1984).

NO_x emissions from soils occur through biotic and abiotic processes. Numerous micro-organisms produce NO_x (almost entirely NO) through both nitrifying and denitrifying pathways. In addition, a number of abiotic pathways, collectively termed chemodenitrification, can contribute to NO production (Williams et al., 1992). An example of such an abiotic processes is the disproportionation of nitrous acid. Williams et al. (1992) have summarized the factors that influence the emission of NO from soils, i.e., temperature, moisture, and total N content. Estimates of this source of NO to the atmosphere are based on measured NO fluxes from soils and extrapolation to the globe using algorithms that seek to account for dependences on the above factors. The estimate in Table 6.1 is based on the work of Davidson (1991), which is heavily weighted by high-emission-rate sites in the savanna. Many more measurements are needed to refine these estimates.

The production of NO_x from ammonia oxidation is somewhat uncertain, because it is the net balance between gas phase chemical processes that both produce and destroy NO_x. The chemistry also depends on the relative concentrations of O_3 and NO_x and their global distribution versus NH_3 sources. A final factor involves the rate of a key reaction involving O_3 that has been estimated to be uncertain by at least a factor of 3 (DeMore et al., 1992).

Oceanic sources of NO_x have not received a great deal of attention. Zafiriou and McFarland (1981) measured the NO in sea-water in the equatorial Pacific and ascribed it to photolytic production of NO from NO_2^-. These measurements imply an estimated the global source from the oceans that is quite small.

NO_x is produced in the stratosphere through the reaction of N_2O with $O(^1D)$ atoms. The mixing time between the stratosphere and troposphere is long compared with the chemical lifetime of NO_x. As a result, most of the NO_x is converted to HNO_3. The downward flux of NO_y is fairly well known, since it must roughly balance the upward flux of N_2O minus some small fraction that is thought to be destroyed through reactions in the upper stratosphere. Assessments of the downward NO_y flux have found it to be quite small, only a fraction of that required to explain mid-oceanic HNO_3 and NO_3^- observations (Logan, 1983). This has resulted in a fairly small estimate of the stratospheric NO_x source.

6.2.2. HONO, CH₃ONO

Direct emissions of nitrous acid to the atmosphere are not well established, but there is evidence that they might occur. Nitrous acid has been shown to form in surface reactions of NO_x and H_2O (Akimoto et al., 1987; Jenkins, et al., 1988) and, by extension, may form in exhaust streams. The efficiency and magnitude of this source have not been extensively investigated. The potential impact of a direct emission of HONO on the radical balance of the atmosphere (see below) warrants further work in this area.

The emission of methyl nitrite from methanol-fueled internal combustion engines has been observed and somewhat quantified. This is thought to occur through a surface reaction of CH_3OH with NO_x in a manner analogous to that of HONO. Jonsson et al. (1979) and Jonsson and Bertilsson (1982) measured the emission of methyl nitrite in the exhaust of vehicles fueled by methanol and methanol–gasoline blend. The results showed that methyl nitrite emissions were at most a few per cent of NO_x emissions and were often lower than that. It was found that exhaust catalysts efficiently removed methyl nitrite and that the concentration of methyl nitrite measured depended on the absolute concentrations of methanol and NO_2, and temperature. Minimal methyl nitrite was observed under realistic conditions in which the exhaust gases were cooled and diluted with air.

6.2.3. Compounds of the Type RONO₂

Not a lot is known about direct emission of alkyl nitrates from internal combustion engines. While low oxygen content and high NO_2 content favor $RONO_2$ formation (see below), the higher temperature characteristic of exhaust streams will promote their thermal decomposition. Not many studies have been performed where $RONO_2$ compounds were explicitly targeted or where the techniques used were even amenable to their detection. Chief among these are the studies of Jonsson et al. (1979) and Jonsson and Bertilsson (1982), described above, wherein methanol-fueled vehicle emissions were sampled. There was no report of methyl nitrate occurrence in their analyses, which would clearly have been able to detect it. Higher-chain-length alkyl nitrates have been proposed as additives for the improvement of diesel engine performance. The extent to which they might be present in the exhaust from such engines is unknown, but it is likely to be quite small, especially for engines fitted with catalytic converters.

6.3. ATMOSPHERIC CHEMISTRY OF NO_y

6.3.1. The Role of NO_x in Photochemical Ozone Production

The chemical transformations occurring in the NO_y family can be thought of as a system of reactions wherein the major source species, NO and NO_2 are eventually oxidized to HNO_3. Along the way are formed a number of intermediates having a variety of chemical properties and undergoing many different reactions, some of which reform NO_x. This system of reactions is shown schematically in Figure 6.1. What follows in this section are brief descriptions of the chemistry pertaining to the troposphere and detailed discussions of photolytic, radical, and thermal decomposition reactions. Some attention will also be given to heterogeneous reactions and aspects of the chemistry that are not well known and require further investigation.

The conversion of NO to NO_2 has a special significance to the chemistry of the troposphere; for this reason NO_x is sometimes referred to as "active odd-nitrogen". Tropospheric NO_x–O_3 photochemistry is often thought of as a "photostationary state" (Leighton, 1961) involving the following three reactions;

$$NO + O_3 \rightarrow NO_2 + O_2 + \Delta H \qquad (6.1)$$

$$NO_2 + h\nu \rightarrow NO + O(^3P) \qquad (6.2)$$

$$\underline{O + O_2 + M \rightarrow O_3 + M + \Delta H} \qquad (6.3)$$

$$\text{net: } h\nu \rightarrow \Delta H$$

the net effect of which is to convert sunlight to heat. Reactions (6.1) and (6.2) have characteristic times ranging from 1 min to several minutes, with respect to NO_x interconversion, at typical midday tropospheric conditions.

It has been demonstrated from simultaneous measurements of NO, NO_2, O_3, and solar UV (Parrish et al., 1986; Cantrell et al., 1992) that the cycle represented by reactions (6.1)–(6.3) is not in balance. The ratio of NO_2 to NO in the daytime is always larger than can be accounted for by this simple cycle. The significance of this to tropospheric ozone chemistry comes in recognizing that there are a number of reactions, in addition to (6.1), that convert NO to NO_2. Any such reaction will result in a net production of ozone through reactions (6.2) and (6.3). Thus the reactions

$$HO_2 + NO \rightarrow NO_2 + HO \qquad (6.4)$$

$$RO_2 + NO \rightarrow NO_2 + RO \qquad (6.5)$$

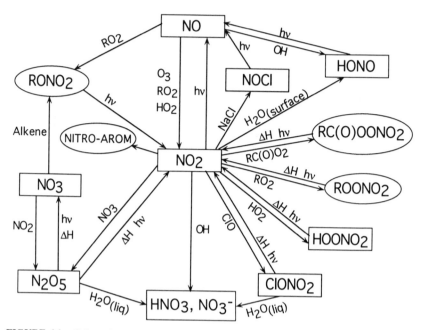

FIGURE 6.1.   Schematic representation of the NO_y chemistry pertaining to the troposphere.

will result in net ozone production. The radicals HO_2 and RO_2 (where R is an organic group) are formed from the photo-oxidation of carbon monoxide and organic compounds (see Chapter 10 for more details).

   The ultimate fate of NO_y in the troposphere is oxidation to HNO_3 and aerosol NO_3^- and deposition either by wet or dry mechanisms. This oxidation can occur simply and rapidly through the reaction

$$NO_2 + OH + M \rightarrow HNO_3 + M \qquad (6.6)$$

Other significant reactions for this conversion are the reaction of N_2O_5, or $ClONO_2$ with liquid water:

$$N_2O_5 + H_2O(aq) \rightarrow 2HNO_3 \qquad (6.7)$$

$$ClONO_2 + H_2O(aq) \rightarrow HNO_3 + HOCl \qquad (6.8)$$

The reaction of NO_3 with phenols has also been shown to result in nitric acid (Carter et al., 1981):

$$PhOH + NO_3 \rightarrow PhO + HNO_3 \qquad (6.9)$$

although this is probably of minor importance. The liquid phase reaction of nitrite to form nitrate should also be considered, since the formation of nitrite from the reaction of PAN (Stephens, 1969) and peroxynitric acid (Park and Lee, 1987) is known:

$$CH_3C(O)OONO_2 + H_2O(aq) \rightarrow NO_2^- + CH_3C(O)OH + O_2 + H^+ \quad (6.10)$$

$$HOONO_2 + H_2O(aq) \rightarrow NO_2^- + O_2 + H_3O^+ \quad (6.11)$$

Reactions of NO_x to intermediate oxidation states and the chemistry of these intermediate species are critical aspects of NO_y chemistry. These intermediate species serve as reservoirs for the storage and transport of NO_x to the regional and global atmosphere. In discussing these intermediate NO_y species, it is useful to distinguish between inorganic and organic compounds.

6.3.2. Inorganic NO_y Compounds

Formation of intermediate inorganic NO_y compounds occurs for the most part through fairly minor reactions or through surface reactions of unknown importance. The compounds HONO and NOX (X = Br, Cl) are hypothesized to form on surfaces containing H_2O and X^- (Finlayson-Pitts and Johnson, 1988; Junkerman and Ibusuki, 1992):

$$2NO_2 + H_2O(aq) \rightarrow HONO + HONO_2 \quad (6.12)$$

$$2NO_2 + X^-(aq) \rightarrow NOX + NO_3^- \quad (6.13)$$

These reactions are not well characterized, although both have been observed in laboratory systems. Reaction (6.13) has been observed in reactions of NO_2 with dry NaX salts; however, no direct atmospheric observations of NOX have been reported. The case for HONO is a bit stronger in that surface formation of HONO has been established under realistic atmospheric conditions, and direct ambient measurements have been made. HONO and NOX compounds photolyze rapidly under tropospheric conditions.

The nitrate radical (NO_3) is a potentially important member of the NO_y family. It plays a role in the night-time formation of nitric acid through its formation of N_2O_5 and it reacts rapidly with alkenes, phenols, and some sulfur-containing organic compounds. A good review of NO_3 physics and chemistry has recently been published (Wayne et al., 1991). Formation of NO_3 occurs through reaction of NO_2 with O_3:

$$NO_2 + O_3 \rightarrow NO_3 + O_2 \quad (6.14)$$

Significant loss mechanisms of NO_3 include reaction with NO_2 to form N_2O_5, reaction with NO to reform NO_2, reactions with alkenes to form organic nitrates, and photolysis:

$$NO_3 + NO_2 \rightarrow N_2O_5 \tag{6.15}$$

$$NO_3 + NO \rightarrow 2NO_2 \tag{6.16}$$

$$NO_3 + RR'C = CR''R^* \rightarrow\rightarrow RR'C(O)\!-\!C(ONO_2)R''R^* \tag{6.17}$$

$$NO_3 + h\nu \rightarrow NO_2 + O \quad \alpha = 0.9 \tag{6.18a}$$

$$\rightarrow NO + O_2 \tag{6.18b}$$

The formation of peroxynitric acid and chlorine nitrate occurs through reactions of hydroperoxy and chlorine monoxide with NO_2:

$$HO_2 + NO_2 + M \rightarrow HOONO_2 \tag{6.19}$$

$$ClO + NO_2 + M \rightarrow ClONO_2 \tag{6.20}$$

Both formation reactions are temperature- and pressure-sensitive. Loss of these species occurs mainly through thermal decomposition in the lower troposphere (Table 6.2), via the reverse of reactions (6.19) and (6.20), and by photolysis in the upper troposphere and stratosphere (DeMore et al., 1992):

$$HOONO_2 + h\nu \;(\lambda = 248\,\text{nm}) \rightarrow HO_2 + NO_2 \quad \alpha = 0.66 \tag{6.21a}$$

$$\rightarrow HO + NO_3 \tag{6.21b}$$

TABLE 6.2   Thermal Decomposition of NO_y Species

Compound	A, $\log_{10}(\text{s}^{-1})$	E_a (kcal mol^{-1})	Rate at 1 atm, 298 K, (s^{-1})	Reference
HNO_3	16	49.3	5×10^{-20}	Roberts (1990)
$RONO_2$	16	40	2×10^{-14}	Roberts (1990)
$ClONO_2$	13.3	21.7	2.1×10^{-3}	Anderson and Fahey (1990)
N_2O_5	14.8	21.8	0.05	Cantrell et al. (1993)
$HOONO_2$	14.15	20.7	8.2×10^{-2}	Graham et al. (1978)
CX_3OONO_2[a]	16.2–16.8	23.5–24.4	0.041–0.2	Koppenkastrop and Zabel (1991)
$ROONO_2$[b]	15.7–16	20.6–21	3.3–3.8	Zabel et al. (1989)
PANs[c]	16.3 ± 0.9	26.9 ± 1.3	3.2×10^{-4}	Roberts and Bertman (1992)

[a]Where X is either a Cl or an F atom.
[b]For a range of alkyl groups R from CH_3 to C_8H_{17}.
[c]A number of PAN-type compounds have been studied and all exhibit roughly the same thermal decomposition rates.

$$\text{ClONO}_2 + h\nu \ (\lambda = 266, 355 \, \text{nm}) \rightarrow \text{Cl} + \text{NO}_3 \quad \alpha = 0.9 \quad (6.22a)$$

$$\rightarrow \text{O} + \text{ClONO} \quad (6.22b)$$

The fate of ClONO is fairly rapid photolysis:

$$\text{ClONO} + h\nu \rightarrow \text{Cl} + \text{NO}_2 \quad (6.23)$$

The compound N_2O_5 can be thought of as the anhydride of nitric acid. It is formed in the atmosphere through the reaction of NO_2 with NO_3:

$$\text{NO}_2 + \text{NO}_3 + \text{M} \rightarrow \text{N}_2\text{O}_5 + \text{M} \quad (6.24)$$

N_2O_5 is unreactive with water vapor ($k < 2 \times 10^{-21}$ cm^{3}molecule^{-1}s^{-1}), and appears to be only slightly soluble in water (Table 6.3). However, the hydrolysis of N_2O_5 in liquid water has been estimated to be quite rapid (5×10^3 s^{-1}):

$$\text{N}_2\text{O}_5 + \text{H}_2\text{O} \rightarrow 2\text{HNO}_3 \quad (6.25)$$

The thermal decomposition of N_2O_5 has been studied numerous times, the latest study having expanded the temperature and pressure range data (Table 6.2; Cantrell et al., 1993):

TABLE 6.3 Henry's Law Coefficients H for NO$_y$ Species

Compound	Temperature (°C)	H (M atm^{-1})	Reference
NO	22	1.9×10^{-3}	Park and Lee (1988)
NO$_2$	22	7×10^{-3}	Lee and Schwartz (1981)
HONO	22	59	Park and Lee (1988)
ClONO$_2$	–	N/A	
N$_2$O$_5$	22	5×10^3	Lee (1990)
HNO$_3$	22	2×10^5	Lee (1990)
HOONO$_2$	22	2×10^4	Park and Lee (1987)
PAN	22	3.6	Lee (1990)
	22	3.5	Kames et al. (1991)
	10.2	8.3	Lee (1984)
RONO$_2$[a]	21–22	0.5 – 2.6	Luke et al. (1989), Lee (1990), Kames and Schurath (1992)
1,2-Ethane dinitrate	21	640	Kames and Schurath (1992)
2-Nitrooxy-ethanol	21	4×10^4	Kames and Schurath (1992)
Nitrooxy-acetone	21	1×10^3	Kames and Schurath (1992)
1,2,-Propane dinitrate	21	175	Kames and Schurath (1992)
Nitrooxy-propanol[b]	21	$6.7–7.3 \times 10^3$	Kames and Schurath (1992)

[a]Simple alkyl nitrates, C$_1$–C$_5$.
[b]A mixture of the two vicinal isomers, separated but not identified in the analysis method.

$$N_2O_5 + M \rightarrow NO_2 + NO_3 + M \qquad\qquad (-6.24)$$

The resulting thermal lifetimes in the troposphere range from 20 s at 298 K and 1 atm to about 1.2 years at the tropopause. The UV absorption and photolysis of N_2O_5 have been studied as a function of wavelength and temperature. Recent studies of UV absorption (e.g., Harwood et al., 1993) are in fairly good agreement, and the following expression can be used for the absorption cross-section between 285 and 380 nm wavelength and between 300 and 225 K temperature:

$$\sigma = \exp[2.735 + ((4728.5 - 17.127\lambda)/T)] \times 10^{20} \qquad (6.26)$$

where λ is in nanometers, T is in Kelvin, and σ is in cm^2 molecule^{-1}. The quantum yield for various products has been reported in several studies (DeMore et al., 1992). N_2O_5 appears to photolyze according to the following processes:

$$N_2O_5 + h\nu \rightarrow NO_3 + NO_2 \qquad\qquad (6.27a)$$
$$\rightarrow NO_3 + NO + O \qquad\qquad (6.27b)$$

Nitrate radical appears to be formed with unity or near-unity efficiency at all wavelengths, while the process (6.27b) varies in importance from 0.72 at 248 nm to less than 0.1 above 290 nm (DeMore et al., 1992).

The heterogeneous reaction of N_2O_5 with sea-salt aerosol has been the subject of a number of studies (see Finlayson-Pitts, 1993, for an extensive reference list). The net reaction appears to be

$$N_2O_5 + NaCl(s) \rightarrow ClNO_2 + Na^+ + NO_3^- \qquad (6.28)$$

The interest in this reaction is not only due to the conversion to inactive NO_y (NO_3^-) but also because of the potential for the formation of reactive chlorine that would result (Reaction (6.23)). A more in-depth discussion on the possible presence of chlorine atoms in the troposphere can be found in Chapter 7.

In summary inorganic compounds HONO, NOCl, and NO_3 are photolytically unstable, while $HOONO_2$, $ClONO_2$, and N_2O_5 are thermally and to some extent photolytically unstable. In addition, they are all fairly soluble in liquid water and react rapidly to yield NO_2^- or NO_3^-. The combination of slow or unproven formation mechanisms and fairly rapid loss mechanisms serves to limit the importance of these species in current tropospheric chemistry models.

6.3.3. Ammonia

Ammonia is not usually considered an NO_y compound; however, its gas phase chemistry can make it a source or sink for NO_x. The principal fate of ammonia in the atmosphere is incorporation into aerosol and subsequent deposition. A small ($\approx 10\%$) fraction of ammonia is oxidized via OH:

$$NH_3 + OH \rightarrow NH_2 + H_2O \tag{6.29}$$

The resulting amide radical can react with either NO or NO_2 or with O_3:

$$NH_2 + NO \rightarrow N_2 + H_2O \tag{6.30}$$

$$NH_2 + NO_2 \rightarrow N_2O + H_2O \tag{6.31}$$

$$NH_2 + O_3 \rightarrow \rightarrow \rightarrow NO_x + H_2O \tag{6.32}$$

Reaction with either nitrogen oxide leads to loss of NO_x, while reaction with O_3 leads to NO_x formation. The uncertainty in the above reaction rates coupled with the spatial variability of O_3 and NO_x makes estimation of the NO_x source rather difficult. A global NH_3 source strength of $82 \, Tg \, N \, year^{-1}$ has been estimated by Stedman and Shetter (1983); the source of NO_x from NH_3 is at most 10% of that and probably smaller still, hence the estimate in Table 6.1.

6.3.4. Organic NO_y Chemistry

In contrast with inorganic NO_y chemistry where only a few compounds are thought to be important, organic NO_y chemistry is recognized to involve many different compounds, some having a high degree of complexity. Organic NO_y compounds can be roughly divided between compounds having a covalent C—N bond (Table 6.4) and those having a C—O—N bond (Table 6.5). The following discussion will outline the chemistries attendant to each type and subtype of organic N compound.

TABLE 6.4   Organic NO_y Compounds of the R—N Type

Name	Structure
Amines	R—NH_2
Amino acids	R—$CH(NH_2)COOH$
Hydrazines	R—$NHNH_2$
Nitramines	R—$NHNO_2$
Nitrosamines	R—NHNO
Nitro compounds	R—NO_2
Nitriles	R—CN

TABLE 6.5   Organic NO$_y$ Compounds of the
C—O—N Type

Name	Structure
Alkyl nitrites	R—ONO
Alkyl nitrates	R—ONO$_2$
Alkyl peroxynitrates	R—OONO$_2$
Peroxycarboxylic nitric anhydrides[a]	R—C(O)OONO$_2$

[a]Also known as peroxyacylnitrates.

6.3.4.1. Compounds of the Type R—N

Reduced nitrogen species such as amines, hydrazines, and amino acids have never been found to constitute significant sources of nitrogen to the atmosphere; however, they are potentially important because of their toxicity. The chemistry of these species is very similar to that of other organic compounds. The organic group is subject to OH attack, and ozone and NO$_3$ attack if it is unsaturated, and oxidation proceeds along the same lines as that of other organic species (Atkinson, 1990). The nitrogen-centered photo-oxidation of reduced nitrogen species is initiated by OH (Pitts et al., 1978):

$$R—NH_2 + OH \rightarrow R—NH + H_2O \tag{6.33}$$

In contrast with NH$_2$, the reaction of R—NH radicals with NO and NO$_2$ appears to proceed through addition to form nitrosamines or nitramines:

$$R—NH + NO \rightarrow R—NHNO \tag{6.34}$$

$$R—NH + NO_2 \rightarrow R—NHNO_2 \tag{6.35}$$

These products are significant because of there biological effects, but are of only minor consequence to oxidant photochemistry. Both nitrosamines and nitramines are likely to deposit to ground or aerosol surfaces and are probably fairly soluble. Nitrosamines photolyze rapidly (5–10 min), while nitramines photolyze less rapidly (2 days; Tuazon et al., 1984), to re-form R—NH radicals. Other reactions of R—NH radicals are not well known, but oxidation of the nitrogen atom by reaction with O$_3$ is likely. The ultimate fate of this nitrogen is liberation as NO or NO$_2$.

Nitro compounds have different chemistry depending on whether the organic group is an aromatic or an alkyl group. Nitroalkyl compounds have some minor combustion sources and are rapidly photolyzed under tropospheric conditions:

$$R\text{—}NO_2 + h\nu \rightarrow R + NO_2 \qquad (6.36)$$

There are no gas phase reactions that can form nitroalkyls because of the virtually instantaneous reaction of any alkyl carbon-centered radical with O$_2$.

The chemistry of nitro-aromatic compounds is somewhat different from that of nitroalkyls. Although primary sources cannot be ruled out, the chief source of nitro-aromatics is through the OH-initiated photo-oxidation of aromatic compounds (Grosjean, 1991; Atkinson and Aschmann, 1994). Reaction of simple aromatic hydrocarbons with OH proceeds mainly through ring addition, illustrated in Figure (6.2) for toluene. The OH-toluene adduct reacts either with NO$_2$ or with O$_2$ (not shown), the fraction going to the nitro adduct being directly

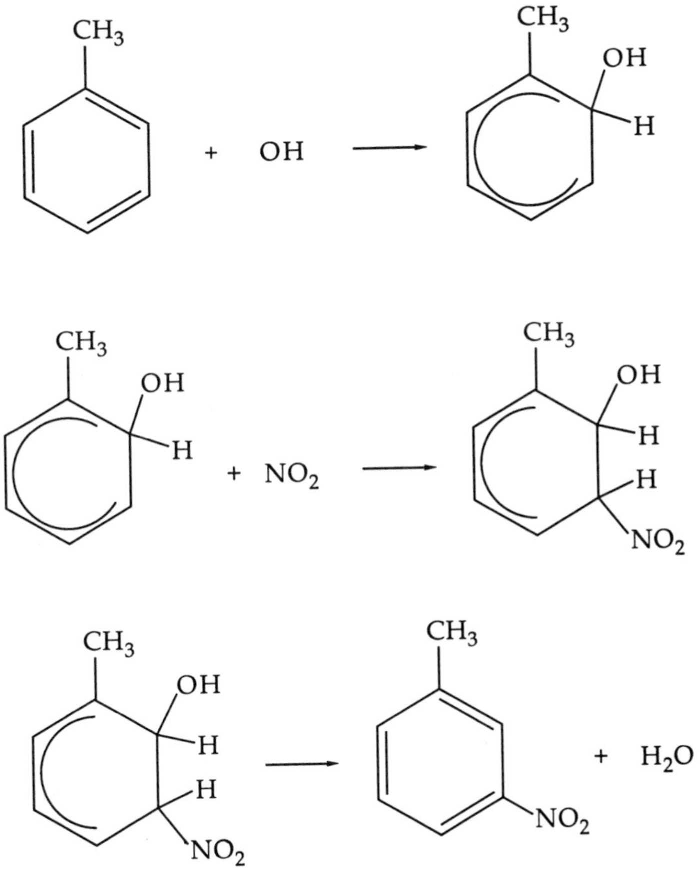

FIGURE 6.2. Formation of nitro aromatics from OH-initiated chemistry in the presence of NO$_x$.

dependent on NO_2 concentration. The formation of the nitro-aromatic occurs when H_2O is eliminated. Results from the recent kinetic studies by Atkinson and Aschmann (1994) can be interpreted to suggest that atmospheric concentrations of NO_2 in general are too low to compete favorably with O_2, and significant synthesis of nitro-aromatics may be possible only very close to combustion sources. Similar chemistry is thought to take place with OH-substituted aromatics (cresols). There is no doubt that many of these reactions have competing channels; however, the evidence for this chemistry is substantial, nitro compounds having been observed in both smog chamber studies (high NO_2 concentrations) and atmospheric measurements. Nitro products are not very volatile, so most of these nitro-aromatics and nitro-cresols are found in the particle phase.

In addition to simple nitro-aromatics, polycyclic aromatic hydrocarbons (PAHs) have been observed to form nitro-PAHs under laboratory conditions. This chemistry is still somewhat uncertain; however, in addition to the OH-initiated chemistry above, there appears to be chemistry initiated by NO_3 or N_2O_5 (Wayne et al., 1991, and references therein). The kinetics of the direct nitrogen chemistry is consistent with attack by N_2O_5; however, stepwise reaction with NO_3 and then NO_2 would yield the same kinetics because of the equilibrium between NO_3, NO_2, and N_2O_5. It is safe to say that these nitro-PAHs would remain in the particle phase and eventually deposit. The existence of these compounds could present a threat to human health.

The chemistry of nitriles in the atmosphere has not been extensively studied, since their sources appear to be relatively minor. The reaction of acetonitrile with OH is very slow and appears to result either in hydrogen abstraction or in addition, probably to the nitrogen (DeMore et al., 1992):

$$CH_3CN + OH \rightarrow CH_2CN + H_2O \tag{6.37}$$

$$+ M \rightarrow CH_3CN{-}OH \tag{6.38}$$

The temperature and pressure dependences of this reaction are not well known. Likewise, the eventual products are not clear, but could result in NO_x production.

6.3.4.2. Compounds of the Type R—O—N

Compounds of this type have fundamentally different chemistry from those of the C—N type. This is because there are many more pathways for atmospheric formation of them and because there are a number of decomposition pathways of them that regenerate NO_x. These features of the chemistry result in the partitioning of a significant fraction of NO_y in the form of R—O—N compounds, particularly organic nitrates.

Alkyl nitrites are esters of nitrous acid. The atmospheric formation of these types of compounds can occur through the reaction of alkoxy radicals with NO:

$$RO + NO + M \rightarrow RONO \qquad (6.39)$$

However, it must be recognized that alkoxy reactions undergo many competing reactions, including oxidation by O$_2$, decomposition, and 1,4 and 1,5 hydrogen transfer (Atkinson, 1990):

$$RO + O_2 \rightarrow carbonyl + HO_2 \qquad (6.40)$$

$$RR'C(O^{\bullet})H + M \rightarrow RC(O)H \text{ (aldehyde) } + R' \qquad (6.41)$$

$$^{\bullet}OCH_2(CH_2)_2CH_3 + M \rightarrow HOCH_2(CH_2)_2CH_2 \qquad (6.42)$$

This renders the gas phase formation of RONO compounds unimportant. The only potentially important source of RONO compounds is the surface formation in exhaust streams of alcohol-fueled vehicles.

The photolysis of RONO compounds is very rapid in every region of the atmosphere:

$$RONO + h\nu \rightarrow RO + NO \qquad (6.43)$$

The net effect of RONO chemistry may be to provide a small radical source in the early morning hours in a manner similar to HONO.

Alkyl nitrates are esters of nitric acid. Several features of their chemistry make them potentially important members of the NO$_y$ family. Several formation pathways are possible for RONO$_2$ compounds:

$$RO + NO_2 + M \rightarrow RONO_2 + M \qquad (6.44)$$

$$RO_2 + NO + M \rightarrow RO + NO_2 + M \qquad (6.45)$$

$$\rightarrow RONO_2 + M \qquad (6.45a)$$

$$NO_3 + alkene \rightarrow \rightarrow difunctional nitrate \qquad (6.46)$$

The first reaction of the two is unimportant for the same reasons as for RONO formation, i.e., other pathways of RO reaction are faster. The second reaction is currently thought to be responsible for RONO$_2$ formation. The reaction has two channels as shown above, the main channel resulting in oxidation of NO to NO$_2$. The other channel is thought to involve a complex mechanism with initial formation of an alkyl peroxynitrite (ROONO) and rearrangement through a three-centered intermediate to form $RONO_2$ (Atkinson et al., 1983). The branching ratio of reaction (6.45) is dependent on temperature, pressure, carbon number, and degree of substitution of the alkyl group. Branching ratios vary from below 0.004 for ethyl to up to 0.40 for some of the octyl compounds.

Nitrate radical reactions with alkenes proceed though addition to the double bond (Wayne et al., 1991):

$$NO_3 + RHC = CR'H \rightarrow (O_2NO)RHC\!\!-\!\!CR'H^\bullet \qquad (6.47)$$

The resulting carbon-centered radical reacts with O_2 to form a peroxy radical:

$$(O_2NO)RHC\!\!-\!\!CR'H^\bullet + O_2 \rightarrow (O_2NO)RHC\!\!-\!\!CR'HO_2 \qquad (6.48)$$

The fate of these peroxy radicals is probably self-reaction or reaction with HO_2, since reaction with NO is not possible at night in general (NO_3 and NO will not coexist in significant amounts because of reaction (6.16)). The resulting stable products are α-difunctional nitrates such as hydroxy or carbonyl nitrates.

An interesting problem comes up when considering methyl nitrate formation. The MeO_2 + NO reaction branch to $MeONO_2$ has not been measured but is probably quite low. Formation of $MeONO_2$ has been attributed (Stephens, 1969; Senum et al., 1986) to the reaction

$$PAN + M \rightarrow CH_3ONO_2 + CO_2 \qquad (6.49)$$

which was thought to proceed by thermal decomposition via a cyclic intermediate. The work of Senum et al. (1986) on the thermal decomposition of PAN in the absence of NO seemed to indicate a pathway for reaction (6.49) that was about 0.5% that of the bond homolytic rate. Recent work (Orlando et al., 1992) in which NO_2 and O_2 concentrations have been systematically varied has shown this cyclic pathway to be at least several orders of magnitude less than that of Senum et al. (1986), if it occurs at all. As a result of these studies, there is no demonstrated source for $MeONO_2$ that can explain measured atmospheric levels.

The loss of alkyl nitrates from the atmosphere is in most cases slow relative to other NO_y species. Important processes include reaction with OH, photolysis, and deposition; thermal decomposition is unimportant under atmospheric conditions (Table 6.2). The $-\!ONO_2$ group is electronegative enough that it lowers the OH reaction rates of alkyl nitrates relative to the parent hydrocarbon, an effect that diminishes with distance from the $-\!ONO_2$ group and therefore with carbon number. The photolysis of simple alkyl nitrates is not very rapid, but is as fast as or faster than OH reaction for compounds of four carbons or less

(Roberts and Fajer, 1989; Talukdar et al., 1995). The pathway of photolysis is the breaking of the RO—NO_2 bond:

$$RONO_2 + h\nu \rightarrow RO + NO_2 \qquad (6.50)$$

Modern measurements of the quantum efficiencies for this process are not available, but it is highly probably that these are close to unity. The reasons for this are fairly simple: the bond strength is far lower than the energy of absorbed photons, there is no structure in the absorption spectra at wavelengths higher than 250 nm, and the quantum efficiency for the process

$$HONO_2 + h\nu \rightarrow HO + NO_2 \qquad (6.51)$$

has been found to be 1 ± 0.05 between 200 and 315 nm (Johnston et al., 1974). The overall photolysis of alkyl nitrates has been observed to be temperature-dependent by Luke et al. (1989). This has been attributed to the temperature dependences of the absorption cross-sections which have been measured by Rattigan et al. (1992) and Talukdar et al. (1995). The absorption by alkyl nitrates drops roughly a factor of 2 in 10 K. Less is known about the photolysis of difunctional nitrates, but indications are that α-hydroxy nitrates have lower photolysis rates and α-keto nitrates have much higher photolysis rates than the corresponding simple alkyl nitrate (Roberts and Fajer, 1989). The solubility of $RONO_2$ compounds has been investigated for some simple alkyl nitrates and difunctional nitrates. Simple alkyl nitrates are not very soluble, having Henry coefficients-between 0.74 and 4 M atm^{-1} (see Table 6.3). Hydroxy nitrates have much higher solubilities, with H up to 4×10^4 M atm^{-1} for nitrooxy-ethanol (Kames and Schurath, 1992).

Alkyl peroxynitrates can be thought of as esters of peroxynitric acid. These compounds are formed in the reaction of peroxy radicals with NO_2:

$$RO_2 + NO_2 + M \rightarrow ROONO_2 + M \qquad (6.52)$$

These reactions appear to be relatively rapid, $(5-8) \times 10^{-12}$ cm^3 molecule^{-1} s^{-1} but they must compete with other reactions of RO_2 radicals such as self-reactions, and reactions with HO_2.

The removal of $ROONO_2$ compounds can occur through reaction with OH, photolysis, or thermal decomposition. Rates of reactions with OH have not been measured but are likely to follow the pattern of those of the $RONO_2$ compounds because of the similar electron-withdrawing properties of the groups. The photolysis of $ROONO_2$ compounds is likely to be a major loss process given that measured cross-sections appear to be larger than those of the corresponding $RONO_2$ compounds (Roberts, 1990, and references therein). As with the $RONO_2$

compounds, the quantum efficiencies of these processes are not well known but are probably close to unity. The thermal decomposition of $ROONO_2$ compounds occurs at significant rates (Table 6.2). This feature restricts the presence of significant concentrations of $ROONO_2$ compounds to the upper troposphere and stratosphere.

Peroxycarboxylic nitric anhydrides are anhydrides of peroxycarboxylic and nitric acids. The most common and abundant such compound is peroxyacetic nitric anhydride, often called peroxyacetyl nitrate, but abbreviated here as PAN. As far as can be determined, this class of compound only exists in the atmosphere. The source of PAN-type compounds is the reaction of peroxyacyl radicals with NO_2, shown here for the compound PAN:

$$CH_3C(O)OO^{\bullet} + NO_2 + M \rightarrow CH_3C(O)OONO_2 + M \qquad (6.53)$$

This reaction is somewhat pressure dependent but essentially temperature-independent (Bridier et al., 1991). Loss of PAN-type compounds from the atmosphere occurs by reaction with OH, photolysis, thermal decomposition, and deposition. Thermal decomposition rates have been measured for a number of PAN-type compounds and to a good approximation have been found to be the same (Roberts and Bertman, 1992). The UV absorption cross-sections of PAN and PPN (peroxypropionic nitric anhydride) were measured by Senum et al. (1984) and that of PAN has recently been measured as a function of temperature (Talukdar et al., 1995). The results of Senum et al. (1984) showed the room temperature cross-sections of PAN and PPN to be similar. Talukdar et al. (1995) found a room temperature cross-section significantly higher than that of Senum et al. (1984), and a lower absorption cross-section with lower temperature. Similar cross-sections can be expected for simple alkyl PAN compounds, but different behavior can be expected of compounds such as PBzN (peroxybenzoic nitric anhydride) and $ClC(O)OONO_2$. The reaction rate of PAN with OH has been measured in several studies, and the temperature dependence recently determined by Talukdar et al. (1995), who found significantly lower rate constants for PAN. The reactions of PAN-type compounds with OH will likewise depend on the structure of the organic substituent, and faster rates can be expected for larger alkyl PANs. Grosjean et al. (1993b) have recently measured the rate of OH reaction with MPAN to be $(3.6 \pm 0.4) \times 10^{-12}\,cm^3\,molecule^{-1}\,s^{-1}$ at 298 K. The net result of the studies on removal rates of PAN is that thermal decomposition is the dominant loss process in the lower atmosphere ($<7\,km$), and photolysis in the upper atmosphere. The reaction of PAN with OH is not competitive with these processes in any region of the atmosphere.

The solubility of PAN in distilled water is quite low (Table 6.3) and somewhat temperature-dependent (Kames et al., 1991). The deposition of PAN on natural pasture and sea water surfaces has been found to be slow (Garland and Penkett,

1976). However, measurements of PAN and O_3 at several surface sites have shown PAN to deposit more rapidly than O_3 (Shepson et al., 1992). This implies that the dry deposition of PAN can be rapid under some circumstances, but the nature of surfaces that might cause this behavior is unknown.

6.4. ATMOSPHERIC OBSERVATIONS AND THEORETICAL UNDERSTANDING

A great deal has been learned about the atmospheric content of NO_y and NO_y species over the past 25 years. This knowledge has been gained through tremendous improvement and innovation in measurement techniques and through a large growth in the geographic area that has been studied. The atmospheric data have been used together with laboratory data to further the theoretical understanding of NO_y chemistry. An example of a theoretical prediction is shown in Figure 6.3, in which the mixing ratios of NO_y species are shown as a function of altitude for conditions characteristic of the mid-latitude marine environment, i.e., Cl chemistry included (Singh and Kasting, 1988). The comparison of such model results with actual measurements is a critical exercise in assessing the state of our understanding of this chemistry. What follows is a summary of the major features of atmospheric NO_y species measurements and discussion of their theoretical understanding.

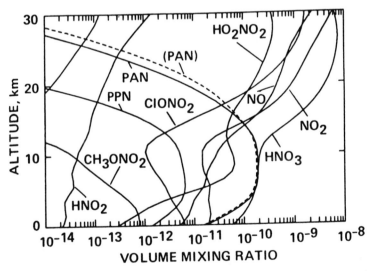

FIGURE 6.3. Model predictions of NO_y species concentrations as a function of altitude for the marine troposphere and lower stratosphere. (From Singh and Kasting, 1988.)

For the purposes of this discussion, the troposphere will be divided into three geographic regimes: urban, regional, and remote. This division is somewhat arbitrary and must be used with care; however, it will be helpful in explaining the current state of knowledge. The following discussion is meant to be a summary of the major issues in tropospheric NO_y chemistry as reflected in ambient measurements and as accounted for in current theoretical models. As such, it is not an exhaustive review of all measurement techniques, measurements, and models. More detailed accounts of aspects of the measurements and models can be found in Fehsenfeld et al. (1988), Hough (1988), Roberts (1990), and Altshuller (1993).

6.4.1. NO_x

The measurement of NO_x has been most often accomplished by the $NO—O_3$ chemiluminescence method, wherein ambient air is reacted with high concentrations of O_3 at low pressure in front of a red-sensitive photomultiplier tube (Fontijn et al., 1970). NO_2 is then determined via selective conversion to NO by photolytic, thermal, or chemical ($FeSO_4$) means. Both the thermal and $FeSO_4$ converters have been found to have significant interferences from other NO_y species, particularly PAN (Fehsenfeld et al., 1987). Several other methods have been developed for NO (laser-induced fluorescence, LIF) or NO_2 (luminol chemiluminescence, tuneable diode laser absorption spectroscopy (TDLAS), LIF) measurement and have been extensively intercompared (Fehsenfeld et al., 1987, 1990).

A good summary of NO_x measurement results in the non-urban troposphere can be found in Fehsenfeld et al. (1988). Mixing ratios vary from several hundred parts per billion (ppb) in urban areas down to only a few parts per trillion (ppt) in remote areas. Overall, NO_x has a relatively constant source with season, but a much longer lifetime in the winter due to reduced photochemical activity. Accordingly, higher mixing ratios are expected in remote areas of North America during the winter. Figure 6.4 shows an example of the NO_x distribution in the remote free troposphere of the western Pacific ocean for 25–45° N and 0–25° N latitude bands (Singh et al., 1994a). Mean NO_x mixing ratios are generally below 100 ppt and are highest in the upper troposphere. The global reservoir of NO_x is balanced by its surface sources, lightning, stratospheric injections, and also its *in situ* formation from NO_y reservoir species (e.g. PAN, HNO_3). It is important to point out that the lifetime of atmospheric NO_x is significantly longer in the upper troposphere (≈ 6 days) compared with the boundary layer (≈ 1 day).

The $NO—NO_2—O_3$ photostationary state relationship has been investigated in a number of studies (Parrish et al., 1986; Cantrell et al., 1992). The ratio of NO_2 to NO has been found to be larger than can be accounted for by the set of

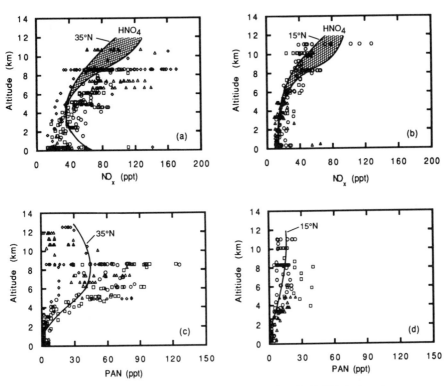

FIGURE 6.4. Vertical distribution of NO$_x$ and PAN over subtropical/mid-latitude and tropical/equatorial western Pacific. Solid curves represent results from a 3D photochemical model. The dashed area in (a) and (b) shows the model calculated abundance of HO$_2$NO$_2$. (From Singh et al., 1994a.)

reactions (6.1)–(6.3), during photochemically active periods. This has been attributed to the presence of peroxy radicals RO$_2$, and HO$_2$ that result from organic photo-oxidation. Implicit in this NO$_x$ photostationary state imbalance is the production of ozone. A comparison of NO$_x$ photostationary state imbalance, as an indirect measure of radicals, with a chemical amplifier measurement of total radicals at a regional site in Alabama showed good agreement at times and disagreement at other times (Cantrell et al., 1992).

6.4.2. Total NO$_y$

The measurement of total NO$_y$ provides an important means of following the chemistry of the atmosphere. The concentrations and systematics of individual

NO_y species relative to the total provide a way to assess the importance of each species and the interconversion of one species to another. Moreover, NO_y is more of a conserved quantity than any of its constituent species.

The measurement of NO_y is accomplished through catalytic conversion to NO and detection by one of the above methods for NO. Both molybdenum and gold have been used for catalytic conversion and work roughly equally well, except that molybdenum appears to be more sensitive to poisoning by materials (probably organic carbon) found in highly polluted air masses (Fehsenfeld et al., 1987). Both catalytic methods have been found to convert NO_y species with good efficiency and not to convert possible interfering species such as NH_3, HCN, and N_2O.

Total NO_y concentrations have varied in a manner similar to those of NO_x, depending on the vicinity of large sources and the degree of photochemical processing and deposition that an air mass has undergone. Thus NO_y levels of several hundred ppb can be found in urban areas and concentrations below 100 ppt have been observed in remote air masses. Figure 6.5 shows the measured NO_y distribution for the northern Arctic/sub-Arctic (50–70° N) atmosphere (Singh et al., 1992a,b). In remote regions it is common to see higher concentrations aloft than near the surface. Both free tropospheric sources of NO_y and its more rapid removal, due to deposition, near the surface of the Earth are responsible for this profile. The high NO_y in the upper troposphere has been largely attributed to the long-range impact of anthropogenic sources.

There is tremendous interest in quantitative assessments of the dependence of ozone formation on NO_x, especially on urban and regional scales. Lin et al. (1988) formulated their model in terms of the efficiency of ozone produced per NO_x converted to products by reactions (6.6) and (6.53). This was found to depend on the concentration of NO_x; fewer ozone molecules are produced at high NO_x than at low NO_x. The nature of NO_x and NO_y on urban to regional scales is that NO_x is converted to products; hence the fraction of NO_y that is NO_x is steadily decreasing as photochemical processing takes place. Trainer et al. (1993) have used the quantity $[NO_y - NO_x]$ as a measure of NO_x products and therefore conversion. The correlation of measured O_3 with $NO_y - NO_x$ has shown a non-linear relationship which is in broad agreement with a regional photochemistry model that simulates the NO_x chemistry (Figure 6.6). The O_3 versus $NO_y - NO_x$ relationship has a slope that varies from roughly 17 at low $NO_y - NO_x$ to 5 at high $NO_y - NO_x$. This slope can be roughly thought of as the O_3 production efficiency per NO_x converted. The addition of more non-methane hydrocarbons (NHHCs) to the model changes the resulting relationship somewhat, but most of the ozone increase is accounted for by the quantity $NO_y - NO_x$, strongly implying that ozone production is NO_x-limited in these air masses.

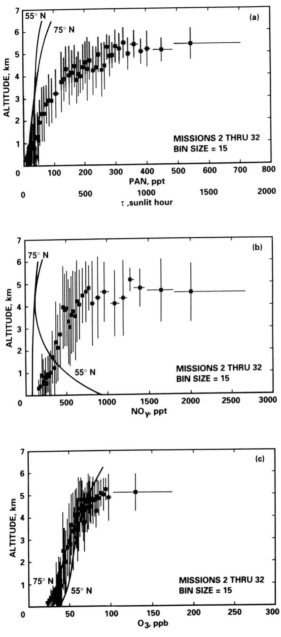

FIGURE 6.5. Mean vertical structure of PAN, NO$_y$ and O$_3$ in the high-latitude troposphere. Solid curves represent results from a 2D photochemical model. (From Singh et al., 1992b)

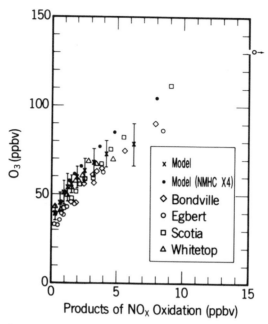

FIGURE 6.6. Relationship between O_3 and NO_x products at four sites in North America and for two model cases. (From Trainer et al., 1993.)

6.4.3. PAN and PAN-Type Compounds

PAN was discovered in the late 1950s in long-path IR measurements of the Los Angeles smog (see Stephens, 1987, for an interesting discussion of this and other early smog studies). Darley et al. (1963) subsequently developed a gas chromatography/electron capture detector (GC/ECD) technique that has become the standard. Developments and improvements since then have included the application of cryogenic collection, capillary GC columns, and the luminol chemiluminescence detector (Singh and Salas, 1983; Burkhardt et al., 1988; Roberts et al., 1989; Bertman et al., 1993; Blanchard et al., 1993). Calibrations have been performed in a variety of ways; synthesis of PAN and PAN-type compounds has been done either photochemically (Stephens, 1969) or in solution (Nielsen et al., 1982; Gaffney et al., 1984), and measurement has been done either by IR absorption or by catalytic conversion to NO and measurement by $NO–O_3$ chemiluminescence.

Ambient levels of PAN vary a great deal, depending on the proximity to NO_x and NMHC sources and the degree of photochemical activity. The highest levels of PAN are found in urban atmospheres in summer and the lowest are found in the remote troposphere in late fall or winter. The systematics of these observations have been discussed by Roberts (1990) and are only briefly summarized here. Urban photochemistry produces afternoon maxima of a few to over 30 ppb of PAN (see Altshuller, 1993, and Roberts, 1990, for summaries). These PAN profiles broadly coincide with O$_3$ maxima and the most intense episodes are observed in summer. The concentrations of PAN found at sites that are regionally representative range from very low (a few ppt) to up to a few ppb. The levels of PAN are highly dependent on the NO$_x$ levels and to a certain extent on the NMHC levels at these rural locations. Diurnal profiles of PAN are again similar to those of O$_3$ in that there are typically afternoon maxima; however, there are some qualitative differences. PAN levels often peak at or around solar noon and drop in the mid- to late afternoon, while O$_3$ stays high. This is due to lower NO$_x$ levels in these air masses, effectively allowing more thermal decomposition to take place in these air masses relative to urban air masses. PAN concentrations in air masses that are characteristic of remote areas do not show diurnal profiles with afternoon maxima, but rather more episodic behavior where PAN levels are determined by the long-range history of the air mass. A characteristic of PAN in remote air masses is a springtime maximum, explained by Penkett and Brice (1986) as being due to "storage" of PAN precursors throughout the winter months and hence increased production at the onset of photochemistry in the spring. This springtime maximum has been observed in several other data sets along with an increase in PAN in remote air masses over the past 15 years (Roberts, 1990). PAN has been proposed as an important reservoir of reactive nitrogen in the global atmosphere (Singh and Hanst, 1981). Figure 6.5 shows the abundance of PAN in the Arctic/sub-Arctic atmosphere, where PAN may have a long lifetime owing to the cold temperatures. In the remote regions of the western Pacific (Figure 6.4) these concentrations are much lower. Comparison with models (solid curves in Figures 6.4 and 6.5) shows that the differences between measurements and simulations are greatest at high latitudes.

Other PAN-type compounds have been observed and quantified. Chief among these is PPN, with relative abundances that vary from a few per cent to 30% of PAN (Singh and Salas, 1989; Altshuller, 1993; Grosjean et al., 1993a). The ratio PPN/PAN has been found to vary systematically with the air mass age, probably owing to the shorter lifetimes of PPN precursors relative to those of PAN (Singh and Salas, 1989; Ridley et al., 1990). Peroxybutyric nitric anhydride (PBN) has been observed in and around the Los Angeles air basin approximately 70% of the time (Grosjean et al., 1993a) at a fractional abundance of about 8% of PAN. The compound peroxymethacrylic nitric anhydride (MPAN) has also been occasionally observed in areas impacted by isoprene emissions (Bertman and Roberts, 1991). The relative abundance of MPAN was small at Niwot Ridge, Colorado (5–7%, Bertman and Roberts, unpublished results, 1991) compared with the 15–25% of PAN predicted by models incorporating isoprene photochemistry (Atherton and Penner, 1990). Measurements in Atlanta found MPAN above detection limits only about 1.5% of the time, but the abundance relative to PAN was 10–30% on those occasions (Williams et al., 1993). The compound peroxybenzoic nitric anhydride (PBzN) has been observed only a few times, often at very low concentration, one study reported PBzN concentrations in the range of 10–20% of PAN (Meijer and Nieboer, 1977).

6.4.4. Nitric Acid and Nitrate Aerosol

Few measurements in atmospheric chemistry have been as vexing as that of HNO_3. The reasons for this difficulty are the extreme solubility of HNO_3 and its equilibration with the particle phase. A number of intercomparisons of HNO_3 and particle nitrate measurements have been conducted (Spicer et al., 1982; Hering et al., 1988; Tanner et al., 1989; Gregory et al., 1990). The broad conclusion from these studies is that HNO_3 measurements must be viewed with caution and that this caution must be carried over to other quantities such as ΣNO_{y_i}, that are derived from them (see below).

Measurements of HNO_3 have revealed a number of features. The production of HNO_3 appears to be driven by photochemistry, consistent with $OH + NO_2$ being the major source. It is apparent from diurnal profiles that HNO_3 is rapidly deposited on surfaces and in water droplets. The HNO_3–NH_3–H_2SO_4 aerosol system has been found to involve an equilibrium between gas phase HNO_3 and particle NO_3^- that is a function of NH_3/NH_4^+ acidity and temperature (Stelson and Seinfeld, 1982a,b). HNO_3 (along with H_2SO_4) has been found to be an important component of acid deposition (see Chapter 12; Schwartz, 1989). Measurements of HNO_3 along with other NO_y species have shown it to be an important fraction of total NO_y (see below) and sometimes the most abundant single component. Models of HNO_3 have had trouble reconciling the relative concentrations of HNO_3 and NO_x; usually the measured ratio HNO_3/NO_x is much smaller than that predicted by models (Liu et al., 1992). The deposition of HNO_3, traditionally a difficult feature for models to simulate, and sources of NO_x in remote areas have been identified as two areas potentially responsible for this discrepancy.

6.4.5. Minor Inorganic NO_y Constituents

Nitrate radical and nitrous acid are the only two other inorganic NO_y species that have been observed in the troposphere. Nitrate radical has been measured by differential optical absorption spectroscopy (DOAS) at 623 and 662 nm (Noxon et al., 1978; Platt et al., 1980), and by matrix isolation ESR spectroscopy (Mihelcic et al., 1985). These measurements have indicated that NO_3 is only present at night at concentrations ranging up to 300 ppt in the boundary layer (Platt et al., 1980), as expected from its absorption cross-section, and simultaneous measurements have shown NO_3 to be a small fraction of NO_2. Measurements of NO_3 and NO_2 at Fritz Peak Observatory (40° N) in the mountains west of Boulder, Colorado implied that there was a loss process for NO_3 that was not attributable to known chemistry and did not appear to be due to reactive organics such as terpenes or the deposition of N_2O_5 on hydrometeors. The resolution of this quandary resulted from the work of Solomon et al. (1989), wherein it was shown that a faint absorption feature of water vapor could interfere with the NO_3 absorption feature. The method of background subtraction of Noxon (1983) did not account for the changing amount of water vapor in the absorption path, thus yielding an apparent rapid change in NO_3. It is clear from these measurements and from known rates of reactions that NO_3 can be a significant reactant for some organic compounds such as biogenic hydrocarbons and sulfur compounds (Winer et al., 1984).

Measurements of HONO have been made both by DOAS and by high-volume sampling of air, with chemical analysis of the resulting NO_2^- ion. In contrast with nitric acid, which is found everywhere, nitrous acid has been observed only relatively near high-NO_x sources. Measurements made in conjunction with NO_2 measurements have shown HONO to be at most 10% of NO_2 (Harris et al., 1982). Measurements made by collection on Na_2CO_3-coated media are compromised by the collection of NO_2 and PAN on these surfaces. The importance of HONO in the understanding of tropospheric chemistry has mostly to do with its behavior as a radical source immediately after sunrise in urban or heavily impacted areas, an aspect that regional photochemical models do not currently address (M. Trainer, personal communication).

6.4.6. Total NO_y versus ΣNO_{y_i}

The relative contribution of individual NO_y species to the total has been an important subject of research in the past few years. Table 6.6 summarizes reported measurements of total NO_y and constituent species. The studies summarized in Table 6.6 were all performed at sites or in air masses of regional or remote character. Under these conditions NO_x is often not the major component of NO_y, but rather the product species PAN and HNO_3 are. This is presumably different in urban areas, but studies in which NO_2 was measured independently are lacking. The trend evident in Table 6.6 is that the sites receiving air masses of the most remote character are the lowest in NO_x/NO_y and the air masses found at higher altitudes have the largest PAN/NO_y. Indeed, PAN is the most abundant NO_y species in the remote free troposphere. It is also apparent from Table 6.6 that ΣNO_{y_i} is less than total NO_y at many sites. This "shortfall" in NO_y is one of the more intriguing problems in tropospheric chemistry. The shortfall appears not to show a day–night systematic difference (Fahey et al., 1986), implying that it cannot be HONO or NO_3. These features of the measurements have led to speculation that organic nitrates are responsible for the shortfall.

TABLE 6.6 Distribution of NO_y among Chemical Species

Location	Sampling Period	Average Ratio					Reference
		NO_x/NO_y	PAN/NO_y	HNO_3/NO_y	NO_3^-/NO_y	$\Sigma NO_y/NO_y$	
Niwot Ridge, CO	7/5–8/31/84	0.29	0.16	0.10	0.02	0.58	Fahey et al. (1986)
	10–11/84	0.40	0.3	–	–	0.88	Unpublished
	Summer 1987	0.32	0.24	0.10	0.04	0.73	Ridley (1991)
Point Arena, CA	4–5/85	0.6	0.1	0.05	–	0.75[a]	Unpublished
Scotia, PA	6–7/86	0.59	0.14	0.16	0.04	0.93	Ridley (1991)
Western Atlantic, PBL	1/4/86	0.32	0.08	–	0.30[b]	0.70	Bottenhiem and Gallant (1989)
FT	1/4/86	0.27	0.11	–	0.18[b]	0.56	Bottenhiem and Gallant (1989)
Continental US, FT	8–9/86	0.14	0.42	0.16–0.32[c]	0.07	0.7–0.9[c]	Ridley (1991)
Pacific, FT	8–9/86	0.09	0.33	0.32–0.49[d]	0.06	0.78–0.9[d]	Ridley (1991)
PBL	8–9/86	0.11	0.04	0.33–0.5	–	–	Ridley (1991)
Mauna Loa, HI, DS[e]	5–6/88	0.14	0.07	0.43	0.15	0.79	Atlas et al. (1992a,b)
US[e]	5–6/88	0.17	0.11	0.51	0.24	1.03	Atlas et al. (1992a,b)
Scotia, PA	7–8/88	0.38	0.2	0.29	–	0.90	Parrish et al. (1993)
Whitetop, VA	8–9/88	0.56	0.17	0.34	–	1.09	Parrish et al. (1993)
Bondville, IL	8/88	0.38	0.12	0.13	–	0.80	Parrish et al. (1993)
Egbert, Ont.	8–9/88	0.48	0.19	0.28	–	0.94	Parrish et al. (1993)
N. America, Greenland, 0–2 km	7–8/88	0.10	0.09	0.21	–	0.38	Singh et al. (1992)
2–4 km	7–8/88	0.05	0.27	0.15	–	0.50	Singh et al. (1992)
4–6 km	7–8/88	0.04	0.33	0.08	–	0.50	Singh et al. (1992)
North Bay, Ont., 0–2 km	7–8/90	0.17	0.22	0.33	0.05	0.70	Singh et al. (1994b)
2–4 km	7–8/90	0.09	0.43	0.22	0.05	0.77	Singh et al. (1994b)
4–6 km	7–8/90	0.06	0.38	0.09	0.04	0.54	Singh et al. (1994b)
Goose Bay, Labrador, 0–2 km	7–8/90	0.21	0.39	0.52	0.06	1.13	Singh et al. (1994b)
2–4 km	7–8/90	0.15	0.77	0.29	0.04	1.19	Singh et al. (1994b)
4–6 km	7–8/90	0.11	0.76	0.11	0.02	1.00	Singh et al. (1994b)

[a] A Teflon filter was placed in front of the NO_y catalyst, so the total NO_y measurement did not include particle NO_3^-.
[b] Total inorganic nitrate, $HNO_3 + NO_3^-$.
[c] Ranges indicate the results of two different HNO_3 measurements, the lower from a tungsten oxide denuder technique, the higher from a filter technique.
[d] Ranges indicate the results of two different HNO_3 measurements, the lower from a filter technique, the higher from a tungsten oxide denuder technique.
[e] DS, down-slope winds; US, up-slope winds.

6.4.7. Alkyl and Multifunctional Nitrates

There are not many reported observations of this class of compounds in the atmosphere. The methods used for these measurements have employed gas chromatography with either ECD or NO_y detection. There have also been reports of the use of mass spectrometry (Atlas, 1988) and nitrogen-selective GC detectors (Flocke et al., 1991; Blanchard et al., 1993). There have been early, sporadic reports of methyl nitrate measurements (Grosjean, 1983; Ciccioli et al., 1986) made in conjunction with PAN on the same system; however, caution must be applied to some of these, since co-elution with chlorinated hydrocarbons can be a problem in these systems under some conditions. Larger alkyl nitrates have also been observed in PAN measurement systems under conditions of high sensitivity (Buhr et al., 1990) or with cryogenic collection (Ridley et al., 1990; Bertman et al., 1993). A charcoal cartridge sampling technique has been developed by Atlas for the collection, transportation, and storage of alkyl nitrates (Atlas and Schauffler, 1991; Muthuramu et al., 1993). These studies have shown that α,β-hydroxy-nitrates and dinitrates can be stored and recovered from charcoal cartridges and that GC analysis of these materials can be done provided that consideration is given to loss of materials on injection port and column surfaces.

Ambient measurements of $RONO_2$ compounds have shown that virtually every possible compound in the size range from C_1 to C_8 can be found, depending on the degree of NO_x and NMHC impact (Flocke et al., 1991; Buhr et al., 1990; Ridley et al., 1990; Atlas et al.. 1993; Bottenhiem et al., 1993; Shepson et al., 1993). Walega et al. (1992) reported that methyl nitrate was found to be up to 10% of NO_y in air masses of tropical origin as determined during the Mauna Loa Observatory Photochemistry Experiment (MLOPEX I) in 1988. In other studies where alkyl nitrates have been measured along with total NO_y or the principal NO_y species (NO_x, HNO_3, particle NO_3^-, PAN), alkyl nitrates have been found to contribute only a few per cent of total NO_y (Buhr et al., 1990; Ridley et al., 1990; Shepson et al., 1993; Flocke et al., 1991). One study conducted in the Arctic during the spring (Bottenhiem et al., 1993) implies that $RONO_2$ might constitute a higher percentage of NO_y under these circumstances, since $\Sigma RONO_2$/PAN was higher than in the above studies, but total NO_y was not measured in this study.

The results of Flocke et al. (1991) at Schauinsland and Julich in Germany indicated the presence of significant concentrations of methyl and ethyl nitrates relative to the C_3 and C_4 nitrates. The branching ratio for reaction (6.45) is uncertain for methyl, but is undoubtedly low ($\ll 1\%$), and is fairly low for ethyl (≤ 0.014). It is difficult to reconcile the observations with this reaction as the sole source of methyl and ethyl nitrates.

Several sets of measurements have been made in the remote Pacific by Atlas and co-workers (Atlas 1988; Atlas et al., 1992a,b, 1993). Initial work by Atlas

(1988) showed the presence of alkyl nitrates of three carbons or more in samples taken ship-board in the mid-Pacific, the sum of 2-butyl and 2- and 3-pentyl nitrates averaged 9 ppt and ranged from 0.8 to 37 ppt. Alkyl nitrate measurements were also made as part of the MLOPEX I study. It was found that the C_3 and C_4 alkyl nitrates correlated with each other and with perchloroethylene, but not with methyl nitrate and only slightly with PAN. In addition, the correlation of higher nitrates with methane was excellent. This strongly implies that the higher alkyl nitrates are of northern hemisphere continental origin, while methyl nitrate appears to be of tropical origin (Walega et al. 1992). The sum of the higher alkyl nitrates averaged about 1.4% of total NO_y as did methyl nitrate. Atlas et al. (1993) recently reported the results of the Soviet–American Gases and Aerosols (SAGA 3) experiment, in which the C_2–C_5 alkyl nitrates were measured. Ethyl and isopropyl nitrates correlated somewhat with brominated hydrocarbons and seemed to show maximum concentrations near the equator. This prompted speculation that these nitrates might be produced in part by surface water biota. 2-Butyl nitrate did not show this behavior, but rather was slightly higher in the northern hemisphere relative to the southern. The most abundant nitrate was isopropyl nitrate, with a mean of 5 ppt, followed by 2- and iso-butyl nitrates, 2.9 ppt, ethyl nitrate, 2 ppt, and 2-methyl-3-butyl nitrate, 0.3 ppt. No measurements of other NO_y species were reported for this study. The lifetime of $RONO_2$ species is long enough (> 1 week) for long-range transport to play an important role in defining their concentrations in remote areas (Atherton, 1989).

6.5. CONCLUSIONS

Odd-nitrogen has a pivotal role in tropospheric chemistry. At least half of the sources of NO_y to the atmosphere can be attributed to anthropogenic activities, mostly combustion-related. There is substantial growth projected in NO_y emissions over the next decades that threatens to further perturb the atmosphere. This growth will be primarily in the developing countries and hence presents a problem similar to that of CO_2 emission control. It is therefore crucial to understand the nature and magnitude of NO_y-related impacts on the global atmosphere.

The involvement of nitrogen oxides in the production of ozone and in acid deposition has been well established. Urban air shed models of NO_x-NMHC chemistry show that NO_x reduction can lead to ozone increases in some areas under some conditions. However, it must be kept in mind that using this as a basis for not restricting NO_x emissions ignores other impacts such as acid aerosol formation and regional and global ozone formation. It is clear that at some sites under some circumstances there is a substantial fraction of total NO_y (up to 40%) that is not identified. Research in urban and regional chemistry is currently aimed at obtaining sufficient understanding of the chemistry and transport so that effective and rational emission control policies can be instituted. Some areas of uncertainty in the endeavor are as follows. Are measured concentrations consistent with source inventories? Is there heterogeneous chemistry that converted inactive NO_y (HNO_3, NO_3^-) to NO_x? How much NO_y is transported to the global atmosphere relative to that deposited?

Research on the global effects of NO_y has focused on the role of NO_x in the production of O_3. This has necessitated a detailed study of the chemistry of the NO_y family so that the fate of NO_x and possible transport mechanisms can be identified. Field studies are relatively sparse in number and have not covered the geographic or temporal range needed to understand this global chemistry. Specific uncertainties in the global area are as follows. How is NO_y distributed among chemical species, geographically, and with season? Are stable NO_y species such as PAN and $RONO_2$ compounds important in transporting NO_x to the global atmosphere? What is the nature of unidentified NO_y and will it contribute to global NO_y? Are there as yet unknown processes that convert particle NO_3^- to NO_x on long time scales? Is there important wintertime chemistry involving NO_3 and perhaps N_2O_5? What long-term changes are taking place in global NO_y?

Acknowledgments

I thank Drs R. Talukdar, J. Burkholder, A.R. Ravishankara, and H.B. Singh for the use of their data prior to publication. I thank Drs M. Trainer and S.C. Liu for helpful discussions concerning this work.

References

Akimoto, H., Takagi, H. and Sakamaki, F. (1987). Photoenhancement of the nitrous acid formation in the surface reaction of nitrogen dioxide and water vapor: extra radical source in smog chamber experiments. *Int. J. Chem. Kinet.*, **19**, 539–551.

Albritton, D.L., Liu, S.C. and Kley, D. (1984). Global nitrate deposition from lightning. in Aneja, V.P., Ed., *Environmental Impact of Natural Emissions*, Air Pollution Control Assoc., Pittsburgh, PA, pp. 100–112.

Altshuller, A.P. (1993). PANs in the atmosphere. *J. Air Waste Manag. Assoc.*, **43**, 1221–1230.

Anderson, L.C. and Fahey, D.W. (1990). Studies with $ClONO_2$: thermal dissociation rate and catalytic conversion to NO using an NO/O_3 chemiluminescence detector. *J. Phys. Chem.*, **94**, 644–652.

Atherton, C. (1989). Organic nitrates in remote marine environments: evidence for long-range transport. *Geophys. Res. Lett.*, **16**, 1289–1292.

Atherton, C.S. and Penner, J.E. (1990). The effects of biogenic hydrocarbons on the transformation of nitrogen oxides in the troposphere. *J. Geophys. Res.*, **95**, 14,027–14,038.

Atkinson, R. (1990). Gas-phase tropospheric chemistry of organic compounds: a review. *Atmos. Environ.*, **24A**, 1–41.

Atkinson, R. and Aschmann, S.M. (1994). Products of the gas-phase reactions of aromatic hydrocarbons: effect of NO_2 concentrations. *Int. J. Chem. Kinet.*, **26**, 929–944.

Atkinson, R., Carter, W.P.L. and Winer, A.M. (1983). Effects of temperature and pressure on alkyl nitrate yields in the NO_x photooxidations of *n*-pentane and *n*-heptane. *J. Phys. Chem.*, **87**, 2012–2018.

Atlas, E.L. (1988). Evidence for $\geqslant C_3$ alkyl nitrates in rural and remote atmospheres. *Nature*, **331**, 426–428.

Atlas, E., Pollock, W., Greenberg, J., Heidt, L. and Thompson, A.M. (1993). Alkyl nitrates, nonmethane hydrocarbons, and halocarbon gases over the equatorial Pacific ocean during SAGA 3. *J. Geophys. Res.*, **98**, 16,933–16,947.

Atlas, E.L., Ridley, B.A., Hübler, G., Walega, J.G., Carroll, M.A., Montzka, D.D., Huebert, B.J., Norton, R.B., Grahek, F.E. and Schauffler, S. (1992a). Partitioning and budget of NO_y species during the Mauna Loa Observatory photochemistry experiment. *J. Geophys. Res.*, **97**, 10,449–10,462.

Atlas, E. and Schauffler, S.M. (1991). Analysis of alkyl nitrates and selected halocarbons in the ambient atmosphere using a charcoal preconcentration technique. *Environ. Sci. Technol.*, **25**, 61–67.

Atlas, E., Schauffler, S.M., Merrill, J.T., Hahn, C.T., Ridley, B., Walega, J., Greenberg, J., Heidt, L. and Zimmerman, P. (1992b). Alkyl nitrate and selected halocarbon measurements at Mauna Loa observatory, Hawaii. *J. Geophys. Res.*, **97**, 10331–10348.

Bertman, S.B., Buhr, M.P. and Roberts, J.M. (1993). Automated cryogenic trapping technique for capillary GC analysis of atmospheric trace compounds requiring no expendable cryogens: application to the measurement of organic nitrates. *Anal. Chem.*, **65**, 2944–2946.

Bertman, S.B. and Roberts, J.M. (1991). A PAN analog from isoprene photooxidation. *Geophys. Res. Lett.*, **18**, 1461–1464.

Blanchard, P., Shepson, P.B., Schiff, H.I. and Drummond, J.W. (1993). Development of a gas chromatograph for trace level measurement of peroxyacetyl nitrate using chemical amplification. *Anal. Chem.*, **65**, 2472–2477.

Bottenhiem, J.W., Barrie, L. and Atlas, E., (1993). The partitioning of nitrogen oxides in the lower Arctic troposphere during spring, 1988. *J. Atmos. Chem.*, **17**, 15–27.

Bottenhiem, J.W. and Gallant, A.J. (1989). PAN over the acrctic: observations during AGASP-2 in April 1986. *J. Atomos. Chem.*, **9**, 301–306.

Bridier, I., Caralp, F., Loirat, H., Lesclaux, R., Veyret, B., Becker, K.H., Reimer, A. and Zabel, F. (1991). Kinetic and theoretical studies of the reactions $CH_3C(O)O_2 + NO_2 + M \rightarrow CH_3C(O)O_2NO_2 + M$ between 248 and 393 K and between 30 and 760 torr. *J. Phys. Chem.*, **95**, 3594–3600.

Buhr, M., Fehsenfeld, F.C., Parrish, D.D., Sievers, R.E. and Roberts J.M. (1990). Contribution of organic nitrates to the total odd-nitrogen budget at a rural, eastern U.S. site. *J. Geophys. Res.*, **95**, 9809–9816.

Burkhardt, M.R., Maniga, N.I., Stedman, D.H. and Paur, R.J. (1988). Gas chromatographic method for measuring nitrogen dioxide and peroxyacetyl nitrate in air without compressed gas cylinders. *Anal. Chem.*, **60**, 816–819.

Cantrell, C.A., Lind, J.A., Shetter, R.E., Calvert, J.G., Goldan, P.D., Kuster, W., Fehsenfeld, F.C., Montzka, S.A., Parrish, D.D., Williams, E.J., Buhr, M.P., Westberg, H.H., Allwine, G. and Martin, R. (1992). Peroxy radicals in the ROSE experiment: measurement and theory. *J. Geophys. Res.*, **97**, 20,671–20,686.

Cantrell, C.A., Shetter, R.E., Calvert, J.G., Tyndall, G.S. and Orlando, J.J. (1993). Measurement of rate coefficients for the unimolecular decomposition of N$_2$O$_5$. *J. Phys. Chem.*, **97**, 9141–9148.

Carter, W.P.L., Winer, A.M. and Pitts, J.N., Jr. (1981). Major atmospheric sink for phenol and cresols. Reaction with the nitrate radical. *Environ. Sci. Technol.*, **15**, 829–831.

Ciccioli, P., Brancaleoni, E., DiPalo, V., Liberti, M. and DiPalo, C. (1986). Misura delle alterazoni della qualita dell'aria da fenomeni di smog fotochimico. *Acqua-Aria*, **7**, 675–683.

Crutzen P.J. (1979). The role of NO and NO$_2$ in the chemistry of the troposphere and stratosphere. *Ann. Rev. Earth Planet. Sci.*, **7**, 443–472.

Darley, E.F., Kettner, K.A. and Stephens, E.R. (1963). Analysis of peroxyacyl nitrates by gas chomatography with electron capture detection. *Anal. Chem.*, **35**, 589–591.

Davidson, E.A. (1991). Fluxes of nitrous oxide and nitric oxide from terrestrial ecosystems. In Rogers, J.E. and Whitman, W.B., Eds, *Microbial Production and Consumption of Greenhouse Gases: Methane, Nitrogen Oxides, and Halomethanes*, American Society of Microbiology, Washington, DC, pp. 219–235.

DeMore, W.B., Sander, S.P., Golden, D.M., Hampson, R.F., Kurylo, M.J., Howard, C.J., Ravishankara, A.R., Kolb, C.E. and Molina, M.J., (1992). *Chemical Kinetics and Photochemical Data for Use in Stratospheric Modeling*, Jet Propulsion Laboratory, NASA, Pasadena, CA.

Dignon, J. and Hameed, S. (1989). Global estimates of nitrogen and sulfur oxides from 1860 to 1980. *J. Air Pollut. Control Assoc.*, **39**, 180–186.

Fahey, D.W., Hübler, G., Parrish, D.D., Williams, E.J., Norton, R.B., Ridley, B.A., Singh, H.B., Liu, S.C. and Fehsenfeld, F.C. (1986). Reactive nitrogen species in the troposphere: measurements of NO, NO$_2$, HNO$_3$, particulate nitrate, peroxyacetyl nitrate (PAN), O$_3$, and total reactive odd nitrogen (NO$_y$) at Niwot Ridge, Colorado. *J. Geophys. Res.*, **91**, 9781–9793.

Fehsenfeld, F.C., Dickerson, R.R., Hübler, G., Luke, W.T., Nunnermaker, L.J., Williams, E.J., Roberts, J.M., Calvert, J.G., Curran, C.M., Delaney, A.C., Eubank, C.S., Fahey, D.W., Fried, A., Gandrud, B.W., Langford, A.O., Murphy, P.C., Norton, R.B., Pickering, K.E. and Ridley, B.A. (1987). A ground-base intercomparison of NO, NO$_x$ and NO$_y$ measurement techniques. *J. Geophys. Res.*, **92**, 14,710–14,722.

Fehsenfeld, F.C., Drummond, J.W., Roychowdhury, U.K., Galvin, P.J., Williams, E.J., Buhr, M.P., Parrish, D.D., Hübler, G., Langford, A.O., Calvert, J.G., Ridley, B.A., Grahek, F., Heikes, B.G., Kok, G.L., Shetter, J.D., Walega, J.G., Elsworth, C.M., Norton, R.B., Fahey, D.W., Murphy, P.C., Hovermale, C., Mohnen, V.A., Demerjian, K.L., Mackay, G.I. and Schiff, H.I. (1990). Intercomparison of NO$_2$ measurement techniques. *J. Geophys. Res.*, **95**, 3579–3597.

Fehsenfeld, F.C., Parrish, D.D. and Fahey, D.W. (1988). The measurement of NO$_x$ in the non-urban troposphere. in Isaksen, I.S.A., Ed., *Tropospheric Ozone*, Reidel, Dordrecht, pp. 185–215.

Finlayson-Pitts, B.J. (1993). Comment on "Indications of photochemical histories of

Pacific air masses from measurements of atmospheric trace species at Point Arena, California" by D.D. Parrish et al. *J. Geophys. Res.*, **98**, 14991–14993.

Finlayson-Pitts, B.J. and Johnson, S.N. (1988). The reaction of NO_2 with NaBr: possible source of $BrNO_3$ in polluted marine atmospheres. *Atmos. Environ.*, **22**, 1107–1112.

Finlayson-Pitts, B.J. and Pitts J.N., Jr. (1986). *Atmospheric Chemistry: Fundamentals and Experimental Techniques*, Wiley, New York, NY.

Flocke, F., Volz-Thomas, A. and Kley, D. (1991). Measurements of alkyl nitrates in rural and polluted air masses. *Atmos. Environ.*, **25A**, 1951–1960.

Fontijn, A., Sabadell, A.J. and Ronco, R.T. (1970). Homogeneous chemiluminescent measurement of nitric oxide with ozone. *Anal. Chem.*, **42**, 575–579.

Gaffney, J.S., Fajer, R. and Senum, G.I. (1984). An improved procedure for high purity gaseous peroxyacyl nitrate production: use of heavy lipid solvents. *Atmos. Environ.*, **18**, 215–218.

Garland, J.A. and Penkett, S.A. (1976). Absorption of peroxyacetyl nitrate and ozone by natural surfaces. *Atmos. Environ.*, **10**, 1127–1131.

Goldenbaum, G.C. and Dickerson, R.R. (1993). Nitric oxide production by lightning discharges. *J. Geophys. Res.*, **98**, 18,333–18,338.

Graham, R.A., Winer, A.M. and Pitts, J.N., Jr. (1978). Pressure and temperature dependence of the unimolecular decomposition of HO_2NO_2. *J. Chem. Phys.*, **68**, 4505–4510.

Gregory, G.L., Hoell, J.M., Jr., Huebert, B.J., Van Bramer, S.E., LeBel, P.J., Vay, S.A., Marinaro, R.M., Schiff, H.I., Hastie, D.R., Mackay, G.I. and Karecki, D.R. (1990). An intercomparison of airborne nitric acid measurements. *J. Geophys. Res.*, **95**, 10,089–10,102.

Grosjean, D. (1983). Distribution of atmospheric nitrogenous pollutants at a Los Angeles area smog receptor site. *Environ. Sci. Technol.*, **17**, 13–19.

Grosjean, D. (1991). Atmospheric fate of toxic aromatic compounds. *Sci. Total Environ.*, **100**, 367–414.

Grosjean, D., Williams II., E.L. and Grosjean, E. (1993a). Peroxyacyl nitrates at southern California mountain forest locations. *Environ. Sci. Technol.*, **27**, 110–121.

Grosjean, D., Williams II, E.L. and Grosjean, E. (1993b). Gas phase reaction of the hydroxyl radical with the unsaturated peroxyacyl nitrate $CH_2 = C(CH_3)C(O)OONO_2$. *Int. J. Chem. Kinet.*, **25**, 921–929.

Hameed, S. and Dignon, J. (1992). Global emissions of nitrogen and sulfur oxides in fossil fuel combustion 1970–1986. *J. Air Pollut. Control Assoc.*, **42**, 159–163.

Harris, G.W., Carter, W.P.L., Winer, A.M., Pitts, J.N., Jr., Platt, U. and Perner, D. (1982). Observations of nitrous acid in the Los Angeles atmosphere and implications for predictions of ozone–precursor relationships. *Environ. Sci. Technol.*, **16**, 414–419.

Harwood, M.H., Jones, R.L., Cox, R.A., Lutman, E. and Rattigan, O.V. (1993). Temperature-dependent absorption cross-sections of N_2O_5. *J. Photochem. Photobiol. A: Chem.*, **73**, 167–175.

Hering, S.V., et al. (1988). The nitric acid shootout: field comparison of measurement methods. *Atmos. Environ.*, **22**, 1519–1539.

Hough, A. (1988). An intercomparison of mechanisms for the production of photo-chemical oxidants. *J. Geophys. Res.*, **93**, 3789–3812.

Jenkins, M.E., Cox, R.A. and Williams, D.J. (1988). Laboratory studies of the kinetics of formation of nitrous acid from the thermal reaction of nitrogen dioxide and water vapor. *Atmos. Environ.*, **22**, 487–489.

Johnston, H.S., Chang, S.-G. and Whitten, G. (1974). Photolysis of nitric acid vapor. *J. Phys. Chem.*, **78**, 1–7.

Jonsson, A., Berg, S. and Bertilsson, B.M. (1979). Methylnitrite in the exhaust from a methanol gasoline fueled automobile. *Chemosphere*, **11/12**, 835–841.

Jonsson, A. and Bertilsson, B.M. (1982). Formation of methyl nitrite in engines fueled with gasoline/methanol and methanaol/diesel. *Environ. Sci. Technol.*, **16**, 106–110.

Junkerman, W. and Ibusuki, T. (1992). FTIR spectroscopic measurements of surface bond products of nitrogen oxides on aerosol surfaces – implications for heterogeneous HNO$_2$ production. *Atmos. Environ.*, **26A**, 3099–3103.

Kames, J. and Schurath, U. (1992). Alkyl nitrates and bifunctional nitrates of atmospheric interest: Henry's law constants and their temperature dependencies. *J. Atmos. Chem.*, **15**, 79–95.

Kames, J., Schweighoefer, S. and Schurath, U. (1991). Henry's law constant and hydrolysis of peroxyacetyl nitrate (PAN). *J. Atmos. Chem.*, **12**, 169–180.

Koppenkastrop, D. and Zabel, F. (1991). Thermal decomposition of chlorofluoromethyl peroxynitrates. *Int J. Chem. Kinet.*, **23**, 1–15.

Kuntasal, G. and Chang, T.Y. (1987). Trends and relationships of O$_3$, NO$_x$, and HC in the south coast air basin of California. *J. Air Pollut. Contr. Assoc.*, **37**, 1158–1163.

Lee, Y.-N. (1984). Kinetics of some aqueous-phase reactions of peroxyacetyl nitrate. *Conf. on Gas–Liquid Chemistry of Natural Waters, Brookhaven National Laboratory, Upton, NY, April 1–4, 1984.*

Lee, Y.-N. (1990). Chemical rate and equilibrium measurements in the aqueous medium. In *NAS Review of DOE/OHER Atmospheric Chemistry Program*, presentation summaries, Washington, DC.

Lee, Y.-N. and Schwartz, S.E. (1981). Reaction kinetics of nitrogen dioxide with liquid water at low partial pressure. *J. Phys. Chem.*, **85**, 840–848.

Leighton, P.A. (1961). *Photochemistry of Air Pollution*, Academic Press, New York, NY.

Lin, X., Trainer, M. and Liu, S.C. (1988). On the nonlinearity of the tropospheric ozone production. *J. Geophys. Res.*, **93**, 15,879–15,888.

Liu, S.C., Trainer, M., Carroll, M.A., Hubler, G., Montzka, D.D., Norton, R.B., Ridley, B.A., Walega, J.G., Atlas, E.L., Heikes, B.G., Huebert, B.J. and Warren, W. (1992). A study of the photochemistry and ozone budget during the Mauna Loa observatory photochemistry experiment. *J. Geophys. Res.*, **97**, 10,463–10,471.

Lobert, J.M., Scharffe, D.H., Hao, W.M. and Crutzen, P.J. (1990). Importance of biomass burning in the atmospheric budgets of nitrogen-containing gases. *Nature*, **346**, 552–554.

Logan, J.A. (1983). Nitrogen oxides in the troposphere: global and regional budgets. *J. Geophys. Res.*, **88**, 10,785–10,807.

Luke, W.T., Dickerson, R.R. and Nunnermaker, L.J. (1989). Direct measurements of the photolysis rates coefficients and Henry's law constants of several alkyl nitrates. *J. Geophys. Res.*, **94**, 14,905–14,921.

Martinez, R.I. (1980). A systematic nomenclature for the "peroxyacyl nitrates", the functional and structural misnomers for anhydride derivatives of nitrogen oxo acids. *Int.*

J. Chem. Kinet., **12**, 771–775.

Meijer, G.M. and Nieboer, H. (1977). Determination of peroxy benzoyl nitrate in ambient air. *VDI Berichte*, **270**, 55–56.

Mihelcic, D., Müsgen, P. and Ehhalt, D.H. (1985). An improved method of measuring tropospheric NO_2 and RO_2 by matrix isolation and electron spin resonance. *J. Atmos. Chem.*, **3**, 341–361.

Muthuramu, K., Shepson, P.B. and O'Brien, J.M. (1993). Preparation, analysis, and atmospheric production of multifunctional organic nitrates. *Environ. Sci. Technol.*, **27**, 1117–1124.

Nielsen, T., Hansen, A.M. and Thomsen, E.L. (1982). A convenient method for the preparation of pure standards of peroxyacetyl nitrate for atmospheric analysis. *Atmos. Environ.*, **16**, 2447–2450.

Noxon, J.F. (1983). NO_3 and NO_2 in the mid-Pacific troposphere. *J. Geophys. Res.*, **88**, 11,017–11,021.

Noxon, J.F., Norton, R.B. and Henderson, W.R. (1978). Observation of atmospheric NO_3. *Geophys. Res. Lett.*, **5**, 675–678.

Orlando, J.J., Tyndall, G.S. and Calvert, J.G. (1992). Thermal decomposition pathways for peroxyacetyl nitrate (PAN): implications for atmospheric methyl nitrate. *Atmos. Environ.*, **26A**, 3111–3118.

Park, J.-Y. and Lee, Y.-N. (1987). Aqueous solubility and hydrolysis kinetics of peroxynitric acid. *193rd ACS National Meeting, Denver, CO, April 5–10, 1987*, Phys. Chem. Sec. Paper No. 67A.

Park, J.-Y. and Lee, Y.-N. (1988). Solubility and decomposition kinetics of nitrous acid in aqueous solution. *J. Phys. Chem.*, **92**, 6294–6302.

Parrish, D.D., Buhr, M.P., Trainer, M., Norton, R.B., Shimshock, J.P., Fehsenfeld, F.C., Anlauf, K.G., Bottenheim, J.W., Tang, Y.Z., Wiebe, H.A., Roberts, J.M., Tanner, R.L., Newman, L., Bowersox., V.C., Olszyna, K.J., Bailey, E.M., Rodgers, M.O., Wang, T., Berresheim, H., Roychowdhury, U.K. and Demerjian, K. (1993). The total reactive oxidized nitrogen levels and partitioning between the individual species at six rural sites in Eastern North America. *J. Geophys. Res.*, **98**, 2927–2939.

Parrish, D.D., Trainer, M., Williams, E.J., Fahey, D.W., Hübler, G., Eubank, C.S., Liu, S.C., Murphy, P.C., Albritton, D.L. and Fehsenfeld F.C. (1986). Measurements of the NO_x-O_3 photostationary state at Niwot Ridge, Colorado. *J. Geophys. Res.*, **91**, 5361–5370.

Penkett, S.A. and Brice, K.A. (1986). The spring maximum in photo-oxidants in the Northern Hemisphere troposphere. *Nature*, **319**, 655–657.

Pitts, J.N., Jr., Grosjean, D., Van Cauwenberge, K., Schmid, J.P. and Fritz, D.R. (1978). Photooxidation of aliphatic amines under simulated atmospheric conditions: formation of nitrosamines, nitramines, amides, and photochemical oxidant. *Environ. Sci. Technol.*, **12**, 946–953.

Platt, U., Perner, D., Winer, A.M., Harris, G.W. and Pitts, J.N., Jr. (1980). Detection of NO_3 in the polluted troposphere by differential optical absorption. *Geophys. Res. Lett.*, **7**, 89–92.

Rattigan, O., Lutman, E., Jones, R.L., Cox, R.A., Clemitshaw, K. and Williams, J. (1992). Temperature-dependent absorption cross-sections of gaseous nitric acid and methyl nitrate. *J. Photochem. Photobiol. A: Chem.*, **66**, 313–326

Ridley, B.A. (1991). Recent measurements of oxidized nitrogen compounds in the troposphere. *Atmos. Environ.*, **25A**, 1905–1926.

Ridley, B.A., Shetter, J.D., Walega, J.G., Madronich, S., Elsworth, C.M., Grahek, F.E., Fehsenfeld, F.C., Norton, R.B., Parrish, D.D., Hübler, G.H., Buhr, M., Williams, E.J., Allwine, E.J. and Westberg, H.H. (1990).The behavior of some organic nitrates at Boulder and Niwot Ridge, Colorado. *J. Geophys. Res.*, **95**, 13,949–13,961.

Roberts, J.M. (1990).The atmospheric chemistry of organic nitrates. *Atmos. Environ.*, **24A**, 243–287.

Roberts, J.M. and Bertman, S.B. (1992). The thermal decomposition of peroxyacetic nitric anhydride (PAN) and peroxymethacrylic nitric anhydride (MPAN). *Int. J. Chem. Kinet.*, **24**, 297–307.

Roberts, J.M. and Fajer, R.W. (1989). UV absorption cross sections of organic nitrates of potential atmospheric importance and estimation of atmospheric lifetimes. *Environ. Sci. Technol.*, **23**, 945–951.

Roberts, J.M., Fajer, R.W. and Springston, S.R. (1989). The capillary gas chromatographic separation of alkyl nitrates and peroxycarboxylic nitric anhydrides. *Anal. Chem.*, **61**, 771–772.

Schwartz, S.E. (1989). Acid deposition: unraveling a regional phenomenon. *Science*, **243**, 753–763.

Senum, G.I., Fajer, R. and Gaffney, J.S. (1986). Fourier transform infrared spectroscopic study of the thermal stability of peroxyacetyl nitrate. *J. Phys. Chem.*, **90**, 152–156.

Senum, G.I., Lee, Y.-N. and Gaffney, J.S. (1984). The ultraviolet absorption spectrum of peroxyacetyl nitrate. *J. Phys. Chem.*, **88**, 1269–1271.

Shepson, P.B., Anlauf, K.G., Bottenhiem, J.W., Wiebe, H.A., Gao, N., Muthuramu, K. and Mackay, G.I. (1993). Alkyl nitrates and thier contribution to reactive nitrogen at a rural site in Ontario. *Atmos. Environ.*, **27A**, 749–757.

Shepson, P.B., Bottenheim, J.W., Hastie, D.R. and Venkatram, A. (1992). Determination of the relative ozone and PAN deposition velocities at night. *Geophys. Res. Lett.*, **19**, 1121–1124.

Singh, H.B. (1987). Reactive nitrogen in the troposphere. *Environ. Sci. Technol.*, **21**, 320–327.

Singh, H.B. and Hanst, P.L. (1981). Peroxyacetyl nitrate (PAN) in the unpolluted atmosphere: an important reservoir for nitrogen oxides. *Geophys Res. Lett.*, **8**, 941–944.

Singh, H.B., O'Hara, D., Herlth, D., Bradshaw, J.D., Sandholm, S.T., Gregory, G.L., Sachse, G.W., Blake, D.R., Crutzen, P.J. and Kanakidou, M.A. (1992a). Atmospheric measurements of peroxyacetyl nitrate and other organic nitrates at high latitudes: possible sources and sinks. *J. Geophys. Res.*, **97**, 16,511–16,522.

Singh, H.B., Herlth, D., O'Hara, D., Zahnle, K., Bradshaw, J.D., Sandholm, S.T., Talbot, R., Crutzen, P.J. and Kanakidou, M.A. (1992b). Relationship of peroxyacetyl nitrate to active and total odd-nitrogen at northern high latitudes: influence of reservoir species on NO$_x$ and O$_3$, *J. Geophys. Res.*, **97**, 16523–16530.

Singh, H.B., Herlth, D., Kolyer, R., Sala, L., Bradshaw, J.D., Sandholm, S.T., Davis, D.D., Kondo, Y., Koike, M., Talbot, R., Gregory, G.L., Sachse, G.W., Browell, E., et al. (1994a). Reactive nitrogen and ozone over the western Pacific: distribution, partitioning and sources based on the 1991 PEM-west (A) expedition. *J. Geophys. Res.*, in press.

Singh, H.B., Herlth, D., O'Hara, D., Zahnle, K., Bradshaw, J.D., Sandholm, S.T., Talbot, R., Gregory, G.L., Sachse, G.W., Blake, D.R. and Wofsy, S.C. (1994b). Summertime distribution of PAN and other reactive nitrogen species in the northern high latitude atmosphere of eastern Canada. *J. Geophys. Res.*, **99**, 1821–1835.

Singh, H.B. and Kasting, J.F. (1988). Chlorine–hydrocarbon photochemistry in the marine troposphere and lower stratosphere. *J. Atmos. Chem.*, **7**, 261–285.

Singh, H.B. and Salas, L.J. (1983). Methodology for the analysis of peroxyacetyl nitrate (PAN) in the unpolluted atmosphere. *Atmos. Environ.*, **17**, 1507–1516.

Singh, H.B. and Salas, L.J. (1989). Measurements of PAN and PPN at selected urban, suburban and rural sites. *Atmos. Environ.*, **23**, 231–239.

Solomon, S., Miller, H.L., Smith, J.P., Sanders, R.W., Mount, G.H. and Schmeltekopf, A.L. (1989). Atmospheric NO_3. 1. Measurement technique and the annual cycle at 40° north. *J. Geophys. Res.*, **94**, 11,041–11,048.

Spicer, C.W., Howes, J.E., Jr., Bishop, T.A., Arnold, L.H. and Stevens, R.K. (1982). Nitric acid measurement methods: an intercomparison. *Atmos. Environ.*, **16**, 1487–1500.

Stedman, D.H. and Shetter, R.E. (1983). The global budget of atmospheric nitrogen species. In Schwartz, S.E., Ed. *Trace Atmospheric Constituents*, Wiley, New York, NY. pp. 411–454.

Stelson, A.W. and Seinfeld, J.H. (1982a). Relative humidity and temperature dependence of the ammonium nitrate dissociation constant. *Atmos. Environ.*, **16**, 983–992.

Stelson, A.W. and Seinfeld, J.H. (1982b). Relative humidity and pH dependence of the vapor pressure of ammonium nitrate-nitric acid solutions at 25 °C. *Atmos. Environ.*, **16**, 993–1000.

Stephens, E.R. (1969). The formation, reactions, and properties of peroxyacyl nitrates (PANs) in photochemical air pollution. *Adv. Environ. Sci.*, **1**, 119–146.

Stephens, E.R. (1987). Smog studies of the 1950's. *EOS Trans.*, **68**, 89–92.

Talukdar, R., Burkholder, J.B., Schmoltner, A.-M., Roberts, J.M. and Ravishankara, A.R. (1995). The atmospheric fate of PAN : UV absorption cross section and rate coefficient for reaction with OH, *J. Geophys. Res.*, in press.

Tanner, R.L., Kelly, T.J., Dezaro, D.A. and Forrest, J. (1989). A comparison of filter, denuder, and real-time chemiluminescence techniques for nitric acid determination in ambient air. *Atmos. Environ.*, **23**, 2213–2222.

Trainer, M., Parrish, D.D., Buhr, M.P., Norton, R.B., Fehsenfeld, F.C., Anlauf, K.G., Bottenheim, J.W., Tang, Y.Z., Wiebe, H.A., Roberts, J.M., Tanner, R.L., Newman, L., Bowersox., V.C., Meagher, J.F., Olszyna, K.J., Rodgers, M.O., Wang, T., Berresheim, H. and Demerjian, K. (1993). Correlation of ozone with NO_y in photochemically aged air. *J. Geophys. Res.*, **98**, 2917–2925.

Tuazon, E.C., Carter, W.P.L., Atkinson, R., Winer, A.M. and Pitts, J.N., Jr. (1984). Atmospheric reactions of N-nitrosodimethylamine and dimethylnitramine. *Environ. Sci. Technol.*, **18**, 49–54.

Walega, J.G., Ridley, B.A., Madronich, S., Grahek, F.E., Shetter, J.D., Sauvain, T.D., Hahn, C.J., Merrill, J.T., Bodhaine, B.A. and Robinson, E. (1992). Observations of peroxyacetyl nitrate, peroxypropionyl nitrate, methyl nitrate and ozone during the Mauna Loa Observatory Photochemistry Experiment. *J. Geophys. Res.*, **97**, 10,311–10,330.

Wayne, R.P., Barnes, I., Biggs, P., Burrows, J.P., Canosa-Mas, C.E., Hjorth, J., LeBras, G.,

Moortgat, G.K., Perner, D., Poulet, G., Restelli, G. and Sidebottom, H. (1991). The nitrate radical: physics, chemistry and the atmosphere. *Atmos. Environ.*, **25A**, 1–203.

Williams, E.J., Hutchinson, G.L. and Fehsenfeld, F.C., (1992). NO$_x$ and N$_2$O emissions from soil. *Global Biogeochem. Cycles*, **6**, 351–388.

Williams II, E.L., Grosjean, E. and Grosjean, D. (1993). Ambient levels of the peroxyacyl nitrates PAN, PPN, and MPAN, in Atlanta, Georgia. *J. Air Waste Manag. Assoc.*, **43**, 873–879.

Winer, A.M., Atkinson, R. and Pitts, J.N., Jr. (1984). Gaseous nitrate radical: possible night-time atmospheric sink for biogenic organic compounds. *Science*, **224**, 156–159.

Zabel, F., Reimer, A., Becker, K.H. and Fink, E.H. (1989). Thermal decomposition of alkyl peroxynitrates. *J. Phys. Chem.*, **93**, 5500–5507.

Zafiriou, O.C. and McFarland, M. (1981). Nitric oxide from nitrite photolysis from the central equatorial Pacific. *J. Geophys. Res.*, **86**, 3173–3182.

7

Halogens in the atmospheric environment

Hanwant B. Singh

7.1. INTRODUCTION AND OVERVIEW

Halogenated chemicals of man-made and natural origin are present as gases and aerosols in the Earth's atmosphere. Interest in the study of inorganic halogens began with attempts to understand the geochemical cycling of sea salt and picked up considerably in the early 1950s when it was recognized that sea salt aerosols could release large amounts of gaseous inorganic halogen species (such as hydrogen chloride) through processes involving dehydration and acidification (Cauer, 1949; Junge, 1957; Eriksson, 1959; Robbins et al., 1959; Cicerone, 1981). Estimates are that some 100–600 Tgyear^{-1} of HCl is released to the atmosphere from this marine source. Synthetic organic halocarbons have been known since the 19th century and some of the early refrigeration systems used methyl chloride (CH_3Cl) and dichloromethane (CH_2Cl_2) as cooling fluids. Chlorofluorocarbons (CFCs) were first synthesized in the late 1890s and their use as ideal refrigerants was recognized in the 1930s. Since then halocarbons as a group have found widespread use in commerce and industry as propellants, refrigerants, and solvents. Because many of these chemicals were thought to be non-toxic and otherwise environmentally benign, usages were such that large fractions were directly released to the environment. Global emissions of all man-made halocarbons grew from virtually negligible quantities in 1940 to about 2.5 Tg year^{-1} in 1990.

The development of the electron capture detector in the 1960s permitted exquisite measurement sensitivity for electron-absorbing halogens. Lovelock (1971) first detected SF_6 and $CFCl_3$ in the atmosphere and, recognizing their inertness, suggested their use as tracers of atmospheric motions. Intensive studies of halogens received a significant boost when Molina and Rowland (1974) proposed that long lived CFCs could decompose in the stratosphere to release Cl atoms which could catalytically deplete the protective ozone layer (see Chapter 11 for more details). It was also recognized that many of the halocarbons were efficient absorbers of infrared radiation and potentially important greenhouse gases (Ramanathan et al., 1985). These concerns, aided by the discovery of large ozone depletion in the stratosphere over the Antarctic continent (Farman et al., 1985), culminated in international agreements (Montreal Protocol) to achieve significant reductions in the future emissions of ozone-destroying chemicals. Indeed new compounds that are either substantially destroyed in the troposphere or do not contain ozone-destroying halogen atoms (e.g. Cl or Br) are being developed as substitutes for CFCs. Toxicity studies over the last two decades have also shown that many halogenated hydrocarbons are mutagens and potential carcinogens. Public health concern resulting from the high concentrations of trihalomethanes in drinking-water of many communities led to the enactment of the Safe Drinking Water Act of 1974, which was further strengthened in a 1986 amendment.

In addition to their environmental significance, this group of chemicals has played a key role in advancing our knowledge of atmospheric chemistry and dynamics. Reaction with hydroxyl radical (OH) is known to provide a principal pathway for the degradation of many organic and inorganic species in the atmosphere. Recognition that some man-made halocarbons were almost exclusively removed by reaction with OH radicals led to the development of methodologies that provided a reliable means of estimating OH radical abundances and validating global photochemical models (Singh, 1977a; Spivakovsky et al., 1990; Prinn et al., 1992). Because of their inertness and the exquisite sensitivity with which they can be measured, select halocarbons and SF_6 have found widespread use as tracers of atmospheric motions and in studies of transport and dispersion (Dabberdt and Dietz, 1986; Prather et al., 1987).

Halogens form a unique group of atmospheric chemicals with a wide variety of man-made and natural sources as well as rather distinct removal mechanisms. They are produced via biological processes in the oceans, from particle to gas conversion in the atmosphere, from biomass burning, and from exclusive man-made synthesis. Some are removed almost exclusively in the troposphere, others largely in the stratosphere, and yet others have no known removal mechanisms and are practically inert. Their atmospheric lifetimes span from a few days to many centuries. Organic halogens are also molecules that can be measured with great sensitivity ($\approx 10^{-14}$ v/v) and specificity. Lack of sensitive measurement techniques is a major impediment to our understanding of the inorganic halogen cycle. Collectively halogens are molecules of both great environmental concern and great utility in understanding the chemistry and dynamics of the atmosphere.

7.2. HALOCARBONS, HYDROXYL RADICALS, AND ATMOSPHERIC LIFETIMES

7.2.1. Halocarbons of Atmospheric Interest

Table 7.1 provides a list of halogenated organics (and SF_6) that have been measured in the global atmosphere. Most of these are synthetic organic chemicals that are produced for commercial and industrial applications. Salient among these is the group known as chlorofluorocarbons (CFCs), used as refrigerants (CFC-12, HCFC-22), blowing agents (CFC-11, HCFC-22), and cleaning agents (CFC-113). Halons are similar bromine-containing chemicals which find wide use as fire extinguishers. During the last two decades the use patterns of CFCs have changed greatly (Figure 7.1), largely in response to regulatory requirements designed to protect the ozone layer. The use of the entire CFC production in propellant applications declined from 69% in 1974 to 18% in 1991. Methyl chloroform (CH_3CCl_3), CH_2Cl_2, C_2Cl_4, and C_2HCl_3 are largely used as degreasers and as dry-cleaning and industrial solvents. Carbon tetrachloride, $CHCl_3$, and CH_2ClCH_2Cl are mainly intermediates used in the manufacture of chemicals such as CFC-11 and CFC-12, HCFC-22, and vinyl chloride monomer, respectively. Methyl chloride is manufactured in relatively small quantities and used almost exclusively for the production of silicones and tetramethyl lead intermediates. Methyl bromide is a commonly used agricultural and space fumigant. All monomethyl halides and many other brominated and iodated species listed in Table 7.1 are produced

TABLE 7.1 Year 1990 Atmospheric Concentrations, Trends, and Lifetimes of Halogenated Organic Species and SF₆

Compound	1990 Global Av. Conc. (ppt)	1990 Trend (ppt year⁻¹)	North/South Ratio	Lifetime (years)	Sources	Sinks
CF_4(CFC-4)	75	1.5		> 50000	MM	Mes./Lyman α
C_2F_6(CFC-116)	5			> 10000	MM	Mes./Lyman α
SF_6	2	0.2	1.05	3000	MM	Electrons/mes.
C_2ClF_5(CFC-115)	5			17000	MM	Strat. O(^{1}D)/mes.
$C_2Cl_2F_4$(CFC-114)	15	1		300	MM	Strat. hv/O(^{1}D)
$C_2Cl_3F_3$(CFC-113)	70	6	1.14	90 ± 10	MM	Strat. hv
CF_3Cl(CFC-13)	3			640	MM	Strat. O(^{1}D)
CCl_2F_2(CFC-12)	480	17	1.03	110 ± 20	MM	Strat. hv/O(^{1}D)
$CHClF_2$(HCFC-22)	90	5	1.15	14 ± 2	MM	Trop. OH
CCl_3F(CFC-11)	260	7	1.05	50 ± 10	MM	Strat. hv/O(^{1}D)
CH_3Cl	610	NDT	1.0	1.4	N(O), BB	Trop. OH
CH_2Cl_2	30	NDT		0.4	MM	Trop. OH
$CHCl_3$	15	5.5		0.6	MM, N(O)	Trop. OH
CH_3CCl_3	155	1.5	1.42	6 ± 1	MM	Trop. OH
CCl_4	110		1.06	45 ± 10	MM	Strat. hv/O(^{1}D)
CH_2ClCH_2Cl	25	NDT		0.4	MM	Trop. OH
C_2Cl_4	12	0.4	4	0.4	MM	Trop. OH
$CBrF_3$ (Halon-1301)	2	0.2	1.05	65 ± 10	MM	Strat. hv/O(^{1}D)
$CBrClF_2$ (Halon-1211)	2			21 ± 5	MM	Trop. hv/Strat
$C_2Br_2F_4$ (Halon-2402)	1.5			26 ± 5	MM	Trop. hv/strat.
CH_3Br	12	0.2	1.20	1.7	N(O), MM, BB	Trop. OH
CH_2Br_2	2			0.2	N(O)	Trop. OH + hv
$CHBr_3$	1–2			0.03–0.1	N(O)	Trop. hv
$CHBrCl_2$ + $CHBr_2Cl$	1			0.4	N(O)	Trop. OH + hv
CH_2BrCH_2Br	1		1.0	0.4	MM	Trop. OH
CH_3I	1			0.02	N(O)	Trop. hv

ΣF = 1800 ppt (100% man-made)
ΣCl = 3800 ppt (84% man-made)
ΣBr = 30 ppt (45% man-made)
ΣI = 1 ppt (0% man-made)

Lifetime estimates are based on global turnover rates and include all known removal processes.
NDT, no detectable trend; MM, man-made; N(O), natural (oceanic); BB, biomass burning; hv, photolytic decomposition; mes., mesosphere; strat., stratosphere; trop., troposphere. OH, O(^{1}D), and hv imply that loss by reaction via these processes is important.

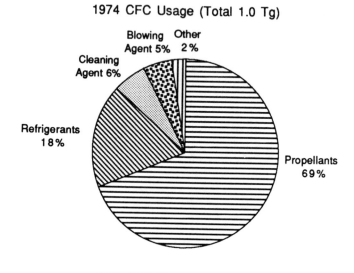

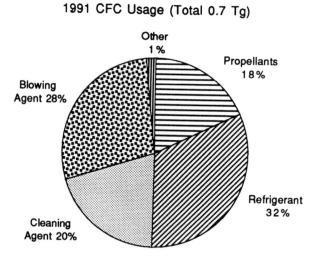

FIGURE 7.1. Changes in the use pattern of CFCs over the last two decades.

biologically in nature. Methyl chloride and CH_3Br have also been identified as products of biomass combustion. Phosgene ($COCl_2$) is the only oxygenated halocarbon that has been identified in the troposphere (Singh, 1976), but it is likely that other such chemicals are present. Both $COCl_2$ and COF_2 have also been identified in the stratosphere (Wilson et al., 1988; Rinsland et al., 1986). The composition of the polluted urban atmosphere is even more complex and many additional chlorinated ethanes, propanes, and aromatics have already been identified (Shah and Singh, 1988).

In order to quantitatively understand the behavior of chemicals in the atmosphere and to develop strategies for control, global emission inventories for a number of CFCs, halons, and other halocarbons have been compiled (Midgley,

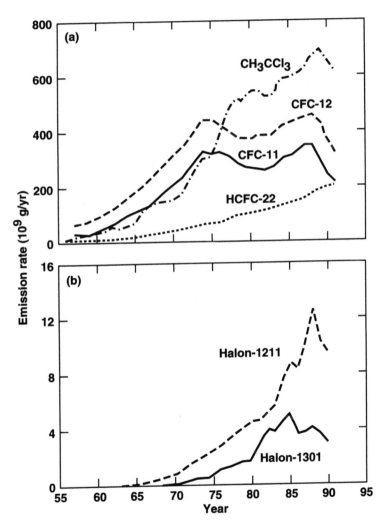

FIGURE 7.2.   Global annual emissions of selected halocarbons. Nearly 95% of these emissions occur in the northern hemisphere.

1989; McCulloch, 1992; Midgley and Fisher, 1993; NASA, 1994). Figure 7.2 shows the global annual release rates of a select group of halocarbons (CFC-11, CFC-12, HCFC-22, and CH_3CCl_3) and halons. For purposes of comparison, it is noted that global release rates of C_2Cl_4 and CH_2Cl_2 for the year 1990 are 350 Gg and 555 Gg, respectively. About 95% of the emissions take place in the northern hemisphere (NH) and about 5% in the southern hemisphere (SH). The decline in emissions after 1987 can largely be attributed to the enactment of the Montreal Protocol designed to scale back the production and use of ozone-destroying chemicals.

7.2.2. Hydroxyl Radicals and Atmospheric Lifetimes

Hydroxyl radical (OH) is produced in the atmosphere by reactions involving ozone, sunlight, and water vapor: $O(^1D) + H_2O \rightarrow 2\ OH$. Reaction with OH is the principal means by which a number of halogenated species (organic and inorganic) are destroyed in the atmosphere. The rate constants of a large number of atmospheric chemicals with OH have been measured over the last two decades and selective compilations are available (Demore et al., 1992). In principle, if the OH field and the rate constant were known, then the lifetime (τ) of a species could be easily calculated: $\tau = 1/k_{OH}[OH]$. In reality, determination of reliable OH concentrations has been possible only through the calculation of τ from the budgets and trends of select halocarbon species. A first demonstration of this technique was presented by Singh (1977a,b) and Lovelock (1977), who selected methyl chloroform (CH_3CCl_3) as the most suitable OH tracer for the following reasons.

- Its principal recognized removal process was due to reaction with OH ($CH_3CCl_3 + OH \rightarrow CH_3CCl_2 + H_2O$) and the OH rate constant was well known.
- Its atmospheric abundance was substantial and could be easily measured in both hemispheres.
- No natural sources existed and reliable annual industrial production/ emission data were available.

It was proposed that a comparison of the amount emitted to the atmosphere and that actually present would provide a direct measure of its removal rate and an indirect measure of the OH abundance. Analysis of CH_3CCl_3 data to estimate OH has been used extensively, but other molecules such as $CHClF_2$ (HCFC-22) can also be used with equal effectiveness. In most cases box models of various kinds have been used to estimate these lifetimes. For the purposes of illustration we describe a simple two-box model that can provide an analytical solution to the atmospheric lifetime (τ) for an exponentially growing emission scenario (Singh, 1977b):

$$dC_n/dt = F_n - (C_n - C_s)/\tau_e - C_n/\tau \tag{7.1}$$

$$dC_s/dt = F_s - (C_n - C_s)/\tau_e - C_s/\tau \tag{7.2}$$

$$F_n = (1 - f) \, m \, \exp(pt)$$

$$F_s = f \, m \, \exp(pt)$$

where C_n and C_s are mean (annual averages) atmospheric loadings and F_n and F_s are annual emissions in the NH and SH, respectively, f is the fraction of the total man-made source emitted in the SH, p is a measure of temporal growth rate, and τ_e is the mean interhemispheric exchange rate. A steady state solution to the above equations leads to:

$$\tau \equiv \tau_c = \gamma/[p(1 - \gamma)] \tag{7.3}$$

$$\tau \equiv \tau_g = \tau_e(1 - f)/(C_n/C_s - p\tau_e - 1) \tag{7.4}$$

where γ is the ratio of the total atmospheric content at time t to the cumulative amount emitted up to that time. Equations (7.3) and (7.4) thus provide two independent ways of calculating atmospheric residence time based on the global content (τ_c) and the north/south gradient (τ_g) method. Equation (7.4) can be rearranged to calculate τ_e from the distribution of nearly inert species such as CFC-11 or CFC-12 ($\tau > 40$ years) and a mean τ_e of 1.1 years has been calculated. For the case of CH_3CCl_3 and HCFC-22, emissions (Figure 7.2) fit an exponential function up to the years 1979 and 1991 respectively ($R^2 = 0.995$). Table 7.2 presents the input parameters to equations (7.3) and (7.4) and the calculated lifetimes for both and HCFC-22 and CH_3CCl_3. The atmospheric data for

TABLE 7.2   Estimation of CH_3CCl_3 and HCFC-22 Atmospheric Lifetimes Using a Two-Box Model

Chemicals	τ_e (years)	f	p	γ	C_n/C_s	Residence Times (years)	
						τ_c [a]	τ_g [b]
CH_3CCl_3 [c]	1.1	0.04	0.1591 (1955–1979)	0.45 (1979)	1.40 (1979)	5.1	4.7
HCFC-22 [d]	1.1	0.05	0.0747 (1955–1990)	0.49 (1990)	1.15 (1990)	12.9	15.4

[a] Global content method, equation (7.3).
[b] Gradient method, equation (7.4).
[c] CH_3CCl_3 emission up to 1979: $Gg \, year^{-1} = 0.00236 \exp[0.1591 (year - 1900)]$; $R^2 = 0.995$.
[d] HCFC-22 emission up to 1991: $Gg \, year^{-1} = 0.24452 \exp[0.0747 (year - 1900)]$; $R^2 = 0.995$.

CH_3CCl_3 are those collected during ALE/GAGE (Prinn et al., 1992), while those of HCFC-22 are taken from Montzka et al. (1993). Lifetimes of about 5 years for CH_3CCl_3 and 14 years for HCFC-22 are calculated (Table 7.2).

The above two-box model is described because of its simplicity. More sophisticated multibox models have also been used to fit the atmospheric data with those of emissions and assumed lifetimes. The most recent of these approaches is due to Prinn et al. (1992), who have used a 12-box time-varying computer model with corrected CH_3CCl_3 ALE/GAGE data over a 12 year period. Based on the analysis of data by the global content method, the north/south gradient method, and also the trend method, they calculate a best value of 5.7 ± 0.7 years.

Small amounts of CH_3CCl_3 are lost to the stratosphere ($\tau_{strat} \approx 45$ years) and by hydrolysis ($\tau_{ocean} \approx 75$ years) in the oceans (WMO, 1989, 1991; Butler et al., 1991). The tropospheric lifetime of CH_3CCl_3 due to OH can be calculated as follows:

$$1/\tau = 1/\tau_{OH} + 1/\tau_{strat} + 1/\tau_{ocean} \qquad (7.5)$$

Given the most recent value of k_{OH} (Table 7.3), a global mean τ_{OH} of 7 ± 1 years and an OH concentration of $(8 \pm 2) \times 10^5$ molecules cm^{-3} provide the best fit. The OH abundance is high in the tropics, with a mean value of $(11 \pm 2) \times 10^5$ molecules cm^{-3} (0–30° N and 0–30° S), and $(6 \pm 1) \times 10^5$ molecules cm^{-3} at mid-latitudes (30–90° N and 30–90° S). Prinn et al. (1992) also infer from the ALE/GAGE data set that OH levels over the period 1978–1990 may have

TABLE 7.3   CFC Replacements and Their Ozone Depletion Potentials (ODPs)

Compound	$k_{OH} \times 10^{12}$ $(cm^3 \, molecule^{-1} \, s^{-1})$[a]	Estimated Mean Lifetime (years)[b]	ODP
$CFCl_3$ (CFC-11)		50	1.00
CHF_2Cl (HCFC-22)	$1.2 \exp(-1650/T)$	14.6	0.05
CF_3CHCl_2 (HCFC-123)	$0.77 \exp(-850/T)$	1.5	0.02
CF_3CHFCl (HCFC-124)	$0.66 \exp(-1250/T)$	6.3	0.02
CF_3CH_2F(HFC-134a)	$1.7 \exp(-1750/T)$	14.8	0.00
CH_3CFCl_2 (HCFC-141a)	$1.3 \exp(-1600/T)$	11.3	0.10
CH_3CF_2Cl (HCFC-142b)	$1.4 \exp(-1800/T)$	21.6	0.06
$CF_3CF_2CHCl_2$ (HCFC-225ca)	$1.5 \exp(-1250/T)$	2.8	0.03
CF_2ClCF_2CHClF (HCFC-225cb)	$0.55 \exp(-1250/T)$	7.6	0.03
CH_3CCl_3	$1.75 \exp(-1550/T)$	5.7	0.10

[a]Demore et al. (1992).
[b]Normalized to CH_3CCl_3 OH lifetime of 7.0 years. Oceanic loss for all HCFCs is expected to be negligible. Stratospheric loss is calculated to be about 50 years for CFC-11, CH_3CCl_3, and HCFC-123. For all other HCFCs it is 100–200 years and does not compete favorably with tropospheric removal (WMO, 1989).

increased at a rate of $(1.0 \pm 0.8)\%$ year^{-1}. This positive trend may be explained by the increased emissions of NO$_x$ and the associated increases in tropospheric ozone. The biggest uncertainty in these calculations comes from atmospheric measurements, which may still be uncertain to $\pm 20\%$, and to a perhaps lesser degree from inaccurate knowledge of emissions.

The best use of the CH$_3$CCl$_3$ (or HCFC-22) mean lifetime estimate is to validate the predicted OH fields by global models (e.g., Spivakovsky et al., 1990) and to estimate lifetimes of other organic chemicals through a process of normalization assuming that their rate constants with OH are accurately known. Only a small error is incurred when lifetimes are estimated by normalizing to a mean temperature of 277 K (Prather and Spivakovsky, 1990). With these assumptions, the lifetime of an unknown chemical X with an OH rate constant of k_{OH-X} can be easily calculated:

$$\tau_X \text{ (years)} = \tau_{OH} \left(k_{OH-CH_3CCl_3}/k_{OH-X} \right)_{277} = 4.55 \times 10^{-14}/(k_{OH-X})_{277} \quad (7.6)$$

The above atmospheric lifetime is only due to removal by OH, but additional removal due to photolysis in the stratosphere or hydrolysis in the oceans may occur. These processes typically occur on time scales that are longer than 40 years. For chemicals where a hydrogen atom is available for abstraction, these losses are quite small in comparison with removal by OH.

Other anthropogenic trace gases such as dichloromethane, 1,2 dichloroethane and tetrachloroethene may also be used to estimate OH values, but accurate atmospheric concentration and emission data are not currently available. In the future it may be possible to add controlled quantities of tailor-made species to act as precise tracers of OH. Unfortunately, detection limits need to improve by three or more orders of magnitude before controlled tracer experiments on the global scale are feasible. Indirect estimation of OH will be greatly facilitated in the future if the emissions of CFC replacements, which are specially designed to be removed by reaction with OH, are accurately monitored.

7.2.3. Chlorofluorocarbon (CFC) Replacements: Hydrochlorofluorocarbons

Phase-out of the CFCs proposed under the Montreal Protocol has led to the development of new compounds that are expected to be more benign to the environment (Manzer, 1990). These new compounds, in addition to being acceptable and non-toxic substitutes for CFC applications, are selected so that they are more easily removed from the troposphere and contain fewer Cl atoms. The OH radical is able to abstract hydrogen atoms from a variety of organic molecules and the inclusion of one or more hydrogen atoms is thus an effective means by which tropospheric removal can be accomplished. These chemicals are commonly called hydrofluorocarbons (HFCs) or hydrochlorofluorocarbons (HCFCs). In this regard it is often convenient to consider an arbitrary scale called the ozone depletion potential (ODP), which is a measure of the ability of a chemical to deplete stratospheric ozone relative to that of CFC-11 on an equivalent weight basis:

$$\text{ODP of X} = \frac{\text{global } \Delta O_3 \text{ due to X}}{\text{global } \Delta O_3 \text{ due to CFC-11}}$$

The quantity ODP depends both on the ability of the molecule to enter the stratosphere (dictated by its rate of tropospheric removal) and on the number of Cl (or Br) atoms it may release in the stratosphere. Chemicals with an ODP of less than 0.2 are currently exempted from regulations. Table 7.3 provides a list of selected new and old chemicals that may be employed as CFC substitutes in the future. Also shown in this table are their atmospheric lifetimes (scaled to CH_3CCl_3) and their ODP values (Fisher et al., 1990; aso see Chapter 11). Compared with CFC-11, these are much safer chemicals as far as ozone depletion is concerned.

HFC-134a (CF_3CH_2F)is a possible replacement for CFC-12 in refrigeration applications, while HCFC-141b (CH_3CFCl_2) and HCFC-123 (CF_3CHCl_2) are candidates for CFC-11 replacement as a blowing agent. In many instances blends and azeotropes involving HCFC-22 and the new products are anticipated. Substantial growth in the use of methyl chloroform and HCFC-22 will probably occur in the future as these compounds find new markets as CFC replacements. A number of tests, however, must be completed before new replacement chemicals can find widespread acceptability. As a caution it is noted that HCFC-132b (CH_2ClCF_2Cl), a highly effective replacement for CFC-113 as a solvent, was found to render male rats sterile and was discarded. Another concern is that HFCs and HCFCs will likely produce many products upon oxidation in the atmosphere, e.g. COF_2, $COClF$, CF_3COCl, $CClF_2CHO$, $CFCl_2C(O)OONO_2$, and some of these may not be environmentally benign. The atmospheric oxidation of HCFC-123 is considered here as an example:

$$
\begin{array}{lcl}
CF_3CHCl_2 + OH \ (+O_2) & \rightarrow & CF_3CCl_2O_2 + H_2O \\
CF_3CCl_2O_2 + NO_2 & \leftrightarrow & CF_3CCl_2OONO_2 \\
CF_3CCl_2O_2 + HO_2 & \rightarrow & CF_3CCl_2\,OOH + O_2 \\
CF_3CCl_2O_2 + NO & \rightarrow & CF_3CCl_2O + NO_2 \\
CF_3CCl_2OOH + OH & \rightarrow & CF_3CCl_2O_2 + H_2O \\
CF_3CCl_2OOH + h\nu & \rightarrow & CF_3CCl_2O + OH \\
CF_3CCl_2OONO_2 + h\nu & \rightarrow & CF_3CCl_2O + NO + O_2 \\
CF_3CCl_2O & \rightarrow & CF_3COCl + Cl \\
RH + Cl & \rightarrow & HCl + R
\end{array}
$$

CF_3COCl does not react with OH. Physical removal following incorporation into rain- and cloud/fog-water and decomposition by photolysis is probably the major tropospheric loss process for CF_3COCl. Other HCFCs such as HCFC-142b (CH_3CF_2Cl) can produce halogenated aldehydes (CF_2ClCHO) that can further react with OH to produce PAN-like compounds ($CFCl_2C(O)OONO_2$). For many of the carbonyl-type species the best protocol molecule may be phosgene ($COCl_2$), which has been measured in both the troposphere and stratosphere (Singh, 1976; Wilson et al., 1988). In the troposphere its likely source is the oxidation of chlorinated organics (principally C_2Cl_4 and C_2HCl_3), while in the stratosphere it is a breakdown product of CCl_4. Phosgene in the troposphere is expected to be removed primarily by rain-out/wash-out processes (removal by photolysis is extremely slow), resulting in an atmospheric residence time of several weeks to months (Singh, 1976; Wilson et al., 1988, Helas and Wilson, 1992). Many of the new HCFC products will be introduced on large scales within the next 5 years or so.

7.3. ORGANIC HALOGENATED COMPOUNDS IN THE ATMOSPHERE

A large number of halogenated organics have been detected in the global atmosphere. In most cases measurements have been made using electron capture gas chromatographic techniques. In some instances data have been collected over long periods of time and atmospheric trends have been measured. Table 7.1 provides 1990 global average concentrations, north/south gradients, and atmospheric trends for these species. In addition, estimated total atmospheric lifetimes and an idea of sources and sinks are also provided. The sources of data presented in Table 7.1 are too numerous to cite but can be found in the published literature (NASA, 1994; WMO, 1989, 1991).

7.3.1. Chlorofluorocarbons (CFCs) and SF_6

Chlorofluorocarbons (CFCs) are a group of halogenated methanes and ethanes that do not exist in nature as such and are synthesized by the chemical industry for commercial applications. Unless they contain a hydrogen atom (HCFCs) which can be abstracted by OH, CFCs do not have any known sinks in the troposphere. Chlorine-containing CFCs decompose in the stratosphere (10–50 km) owing to photolysis and to a very small degree ($\approx 1\%$) by reaction with $O(^1D)$ atoms. Fully fluorinated species are almost entirely lost in the mesosphere and thermosphere (i.e., above 60 km), resulting in lifetimes that can be on the order of thousands of years (Ravishankara et al., 1993).

Only a few fully fluorinated species have been measured in the atmosphere. CF_4 and C_2F_6 are inadvertently produced and emitted to the atmosphere during the production of aluminum ($4AlF_3 + 3C \rightarrow 4Al + 3CF_4$) and of iron and steel ($2CaF_2 + C \rightarrow Ca + CF_4$), involving high temperatures (1500–2000 K). Alumina reduction is the dominant anthropogenic source of CF_4 (0.5–2.5 kg of CF_4 per 1000 kg of aluminum produced) and may account for two-thirds (≈ 20 Gg year^{-1}) of its global source (≈ 30 Gg year^{-1}). Emission rates of C_2F_6 during aluminum production are 6–10 times lower than those for CF_4. Small amounts of CF_4 (≈ 0.35 Gg year^{-1}) and C_2F_6 (≈ 0.2 Gg year^{-1}) are also made intentionally (mostly in Japan and the US). CF_4 is principally used in the semiconductor industry, where cryogenic recovery procedures are often utilized to minimize emissions. C_2F_6 is used in various types of electrical equipment as a dielectric and a coolant gas. SF_6 is a synthetic chemical that finds applications as a dielectric in the electrical industry, as a degassing agent in the metal industry, and as an intentional atmospheric tracer. Much of the production ($> 90\%$) is eventually released to the atmosphere. An increasing trend in atmospheric concentrations of fully fluorinated species has been detected (Table 7.1). These fully fluorinated chemicals are non-toxic and chemically inert. They will, however, accumulate in the atmosphere, making a minor contribution as greenhouse gases (Ko et al., 1993). It is not known whether any natural sources exist.

CFC-11, CFC-12, and CFC-113 are three of the most important fully halogenated CFCs that have found wide applications as propellants, refrigerants, and solvents. Over the last 20 years or so their atmospheric abundances have increased at average rates of 9, 18, and 6 ppt year^{-1}, respectively. Figure 7.3 shows the measured atmospheric growth of CFC-11 at sites in the northern and southern hemispheres. Other CFCs (Table 7.1) have grown similarly and in all cases their increased emissions have been reflected in their increased atmospheric concentrations. CFC-11 is one molecule for which the world emissions are best described (Figure 7.2) and the atmospheric data are consistent among many groups (Figure 7.3). Therefore this molecule provides an excellent test for comparison of atmospheric and emission data. One can fit the emission data to the estimated global content and trend to calculate a CFC-11 lifetime of about 42 ± 6 years (Cunnold et al., 1993), in good agreement with theoretical calculations of stratospheric loss rates due to photolysis. These comparisons continue to be refined (NASA, 1994) but are largely consistent with the view that for molecules such as CFCs no important tropospheric removal processes are operative.

The recent decline in the emissions of several CFCs resulting from the adoption of the 1987 Montreal Protocol (Figure 7.2) has also been spectacular and easily detected in the atmosphere (Cunnold et al., 1994, Elkins et al., 1993). Figures 3.12(a) and 3.12(b) in Chapter 3 show this dramatic decline during the period 1988–1993, the atmospheric growth rate of CFC-11 declined from 10–12 to 3–5 ppt year^{-1}, while that of CFC-12 registered an equally impressive decrease

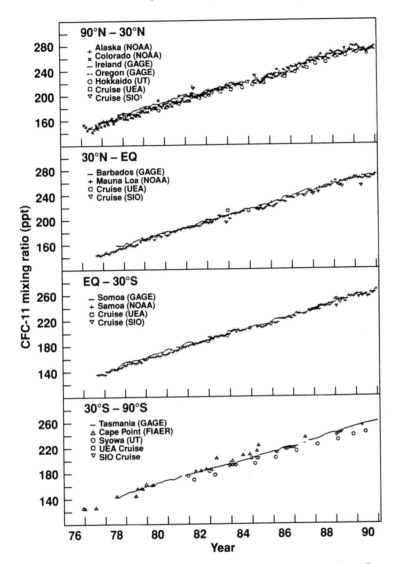

FIGURE 7.3. CFC-11 (CCl_3F) observations (ppt) in the four semi-hemispheres. Sources are summarized in WMO (1991) and NASA (1994).

from 17–22 to 7 ppt year^{-1}. In the future, atmospheric measurements may be used to monitor compliance with the emission reductions required by the Montreal Protocol. Unlike fully halogenated CFCs, HCFC-22 is removed in the troposphere by reaction with OH, resulting in a lifetime of about 14 years.

7.3.2. Chlorocarbons

Chlorinated hydrocarbons have much shorter lifetimes than CFCs (Table 7.1) and more complex sources. As stated earlier, these chlorocarbons (except CCl_4) break down in the troposphere upon reaction with the hydroxyl radical. The most abundant halocarbon in the atmosphere, CH_3Cl, is almost uniformly distributed in the globe at a concentration of about 600 ppt. Given its estimated lifetime of 1.4 years, a global source of about 3.5 Tgyear^{-1} is required. Man-made production is extremely small ($< 1\%$), but oceans and biomass burning are found to be important sources (Singh et al., 1983; Lobert et al., 1991). However, sources are only roughly known, because oceanic data are extremely sparse and the biomass burning emission is a function of the chlorine content of the vegetation and hence highly variable. It is estimated that the marine source represents 70–80% of the total and combustion sources account for the remaining 20–30%. The mechanisms by which CH_3Cl is produced in the oceans are not known, but it has been suggested that CH_3I (a product of marine algae) could react with Cl^- ions in sea-water to produce CH_3Cl (Zafiriou, 1975). The possibility of direct release of CH_3Cl from marine micro-organisms or phytoplankton cannot be ruled out.

Carbon tetrachloride (CCl_4) has been increasing slowly at a rate of 1–2 ppt year^{-1}. Despite early concerns, there is little doubt that CCl_4 is of near exclusive man-made origin (Simmonds et al., 1988). Methyl chloroform emissions have grown enormously (Figure 7.2), because it is a preferred substitute for the more

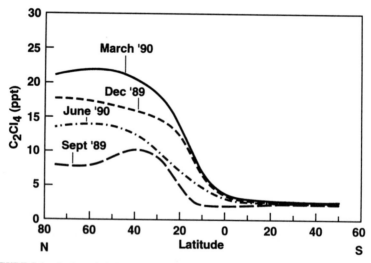

FIGURE 7.4. Surface air latitudinal distribution of tetrachloroethene (C_2Cl_4) in four seasons. (Adapted from Wang et al., 1994.)

toxic and reactive chloroethene solvents (e.g., C_2HCl_3 and C_2Cl_4) and in some cases for CFCs. Among the chlorocarbons listed in Table 7.1, CH_2Cl_2 and $CHCl_3$ sources are not well quantified and both man-made and natural sources (e.g., oceanic) are probably important. Species with atmospheric lifetimes of less than 1 year will show strong seasonal variations associated with the seasonal variation in OH. Figure 7.4 shows this latitudinal profile of C_2Cl_4 in four seasons (Wang et al., 1994). This seasonal behavior is largely consistent with the view that OH radicals are the dominant removers of these species.

7.3.3. Bromocarbons

Table 7.1 also summarizes the concentrations and north/south distribution of bromine-containing species that have been identified in the troposphere. In general, brominated species are far less abundant in the atmosphere than chlorine species. Molecules with two or more bromine atoms generally absorb ultraviolet radiation in the troposphere and are easily photodissociated. In some cases (e.g., $CHBr_3$) this process is far more efficient than removal via OH radicals. For brominated chemicals, both man-made and natural sources are known to exist, with oceans being an important natural source. A knowledge of the sources and sinks of bromine-containing species is important, because bromine is 40 (30–120) times more effective than chlorine in stratospheric ozone destruction (WMO, 1991). This is a result of synergistic chemistry between chlorine and bromine atoms ($ClO + BrO \rightarrow Br + Cl + O_2$).

Three man-made halons have been measured in the atmosphere at concentrations of about 2 ppt each. Halon-1301 ($CBrF_3$) and Halon-1211 ($CBrClF_2$) have been increasing at a rapid rate of about 10% $year^{-1}$. The atmospheric burdens and growth rates of these are consistent with their exclusive man-made sources (McCulloch, 1992). Man-made halons are a potent source of stratospheric Br because they have the longest lifetimes among the group of known brominated species. According to the provisions of the Montreal Protocol, further manufacturing of halons is to cease by the end of this decade.

Methyl bromide (CH_3Br) is a ubiquitous component of the atmosphere and has been measured at a concentration of about 15 ppt in the NH and 12 ppt in the SH. This north/south gradient is thought to be a result of its man-made emissions. As has been summarized in Singh and Kanakidou (1993), man-made production is estimated to be about 65 Gg $year^{-1}$ for 1990 and a large fraction is expected to be released to the atmosphere. Nearly 95% of all CH_3Br sold is used in various soil, grain, and space fumigation applications, with a very small fraction ($\approx 5\%$) finding use as a chemical intermediate. It decomposes in soils at a rapid rate (5–10% day^{-1}) via processes of hydrolysis and demethylation. Numerical models of soil vapor migration, which take into account current use practices and typical soil types, have been used to calculate that some 30–60% of the CH_3Br applied may be ultimately released to the atmosphere. Sea-water measurements also show that ocean waters are supersaturated with CH_3Br (Singh et al., 1983; Khalil et al., 1993). Because CH_3Br is rapidly removed from sea-water by hydrolysis and reactions with Cl ions ($\approx 4-6\%$ day^{-1}), a large source of 200–300 Gg year^{-1} within the oceans has been postulated (Singh and Kanakidou, 1993). Atmospheric abundances, interhemispheric gradients, oceanic concentrations, man-made emissions, and removal processes have been analyzed to estimate a CH_3Br global source of ≈ 100 Gg year^{-1} of which approximately 35% (≈ 35 Gg year^{-1}) is man-made. Indirect emissions from biomass burning are about 1% (v/v) of the corresponding CH_3Cl emissions, resulting in an estimated source of 10–50 Gg year^{-1} (Manö and Andreae, 1994). The remainder is largely oceanic. A global trend of 0.1–0.2 ppt year^{-1} is predicted and has been detected (Singh and Kanakidou, 1993, Khalil et al., 1993). Substantial uncertainties in calibrations, source estimates, and deposition processes are present. It is possible that both the sources and sinks of CH_3Br are considerably larger than those stated above.

Bromoform ($CHBr_3$) has been identified as a marine product of biological origin with high concentrations in the cold waters of the Arctic Ocean (Dyrssen and Fogelqvist, 1981). Arctic and Antarctic ice algae have been found to contain and release significant quantities of $CHBr_3$ (Sturges et al., 1992). Anthropogenic activities involving the practice of chlorinating sea-water (to prevent fouling by marine organisms) prior to use as a coolant in coastal power plants can also release $CHBr_3$ (Fogelqvist and Krysell, 1991). The atmospheric distribution of both $CHBr_3$ and CH_2Br_2 has been measured and no north/south gradients are observed (Penkett et al., 1985; Class and Ballschmiter, 1988). Concentrations within the boundary layer are 0.5–2 ppt but are lower aloft. A strong seasonal cycle for $CHBr_3$ with high winter/spring (9–11 ppt) and low summer (2–3 ppt) mixing ratios has been observed at northerly high latitudes (Cicerone et al., 1988). This cycle is due to the combination of a strong winter/spring source in the cold waters of the Arctic Ocean and a weak sink due to photolysis. Because of uncertainties in its absorption cross-sections, a global mean lifetime of as little as 10 days and as large as 30 days can be calculated. The lifetime of CH_2Br_2 is considerably longer because of its much slower rate of photolysis. Other molecules such as CH_2BrCl, $CHBr_2Cl$, and $CHBrCl_2$ have also been detected. Extremely few sea-water data are available, but the atmospheric abundance of these brominated species is largely consistent with a dominant oceanic source.

7.3.4. Iodocarbons

Iodine-containing species are the least abundant in the atmosphere. Methyl iodide (CH_3I) is present in the marine boundary layer at a mean concentration of about 2 ppt (Lovelock et al., 1973; Singh et al., 1983) and is likely to possess a strong vertical gradient owing to its short lifetime of only about 1 week resulting from rapid photodissociation. Sea-water is significantly supersaturated with CH_3I and film models have been used to estimate sea-to-air fluxes of 0.5–1.0 Tg year^{-1} (Singh et al., 1983; Reifenhäuser and Heumann, 1992). Other iodine-containing species such as CH_2ICl and CH_2I_2 have also been identified in sea-water, but their atmospheric abundances are not known (Class and Ballschmiter, 1988; Moore and Tokarczyk, 1993). Indications are that the sea-to-air flux of I resulting from these species is not negligible. It is estimated that a global background of about 1 ppt of organic I originating almost exclusively from the oceans is present. Attempts have been made to link the presence of CH_3I decomposition products (such as IO) to ozone destruction in the tropical marine boundary layer (Chameides and Davis, 1980) and to dimethyl sulfide oxidation (Chatfield and Crutzen, 1990); however, given the very low abundance of available I and the much slower revised reaction rate of IO with dimethyl sulfide, it now appears that these proposed mechanisms are probably not important in the atmosphere.

Mechanisms by which organohalides are produced in sea-water are not well known. It is known that kelp and other microbes contain and release methyl iodide as a direct product (Manley and Dastoor, 1988). Geschwend et al. (1985) have found that a variety of macroalgae can release significant amounts of $CHBr_3$, CH_2Br_2 and other chemicals to sea-water. Zafiriou (1975) has proposed that Cl^- and Br^- ions in sea-water may react with CH_3I, to rapidly (over several days) form CH_3Cl and CH_3Br ($CH_3I + Cl^- \rightarrow CH_3Cl + I^-$). Singh et al. (1983) were unable to find any correlation between sea-water concentrations of CH_3I and those of CH_3Cl or CH_3Br, suggesting that this was probably not a dominant mechanism. Another possibility (White, 1982) may be either direct emission from organisms or reaction between dimethyl sulfonium and halide ions in sea-water ($(CH_3)_2S^+$—$R + Br^- \rightarrow CH_3Br + CH_3SR$). Proposed Cl^- ion reactions may also take place with $CHBr_3$ to produce a variety of substituted chloro/bromo organics ($CHBr_3 (+Cl^-) \rightarrow CHBr_2Cl (+Cl^-) \rightarrow CHBrCl_2 (+Cl^-) \rightarrow CHCl_3$). It is possible that phytoplankton, algae, and ion reactions all contribute to sea-water formation of organohalides. In most cases specific mechanisms are yet to be understood. Present estimates of the source strength of naturally occurring halocarbons are largely based on their removal rate by reaction with OH. Removal by deposition processes, if found to be present, would require a significant upward revision in these estimates.

7.3.5. Organic Halogen Budget

By adding halogen atoms from individual species, the total organic F, Cl, Br, and I burden in the troposphere is provided in Table 7.1. Comparison of individually measured species with bulk measurements of total Cl and Br (Berg et al., 1980) suggests that most of the organic chlorine is probably accounted for by chemicals of Table 7.1. This cannot be said with certainty for Br and I species. In 1990

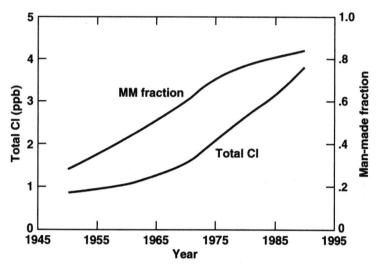

FIGURE 7.5.   Increasing abundance of organically bound chlorine and its man-made (MM) fraction in the troposphere.

about 3800 ppt of Cl was present in the troposphere and about 84% was of man-made origin. Similarly, about 1800 ppt of F was present, virtually all of which was of man-made origin. Of the 30 ppt of Br present, nearly 45% is estimated to be of man-made origin. The iodine cycle, on the other hand, is nearly unperturbed, with almost 100% of I attributable to its marine source. There is little doubt that the Cl and Br cycles of the troposphere have been strongly impacted by human activities. Figure 7.5 shows the total organic Cl burden of the atmosphere over the period from 1950 to 1990. Extrapolations for man-made species to earlier years (1950–1970) are based on available emission data. The Cl burden in the atmosphere increased from about 0.8 ppb in 1950 (mostly CH_3Cl, CCl_4, CFC-12, and CFC-11) to 3.8 ppb in 1990; during this period the man-made contribution to atmospheric Cl increased from 30% to 84%. If Montreal Protocol objectives are adhered to, the atmospheric burden of Cl is expected to peak within this decade and then slowly decline over the next century.

7.4. INORGANIC HALOGENS IN THE TROPOSPHERE

Sea-salt aerosol, produced by the bursting of small air bubbles near the sea surface, constitutes one of the largest inputs of halogen substances ($\approx$ 6000 Tg year^{-1} as particulate Cl$^-$) to the atmosphere (Woodcock, 1953; Blanchard and Woodcock, 1980). Bulk atmospheric measurements have shown that inorganic halogens are present in gaseous and aerosol forms (e.g., Rahn et al., 1976). Almost all the inorganic gaseous and aerosol data were collected by passing large quantities of air through filters for particulate collection, followed by impregnated filters (generally LiOH or K_2CO_3) for inorganic halogen collection. In some studies activated charcoal beds are inserted after the filter to collect organic halocarbons as well. These samples are then generally analyzed by standard techniques of neutron activation. Artifact Cl^- deficits can be produced on filters owing to reactions between acidified fine aerosols (generally sulfates) and sea-salt aerosols. It is possible to speculate that gaseous species such as HCl, HBr, HI, Cl_2, HOCl, NOCl, NOBr, NO_2Cl, IO, $ClONO_2$, and $BrONO_2$ may be present, especially in the marine boundary layer, but HCl is the only species that has been specifically identified. Compared with organic halogen species, for which extensive sensitive and specific measurements have been made, the situation is much less satisfactory for inorganic species. Inorganic halogens can be expected to be quite reactive in the troposphere, with lifetimes of a few hours to several days.

7.4.1. Inorganic Chlorine and Chlorine Atoms

Experiments have indicated that marine aerosols show depletion of chloride and excess of sulfate compared with bulk sea-water. In marine aerosol the Cl^-/Na ratio is typically found to be 1.0–1.7, compared to 1.8 in sea-water or in newly formed particles (Chesselet et al., 1973; Kritz and Rancher, 1980). In highly acidified atmospheres, generally in coastal regions, Cl^-/Na ratios of less than 0.5 are frequently observed. It has also been observed that this deficit is more pronounced in smaller particles compared with larger ones. This Cl^- deficit is attributed to the loss of gaseous inorganic chlorine (GIC) from the acidification of marine aerosol (Cauer, 1949; Eriksson, 1959; Robbins et al., 1959; Duce, 1969). Thermodynamic considerations have led people to conclude that this GIC must largely be HCl. Equilibrium model calculations in $H_2SO_4/NaCl$ droplet systems (HCl(g) = H^+(aq) + Cl^-(aq)) confirm that most of the hydrogen ions in an acidic sodium chloride droplet will be lost to the atmosphere as gaseous HCl (Clegg and Brimblecombe, 1985). The mechanism postulated for this particle-to-gas conversion involves reactions that acidify aerosols to low pH values (pH 2–3):

$$2NaCl(s) + H_2SO_4 \rightarrow Na_2SO_4(s) + HCl(g)$$

$$2NaCl(s) + HNO_3 \rightarrow NaNO_3(s) + HCl(g)$$

Most of the sea-salt aerosol is present as giant particles and is redeposited on the sea surface within 1 or 2 days. The smaller sizes (less than 2 μm) are airborne longer and are the ones that can react with acids to release GIC. A variety of experiments have shown that the fraction (f) of the sea-salt particle Cl^- that is converted to GIC is highly variable (3–50%) and is a function of size, relative humidity, and the degree of local pollution. Experimental artifacts due to reactions between acidified fine aerosols (generally sulfates) and sea-salt aerosols on the collection media tend to artificially enhance the release of GIC. Gross estimates, derived from the product of total sea-salt release and some average Cl^- deficit ($\approx$ $6000 \times f$ Tg year^{-1}), suggest that degassing from sea-salt aerosol could produce some 100–600 Tg year^{-1} of GIC ($\approx$ HCl) in the marine boundary layer. All other sources of HCl combined (anthropogenic emissions, volcanic activity, oxidation of chlorinated hydrocarbons) contribute an additional 10–20 Tg year^{-1} (Legrand and Delmas, 1988). Volcanic sources can vary greatly, especially when large eruptions occur (e.g., Tambora in 1815), but even routine volcanic emissions can provide a locally and regionally important source of HCl. Unlike the marine source of chlorine, volcanic sources may also directly inject this chlorine into the stratosphere, causing ozone depletion in a manner similar to CFCs.

A body of data exists that shows that GIC ($\approx$ HCl) is typically present at concentrations of 0.1–2 ppb in the marine boundary layer and that these concentrations decline with altitude and as the air moves over continental regions (see Singh and Kasting, 1988). It is found that GIC is some five (1–10) times more abundant than particulate inorganic chlorine (Rahn et al., 1976; Kritz and Rancher, 1980). Specific measurements of HCl have been attempted using spectroscopic (Farmer et al., 1976; Harris et al., 1992) and derivative (Vierkorn-Rudolph et al., 1984; Singh and Kasting, 1988, and references therein) techniques. Compared to standard impregnated filter methods, these more specific techniques report generally lower (0.05–0.5 ppb) HCl concentrations. Free tropospheric mixing ratios are definitely below 150 ppt. No latitudinal distributions are available. Assuming that HCl is removed from the atmosphere in 4–8 days, these specific measurements would suggest a somewhat lower global source of 50–150 Tg year^{-1}. It is noted that while the global source of marine HCl is quite large, its magnitude is highly uncertain. A consequence of the increased NO_2 and SO_2 emissions over the last several decades would be to enhance the release of HCl from sea-salt aerosols.

Recent preliminary work by Pszenny et al. (1993) suggests that in addition to HCl, other inorganic halogens such as Cl_2 and HOCl may also be released. While the basic scheme of sea-salt aerosol acidification remains viable, other mechanisms based on laboratory studies have also been suggested. These mechanisms may have some significance in regions where marine air is impacted by episodes of pollution:

$$2NO_2(g) + NaCl(s) \rightarrow NaNO_3(s) + NOCl\ (g)$$
$$NOCl + h\nu \rightarrow NO + Cl$$
$$NOCl + H_2O \rightarrow HCl + HNO_2$$

$$N_2O_5 + NaCl(s) \rightarrow ClNO_2 + NaNO_3(s)$$
$$ClNO_2 + h\nu \rightarrow Cl + NO_2$$

Cl atoms will react with hydrocarbons in the atmosphere to produce HCl. One reason why inorganic chlorine may play an important role in atmospheric chemistry is because reactive chlorine species (HCl, Cl_2, $HOCl$, etc.) may be sources of tropospheric Cl atoms. Cl atoms are some 10–1000 times more reactive than OH toward a variety of hydrocarbons (Demore et al., 1992). There is little doubt that high concentrations of Cl atoms ($> 10^4$ molecules cm^{-3}) could contribute significantly to the oxidizing capacity of the atmosphere. Singh and Kasting (1988) evaluated several possible mechanisms and concluded that only 10^2–10^3 molecules cm^{-3} of Cl atoms are probably present, largely from the oxidation of HCl ($HCl + OH \rightarrow H_2O + Cl$). Even at these low concentrations they may oxidize some 20–30% of select hydrocarbons. Reports based on aerosol chamber studies have suggested that molecular chlorine (Cl_2) was a product of reactions involving sea-salt aerosols, ozone, and sunlight (Zetzsch et al., 1988; Zetzsch and Behnke, 1992). These results, coupled with the observation that the Cl deficit in marine aerosols cannot always be accounted for by the corresponding acidification, have led to suggestions that chlorine species other than HCl, possibly HOCl and Cl_2, may be released in significant quantities (Keene et al., 1990; Pszenny et al., 1993).

An indirect test for the high atmospheric abundance of Cl atoms is possible by looking at the reactivities of a variety of hydrocarbons and molecules such as dimethyl sulfide. The reactions of Cl atoms with C_2H_6, C_3H_8, C_2H_2, C_2Cl_4, and dimethyl sulfide are extremely fast and at 298 K some 237, 145, 64, 241, and 65 times faster than their corresponding OH rates, respectively (Demore et al., 1992). Night/day measurements of these species were performed in the tropical Pacific Ocean and offered an opportunity to assess the presence of Cl atoms in the early morning hours. From these measurements, Singh et al. (1994) infer that Cl atom concentrations in the vicinity of 10^5 molecules cm^{-3} may be present in the marine boundary layer of the tropical Pacific Ocean to explain the observed night/day differences in the mixing ratios of ethane and propane. High early morning HCl mixing ratios, suggested by the mechanism proposed by Keene et al. (1990), were not observed. In the free troposphere Cl atom concentrations are expected to be extremely small. Parrish et al. (1993) have proposed that atmospheric concentration ratios of select hydrocarbons (*i*-butane/*n*-butane versus *i*-butane/propane) can be employed to separate Cl and OH chemistries along orthogonal axes (see Figure 7.6). One consequence of the known kinetics is that the ratio of *i*-butane to *n*-butane in the atmosphere should remain nearly independent of the age of the air mass if OH chemistry dominates but vary greatly if Cl chemistry dominates. Analysis of data from a variety of sources is supportive of the view that OH kinetics fits the data quite well and high

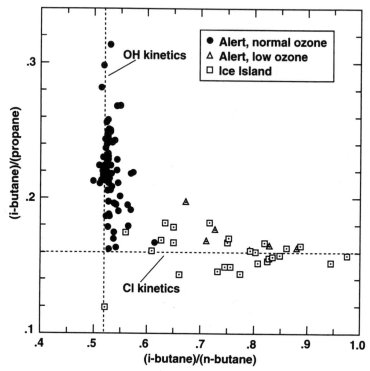

FIGURE 7.6.   Plot of *i*-butane/*n*-butane versus *i*-butane/propane from Alert and Ice Island, Canada. Dashed lines show that data would align vertically under conditions of OH kinetics and horizontally under conditions of Cl kinetics. (From Jobson et al., 1994.)

concentrations of Cl atoms ($> 10^4$ molecules cm^{-3}) are not present in the bulk of the troposphere (Parrish et al., 1993).

In certain special environments, however, halogen radicals may indeed play an important role. One such environment is the Arctic, where very low ozone (Figure 7.7) and hydrocarbon concentrations are encountered during episodes in springtime as the Sun comes up (Barrie et al., 1988; Bottenheim et al., 1990). This phenomenon is not seen in the dark winter, nor in the summer. Hydrocarbon data collected during these episodes of extremely low O_3 and at other times have been plotted in Figure 7.6 to accentuate the role of OH and Cl chemistries (Jobson et al., 1994). These data support the contention that under normal circumstances OH kinetics dominates. Within the depleted O_3 air masses, however, indirect proof can be found to show that Cl atoms are present at an estimated concentration of 10^4–10^5 molecules cm^{-3}. Cl concentrations of this magnitude are sufficient to cause only a very small fraction ($\approx 5\%$) of the

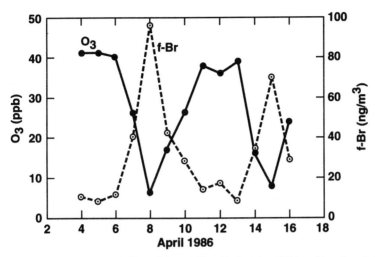

FIGURE 7.7. Surface concentrations of O_3 and filterable bromine (f-Br) at Alert, Canada in April 1986 as reported by Barrie et al. (1988).

observed O_3 destruction. Thus it appears that while Cl atoms may be quite effective in oxidizing non-methane hydrocarbons (NMHC) in special boundary layer situations, their significance in the global troposphere is quite modest.

7.4.2. Inorganic Bromine and Bromine Atoms

The database dealing with inorganic bromine is far more sparse than that of inorganic chlorine. No inorganic Br species have been specifically identified or measured in the troposphere. Filter collection techniques followed by elemental analysis have led to some understanding of inorganic Br behavior. In marine aerosol a Br^- deficit similar to the Cl^- deficit is observed and presumably HBr (or Br_2) is released (Duce et al., 1965; Kritz and Rancher 1980). Gaseous inorganic bromine is present at very low concentrations of about 3 ppt (0–7 ppt) and is some two (one to three) times more abundant than particle inorganic bromine (Rahn et al., 1976). Rancher and Kritz (1980) also observe that marine aerosols contain only about half as much Br during the day compared with the night, while an opposite behavior is seen in the gas phase. This observation is suggestive of the fact that an exchange of Br between particle and gas phases associated with the presence of sunlight is taking place, but few details are available. In laboratory studies, reactions of NO_2 and N_2O_5 (at ppm levels) with sea-salt bromide have been shown to produce NOBr and NO_2Br (Finlayson-Pitts et al., 1990).

In the Arctic spring, extremely high concentrations of particulate and gaseous bromine (some 20 times normal) have been observed (Berg et al., 1983). During spring, episodes of extremely low ozone coincident with high concentrations of filterable Br (f-Br) have also been observed (Figure 7.7). This filterable Br (inorganic bromine) can be defined as the sum of Br on particles and gaseous species such as HBr and $BrNO_3$, as collected by a Whatman 41 cellulose acetate filter or a combination Teflon/nylon filter (Barrie et al., 1988; Bottenheim et al., 1990). High concentrations of total Br are possibly derived from $CHBr_3$ (and other organobromines) synthesized in the cold environs of the Arctic Ocean and the general stagnation following the dark winter. Barrie et al. (1988) have suggested that photolysis of $CHBr_3$ can produce Br radicals which may catalytically deplete this ozone:

$$Br + O_3 \rightarrow BrO + O_2$$
$$BrO + BrO \rightarrow Br + O_2$$

Closer inspection shows that the rate of photolytic decomposition of $CHBr_3$ is too slow to generate enough Br radicals; once formed, they would quickly convert to non-radical species such as HBr, HOBr, $BrNO_3$, and organic bromine compounds, and further ozone loss would cease. Photochemical calculations suggest that daytime Br concentrations of $\approx 10^6$ molecules cm^{-3} would be required to explain the observed O_3 loss. In order to sustain such a concentration of Br radicals, a continuous source of Br_2 is probably required. McConnell et al. (1992) have suggested that inorganic (HBr, HOBr, $BrNO_3$) and some organic (e.g., BrCHO) bromine species could be scavenged by aerosols and photo-chemically converted to Br_2 in the aerosol phase, with subsequent volatilization to the gas phase. Fan and Jacob (1992) contend that such a mechanism is kinetically possible in the highly acidic aerosol environment that is likely to be present under these circumstances. Based on the reaction of Cl and Br atoms with acetylene, Jobson et al. (1994) infer that Br atom concentrations in the vicinity of 10^7–10^8 molecules cm^{-3} are consistent with the observed loss of acetylene. If this is true, then more than 95% of the observed O_3 loss could result from Br atom chemistry alone. It is possible that high bromine particle loading and coincident high acidity of the Arctic haze produce free bromine to sustain substantial Br radical abundances. However, the causes of the extremely high gaseous and particulate Br loading in the Arctic spring are not yet fully understood.

7.4.3. Inorganic Iodine

Elemental analysis techniques have shown that the total gaseous I (organic and inorganic) abundance in the atmosphere is between 1 and 5 ppt. Gaseous inorganic iodine is present at very low concentrations of ≈ 1 ppt (0–3 ppt) and it is some 2–10 times more abundant than particle inorganic iodine (Moyers and

Duce, 1972; Rahn et al., 1976). Unlike Cl^- and Br^-, I^- (and/or IO_3^-) is found to be strongly enriched in sea-salt aerosols compared with bulk sea-water. Measured I^-/Na ratios in the marine aerosol are some 10^2–10^3 times greater than sea-water values (Miyake and Tsunogai, 1963). This has led to suggestions that gaseous iodine is being incorporated into marine aerosols subsequent to their formation. To date, no inorganic iodine species has been specifically identified or measured in the troposphere. It is possible that much of the inorganic I is present largely as a result of the breakdown of organic species (e.g., CH_3I, CH_2I_2).

7.5. HUMAN EXPOSURE TO TOXIC HALOCARBONS

A large number of halogenated chemicals are released in urban environments and high concentrations can be found in outdoor as well as indoor spaces. CFCs as a group are considered non-toxic and low-level exposure does not pose a health hazard. This is not true of other chlorinated hydrocarbons, many of which are suspect in causing human cancer and other disease. Their measurements in the urban environment are an integral part of estimating population exposure, evaluating long-term health effects, and formulating strategies for control. Large databases have been compiled to assess the exposure to such chemicals in outdoor and indoor environments (Shah and Singh, 1988). Yet another source of health concern is the ingestion of chlorinated hydrocarbons from drinking-water. Substantial intake of halogenated toxic chemicals through air and water media is possible recognizing that on average a person breathes $25\,m^3\,day^{-1}$ of air and drinks $1\,l\,day^{-1}$ of liquids.

7.5.1. Urban Air Concentrations

Urban concentrations of CFCs have been measured in many cities and typically do not exceed 5 ppb for an individual CFC. A sampling of such data based on three short term field studies at a site in San Jose, California is provided in Table 7.4 (Singh et al., 1992). As expected, concentrations are a strong function of local meteorology, but mean concentrations of CFCs can be expected to be about two to five times their global background values. However, exceptions do exist. As an example, unusually high concentrations of CFC-113 have been measured in Silicon Valley, California where CFC-113 is widely used as a solvent/degreaser in the electronics industry (Table 7.4). Substantially elevated concentrations of other select halogenated species have also been observed, as shown in Table 7.4. Most abundant of the man-made species are CH_2Cl_2, CH_3CCl_3, and C_2Cl_4 with maximum concentrations in the 3–10 ppb range. Urban sources are sufficient to impact the atmospheric abundances of CH_3Cl and CH_3Br, molecules of

TABLE 7.4 **Atmospheric Measurements of Selected Halocarbons in San Jose, California in Three Seasons of 1985**

Chemicals	Mixing ratios (ppt)		
	April 4–16	August 12–24	December 13–21[a]
CFC-11	529 (252–1613)[b]	450 (244–1330)	585 (259–971)
CFC-12	1020 (458–2751)	881 (427–2058)	1435 (670–2450)
CFC-113	1256 (395–4605)	616 (166–2410)	1211 (476–2321)
CFC-114	59 (19–239)	72 (12–888)	227 (34–967)
CH_3Cl	1060 (673–2508)	–	1118 (594–4870)
CH_3Br	400 (44–4661)	121 (5–1067)	2869 (239–15424)
CH_2Cl_2	1534 (403–4311)	1119 (142–8257)	4181 (1034–10310)
$CHCl_3$	64 (23–138)	58 (11–180)	102 (38–203)
CCl_4	193 (55–398)	144 (85–213)	155 (90–266)
CH_3CCl_3	360 (120–905)	283 (133–518)	1219 (345–3174)
C_2HCl_3	63 (8–266)	68 (10–266)	271 (71–907)
C_2Cl_4	427 (58–1530)	264 (36–767)	1858 (311–6639)
CH_2BrCH_2Br	21 (9–41)	–	7 (2–18)

[a]A period of extreme stagnation.
[b]Arithmetic mean (maximum value–minimum value).

substantially natural origin. Table 7.5 shows annual mean concentrations of four select "toxic chemicals" from three California sites over a 4 year period. Benzene, a suspect carcinogen with a predominantly automobile exhaust source, is included for comparison. High concentrations of chloroform have also been measured in urban areas and in many indoor environments.

7.5.2. Drinking-Water Concentrations

Chlorine has historically been the most common disinfectant in the United States. Trihalomethanes and other halogenated hydrocarbons are formed during chlorination of water. These concerns led to creation of the Safe Drinking Water Act in 1974, which was amended in 1986. These acts promulgated acceptable standards for specific toxic substances in drinking water. In a recent study the occurrence of disinfection by-products in US drinking-water was evaluated at 35 water treatment facilities that had a broad range of source water qualities and treatment processes (Krasner et al., 1989). Table 7.6 shows the median concentrations of halogen species based on measurements conducted over a 3 month period. Some $180 \mu g \, l^{-1}$ (median concentration) of organic halogens were found to be present in treated water. Trihalomethanes were the largest class of disinfection by-products detected on a weight basis, with haloacetic acids being the next most significant. Formaldehyde and acetaldehyde, known by-products of

TABLE 7.5   Annual Average (1987–1990) Mixing Ratios of Key Chemicals at Selected Bay Area Sites

Chemicals	Site	Annual Average Mixing Ratio (ppb)[a]			
		1987	1988	1989	1990
CH_2Cl_2	Mountain View	1.36	0.86	0.70	0.83
	Santa Rosa	0.59	0.46	0.35	0.32
	Fremont	0.69	0.46	0.47	0.42
CH_3CCl_3	Mountain View	0.41	0.66	0.55	0.58
	Santa Rosa	0.25	0.26	0.46	0.57
	Fremont	1.10	2.34	3.82	2.43
C_2Cl_4	Mountain View	0.27	0.29	0.31	0.17
	Santa Rosa	0.20	0.11	0.11	0.11
	Freemont	0.26	0.22	0.26	0.16
C_6H_6	Mountain View	1.60	3.46	1.87	1.52
	Santa Rosa	1.44	1.50	2.03	1.41
	Fremont	1.50	1.73	1.52	1.40

[a]Annual average based on two 24 h samples collected monthly.

ozonation, were also produced by chlorination. Cyanogen chloride was preferentially produced in water where chloroamines were used as pre-disinfectants. Reduction in drinking-water halocarbon concentrations can be achieved by switching to different disinfectants (e.g. ozone, ClO_2) or, in the home, by charcoal filtration.

7.6. SUMMARY AND FUTURE NEEDS

Halocarbons will continue to be important components of the global atmosphere. CFCs already present in the atmosphere will persist for many decades. It is conceivable that new HCFCs will not pose a threat to stratospheric ozone, which will slowly recover over the next 50 years. These new, highly reactive chemicals will produce many intermediate products (e.g., halogenated carbonyls) in the troposphere and their chemistry and fate need to be studied. Since lifetimes of many species and estimates of OH radicals are normalized to those of methyl chloroform, it is important that major calibration uncertainties be removed. In the future it may be possible to release controlled quantities of target chemicals to determine OH concentrations more accurately than is possible today. These procedures, once refined, could also be used to infer seasonal and long-term trends in OH abundances. Mechanisms by which organic halogens are produced in nature, especially in the marine environment, are poorly understood and should be further elucidated. It is evident that the natural source of inorganic

TABLE 7.6   Quarterly Median Concentrations of Halocarbons and Other By-Products of Disinfection in Drinking-Water

Disinfection By-Product	Concentration ($\mu g\, l^{-1}$)
Trihalomethanes	
Chloroform	15
Bromodichloromethane	10
Dibromochloromethane	4.5
Bromoform	0.6
Total	44
Haloacetonitriles	
Trichloroacetonitrile	<0.01
Dichloroacetonitrile	1.1
Bromochloroacetonitrile	0.6
Dibromoacetonitrile	0.5
Total	2.5
Haloketones	
1,1-Dichloropropanone	0.5
1,1-Trichloropropanone	0.4
Total	0.9
Haloacids	
Monochloroacetic acid	1.2
Dichloroacetic acid	6.8
Trichlororacetic acid	5.8
Monobromoacetic acid	<0.5
Dibromoacetic acid	1.5
Total	20
Aldehydes	
Formaldehyde	5.1
Acetaldehyde	2.7
Total	6.9
Miscellaneous	
Chloropicrin	0.1
Chloral hydrate	3.0
Cyanogen chloride	0.6
2,4,6-Trichlorophenol	<0.4
Total organic halide	180

[a]Data from April–June 1988, based on a study of 35 water treatment facilities in the United States.

chlorine (largely HCl) far exceeds all sources (man-made and natural) of organic chlorine. However, it is not known what fraction of the HCl produced in the marine boundary layer actually enters the free troposphere. Techniques employed to measure inorganic halogens are generally non-specific and suffer from sampling and analysis problems. Real progress in our understanding of inorganic halogens must await the development of sensitive and specific measurement techniques. HCl is the only species specifically identified to date and even here significant uncertainties exist. The search for halogen free radicals (Cl, Br) must continue, since these can provide a significant perturbation in our knowledge about the oxidative capacity of the atmosphere.

Acknowledgments

This research was supported by the NASA Global Tropospheric Experiment. Thanks are due to Drs H. Niki (York University) and F.S. Rowland (University of California at Irvine) for providing us with preprints of their unpublished work, and to Dr M. Kritz (State University of New York) for critical review.

References

Barrie, L.A., Bottenheim, J.W., Shnell, R.C., Crutzen, P.J. and Rasmussen, R.A. (1988). Ozone destruction and photochemical reactions at polar sunrise in the lower arctic atmosphere. *Nature*, **334**, 138–140.

Berg, W.W., Crutzen, P.J., Grahek, F.E., Gitlin, S.N. and Sedlacek, W.A. (1980). First measurements of total chlorine and bromine in the lower startosphere. *Geophys. Res. Lett.*, **7**, 937–940.

Berg, W.W., Sperry, P.D., Rahn, K.A. and Gladney, E.S. (1983). Atmospheric bromine in the Arctic. *J. Geophys. Res.*, **88**, 6719–6736.

Blanchard, D.C. and Woodcock, A.H. (1980). The production, concentration and vertical distribution of the sea salt aerosol. *Ann. N. Y. Acad. Sci.*, **338**, 330–347.

Bottenheim, J.W., Barrie, L.A., Atlas, E., Heidt, L., Niki, H., Rasmussen, R.A. and Shepson, P.J. (1990). Depletion of lower tropospheric ozone during arctic spring: the polar sunrise experiment 1988. *J. Geophys. Res.*, **95**, 18,555–18,568.

Butler, J., Elkins, J., Thompson, T., Hall, B., Swanson, T. and Koropalav, V. (1991). Oceanic consumption of CH_3CCl_3: implications for tropospheric OH. *J. Geophys. Res.*, **96**, 22,347–22,355.

Cauer, H. (1949). Ergebnisse chemische-meteorologischeer forschung. *Arch. Meteorol. Geophys. Bioklim. Ser. B.*, **1**, 221–226.

Chameides, W.L. and Davis, D.D. (1980). Iodine: its possible role in tropospheric photochemistry. *J. Geophys. Res.*, **85**, 7383–7398.

Chatfield, R.B. and Crutzen, P.J. (1990). Are there interactions of iodine and sulfur species in marine air photochemistry. *J. Geophys. Res.*, **95**, 22,319–22,341.

Chesselet, R., Morelli, J. and Buat-Menard, P. (1974). Some aspects of the geochemistry of marine aerosols, in Dyrssen, D. and Jagner, D., Eds., *Proc. Nobel Symposium 20, The Changing Chemistry of the Oceans*, pp. 93–114.

Cicerone, R. (1981). Halogens in the atmosphere. *Rev. Geophys. Space Phys.*, **19**, 123–139.

Cicerone, R.J., Heidt, L.E. and Pollock, W.H. (1988). Measurements of atmospheric methyl bromide and bromoform. *J. Geophys. Res.*, **93**, 3745–3749.

Class, T.H. and Ballschmiter, W. (1988). Chemistry of organic traces in air. VIII: Sources and distribution of bromo and bromochloromethanes in marine air and surface water of the Atlantic Ocean. *J. Atmos. Chem.,* **6**, 35–46.

Clegg, S.L. and Brimblecombe, P. (1985). Potential degassing of hydrogen chloride from acidified sodium chloride droplets. *Atmos. Environ.,* **19**, 465–470.

Cunnold, D., Fraser, P., Weiss, R., Prinn, R., Simmonds, P., Alyea, F. and Crawford, A. (1992). Global trends and annual releases of $CFCl_3$ and CF_2Cl_2 estimated from ALE/GAGE and other measurements from July 1978 to June 1991. *J. Geophys. Res.,* **99**, 1107–1126.

Dabberdt, W.F. and Dietz, R.N. (1986). Gaseous tracer technology and application, In Lenschow, D.H., Ed., *Probing the Atmospheric Boundary Layer.* American Meteorological Society, Boston, MA, pp. 103–128.

Demore, W.B., Sander, S.P., Golden, D. Hampson, R.F., Kurylo, M.J., Howard, C J., Ravishankara, A.R., Kolb, C.E. and Molina, M.J. (1992). Chemical kinetics and photochemical data for use in stratospheric modelling. *Evaluation No. 10.*NASA/Jet Propulsion Lab.

Duce, R.A. (1969). On the source of gaseous chlorine in the marine atmosphere. *J. Geophys. Res.,* **74**, 4597–4599.

Duce, R.A., Winchester, J.W. and Van Nahl, T.W. (1965). Iodine, bromine and clorine in the Hawaiian marine atmosphere. *J. Geophys. Res.,* **70**, 1775–1782.

Dyrssen, D. and Fogelgvist, E. (1981). Bromoform Concentrations in the Arctic Ocean in the Svalbard area. *Oceanol. Acta,* **4**, 313–317.

Elkins, J.W., Thompson, T.M., Swanson, T.H., Butler, J.H., Hall, B.D., Cummings, S.O., Fisher, D.A. and Raffo, A.G. (1993). Decrease in the growth rates of atmospheric chlorofluorocarbons 11 and 12. *Nature,* **364**, 780–783.

Eriksson, E. (1959). The yearly circulation of chloride and sulfar in nature, meteorological, geochemical and pedological implications. *Tellus,* **11**, 375–404.

Fan, S.M. and Jacob, D.J. (1992). Surface ozone depletion in arctic spring sustained by bromine reactions on aerosols. *Nature,* **359**, 522–524.

Farman, J.C., Gardiner, V. and Shanklin, J.D. (1985). Large losses of ozone in Antarctica reveal seasonal ClO_x/NO_x interaction. *Nature,* **315**, 207–210.

Farmer, C.B., Raper, V. and Norton, R.H. (1976). Spectroscopic detection and vertical distribution of HCl in the troposphere and stratosphere. *Geophys. Res. Lett.,* **3**, 13–16.

Finlayson-Pitts, B.J., Livingston, F.E. and, Berko, H.N. (1990). Ozone destruction and bromine photochemistry at ground level in the arctic spring. *Nature,* **343**, 622–625.

Fisher, D.A., Hales, C.H., Filkin, D.L., Ko, M.K., Sze, N.D., Connell, P.S., Wuebbles, D.J., Isaksen, I.S. and Strodal, F. (1990). Model calculations of the relative effects of CFCs and their replacements on stratospheric ozone. *Nature,* **344**, 508–512.

Fogelqvist, E. and Krysell, M. (1991). Naturally and anthropogenically produced bromoform in the Kattegatt, a semi-enclosed oceanic basin. *J. Atmos. Chem.,* **13**, 315–324.

Geschwend, P., MacFarlane, J.K. and Newman, K.A. (1985). Volatile halogenated organic compounds released to sea water from temperate marine macroalgae. *Science,* **227**, 1033–1035.

Harris, G.W., Klemp, D. and Zenker, T. (1992). An upper limit on the HCl near-surface mixing ratio over the Atlantic measured using TDLAS. *J. Atmos. Chem.,* **15**,

327–332.

Helas, G. and Wilson, S.R. (1992). On sources and sinks of phosgene in the troposphere. *Atmos. Environ.,* **16**, 2975–2982.

Jobson, B.T., Niki, H., Yokouchi, Y., Bottenheim, J., Hopper, F. and Leaitch, R. (1994). Measurements of C_2–C_6 hydrocarbons in the high arctic: Evidence for Br and Cl chemistry during low ozone episodes at Alert. *J. Geophys. Res.,* in press.

Junge (1957). Chemical analysis of aerosol particles and gas tracers at the island of Hawaii. *Tellus,* **9**, 528–537.

Keene, W.C., Pszenny, A.A.P., Duce, R.A., Jacob, D.J., Galloway, J.N., Schultz-Tokos, J.J., Sievering, H. and Boatman, J.F. (1990). The geochemical cycling of reactive chlorine through the marine troposphere. *Global Biogeochem. Cycles,* **4**, 407–430.

Khalil, M., Rasmussen, R.A. and Gunawardera, R. (1993). Atmospheric methyl bromide: trends and global mass balance. *J. Geophys. Res.,* **98**, 2887–2896.

Ko, M.K., Sze, N.D., Wang, W., Shia, G., Goldman, A., Murcray, F., Murcray, D. and Rinsland, C. (1993). Atmospheric sulfur hexafluoride: sources, sinks and greenhouse warming. *J. Geophys. Res.,* **98**, 10,499–10,507.

Krasner, S.W., McGuire, M.J., Jacangelo, J.G., Patania, N.L., Reagan, K.M. and Marco-Aieta, E. (1989). The occurrence of disinfection by-products in U.S. drinking water. *J. AWWA,* **81**, 41–53.

Kritz, M.A. and Rancher, J. (1980). Circulation of Na, Cl, and Br in the tropical marine atmosphere. *J. Geophys. Res.,* **85**, 1633–1639.

Legrand, M.R. and Delmas, R.J. (1988). Formation of HCl in the Antarctic atmosphere. *J. Geophys. Res.,* **93**, 7153–7168.

Lobert, J., Scharffe, D., Hao, W., Kuhlbusch, T., Seuwen, R., Warneck, P. and Crutzen, P. (1991). Experimental evaluation of biomass burning emissions: nitrogen and carbon containing compounds. In *Global Biomass Burning: Atmospheric, Climatic, and Biospheric Implications*, Levine, J.S., Ed., MIT Press, Cambridge, MA, pp. 289–304.

Lovelock, J.E. (1971). Atmospheric fluorine compounds as indicators of air movements. *Nature,* **230**, 379.

Lovelock, J.E. (1977). Methyl chloroform in the troposphere as an indicator of OH radical abundance. *Nature,* **267**, 32–33.

Lovelock, J.E., Maggs, R.J. and Wade, R.J. (1973). Halogenated hydrocarbons in and over the Atlantic. *Nature,* **241**, 194–196.

Manley, S.L. and Dastoor, M.N. (1988). Methyl iodide (CH_3I) production by kelp and associated microbes. *Marine Biol.,* **98**, 477–482.

McConnell, J.C., Henderson, G.S., Barrie, L., Bottenheim, J., Niki, H., Langford, C.H. and Templeton, E.J. (1992). Photochemical bromine production implicated in arctic boundary-layer ozone depletion. *Nature,* **355**, 150–152.

McCulloch, A. (1992). Global production and emission of bromochlorodifluoro methane and bromotriflouro methane (Halons 1211 and 1301). *Atmos. Environ.* **26A**, 1325–1329.

Manö, S. and Andreae, M.O. (1994). Emission of methyl bromide from biomass burning. *Science,* **263**, 1255–1257.

Manzer, L.E. (1990). The CFC–ozone issue: progress on the development of alternatives to CFCs. *Science,* **249**, 31–35.

Midgley, P.M. (1989). The production and release to the atmosphere of 1,1,1-trichloroethane (methyl chloroform). *Atmos. Environ.,* **23**, 2663–2664.

Midgley, P.M. and Fisher, D.A. (1993). The production and release to the atmosphere of chlorodifluoromethane (HCFC-22). *Atmos. Environ.,* **27A**, 2215–2223.

Miyake, Y. and Tsunogai, S. (1963). Evaporation of iodine from the ocean. *J. Geophys. Res.,* **68**, 3989–3993.

Molina, M.J. and Rowland, F.S. (1974). Stratosphere sink for chlorofluoromethanes: chlorine atom catalyzed destruction of ozone. *Nature,* **249**, 810–812.

Montzka, S.A., Myers, R.C., Butler, J.H., Elkins, J.W. and Cummings, S.O. (1993). "Global tropospheric distribution and calibration scale for HCFC-22. *Geophys. Res. Lett.,* **20**, 703–706.

Moore, R.M. and Tokarczyk, R. (1993). Volatile biogenic halocarbons in the northwest Atlantic. *Global Biogeochem. Cycles,* **7**, 195–210.

Moyers, J.L. and Duce, R. (1972). Gaseous and particulate bromine in the marine troposphere. *J. Geophys. Res.,* **77**, 5330–5338.

NASA (1994). Report on concentrations, lifetimes, and trends of CFCs, halons, and related species, *Reference Publication 1339*. Kaye, J.A., Penkett, S.A., and Ormond, F.M., Eds. Washington, DC.

Parrish, D.D., Hahn, C.J., Williams, E.J., Norton, R.B., Fehsenfeld, F.C., Singh, H.B., Shetter, J.D., Gandrud, B.W. and Ridley, B.A. (1993). Reply to comments by B. Finlayson-Pitts. *J. Geophys. Res.,* **98**, 14,995–14,997.

Penkett, S.A., Jones, B.M.R., Rycroft, M.J. and Simmons, D.A. (1985). An interhemispheric comparison of the concentrations of bromide compounds in the atmosphere. *Nature,* **318**, 550–553.

Prather, M., McElroy, M.J., Wofsy, S., Russell, G. and Rival, D. (1987). Chemistry of the global troposphere: fluorocarbons as tracers of air motions. *J. Geophys. Res.,* **92**, 6579–6613.

Prather, M. and Spivakovsky, C. M. (1990). Tropospheric OH and lifetimes of hydrochlorofluorocarbons. *J. Geophys. Res.,* **95**, 18723–18729.

Prinn, R.G., Cunnold, D.M., Simmonds, P.G., Alyea, F.N., Boldi, R., Crawford, A.J., Frazer, P., Gutzler, D., Hartley, D., Rosen, R. and Rasmussen, R.A. (1992). Global average concentration and trend for hydroxyl radical deduced from ALE/GAGE trichloroethane (methyl chloroform) data for 1978–1990. *J. Geophys. Res.,* **97**, 2445–2461.

Pszenny, A.P., Keene, W.C., Jacob, D.J., Fan, S., Maben, J.R., Zetwo, M.P., Young-Springer, M. and Galloway, J.N. (1993). Evidence of inorganic chlorine gases other than hydrogen chloride in marine surface air. *Geophys. Res. Lett.,* **20**, 699–702.

Rahn, K.A., Borys, R.D. and Duce, R.A. (1976). Tropospheric halogen gases: inorganic and organic components. *Science,* **192**, 549–550.

Ramanathan, V., Cicerone., R.H., Singh, H.B. and Kiehl, J. (1985). Trace gas trends and their potential role in climate change. *J. Geophys. Res.,* **90**, 5547–5566.

Rancher, J. and Kritz, M.A. (1980). Diurnal fluctuations of Br and I in the tropical marine atmosphere. *J. Geophys. Res.,* **192**, 5581–5587.

Ravishankara, A.R., Solomon, S., Turnipseed, A. and Warren, R.F. (1993). Atmospheric lifetimes of long-lived halogenated species. *Science,* **259**, 194–199.

Reifenhäuser, W. and Heumann, L.G. (1992). Determination of methyl iodide in the antarctic atmosphere and the south polar sea. *Atmos. Environ.,* **26A**, 2905–2912.

Rinsland, C.P., Zander, R., Brown, L.R., Farmer, C.B., Park, J.H., Norton, R.H., Russell, J.M. III. and Raper, O.F. (1986). Detection of carbonyl fluoride in the stratosphere. *Geophys. Res. Lett.,* **13**, 769–772.

Robbins, R.C., Codle, R.D. and Eckhardt, D.L. (1959). The conversion of sodium chloride to hydrogen chloride in the atmosphere. *J. Meteorol.,* **16**, 53–56.

Shah, J.S. and Singh, H.B. (1988). Distribution of volatile organic chemicals (VOC) in the outdoor and indoor air environment: a national VOC data base. *Environ. Sci. Technol.,* **22**, 1381–1388.

Simmonds, P., Cunnold, D., Alyea, F., Cardelino, C., Crawford, A., Prinn, R., Fraser, P., Rasmussen, R. and Rosen, R. (1988). Carbon tetrachloride lifetimes and emissions determined from daily global measurements during 1978–1985. *J. Geophys. Res.,* **1**, 35–58.

Singh, H.B. (1976). Phosgene in the ambient air. *Nature,* **264**, 428–429.

Singh, H.B. (1977a). Atmospheric halocarbons: Evidence in favour of reducing average hydroxyl radical concentration in the troposphere. *Geophys. Res. Lett.,* **4**, 101–104.

Singh, H.B. (1977b). Preliminary estimates of average tropospheric OH concentrations in the Northern and Southern Hemispheres. *Geophys. Res. Lett.,* **4**, 453–456.

Singh, H.B. and Kanakidou, M. (1993). An investigation of the atmospheric sources and sinks of methyl bromide. *Geophys. Res. Lett.,* **20**, 133–136.

Singh, H.B. and Kasting, J.F. (1988). Chlorine–hydrocarbon photochemistry in the marine troposphere and lower stratosphere. *J. Atmos. Chem.,* **7**, 261–285.

Singh, H., Salas, L. and Stiles, R. (1983). Methyl halides in and over the Eastern Pacific (40° N–32° S). *J. Geophys. Res.,* **88**, 3684–3690.

Singh, H.B., Salas, L., Viezee, W., Sitton, B. and Ferek, R. (1992). Measurement of volatile organic chemicals (VOC) at selected sites in California. *Atmos. Environ.,* **26A**, 2929–2946.

Singh, H.B., Gregory, G., Anderson, B., Browell, E., Davis, D., Crawford, J., Talbot, R., Bradshaw, J., Blake, D., Sachse, G., Thornton, D., Newell, R. and Merrill, J. (1994). Low ozone in the marine boundary layer of the tropical Pacific ocean: photochemical loss, chlorine atoms and entrainment. *J. Geophys. Res.,* in press.

Spivakovsky, C., Yevich, R., Logan, J., Wofsy, S. and McElroy, M. (1990). Tropospheric OH in a three dimensional chemical tracer model: an assessment based on observation of CH_3CCl_3. *J. Geophys. Res.,* **95**, 1844–1847.

Sturges, W.T., Cota, G. and Buckley, P.T. (1992). Bromoform emission from arctic ice algae. *Nature,* **358**, 660–662.

Vierkorn-Rudolph, B., Bachmann, K., Scharz, B. and Meixner, F.X. (1984). Vertical profiles of hydrogen chloride in the troposphere. *J. Atmos. Chem.,* **2**, 47–63.

Wang, C.D., Blake, D. and Rowland, F.S. (1994) Seasonal variations in the atmospheric distribution in remote surface locations of a reactive chlorine compound, tetrachloroethylene. *Geophys. Res. Lett.,* in press.

White, R.H. (1982). Analysis of dimethyl sulfonium compounds in marine algae. *J. Marine Res.,* **40**, 529–536.

Wilson, S.R., Crutzen, P.J., Schuster, G., Griffith, D. and Helas, G. (1988). Phosgene measurements in the upper troposphere and lower stratosphere. *Nature,* **334**, 689–691.

WMO (1989). Scientific assessment of stratospheric ozone:1989. *World Meteorological Organization, Report No. 20*, Volume II AFEAS report.

WMO (1991). Scientific assessment of ozone depletion: 1991. *World Meteorological Organization, Report No. 25.*

Woodcock, A.H. (1953). Salt nuclei in marine air as a function of altitude and wind force. *J. Meteorol.,* **10**, 362.

Zafiriou, (1975). Reaction of methyl halides with sea water and marine aerosols. *J. Marine Res.,* **33**, 75–81.

Zetzsch, C. and Behnke, W. (1992). Hetrogeneous sources of atomic Cl in the troposphere. *Ber. Bunsenges. Phys. Chem.,* **96**, 488–493.

Zetzsch, C., Pfahler, G. and Behnke, W. (1988). Hetrogeneous formation of chlorine atoms from NaCl in a photosmog system. *J. Aerosol. Sci.,* **19**, 1203–1206.

8

Sulfur in the atmosphere

H. Berresheim, P.H. Wine, and D.D. Davis

8.1. INTRODUCTION

The atmosphere represents an important sulfur reservoir for living organisms. Compared with other sulfur reservoirs such as the oceans or the lithosphere, it acts as a highly mobile medium in which wind distributes sulfate and other essential nutrients over large areas on a relatively short time scale. Biological processes not only make use of atmospheric sulfur, but in turn have a decisive impact on the chemical composition and total burden of sulfur compounds in the atmosphere. Sulfur in various chemical forms is released from many different

251

organisms as an end-product of their metabolism. In other words, the natural sulfur cycle in our planet's atmosphere is uniquely connected with the geochemical sulfur cycle through the existence and action of life, and in turn through a life-supporting balanced exchange of sulfur between each reservoir. Therefore the presence and fate of sulfur in the atmosphere cannot be understood without knowing its relationship with other sulfur reservoirs, in particular the biosphere. This more holistic view of the atmospheric sulfur cycle as part of the global "biogeochemical" sulfur cycle has now been generally accepted by earth scientists.

Sulfur is one of the major elements essential to life on this planet. Living organisms (including plants) ingest sulfur from their environment, mainly in the form of sulfate or amino acid sulfur. In a process called *assimilatory sulfate reduction*, micro-organisms (bacteria, algae, fungi) and plants use sulfate to build certain proteins to store energy and to support cell growth. During food digestion, energy is released by the decomposition of proteins back into their chemical building blocks, the amino acids. Important sulfur-containing amino acids are cysteine, cystine, and methionine. Their further catabolism results in relatively simple volatile sulfur compounds which the organism releases back into its environment. Animals, including man, are unable to produce sulfur amino acids by this process and thus they depend on plants, fungi, and bacteria as sulfur-supplying food.

Certain micro-organisms living in anoxic environments such as the Earth's crust or the deep oceans obtain energy directly from using sulfate (in place of oxygen) as an electron acceptor. This process is called *dissimilatory sulfate reduction*. The reaction of organic substrate with sulfate during this process can be symbolized by the following equation:

$$2 \ CH_2O \ + \ SO_4^{2-} = \xrightarrow{\text{multisteps}} 2 \ HCO_3^- \ + \ H_2S \qquad (8.1)$$

A substantial fraction of the hydrogen sulfide released during this process combines with iron minerals to form pyrite, FeS, and is thus incorporated into the sediment layers of the lithosphere. In general, the turnover of sulfur in dissimilatory processes is several orders of magnitude faster than in assimilatory processes. The biological sulfur cycle is therefore mainly controlled by anaerobic sulfate-reducing bacteria (Thode, 1991). For a more detailed discussion of the biogeochemical mechanisms involved in assimilatory and dissimilatory sulfate reduction the reader is referred to previous reviews given by other authors (e.g., Krouse and McCready, 1979; Andreae and Jaeschke, 1992; and references therein).

An important part of the sulfur cycle in the Earth's crust is microbial decomposition of buried organic matter involving dissimilatory sulfate reduction and the reaction of H_2S with organic matter. This process represents the major source of sulfur in fossil oil and coal deposits (Orr and White, 1990). Since the beginning of the industrial age, man has been exploiting these fossil deposits as a primary energy source. The combustion of fossil fuels and its expansion on a worldwide scale have led to a dramatic perturbation of the global natural sulfur cycle in the atmosphere. Environmental problems due to anthropogenic sulfur pollution range from local- to global-scale phenomena such as health effects, visibility reduction, acid rain, and global climate change.

Today we know that significant "background" levels of anthropogenic sulfur exist even in the marine atmosphere, e.g., over the North Atlantic, owing to long-range horizontal and vertical mixing processes. However, to this date it still remains a challenging task to assess the potential consequences of global anthropogenic sulfur pollution. Such an evaluation depends to a large extent on our accurate knowledge of the natural, i.e., unperturbed, atmospheric sulfur cycle. Compared with anthropogenic sulfur emissions, current estimates of natural sulfur emissions are more uncertain. Additional sulfur flux measurements are required to reduce these uncertainties.

Furthermore, many aspects of atmospheric sulfur chemistry have yet to be studied. Our understanding of both elementary reaction kinetics and reaction mechanisms involving atmospheric sulfur compounds is crucial to improving current models of the global sulfur cycle as well as parameterizations of important atmospheric processes such as cloud formation and albedo effects in global climate models.

8.2. OVERVIEW OF ATMOSPHERIC SULFUR COMPOUNDS

Four stable isotopes of sulfur are known to occur naturally. Their atomic weight distribution is approximately: 32 (95.0%), 33 (0.76%), 34 (4.22%), and 36 (0.014%), yielding an average value of 32.064. Atmospheric sulfur compounds show small but distinct differences in sulfur isotope ratios, depending on their origin (e.g., burning of coal or oil, sea spray, volcanic emissions, biogenic emissions) and corresponding fractionation processes (Thode, 1991).

Sulfur also occurs in various oxidation states in the atmosphere, mainly as $S(-2)$, $S(+4)$, and $S(+6)$. Compounds containing the highest oxidation state, $S(+6)$, are relatively stable under atmospheric conditions. On the other hand, since the atmosphere is an oxidizing medium, both $S(-2)$ and $S(+4)$ species are subject to reactions with atmospheric oxidants such as O_2, OH, H_2O_2, and O_3. Thus the distribution and residence times of individual $S(-2)$ and $S(+4)$ species in the atmosphere are largely controlled by chemical processes. However, physical processes also play an important role in the atmospheric sulfur cycle. Essentially the cycle begins with the injection of sulfur into the atmosphere and ends with sulfur removal by dry or wet deposition to the Earth's surface. The potential of sulfur compounds to influence physical processes in the atmosphere, such as cloud formation and radiation scattering, increases with oxidation state. This results largely from significant differences in *physical properties* between oxysulfur and reduced sulfur compounds, e.g., corresponding vapor pressures and aqueous solubilities. The following two sections present a short overview of the atmospheric sulfur compounds discussed in this chapter, their physicochemical properties, atmospheric concentrations, and their relationships to important physical processes in the atmosphere.

8.2.1. Physical and Chemical Properties

Table 8.1 presents an overview of the most important sulfur compounds identified to date in polluted and/or relatively clean air. Table 8.2 shows a compilation of basic physical and chemical constants for each compound. Only the free gaseous forms are listed in Table 8.2. With the exception of HMSA, all compounds are known to occur in the atmospheric gaseous phase. Not explicitly included are corresponding ionic species of the same oxidation state occurring in aerosol particles or in the aqueous phase (rain-, fog-, cloud-water, etc.). For example, inorganic sulfates (XSO_4) must be mentioned here as the derivatives of sulfuric acid since they represent a major fraction of the total sulfur mass in the atmosphere. The most important sulfate compounds in the troposphere are ammonium sulfate, $(NH_4)_2SO_4$, and ammonium bisulfate, $(NH_4)HSO_4$, which result from the neutralization or partial neutralization reaction, respectively, between H_2SO_4 and NH_3. Biogenic ammonia emitted mainly from continental environments represents the principal source of free alkalinity in the atmosphere. Compounds formed by aqueous phase absorption and dissociation of gas phase sulfur species are implicitly referred to by the pK values given in Table 8.2.

From a comparison of the Henry law constants in Table 8.2 it follows that the solubility of sulfur compounds increases with oxidation state. Therefore reduced sulfur species occur preferentially in the gas phase whereas the $S(+6)$ compounds are found predominantly in atmospheric (moist) particles or droplets. The extremely low volatility of higher-oxysulfur compounds, in particular H_2SO_4 and MSA, also implies that they have a potential for nucleating new "particles" (low- or subnanometer size range) via self-condensation or binary condensation with water vapor under atmospheric conditions. In general, the chemical reactivity of atmospheric sulfur compounds is inversely related to their sulfur oxidation state. Reduced sulfur compounds are rapidly oxidized by the hydroxyl radical, OH, and to a lesser extent by NO_3 and (possibly) Cl atoms. Their atmospheric lifetimes are typically on the order of a few hours to a few

TABLE 8.1   Currently Identified Atmospheric Sulfur Compounds

Structure and Name	Symbol	Structure and Name	Symbol
H–S–H HYDROGEN SULFIDE	(H_2S)	$$HO-\overset{\overset{O}{\|\|}}{\underset{\underset{O}{\|\|}}{S}}-OH$$ SULFURIC ACID	(H_2SO_4)
CH_3–S–CH_3 DIMETHYL SULFIDE	(DMS)	$$CH_3-\overset{\overset{O}{\|\|}}{\underset{\underset{O}{\|\|}}{S}}-OH$$ METHANE SULFONIC ACID	(MSA)
S=C=S CARBON DISULFIDE	(CS_2)	$$CH_3-\overset{\overset{O}{\|\|}}{\underset{\underset{O}{\|\|}}{S}}-CH_3$$ DIMETHYL SULFONE	($DMSO_2$)
O=C=S CARBONYL SULFIDE	(OCS)		
CH_3–S–S–CH_3 DIMETHYL DISULFIDE	(DMDS)	$$CH_3O-\overset{\overset{O}{\|\|}}{\underset{\underset{O}{\|\|}}{S}}-OH$$ MONOMETHYL SULFATE	($MMSO_4$)
CH_3–S–H METHYL MERCAPTAN	(MeSH)	$$CH_3O-\overset{\overset{O}{\|\|}}{\underset{\underset{O}{\|\|}}{S}}-OCH_3$$ DIMETHYL SULFATE	($DMSO_4$)
O=S=O SULFUR DIOXIDE	(SO_2)	$$HOCH_2-\overset{\overset{O}{\|\|}}{\underset{\underset{O}{\|\|}}{S}}-OH$$ HYDROXYMETHANE SULFONIC ACID	(HMSA)
CH_3–S–CH_3 $\overset{\|\|}{O}$ DIMETHYL SULFOXIDE	(DMSO)	$$HOCH_2-\overset{\overset{O}{\|\|}}{\underset{\underset{O}{\|\|}}{S}}-CH_2OH$$ BIS-HYDROXYMETHYL SULFONE	($BHMSO_2$)

days. One exception is OCS which has an estimated lifetime of approximately 4 years (Chin, 1992). The residence times of the S(+6) compounds in the troposphere are mainly determined by their physical removal via dry and wet deposition and are typically on the order of a few days. For some organic S(+6)

TABLE 8.2 Physical and Chemical Properties of Atmospheric Sulfur Compounds

Symbol	Mol. Wt.	M.P. (°C)	B.P. (°C)	K_H (25 °C) (mol l^{-1} atm^{-1})		pK_1		pK_2	
Reduced sulfur									
H_2S	34.08	−85.5	−60.7	0.102	[MH]	6.98	[MH]	17.1	[G]
DMS	62.13	−98.3	37.3	0.559	[DW]				
CS_2	76.14	−111.5	46.2	0.044	[AD]	hb			
OCS	60.08	−138.2	−50.2	0.021	[AD]	hb			
DMDS	94.20	−84.7	109.7	0.909	[AD]				
MeSH	48.11	−123.0	6.2	0.394	[AD]	10.70			
Lower oxysulfur									
SO_2	64.07	−72.7	−10	1.42	[M]	1.86	[HUE]	7.20	[SI]
DMSO	78.13	18.4	189	9.74E4	[WB]	32–33	[SJ]		
Higher oxysulfur									
H_2SO_4	98.08	10.4	330	>1E9	e [RO, A]	−3		1.96	
MSA	96.10	20	167^a	6.5E13/mH$^+$	[CB]	−1.86	[CW]		
$DMSO_2$	94.13	110	238	1E5–1E6	e [W]	st		e	
$MMSO_4$	112.10	<−30	130–140 d	<1E5	e [HE]	dis	[HE]		
$DMSO_4$^{b}	126.13	−31.7	188.5 d	<1E5	e [HE]	hm	[HE]		
HMSA	112.10	?	ust	ug	e [HE]	<0	[JH]	11.28	[BEH]
$BHMSO_2$	126.13	?	ust	>1E5	e [HE]	<0	e	?	

All data from Lide (1991), *CRC Handbook of Chemistry and Physics*, 72 Edn, CRC Press, Boca Raton, FL, and Dean (1992), *Lange's Handbook of Chemistry*, 14th Edn, McGraw Hill, New York, NY, unless specific references are given in brackets. Henry law constants (K_H) refer to physical solubility in pure water. An operational constant is given for MSA [CB] following [BD]; mH$^+$ is the molal H$^+$ concentration.
^{a}At 10 nmHg. ^{b}Mutagen and suspected carcinogen. K_H estimated from solubility and/or vapor pressure data given in associated references; e, estimate based on indicated literature; d, decomposes; hb, hydrolyzes to form bisulfide [J, AC]; hm, hydrolyzes to from $MMSO_4$ and H_2SO_4; dis, disproportionates into $DMSO_4$; ug, occurrence in atmospheric gas phase unknown; st, stable; ust, unstable.

[AD] Adewuyi (1989), in *Biogenic Sulfur in the Environment*, Saltzman and Cooper (Eds), *ACS Symp. Ser.*, Vol. 393 p. 529, Am. Chem. Soc., Washington, DC; [AC] Adewuyi and Carmichael (1987), *Environ. Sci. Technol.*, **21**, 170; [A] Ayers et al. (1980), *Geophys. Res. Lett.*, **7**, 433; [BEH] Betterton et al. (1988), *Environ. Sci. Technol.*, **22**, 92; [BD]Brimblecombe and Dawson (1984), *J. Atmos. Chem.*, **2**, 95; [CW] Clark and Woodward (1966), *Trans. Faraday Soc.*, **62**, 2226; [CB] Clegg and Brimblecombe (1985), *Sci. Technol. Lett.*, **6**, 269; [DW] Dacey et al. (1984), *Geophys. Res. Lett.*, **11**, 991; [G] Giggenbach (1971), *Inorg. Chem.*, **10**, 1333; [HE] Hansen and Eatough (1991); [HUE] Huss and Eckert (1977), *J. Phys. Chem.*, **81**, 2268; [JH] Jacob and Hoffmann (1983), *J. Geophys. Res.*, **88**, 6611; [J] Johnson (1981), *Geophys. Res. Lett.*, **8**, 938; [M] Maahs (1983), *J. Geophys. Res.*, **88**, 10721; [MH] Millero and Hershey (1989), in *Biogenic Sulfur in the Environment*, Saltzman and Cooper, Eds, *ACS Symp. Ser.*, Vol. 393, p. 282, Am. Chem. Soc., Washington, DC; [R] Roedel (1979), *J. Aerosol Sci.*, **10**, 375; [SI] Sillen (1964), *Stability Constants of Metal-Ion Complexes. I. Inorganic Ligands*, 2nd Edn, *Chem. Soc. (London) Spec. Publ. 17*; [SJ] Stewart and Jones (1967), *J. Am. Chem. Soc.*, **89**, 5069; [WB] Watts and Brimblecombe (1987), *Environ. Technol. Lett.*, **8**, 483; [W] Watts et al. (1990).

species, such as $MMSO_4$ and $DMSO_4$, photolysis and/or methylation reactions may also be important removal pathways. The latter two compounds are chemically interdependent in the atmosphere and are important to mention because $DMSO_4$ is a known mutagen and suspected carcinogen (Hansen and Eatough, 1991).

8.2.2. Atmospheric Mixing Ratios

Table 8.3 presents measured atmospheric mixing ratios, in parts per trillion by volume (ppt), of the most important sulfur gases in the marine and remote continental atmosphere. This list may not be complete. We believe, however, that the data provide a representative up-to-date view of geographical distributions and trends of the individual sulfur gases in the corresponding regions. These trends and the distributions of other sulfur compounds not listed in Table 8.3 are summarized in the following text.

8.2.2.1. Reduced Sulfur Compounds

DMS

To our present knowledge the major sources and sinks of atmospheric DMS are assimilatory sulfate reduction and photochemical oxidation by OH, respectively. It is now well established that DMS is the dominant sulfur compound emitted from the world's oceans. A precursor of marine DMS is believed to be dimethylsulfonium propionate (DMSP) which is formed and released by certain phytoplankton species (e.g., *Phaeocystis pouchetii*) and bacteria. Aerobic organic soils and certain land plants (e.g., corn, alfalfa, wheat, and the major salt marsh plant *Spartina Alterniflora*) are significant terrestrial sources of DMS.

Following the early studies by Lovelock et al. (1972), numerous measurements of DMS have been conducted in the marine atmosphere and in seawater. Based on these measurements, the DMS mixing ratio in the marine boundary layer (MBL) varies from < 10 ppt in polluted air to 50–100 ppt over oligotrophic waters and up to about 1 ppb over eutrophic (e.g., coastal, upwelling) waters. The global average DMS mixing ratio at sea surface level is approximately 100 ppt. Typically, DMS mixing ratios decline rapidly with altitude to a few ppt in the free troposphere (FT). However, occasionally significant amounts of DMS are "pumped" into the FT by convective clouds and large-scale storm systems such as hurricanes. The latter occur with an average frequency of about 50 year^{-1} and can last for many days. The diurnal cycle of DMS in the MBL strongly depends on photochemical reactions, air mass origin, MBL height, and DMS emission rates. DMS levels with average night/day ratios between 1.0 (no variation) and 3.7 have been reported. The seasonal variation in DMS roughly follows the primary productivity cycle in the oceans. In temperate and high latitudes, spring and summer DMS mixing ratios are typically one order of magnitude higher than in winter.

In continental environments a number of DMS flux measurements have been conducted. However, relatively few atmospheric mixing ratios have been reported. Typical continental surface DMS levels are on the order of 20 ppt or less in rural areas and forests but can reach high ppt levels over wetlands and downwind from industrial sources such as Kraft pulp and paper mills. Recently

TABLE 8.3 Observed Mixing Ratios of Atmospheric Sulfur Gases (ppt)

Location	Range	Average[a]	Remarks	Reference(s)
Dimethylsulfide				
A. Marine surface layer[b]				
Atlantic Ocean	<1–980	80	≤100 ppt: MC and/or OW; $R = 1.0–3.7$ / >100 ppt: MM and/or EW; $R = 1.4–2.3$ / Peak spring/winter ratio: 112 (North Sea)	1–11
Pacific Ocean	5–670	110	5–400 ppt: MM, North Pacific; $R \approx 1.0–1.6$ / <20–670 ppt: MM, South Pacific	1, 12–17, & 74–76
Indian Ocean (Amsterdam I.)	13–400	108	MM; $R \approx 1.6$ / Summer/winter ratio: 4–23	18–20
Southern Ocean (south of 40°S)	2–1048	100	MM, EW, and OW; $R = 1.4–1.6$ (fall, summer) / Summer/winter ratio: 10–20 (Cape Grim)	1, 21–24
B. Continental surface layer				
Tropical forests (Amazon, Congo)	5–75	20	No difference between dry and wet seasons; $R \leq 2$ (mainly due to mixing)	25–27
Temperate forests and rural areas	<1–20	8	Pine forest and rural areas; $R \approx 2.5$	4,28
	<10–400	45	Wheat field, spring and summer; $R \approx 3.9$	29
Wetlands, temperate NH	<1–500	60	Coastal wetlands, marshes	30, 31
	4–140	45	Swamps, moors	4, 31
City, industry	440–560	500	Downwind from paper mills (10–70 km)	31, 32
C. Free troposphere (2–5 km)				
Marine (Atlantic, Pacific, S. Ocean)	<1–3	1.5	Relatively stratified atmosphere	2–4, 6, 12 & 17, 22, 74
	5–19	15	Strong convection and/or DMS emissions	
Tropical forests	<1–5	1.5		25–27
Germany/Bavaria	<0.1	<0.1		4
Hydrogen sulfide				
A. Marine surface layer[b]				
Atlantic Ocean	<2.5–75	7.5	MM, $R = 1.4–1.7$, North Atlantic	5, 6, 33
Coastal waters	10–260	65	Mainly advection from coastal wetlands	5, 6, 33
Pacific Ocean	<0.4–14	3.6	MM, no diel trend, equatorial Pacific	34, 75

	Range	Value	Location/notes	References
B. Continental surface layer				
Tropical forests	10–60	35	Amazon, Congo; max. day/night ratio: 4–8	25–27
	105–6400	780/3750	Ivory Coast; above oxic/anoxic soils	35
Temperate forests and rural areas	5–150	60	Pine forests and rural areas	35–37
	<10–350	100	Wheat field, spring and summer	29
Wetlands, NH	<3–2740	840	Coastal wetlands, marshes	31, 37, 38
	80–820+	450	Swamps, moors	4, 31, 37, 38
City, industry	80–810+	365		38–41
C. Free troposphere (2–5 km)				
Tropical forests	7–13	8.5	Amazon, Congo	26, 27
	70		Ivory Coast	35
Rural areas, NH	1–7		France, Germany/Bavaria	4, 35
	140	6	Advection from coastal wetland areas	4
Carbon disulfide				
A. Marine surface layer[b]				
Atlantic Ocean	1–15	2	Marine air	4, 5, 7, 9, 42
	5–30	18	MM, with continental background air	4, 7, 42
	4–420	150	MC, advection from pollution sources	4, 42
Pacific Ocean	<6		Marine air	43
	>20		Advection of polluted air	43
Southern Ocean	<7–35+	<7	MM, >7 ppt only in polluted air	24
B. Continental surface layer				
Temperate forest and rural areas	5–106	35	Pine forest and rural areas	4, 28, 44–46
	15–560	80	Wheat field, spring and summer	29
Wetlands, NH	5–390	120	Coastal wetlands, marshes	31
City, industry	65–370	190		44, 47
C. Free troposphere (2–5 km)				
Atlantic Ocean	1–8	5	MM, North Sea; polluted air: ≈30 ppt	4
	65–166	115	At 6.1 km, possibly continental air	48
Pacific Ocean	<6		Marine air	43
Coastal waters	1–21	5.7	At 6 km, MM, mainly SE Pacific coast	49
Continental	<3–18	7	Continental background air	4, 49, 50

TABLE 8.3 (*Continued*)

Location	Range	Average[a]	Remarks	Reference(s)
Carbonyl sulfide				
Total troposphere				
Northern hemisphere	300–1800	512 + 119	All data (n = 1066), incl. pollution	51
10°N–85°N, 52°E–155°W		514 ± 64	Flight data only (n = 464), no trends	51 (& 4, 7, 43, 48)
Arizona and Swiss Alps		530 ± 40	No trend over last decade	52
A. Marine surface layer[b]	400–800	500	Advection of continental air: > 600 ppt	4, 7, 24, 43,
(All oceans)			North Sea: 740 ppt (0.2 km)	& 53–56
B. Continental surface	300–1800	545	Max. values in polluted air and wetlands	29
	500–7000		Wheat field, summer	
Tropopause and	280–520	380	At 10–15 km, declines with altitude	46, 57, 58
stratosphere	<10–30	16	At 27–31 km	57–59
Sulfur dioxide[c]				
Total troposphere				
Pacific and Western	MBL: 54 ± 19,	FT: 85 ± 28	Marine atmosphere	60
N. America (57°S–70°N)	MBL: 112 ± 79,	FT: 160 ± 100	Continental atmosphere	
Coastal Europe/N. &	20–1100	260 ± 257	At 0–4 km, decrease to 4 km; max.: 35°N–65°N	61
S. America/Africa/Atlantic	20–300	120 ± 84	At 4–10 km, increase to 7 km, then decrease	
65°N–60°S, 80°W–15°E	48–138	78 ± 31	Lower stratosphere; higher than below tropopause	
A. Surface layer (All oceans)	<4–160	20	Marine air; summer/winter ratio: 6–9	12, 20–22,
			at Amsterdam I. (Indian Ocean)	& 60–65, 74–76
Spitsbergen (1983–1992)	14–1880	210	Monthly averages, winter/summer ratio: 10	66
Atlantic, E. Pacific	9–770	130	Near coasts, mixed marine/continental air	8, 13, 16
Atlantic, E. Pacific	150–6000	1500	MC, polluted air	8, 67–69
Amazon forest	10–50	25	Up to 1 km, without biomass burning	25, 26
Rural areas, NH	80–7000	≈2000	Varies strongly with air mass	28, 31, 70

B. Free troposphere (2–5 km)

North Pacific	10–300	≥20	Air masses from Asia and/or Mt Pinatubo	12, 64
Southern Ocean	12–20	16 ± 3	Marine air; South Pacific: 10–80 ppt	22
Amazon forest		17	Wet and dry season	25, 26

C. Upper FT (>5 km)/lower stratosphere

Europe, North Sea, and Arctic	10–400	50	Refs 70–72: constant profile Ref. 73: strong decrease to 10 ppt	46, 70–73

Data with potentially significant bias due to use of inadequate oxidant scrubbers (DMS data) or interference by OCS (H$_2$S data) are not included in this table. Surface layer data include measurements up to ca 300 m altitude.

[a] Weighed by number and spatial coverage of data. [b] Coastal wetlands not included. [c] Remote regions only. MBL, marine boundary layer; FT, free troposphere; NH, northern hemisphere; EW, eutrophic waters; OW, oligotrophic waters; R, average night-time/daytime minimum ratio; MM, mainly marine air; MC, mainly continental air.
1 Andreae et al. (1985), J. Geophys. Res., 90, 12,891; 2 Ferek et al. (1986), Nature, 320, 514; 3 Van Valin et al. (1987), Geophys. Res. Lett., 14, 715; 4 Georgii et al. (1987), Ber. Inst. Meteorol. Geophys., 68, Univ. Frankfurt; 5 Saltzman and Cooper (1988), J. Atmos. Chem., 7, 191; 6 Andreae et al. (1990), EOS, 71, 1254; 7 Bates and Johnson, (1990), EOS, 71, 1255; 8 Berresheim et al. (1991), Tellus B, 43, 353; 9 Cooper and Saltzman (1991), J. Atmos. Chem., 12, 153; 10 Bürgermeister and Georgii (1991), J. Atmos. Chem., Environ., 25A, 587; 11 Davison and Hewitt (1992), J. Geophys. Res., 97, 2475; 12 Andreae et al. (1988), J. Atmos. Chem., 6, 149; 13 Bates et al. (1990), J. Atmos. Chem., 10, 59; 14 Quinn et al. (1990), J. Geophys. Res., 95, 16,405; 15 Bandy et al. (1992), EOS, 73, 28; 16 Bates et al. (1992), J. Geophys. Res., 97, 9859; 17 Berresheim et al. (1993), J. Geophys. Res., 98, 12,701; 18 Nguyen et al. (1990), J. Atmos. Chem., 11, 123; 19 Nguyen et al. (1992); J. Atmos. Chem., 15, 39; 20 Putaud et al. (1992); J. Atmos. Chem., 15, 117; 21 Berresheim (1987), J. Geophys. Res., 92, 13,245; 22 Berresheim et al. (1990), J. Atmos. Chem., 10, 341; 23 Ayers et al. (1991), Nature, 349, 404; 24 Staubes and Georgii (1993), Tellus B, 45, 127; 25 Andreae and Andreae (1988), J. Geophys. Res., 93, 1487; 26 Andreae et al. (1990), J. Geophys. Res., 95, 16,813; 27 Bingemer et al. (1992), J. Geophys. Res., 97, 6207; 28 Berresheim and Vulcan (1992), Atmos. Environ., 26A, 2031; 29 Hofmann et al. (1992), in Precipitation Scavenging and Atmosphere–Surface Exchange, Schwartz and Slinn, Eds, Hemisphere, Washington, DC, p. 967; 30 Maroulis and Bandy (1977), Science, 196, 647; 31 Berresheim (1993), Atmos. Environ., 27A, 211; 32 Bürgermeister (1984), M.S. Thesis, Univ. Frankfurt; 33 Andreae et al. (1991), J. Geophys. Res., 96, 18,753; 34 Yvon et al. (1993), J. Geophys. Res., 98, 16,979; 35 Delmas et al. (1980), J. Geophys. Res., 85, 4468; 36 Natusch et al. (1972), Anal. Chem., 44, 2067; 37 Jaeschke et al. (1978), Pure Appl. Geophys., 116, 463; 38 Jaeschke and Herrmann (1981), Int. J. Environ. Anal. Chem., 10, 107; 39 Jaeschke et al. (1980), J. Geophys. Res., 85, 5639; 40 Breeding et al. (1973), J. Geophys. Res., 78, 7057; 41 Slatt et al. (1978), Atmos. Environ., 12, 981; 42 Kim and Andreae (1987), J. Geophys. Res., 92, 14,733 43 Blomquist et al. (1992), EOS, 73, 27; 44 Maroulis and Bandy (1980), Geophys. Res. Lett., 7, 681; 45 Jones et al. (1983), J. Geophys. Res., 90, 10,483; 49 Tucker et al. (1985), Geophys. Res. Lett., 12, 9; 50 Bandy et al. (1981), Geophys. Res. Lett., 8, 1180; 51 Bandy et al. (1992), J. Atmos. Chem., 14, 527; 52 Rinsland et al. (1992), J. Geophys. Res., 97, 5995; 53 Rasmussen et al. (1982), Atmos Environ., 16, 1591; 54 Johnson and Harrison (1986), J. Geophys. Res., 91, 7883; 55 Bingemer et al. (1990), J. Geophys. Res., 95, 20617; 56 Mihalopoulos et al. (1991), J. Atmos. Chem., 13, 73; 57 Inn et al. (1979), Geophys. Res. Lett., 6, 191; 58 Leifer (1989), J. Geophys. Res., 94, 5173; 59 Zander et al. (1988), J. Geophys. Res., 93, 1669; 60 Maroulis et al. (1980), J. Geophys. Res., 85, 7345; 61 Ockelmann (1988), Ber. Inst. Met. Geophys., 75, Univ. Frankfurt; 62 Herrmann and Jaeschke (1984), J. Atmos. Chem., 1, 111; 63 Pszenny et al. (1989), J. Geophys. Res., 94, 9818; 64 Thornton et al. (1992), EOS, 73, 28; 65 Saltzman et al. (1993), J. Atmos. Chem., 17, 73; 66 H. Dovland and E. Joranger (1993), personal communication; 67 Thornton et al. (1987), Global Biogeochem. Cycles, 1, 317; 68 Saltzman et al. (1986), J. Geophys. Res., 91, 7913; 69 Luria et al. (1987), Atmos. Environ., 21, 1631; 70 Georgii and Meixner (1980), J. Geophys. Res., 85, 7433; 71 Jaeschke et al. (1976), Geophys. Res. Lett., 9, 517; 72 Meixner (1984), J. Atmos. Chem., 2, 175; 73 Möhler and Arnold (1992), Geophys. Res. Lett., 19, 1763; 74 Thornton and Bandy (1993), J. Atmos. Chem., 17, 1; 75 Yvon et al. (1992), EOS, 73, 81; 76 Huebert et al. (1993), J. Geophys. Res., 98, 16,985.

DMS levels up to 1 ppb have also been measured in an African rain forest as a result of high DMS emissions from certain tree species (J. Kesselmeier, 1993, personal communication).

H_2S

As discussed earlier, H_2S is the dominant end-product of dissimilatory sulfate reduction. It is rapidly oxidized in the atmosphere by OH. Although marine sediments are a major source of H_2S, very little escapes into the atmosphere (from the open ocean) due to its oxidation in aerobic surface waters. On a global scale H_2S is mainly emitted from terrestrial sources such as anoxic soils, tidal flats, plants, and various industrial sources (e.g., paper and cellulose industry). It is also the dominant reduced sulfur gas emitted by volcanoes.

Average surface-level H_2S mixing ratios over remote oceans are 5–15 ppt and show approximately 1.5 times higher values at night than during daytime. Values ranging between 100 and 300 ppt have been measured over shallow coastal waters, while values up to 1 ppb, with large spatial and temporal fluctuations, have been observed over tidal flats, coastal wetlands, and swamps. In temperate latitudes H_2S levels range between 5 and 150 ppt over vegetated and non-vegetated soils. Relatively moderate H_2S mixing ratios (30–40 ppt at ground level, 1–10 ppt at 3 km) have been measured in and above tropical forests, with the exception of very high levels observed in an Ivory Coast forest (Table 8.3, reference 35). H_2S mixing ratios up to 1 ppb and up to 100 ppb have been measured downwind from industrial sources and in volcanic plumes, respectively.

OCS

Due to its low chemical reactivity, OCS is the most abundant sulfur gas in the global background atmosphere. It is also the only sulfur gas (except during large volcanic eruptions) which can diffuse into the stratosphere where it can be photolyzed at < 388 nm (energetic threshold). Major tropospheric sinks for OCS appear to be uptake by plants and hydrolysis in natural waters.

Measurements of atmospheric OCS and CS_2 mixing ratios and fluxes have been reviewed in detail by Chin and Davis (1993). OCS shows an average tropospheric mixing ratio of approximately 500 ± 50 ppt and no significant variation with altitude up to about 15 km. In the stratosphere it declines rapidly to levels of 10–20 ppt between 25 and 35 km. Long-term measurements over remote marine and continental regions have shown no statistically significant seasonal or secular trends of OCS. However, individual measurements over the Atlantic Ocean have shown elevated OCS levels (up to 800 ppt) in the northern hemisphere, indicating a possible impact of anthropogenic sources on global atmospheric OCS levels.

Other Reduced Sulfur Species

CS_2 is the major chemical precursor for OCS in the atmosphere. Its mixing ratio in the lower troposphere varies from less than 10 ppt over open ocean areas to 15–50 ppt in background continental air and several hundred ppt in polluted air. Most airborne studies have shown a strong decrease in CS_2 levels with altitude (typically 5–7 ppt at 2–5 km in background air) but relatively high levels (up to 166 ppt at 6 km) have also been observed. DMDS and MeSH are probably only important in the vicinity of specific industrial sources such as Kraft pulp mills and refineries. Atmospheric mixing ratios of MeSH over the surface ocean and in tropical forests are typically < 5 ppt.

8.2.2.2. Oxysulfur Compounds

SO_2

Sulfur dioxide is the most abundant anthropogenic sulfur pollutant. It is predominantly emitted through fossil fuel combustion and is also a major product of the oxidation of reduced sulfur compounds in the atmosphere. Boundary layer SO_2 mixing ratios in continental background air range from about 20 ppt to 1 ppb. They can reach tens of ppm in power plant and volcanic plumes. In the marine boundary layer downwind from anthropogenic sources, SO_2 levels can still be as high as a few ppb. Due to its solubility (Table 8.2) and high reactivity in the aqueous phase, atmospheric SO_2 can be rapidly removed by absorption and oxidation in cloud and fog droplets ("rain-out") or during precipitation events ("wash-out").

Average SO_2 values measured in relatively unpolluted MBL air range between 20 and 50 ppt. Airborne studies by Maroulis and co-workers in the marine atmosphere revealed an increase in SO_2 levels from approximately 50 ppt in the MBL to 85 ppt in the free troposphere. Earlier aircraft measurements by Meixner, Jaeschke, Georgii, and others over continental areas showed a vertical decrease in SO_2 levels to about 30–70 ppt at 6 km and no further decline up to and above the tropopause at approximately 12 km (Warneck, 1988). Recent measurements over the Arctic (Table 8.3) reveal yet another vertical trend, with SO_2 levels slightly decreasing up to the tropopause and rapidly decreasing in the lower stratosphere.

H_2SO_4, MSA, and Their Aerosol and Aqueous Phase Ions

Gaseous sulfuric acid in the atmosphere is mainly a product of the reactions of SO_2 with OH and of SO_3 with water vapor. The only known precursor of gaseous methanesulfonic acid, MSA(g), in the atmosphere is DMS. Since both H_2SO_4 and MSA(g) rapidly condense and/or interact with particles in the atmosphere, their vapor mixing ratios are very low in background air. A mass spectrometric detection technique for both gases has only recently been developed (Eisele and Tanner, 1993). It is based on atmospheric pressure, selected ion chemical ionization mass spectrometry (SI/CI/MS), and offers a real-time detection sensitivity of 10^{-3} ppt. Initial field measurements with this technique (Berresheim et al., 1993a; Eisele and Tanner, 1993) have shown a strong diurnal variation in both gases basically tracking the solar radiation flux. H_2SO_4 levels in rural continental and marine air ranged from a few ppq (parts per quadrillion) at night to 1.2 ppt at midday. A relatively smaller range, under mostly cloudy conditions, was observed for MSA(g) in marine air (0.002–0.2 ppt).

In contrast with the above-mentioned measurements in the gas phase, numerous measurements of sulfate and methanesulfonate, MSA(p), in atmospheric particles and droplets have been made by various investigators. Anthropogenic sulfate pollution and the distribution of continental sulfate in the atmosphere are discussed in detail in Chapters 4, 5, and 12. Aerosol sulfate mixing ratios typically range from less than 200 ppt over remote continental areas to about 1–2 ppb in rural areas and $\geqslant 2$ ppb (24 h average) in polluted air. (For aerosol compounds we define mixing ratios in moles of substance/mole of air; e.g., 1 ppt $= 10^{-12}$ mole/mole; 22.4 ppt $= 1$ nmol m^{-3} at STP.) Aqueous phase concentrations over the continents are typically 10–50 μM in rainwater, 20–100 μM in cloudwater, and several hundred μM in fogwater (e.g., NAPAP, 1991).

Generally, less than half of the sulfate mass in the marine atmosphere derives from sea water sulfate. Other sources such as reduced sulfur oxidation, advection of soil dust, and anthropogenic sulfate contribute about 50–80%. The concentration of this so-called non-sea-salt sulfate, nss-SO$_4^{2-}$, is usually calculated from the total sulfate and sodium content of the aerosol:

$$[\text{nss-SO}_4^{2-}] = [\text{SO}_4^{2-}(\text{total})] - 0.0603\,[\text{Na}^+] \tag{8.2}$$

The brackets in equation (8.2) represent mol m^{-3} units and the factor on the right-hand side represents the average [SO$_4^{2-}$]/[Na$^+$] mole ratio in seawater, which is assumed to be constant. Levels of nss-SO$_4^{2-}$ in the remote MBL vary from about 5–50 ppt over the Southern Ocean to 50–200 ppt over most of the Pacific to 100–700 ppt over the northern central Atlantic. In the northern hemisphere significant enhancements of nss-SO$_4^{2-}$ levels in both the MBL and FT have been observed as a result of long-range transport of continental air. A continental influence appears to be negligible over most of the remote Southern Ocean where nearly constant vertical profiles of nss-SO$_4^{2-}$ have been measured. Seasonal variations observed in the MBL over the Pacific and northern Atlantic show a maximum in fall and a minimum in spring (amplitude ratio 1.6–1.7) (Savoie and Prospero, 1989). A much larger seasonal variation (factor of 5–10; 7 ppt in midwinter, 50 ppt in mid-summer) has been recorded at Cape Grim, Tasmania (40° S) (Ayers et al., 1991). However, with respect to all the data reported it must be kept in mind that calculations of nss-SO$_4^{2-}$ levels based on equation (8.2) frequently involve the subtraction of two large numbers (i.e., total SO_4^{2-} minus sea salt SO_4^{2-}) leading to high uncertainties in nss-SO_4^{2-} values.

MSA(p) mixing ratios are mainly determined by the distribution of DMS emission sources and by deposition processes. Values of < 10 ppt and 5–100 ppt have been measured in rural and urban air, respectively. MSA(p) levels in the MBL are on the order of 5–15 ppt, with maximum values occurring at high latitudes. The vertical distribution of aerosol MSA(p) typically shows values of $\leqslant 5$ ppt at 3 km in the marine FT. Aqueous MSA(p) concentrations have also been measured in marine cloudwater, with values ranging between 0.2 and 5.2 μM (Table 8.3, reference 22). The seasonal cycle of MSA(p) in the MBL appears to increase with latitude. Summer/winter ratios of 12–25 and higher have been reported for latitudes $\geqslant 40°$ in both hemispheres. A similar variation has also been observed for MSA(p) in rainwater at Cape Grim (1.0–1.6 μM in summer, 0.1 μM in winter).

DMSO and $DMSO_2$

Dimethylsulfone and its precursor, dimethylsulfoxide, are potentially important oxidation products of DMS. Measurements of both gases in ambient air were first conducted by Harvey and Lang (1986) using an off-line preconcentration-sampling technique (typical sample integration time 8–12 h). For both gases values between 0.6 and 1.6 ppt were found in coastal Miami air. More recent measurements over the North Atlantic (Pszenny et al., 1990) showed a much larger range of mixing ratios for DMSO (<0.7–271 ppt) and $DMSO_2$ (<0.7–56 ppt). The relatively long sampling times required make the above technique unsuitable for mechanistic studies of DMS chemistry in the atmosphere. Recently, a real-time SI/CI/MS method has been developed for DMSO with a detection limit of 0.5 ppt at a statistical signal integration time of 60 s (Berresheim et al., 1993b). First measurements with this technique have been conducted at an elevated coastal site in Washington state (Berresheim et al., 1993a). DMSO showed a significant diurnal variation ranging from <0.5 ppt at night to 3.2 ppt during daytime. The potential implications of these measurements for our understanding of atmospheric DMS chemistry will be discussed in Section 8.3.2.1.

Liquid-based techniques have been used to measure DMSO and $DMSO_2$ in rainwater and aerosol. Coastal rainwater samples showed DMSO concentrations between <0.1 and 100 nM (Harvey and Lang, 1986). Both compounds have also been measured in marine aerosol (Watts et al., 1990), showing summer maxima of 0.1 and 0.03 ppt, respectively. Winter minima were about 30–50 times lower.

Other Oxysulfur Compounds

HMSA is formed in atmospheric hydrosols via addition of formaldehyde to bisulfite. It largely resists oxidation by dissolved O_3 and H_2O_2. Direct measurements of HMSA in urban fog samples have shown concentrations of $0-140\,\mu M$ (Munger et al., 1986). Other studies have inferred HMSA concentrations from measuring the difference between the total $S(+4)$ and "free" sulfite/bisulfite content in hydrosols (e.g., Chapman, 1986). Total $S(+4)$ levels in marine stratus clouds and in precipitation were typically $\leqslant 10\,\mu M$, but values up to 3 mM have been measured in urban fog. The results indicate that HMSA and possibly other aldehyde–bisulfite adducts may represent a substantial or sometimes even dominant fraction of $S(+4)$ in the atmospheric aqueous phase.

Measurements and the atmospheric chemistry of $MMSO_4$, $DMSO_4$, and $BHMSO_2$ have recently been reviewed by Hansen and Eatough (1991). The former two compounds occur predominantly in the gas phase. They are most likely formed from the reaction of methanol with sulfuric acid in the atmosphere. Mixing ratios between 45 and 450 ppt have been measured in urban air and may reach levels two to five times higher than SO_2 downwind from power plant emissions. $BHMSO_2$ has only been found in aerosol particles, typically at 40–350 ppt in urban air. However, under conditions favorable for its production, concentrations may range up to one order of magnitude higher than those of aerosol sulfate. A mechanism proposed by Hansen and Eatough (1991) for $BHMSO_2$ formation in the atmosphere involves reactions of ethene, ozone, and SO_2.

8.3. THE ATMOSPHERIC SULFUR CYCLE

An overview of the most important processes involved in the atmospheric sulfur cycle is given in Figure 8.1. In addition, other more temporary phenomena can also be of importance on both regional and global scales. Noteworthy among these are biomass burning and rapid vertical mixing of air in connection with cumulonimbus clouds or large storm systems such as hurricanes.

Anthropogenic sources emit predominantly SO_2 into the atmosphere. The major sulfur gases released by natural sources are DMS, H_2S, OCS, and CS_2. Only volcanism emits significant amounts of natural SO_2 into the air. Most of the sulfate mass injected through sea spray, soil erosion by wind, volcanism, or industrial activities is concentrated in relatively large micron-size particles. Unless this aerosol is injected into the free troposphere or higher altitudes, e.g., by dust storms or volcanic eruptions, it is typically deposited back to the surface after only a few hundred kilometers of transport due to gravitational settling. On the other hand, fine particle submicron sulfate aerosol has a much longer atmospheric lifetime on the order of a few days in the troposphere and a few years in the stratosphere (see Chapter 5). Gas phase oxidation of reduced sulfur gases and SO_2 represent major indirect sources of fine particle sulfate. As already discussed in Section 8.2.1, submicron sulfate particles are produced from further

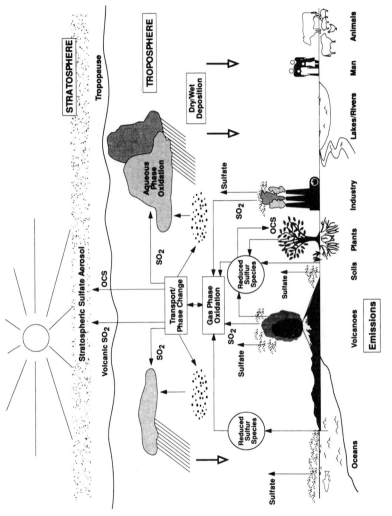

FIGURE 8.1.   Major processes involved in the atmospheric sulfur cycle.

oxidation of $S(+4)$ to H_2SO_4 followed by homogeneous or heterogeneous nucleation of sulfuric acid. These particles are believed to be major sources of cloud condensation nuclei (CCNs) in the atmosphere (Twomey, 1971). The cycling of sulfur through the atmosphere is eventually terminated by dry or wet deposition to the Earth's surface, with all its environmental implications as indicated in Figure 8.1.

8.3.1. Emissions

Table 8.4 presents our current best estimates of global sulfur fluxes into the atmosphere. The corresponding values are based on a critical evaluation of recent estimates published by various authors. Estimates of the anthropogenic sulfur flux are well established by now with a relatively narrow range of uncertainty. The results suggest that natural sources represent roughly 30% of the total sulfur emissions into the atmosphere ignoring contributions from sea-salt and soil dust sulfate. Bates et al. (1992b) estimate about 24% globally based on a somewhat lower value for the oceanic sulfur flux. According to the data in Table 8.4, anthropogenic sources contribute about 80% to the emissions in the northern hemisphere (NH) and 30% in the southern hemisphere (SH).

8.3.1.1. Anthropogenic Emissions

The most recent and most detailed estimates to date of anthropogenic sulfur emissions have been presented by Spiro et al. (1992), Hameed and Dignon (1992), and Dignon and Hameed (1989). Figure 8.2 shows a revised update of the historical trend of anthropogenic sulfur emissions since the beginning of the industrial age. Approximately 90–94% of these emissions are from the NH. Based on the data in Figure 8.2, anthropogenic sulfur emissions surpassed natural sulfur emissions in the NH by about 1910 and on a global scale during the decade of the 1950s. This conclusion is further supported by Greenland ice core studies (Mayewski et al., 1990; Neftel et al., 1985) and model calculations (Langner et al., 1992) indicating a dramatic increase in atmospheric sulfate levels in the NH since about 1890. Figure 8.2 also includes the temporal variation in anthropogenic CO_2 emissions over the same period. The similarity between the two trends clearly indicates a strong correlation between man-made sulfur emissions and fossil fuel combustion. Recent estimates suggest that approximately 87–94% of the anthropogenic sulfur in the atmosphere is contributed from fossil fuel combustion, coal coking, and petroleum uses. Based on earlier work by Warneck (1988) and Brimblecombe et al. (1989), we estimate that sulfate and reduced sulfur compounds each represent about 3% of the total anthropogenic sulfur released into the atmosphere. The bulk portion (94%) is emitted as SO_2.

TABLE 8.4 Global Sulfur Emission Estimates (Tg S year^{-1})

Source	H$_2$S	DMS	CS$_2$	OCS	SO$_2$	SO$_4$	Total[a]
Fossil fuel combustion + industry	<0.01?	—	Total reduced S: 2.2	0.08	70	2.2	71–77 (mid-1980s) (68/6)
Biomass burning	<0.3		<0.01?	0.17	2.8	0.1	2.2–3.0 (1.4/1.1)
Oceans		15–25	0.08		—	40–320	15–25 (8.4/11.6)[b]
Wetlands	0.006–1.1	0.003–0.68	0.0003–0.06	0.0006–0.12		—	0.01–2 (0.8/0.2)
Plants + soils	0.17–0.53	0.05–0.16	0.02–0.05	0.01–0.03	—	2–4	0.25–0.78 (0.3/0.2)[c]
Volcanoes	0.5–1.5	—	—	0.01	7–8	2–4	9.3–11.8 (7.6/3.0)
Anthropogenic (total)							73–80
Natural (total, without sea-salt and soil dust)							25–40
Total							98–120

[a]Numbers in parentheses are fluxes from northern hemisphere/southern hemisphere.
[b]Excluding sea-salt contributions.
[c]Excluding soil dust contributions.

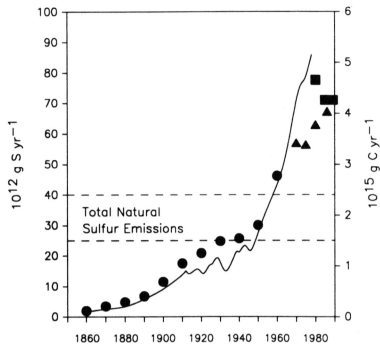

FIGURE 8.2.   Global anthropogenic SO_2 and CO_2 emissions since 1860. Data are from Dignon and Hameed (1989; circles), Hameed and Dignon (1992; triangles), and Spiro et al. (1992; squares). Data for CO_2 (solid curve) are from Bolin et al. (1981). The dashed lines indicate the estimated range for the global natural sulfur flux (Table 8.4).

Figure 8.2 shows recent deviations between sulfur and CO_2 emission trends since about the mid-1960s illustrating the effect of abatement policies on SO_x (SO_2 + SO_4^{2-}) emissions in North America and Western Europe. Spiro et al. (1992) also took into account the improved use of sulfur recovery technologies in the former USSR and estimate that global man-made sulfur emissions have actually decreased during the 1980s. On the other hand, the study by Hameed and Dignon (1992) shows a significant geographical shift in anthropogenic sulfur emissions. Asia (not including the former USSR) has become the leading sulfur polluting continent since about 1981 with a continuing increase in anthropogenic sulfur emissions ($> 20\,Tg\,S\,year^{-1}$ in 1986), while those from North America ($< 15\,Tg\,S\,year^{-1}$ in 1986) have steadily declined since about 1970. The emissions from Asia have nearly doubled between 1975 and 1986 mainly because of the increased use of coal combustion in China. On a country-by-country basis, the largest emitters of SO_x in 1986 were the USSR, USA, China, India, and Poland as shown in Figure 8.3.

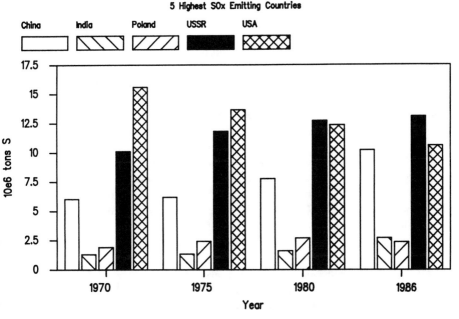

FIGURE 8.3. Recent trends in anthropogenic SO$_x$ emissions from the five highest emitter countries. (Reproduced from Hameed and Dignon 1992, with permission.)

8.3.1.2. Biomass Burning

Approximately 95% of global biomass burning is human-initiated (Bates et al., 1992b). A detailed review of sulfur emissions from biomass burning has been given by Andreae (1991). SO$_2$ and OCS appear to be the only sulfur compounds emitted in significant amounts, while H$_2$S and CS$_2$ emissions are probably on the order of a few per cent of the OCS flux (Chin and Davis, 1993). These minor sulfur compounds (including OCS) are predominantly released from smoldering ash after the end of the active burning phase. Recently Penner et al. (1992) calculated the amount of smoke emitted from biomass burning to be 114 Tg year^{-1}. Using this value and a sulfur/carbon emission ratio of 0.85 g S/ kg C (Andreae, 1991), we estimate that about 0.1 Tg S year^{-1} of sulfate is emitted from biomass burning. The uncertainties in all these estimates are high. Even the total burning area is only known with a certainty of about ± 50% (Robinson, 1989). However, it appears likely that biomass burning is the dominant atmospheric sulfur source in tropical latitudes (25° N–25° S) during the burning season (Bates et al., 1992b), although on a global annual basis it represents a relatively small source (Table 8.4).

8.3.1.3. Natural Emissions

Measurements of sulfur fluxes from different natural sources have been reviewed by a number of authors, most recently by Andreae and Jaeschke (1992), Bates et al. (1992b), and Aneja (1990). Natural sulfur emissions typically show large spatial and temporal variations which make it difficult to extrapolate individual measurements on a global scale. On the other hand, a significant number of field studies have been conducted over the last two decades resulting in continuous improvements in global flux estimates. Although some of the estimated ranges shown in Table 8.4 are still large, it appears that sulfur emissions from wetlands, plants, and soils may only be of local or regional importance. The most significant natural sulfur sources on a global scale are oceanic and volcanic sulfur emissions. However, our present estimate for the total natural sulfur flux still has an uncertainty of at least ± 30% compared with less than ± 10% for anthropogenic emissions. Clearly more studies are needed to further reduce these uncertainties.

Oceans

Current estimates of oceanic DMS emissions range from 11.9–15.4 Tg S year^{-1} (Spiro et al., 1992; Bates et al., 1992b; Erickson et al., 1990) to 19–51 Tg S year^{-1} (Andreae and Jaeschke, 1992). Some of the uncertainties in these estimates are due to the calculation of the oceanic DMS flux, which is usually based on measured DMS concentrations in surface seawater and a parameterization scheme for gas exchange across the air–sea interface (e.g., Liss and Merlivat, 1986). Other major uncertainties arise from the temporal and spatial variability of DMS in seawater. The average estimate by Andreae and Jaeschke (1992; 35 Tg S year^{-1}) is probably too high due to (a) an overemphasis of data from tropical and upwelling regions, (b) the assumption of a single global value for the vertical transfer velocity v_p, of DMS in the sea-water column, and (c) the lack of wintertime measurements in their data set. Bates et al. (1992b) include winter data and a spatial and seasonal variability of v_p in their estimate. Spiro et al. (1992) and Erickson et al. (1990) use field data reported by Bates and co-workers and calculate a global v_p field based on climatological data.

However, Bates et al. (1992b) likely underestimated DMS emissions from the Southern Ocean by at least a factor of 2. Berresheim (1987) calculated the DMS flux from the ocean area south of 40° S to be 6.4 Tg S year^{-1} based on seasonal primary productivity rates and his measurements during the austral fall season in this region. Bates et al. (1992b) estimate a DMS flux of only 3.5 Tg S year^{-1} from the ocean area between 35° S–80° S. Recent measurements by McTaggart and Burton (1992) and Gibson et al. (1990) show very high DMS concentrations in Antarctic surface waters during the austral summer season, on the average 18.5 nM in open ocean waters and as much as 290 nM in coastal waters. In comparison, the global average DMS sea surface concentration is on the order of

2 nM. Based on their measurements, Gibson et al. (1990) estimate the annual DMS flux from the Antarctic shelf area alone to be 1.3 Tg S year^{-1}. This suggests that the global estimate by Bates et al. (1992b) probably represents a lower limit. In summary, we consider the DMS flux range given in Table 8.4 to be the most reasonable estimate at the present time.

The estimated H$_2$S flux in Table 8.4 is based on recent measurements by Andreae et al. over the North Atlantic and by Yvon et al. over the equatorial Pacific (Table 8.3, references 33 and 34). OCS and CS$_2$ flux estimates have been adopted from Chin and Davis (1993). The flux of sea-salt sulfate into the atmosphere represents probably the largest uncertainty in all the estimates listed in Table 8.4. Corresponding values have been adopted from Andreae and Jaeschke (1992) bracketing those estimated by Bates et al. (1992b; 131–275 Tg S year^{-1}).

Wetlands

Biogenic sulfur emissions from wetlands (tidal flats, coastal salt marshes, freshwater marshes, and swamps) can vary by up to six orders of magnitude in time and space (see, e.g., Aneja, 1990; Adams et al., 1980). Overall, the contributions of wetlands to the atmospheric sulfur cycle are now considered to be of regional but not global importance owing to the relatively small area of wetland regions. Bates et al. (1992b) estimate the global wetland area to be 3.4 ×10^6 km^2 in the NH and 0.3 ×10^6 km^2 in the SH. They estimate a corresponding global sulfur flux of only 0.008 Tg S year^{-1} with approximately 84% of the emissions in the NH. Andreae and Jaeschke (1992) estimate a total flux of 1.76 Tg S year^{-1} with relative contributions from H$_2$S (57%), DMS (34%), OCS (6%), and CS$_2$ (3%). We have applied this relative distribution to our present estimate in Table 8.4. Based on the reviews by Andreae and Jaeschke (1992) and Aneja (1990), we consider 2 Tg S year^{-1} as an upper limit.

Soils and Plants

A wide range of reduced sulfur compounds are emitted from soils (Banwart and Bremner, 1975), from crops and other lower plants (Fall et al., 1988), and from trees (Rennenberg, 1991). Extensive studies of these sulfur sources have been conducted in temperate and subtropical latitudes. All these studies have shown a strong positive correlation between sulfur emissions and temperature and light conditions. However, early global estimates based on these observations have drastically overestimated sulfur fluxes from tropical land masses as recent measurements in Amazonia and the Congo rain forest have shown (Table 8.3). In addition to temperature and light intensity, terrestrial sulfur fluxes also depend strongly on plant and soil types and on other factors such as water and nutrient availability. Andreae and Jaeschke (1992) estimate the annual sulfur flux from tropical forests to range between 0.23 and

$0.76\,\mathrm{Tg\,S\,year}^{-1}$ assuming a tropical forest area of $14.5 \times 10^{12}\,\mathrm{m}^2$ and an area-based average flux of $30\text{--}100\,\mathrm{ng\,S\,m}^{-2}\,\mathrm{min}^{-1}$.

For the rest of the total land area ($88 \times 10^{12}\,\mathrm{m}^2$) they calculate a flux of $3.5\,\mathrm{Tg\,S\,year}^{-1}$, resulting in a global estimate of $3.8\text{--}4.3\,\mathrm{Tg\,S\,year}^{-1}$. However, their estimate for the extratropical regions is probably too high or miscalculated. Andreae and Jaeschke (1992) state that their corresponding estimate is based on average flux data reported in the natural sulfur emission survey established recently by Guenther et al. (1989) for the area of the United States. However, a corresponding value of only $0.19\,\mathrm{Tg\,S\,year}^{-1}$ results using the national average flux of $4.0\,\mathrm{ng\,S\,m}^{-2}\,\mathrm{min}^{-1}$ reported by Guenther et al. (1989). This would add up with the tropical flux estimated by Andreae and Jaeschke (1992) to only $0.42\text{--}0.95\,\mathrm{Tg\,S\,year}^{-1}$ emitted from terrestrial biogenic sulfur sources. Using a somewhat higher total area estimate (Bates et al., 1992b; $1.3 \times 10^{14}\,\mathrm{m}^2$), values between 0.48 and $1.01\,\mathrm{Tg\,S\,year}^{-1}$ are obtained. Corresponding results by Bates et al. (1992b; $0.35\,\mathrm{Tg\,S\,year}^{-1}$) and Spiro et al. (1992; $0.91\,\mathrm{Tg\,S\,year}^{-1}$) support this correction of the Andreae and Jaeschke (1992) estimate.

The sulfur emission inventory by Guenther et al. (1989) does not account for a possible recycling or uptake of reduced sulfur gases from the atmosphere by plants and/or soils. Several studies (reviewed by Chin and Davis, 1993) have shown that plant uptake may represent a major sink for atmospheric OCS. In their review Chin and Davis (1993) estimate the global uptake flux of OCS by plants to be $\approx 0.23\,\mathrm{Tg\,S\,year}^{-1}$ assuming that OCS emissions from vegetation are negligible on a global basis. If we correct our estimated range of $0.48\text{--}1.01\,\mathrm{Tg\,S\,year}^{-1}$ by this value, we obtain the final estimate shown in Table 8.4 for the net global sulfur flux from soils and plants. Fluxes for individual sulfur gases have been derived on a percentage basis as proposed by Andreae and Jaeschke (1992) and Aneja (1990): H_2S, 68%; DMS, 21%; CS_2, 7%; OCS (net), 4%. Also, the estimate by Andreae and Jaeschke (1992) for soil dust sulfate emissions has been adopted here. Overall, the total estimate for the terrestrial biogenic sulfur flux (including emissions from wetlands) bears a large uncertainty second only to that for the flux of sea-salt sulfate into the atmosphere.

Volcanoes

The range of global volcanic sulfur emissions given in Table 8.4 is based on previous estimates by Stoiber et al. (1987; $9.3\,\mathrm{Tg\,S\,year}^{-1}$) and Berresheim and Jaeschke (1983; $11.8\,\mathrm{Tg\,S\,year}^{-1}$). Estimated fluxes for individual sulfur compounds are largely based on the work of Berresheim and Jaeschke (1983) and for OCS and CS_2, the review by Chin and Davis (1993). Estimates for the hemispheric contributions (72% from the NH, 28% from the SH) have been adopted from Bates et al. (1992b).

Although Table 8.4 shows a relatively narrow range for the global volcanic sulfur flux, the corresponding year-to-year variation in this flux can be very large. The estimated annual average flux depends very much on the period over which the sulfur flux has been averaged and on the frequency of large explosive eruptions as well as the amount of non-eruptive contributions during the same period. Berresheim and Jaeschke (1983) emphasized that their average flux estimate for the period 1961–1979 was probably somewhat low due to the lack of direct measurements during highly explosive eruptions. However, over the last decade sulfur fluxes from large eruptions have been well documented by airborne and/or remote sensing measurements. The eruptions of El Chichon in 1982 and Pinatubo in 1991 alone contributed about 3 and 7.5 Tg of S to the atmosphere, respectively (A. Krueger, 1992, personal communication). This suggests that annual volcanic sulfur emissions (eruptive plus non-eruptive) are on the order of at least 10% of the total anthropogenic sulfur flux into the atmosphere and may range up to 30% during years with highly explosive eruptions.

8.3.2. Chemical Transformations and Mechanisms

In this section we shall discuss results obtained from laboratory and field studies that have significantly contributed to our understanding of atmospheric sulfur chemistry. Mechanistic models derived from these results will be discussed and modifications will be proposed where deemed appropriate based on our current knowledge. We will mainly focus on the atmospheric transformations of DMS in the remote marine atmosphere, and those of H_2S, CS_2, OCS, and SO_2 in the remote marine and continental atmosphere. By comparison, SO_2 oxidation reactions in the polluted atmosphere will be discussed only briefly. For more details concerning the role of SO_2 as an atmospheric pollutant the reader is referred to Chapters 1, 4, and 12. We would also like to refer the reader to the review by Hansen and Eatough (1991) for a detailed discussion of the atmospheric chemistry of organic S(+ 6) pollutants (HMSA, $MMSO_4$, $DMSO_4$, and $BHMSO_2$), which have not been considered in the following text. Figure 8.4 gives a schematic overview of some of the more important reduced sulfur compounds in the atmosphere and what are thought to be their principal reaction and removal pathways.

8.3.2.1. The Remote Marine and Continental Atmosphere
In the *remote* marine atmosphere the most important reactive sulfur gas is DMS. Other reduced sulfur compounds such as H_2S and CS_2 are believed to contribute less than 20% to S(+ 4) and S(+ 6) levels over the remote oceans. The polluted marine atmosphere would include mainly coastal and open ocean regions where continental air masses predominate. Field measurements have shown that large

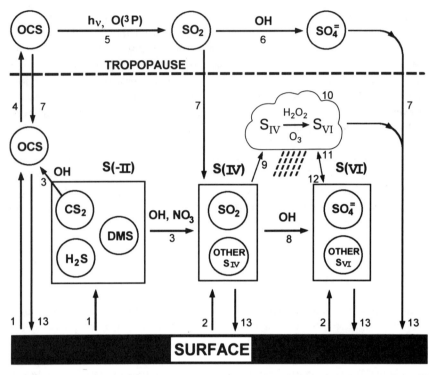

FIGURE 8.4. Major pathways of reduced and oxidized sulfur compounds in the atmosphere: (1) emission of DMS, H$_2$S, CS$_2$, and OCS; (2) emission of S($+4$) and S($+6$); (3) oxidation of DMS, H$_2$S, and CS$_2$ by OH, and DMS by NO$_3$ in the troposphere; (4) transport of OCS into the stratosphere; (5) photolysis of OCS or reaction with O(^{3}P) to form SO$_2$ in the stratosphere; (6) oxidation of SO$_2$ in the stratosphere; (7) transport of stratospheric OCS, SO$_2$, and sulfate into the troposphere; (8) oxidation of SO$_2$ and other S($+4$) products by OH in the troposphere; (9) absorption of S($+4$) – mainly SO$_2$ – into hydrosols (cloud/fog/rain droplets, moist aerosol particles); (10) liquid phase oxidation of S($+4$) by H$_2$O$_2$(aq) and O$_3$(aq) in hydrosols (and by O$_2$ in the presence of elevated levels of catalytic metal ions); (11) absorption/growth of S($+6$) aerosol – mainly sulfate – into hydrosols; (12) evaporation of cloud-water leaving residual S($+6$) aerosol; (13) deposition of OCS, S($+4$), and S($+6$). The SURFACE bar represents anthropogenic and/or natural sources and sinks (e.g., industry, soils, plants, seawater, ice sheets) depending on the type of environment being discussed.

oceanic regions in the northern hemisphere, in particular the North Atlantic, can be periodically influenced by long-range advection of air from bordering industrialized countries (Table 8.3). The major impact of these anthropogenically influenced air masses on marine atmospheric sulfur is to further enhance nss-SO$_4^{2-}$ levels over the oceans via the transport of both anthropogenic SO$_2$ and sulfate. They also increase the percentage contribution of H$_2$S and CS$_2$ to the total reduced sulfur observed in the marine atmosphere.

DMS Oxidation in the Remote Marine Atmosphere

As will be discussed in detail in Section 8.4.1, one of the centrally important issues surrounding oceanic emissions of DMS is its potential role as a climate-modulating species. The general mechanism by which this might occur involves the oxidation of the sulfur atom in DMS to the $+6$ state followed by heteromolecular nucleation which leads to the formation of new CCN particles. However, both theoretical considerations as well as experimental observations suggest that only the fraction of DMS that is oxidized to gas phase H_2SO_4 plays a significant role in new CCN formation (e.g., Kreidenweis and Seinfeld, 1988). The two most important sulfur reactants involved in this production are the species SO_2 and SO_3. In the case of SO_2, gas phase H_2SO_4 is produced via its reaction with OH radicals; for the species SO_3, it is reaction with gas phase H_2O. Since the yields of both SO_2 and SO_3 from DMS are strongly mechanism-dependent, any quantitative assessment of the DMS–CCN linkage is critically dependent on there being available a detailed understanding of the DMS oxidation scheme.

Historically, both field observations and laboratory studies have contributed to our current understanding of atmospheric DMS chemistry. For example, early field measurements of MSA in atmospheric aerosols (Panter and Penzhorn, 1980) gave the first indication that this compound and possibly other organic sulfur compounds were produced in the atmospheric oxidation of DMS. Similarly, since the early work of Cox and Sandalls (1974), a number of photochemical chamber studies involving DMS–NO_x-air mixtures have identified still other DMS oxidation products and have also led to the investigation of the relative yields of these products as function of varying experimental conditions. In general, all studies seem to agree that SO_2 and MSA(g) are two of the major products from the DMS + OH reaction. However, the results from individual research groups have strongly disagreed with respect to the relative yields of these products. The chamber studies by Niki et al. (1983) and Hatakeyama et al. (1985) suggest that MSA is the dominant product, with lesser amounts of SO_2 being formed. By contrast, the studies by Yin et al. (1986) and Grosjean (1984), although qualitatively showing similar products, indicate quite different relative yields of SO_2 and MSA, SO_2 being by far the dominant product. Similar results were also obtained in a more recent study by Yin et al. (1990) involving an outdoor photochemical chamber.

At this time there appears to be no simple explanation for the major differences between the results of Hatakeyama et al. and Yin and co-workers. Differences in chamber temperatures and concentration levels of initial reactants have been most frequently cited as the reason(s) for the observed differences; but no experimental study has yet confirmed this. In spite of these apparent difficulties, chamber studies have provided a basis for examining the larger mechanistic picture for the oxidation of DMS. In this context they have been instrumental in conceptually identifying potentially important reactive intermediates as well as more stable products such as DMSO and $DMSO_2$. Both of the latter species have now been observed in recent field studies (see Section 8.2.2).

With respect to providing "definitive" answers to DMS oxidation product yields in the remote MBL, however, chamber studies as previously configured appear to have severe limitations. The reasons are twofold. (1) The surface-to-volume ratio of a typical irradiation chamber is orders of magnitude greater than that of a real MBL atmosphere, thus leading to abnormally high surface loss rates for some products and heterogeneous surface production of others. (2) The mixing ratios of the DMS–NO_x reactants in most chamber studies exceed those in the remote MBL by several orders of magnitude. For example, NO levels in the remote MBL are typically 2–8 ppt (Davis et al., 1987; McFarland et al., 1979) (the latter NO mixing ratio corresponds to approximately 8–25 ppt of NO_x) and DMS levels are 10–300 ppt (see Section 2.2). By comparison, DMS mixing ratios in chamber studies have ranged from 0.3 to 20 ppm and those of NO_x from 0.2 to 18 ppm. The latter levels of reactants can dramatically alter the competitive kinetics of many of the more critical DMS oxidation steps. Thus they can potentially lead to significantly different final product yields as compared with those found under remote marine atmospheric conditions.

Detailed kinetic studies designed to assess the aforementioned complications in the DMS oxidation scheme have been the focus of several laboratory investigations in recent years. These have been critically reviewed by other authors (e.g., Turnipseed and Ravishankara, 1993; Tyndall and Ravishankara, 1991; Yin et al., 1990) and thus only a brief summary will be presented in this section of the text for purposes of aiding the reader in the follow-on discussions of DMS field studies. Later in the text a more comprehensive discussion of the findings from these earlier kinetic studies will be presented. For reader convenience a summary of recommended rate constants for reactions currently thought to be important in the atmospheric oxidation of DMS is also provided in Table 8.5(a).

Current evidence suggests that the atmospheric oxidation of DMS is initiated by one of two species, OH or NO_3. The OH species dominates during daylight hours and NO_3 at night. However, the relative importance of these two pathways is most likely a strong function of latitude, altitude, and season of the year, reflecting significant variations in the levels of both NO_x and HO_x (see discussion later in text). The rate constant for the DMS + OH reaction alone (Table 8.5(a)) suggests a DMS lifetime of approximately 1 day assuming a daily average OH concentration of 2.5×10^6 molecules cm^{-3}.

A number of details concerning the DMS + OH and DMS + NO_3 oxidation mechanisms have been revealed to date. For example, laboratory studies by Hynes et al. (1986) have shown that the DMS + OH reaction branches into two channels: (1) hydrogen abstraction from a methyl group, forming the radical product

TABLE 8.5(a)   Rate Constants for Some Important Reactions in the Atmospheric Oxidation of DMS

Reaction	A^a	E/R^b	$k(298\ K)^a$	Reference(s)
$OH + (CH_3)_2 \rightarrow CH_3SCH_2 + H_2O$	960	234	440	1
$OH + (CH_3)_2S + M \leftrightarrow (CH_3)_2SOH + M$ $(CH_3)_2SOH + O_2 \rightarrow$ products	c	c	170	1, 2
$NO_3 + (CH_3)_2S \rightarrow CH_3SCH_2 + HNO_3$	19	−520	110	2
$Cl + (CH_3)_2S \rightarrow CH_3SCH_2 + HCl$	18000	0	18000	3
$Cl + (CH_3)_2S + M \rightarrow (CH_3)_2SCl + M$	6600	−260	16000	3
$IO + (CH_3)_2SO \rightarrow (CH_3)_2SO + I$			< 1.2	2
$OH + (CH_3)_2SO \rightarrow$ products			8100	4, 5
$NO_3 + (CH_3)_2SO \rightarrow NO_2 + (CH_3)_2SO_2$			17	5
$Cl + (CH_3)_2SO \rightarrow$ products			7400	5
$CH_3SCH_2 + O_2 + M \rightarrow CH_3SCH_2O_2 + M$			570^d	6
$CH_3SCH_2O_2 + NO \rightarrow CH_3SCH_2O + NO_2$	1010	−130	1550	6, 7
$CH_3SCH_2O + M \rightarrow CH_3S + CH_2O + M$			e	7, 8
$CH_3SCH_2O_2 + HO_2 \rightarrow CH_3SCH_2OOH + O_2$	60	−700	630	2^f
$CH_3S + O_2 + M \rightarrow CH_3SOO + M$	0.14	−1550	25	9^g
$CH_3SOO + M \rightarrow CH_3S + O_2 + M$	17400^h	3430	0.174^h	9^g
$CH_3S + O_2 \rightarrow$ products			< 0.0006	9, 10
$CH_3S + O_3 \rightarrow$ products	200	−290	530	11
$CH_3S + NO_2 \rightarrow CH_3SO + NO$	2060	−320	6030	11
$CH_3SOO + O_3 \rightarrow$ products				11^i
$CH_3SOO + NO_2 \rightarrow$ products	$(2200)^j$	$(0)^j$	$(2200)^j$	11
$CH_3SOO + NO \rightarrow$ products	$(1100)^k$	$(0)^k$	$(1100)^k$	11
$CH_3SO + O_3 \rightarrow$ products			100	10
$CH_3SO + NO_2 \rightarrow$ products			1200	12

^{a}Units are $10^{-14}\ cm^3\ molecule^{-1}\ s^{-1}$ except where otherwise indicated.

^{b}Units are degrees Kelvin.

^{c}The effective rate constant for *irreversible* addition at 298 K is $k = (4.1 \times 10^{-31}[O_2])/(1 + 4.1 \times 10^{-20}[O_2])$.

$^dP = 760\ Torr$.

^{e}Generally assumed to be very fast; recent experimental work (refs 6 and 7) is consistent with this assumption.

^{f}Arrhenius parameters estimated based on known parameters for HO_2 reactions with CH_3O_2, $C_2H_5O_2$, and $CH_3C(O)O_2$. Products are estimated based on results from the $CH_3O_2 + HO_2$ reaction.

^{g}Rate constants are for 760 Torr air; they are estimated to be 10 times faster than those measured (ref. 8) in 80 Torr He.

^{h}Units are $10^7\ s^{-1}$.

$^i k < 8 \times 10^{-13}\ cm^3\ molecule^{-1}\ s^{-1}$ at $T = 227\ K$.

^{j}Based on data obtained over temperature range 227–246 K.

^{k}Based on data obtained over the temperature range 227–256 K.

1 Hynes et al. (1986), *J. Phys. Chem.*, **90**, 4148; 2 Atkinson et al. (1992), *J. Phys. Chem. Ref. Data*, **21**, 1125; 3 Stickel et al. (1992), *J. Phys. Chem.*, **96**, 9875; 4 Hynes et al. (1992), *12th Int. Symp. on Gas Kinetics, Reading*, Abstract G6; 5 Barnes et al. (1989), in *Biogenic Sulfur in the Environment*, Saltzman and Cooper, Eds, *ACS Symp. Ser.*, Vol. 393, pp. 476–488; 6 Wallington et al. (1993), *J. Phys. Chem.*, **97**, 8442; 7 Zhao, Stickel, and Wine (1994), personal communication; 8 A. Ravishankara (1993), personal communication; 9 Turnipseed et al. (1992), *J. Phys. Chem.*, **96**, 7502; 10 Tyndall and Ravishankara (1989), *J. Phys. Chem.*, **93**, 4707; 11 Turnipseed et al. (1993), *J. Phys. Chem.*, **97**, 5926; 12 Domine et al. (1990), *J. Phys. Chem.*, **94**, 5839.

CH_3SCH_2, and (2) addition of OH to the sulfur atom. The abstraction channel dominates in the absence of O_2, whereas, the addition channel involves oxidation of the adduct with O_2 in competition with adduct decomposition back into DMS + OH. In the presence of O_2 both channels have equivalent rates at approximately 285 K, with H abstraction dominating at higher temperatures. Both the addition and abstraction branches as well as the overall observed rate constant, k_{obs}, are shown as a function of temperature in Figure 8.5. Concerning the second DMS initiator species, NO_3, recent laboratory studies suggest that the most likely primary products of this reaction are HNO_3 and CH_3SCH_2. As a result, both the DMS/OH abstraction channel and the DMS/NO_3 reaction lead to the same sulfur-containing free radical, CH_3SCH_2 (e.g., see Figure 8.6(a)). Thus the ensuing chemistry from these two different initiator species is indistinguishable.

Field studies aimed at understanding the oxidation mechanism of atmospheric DMS have typically focused on one of two aspects of this chemistry: (a) the influence of temperature on the relative distributions of measurable stable products; (b) assessments of the diurnal trends in DMS and its associated products in the marine boundary layer. As illustrated in several proposed DMS oxidation schemes (Turnipseed and Ravishankara, 1993; Bandy et al., 1992; Yin et al., 1990; see also Figures 8.6(a) and 8.6(b)), the chemical degradation of DMS involves numerous competitive kinetic steps. Among the more important of these is the aforementioned OH initiation step which can involve either addition or abstraction. Both processes are temperature-dependent, but so also are many

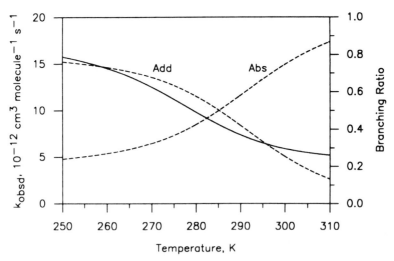

FIGURE 8.5. Temperature dependence of the branching ratio between the addition (Add) and abstraction (Abs) channels in the DMS + OH reaction (dashed curves). Also shown is the overall observed rate constant k_{obsd} (solid curve).

other critical DMS reactions, and still others have a critical dependence on the absolute levels of photogenerated free radicals. As a result, the DMS product distribution for a given set of environmental conditions as well as the diurnal trends in DMS might be expected to have a significant but perhaps complex dependence on both temperature and the level of photochemical activity.

Qualitatively, some field studies seem to confirm aspects of the above cited predictions, while in other cases there seem to be conflicting results. For example, as discussed earlier in the text, two of the known end-products of the OH-initiated degradation of DMS are H_2SO_4 (sulfate; see Figure 8.4) and MSA. Of the two, sulfate formation is now viewed mechanistically as defining the higher-energy pathway. In fact, field studies over remote tropical and subtropical ocean areas have shown that the mole ratio of MSA(p)/nss-SO_4^{2-} in aerosol ranges from about 0.02 to 0.1. By contrast, at higher latitudes this ratio has been found to vary from 0.15 to 1 (Huebert et al., 1993; Bates et al., 1992a; Ayers et al., 1991; Savoie and Prospero, 1989; Berresheim, 1987; Saltzman et al., 1983). Collectively, therefore, these data do suggest an inverse correlation of the MSA(p)/nss-SO_4^{2-} ratio with temperature. Conflicting with this simple interpretation, though, are seasonal data obtained at Cape Grim, Tasmania (40° S) and Norfolk Island (30° S) in the South Pacific Ocean, both of which show an average maximum MSA(p)/nss-SO_4^{2-} ratio of 0.1–0.2 during the austral summer and a minimum of 0.02–0.06 during winter (Ayers et al., 1991; Saltzman et al., 1986). Whether this is indeed a conflict or simply a reflection of uncertainties in the data is not clear at this time, (see discussion on methodology employed in evaluating the quantity nss-SO_4^{2-}, Section 8.2.2). Such data also raise some important questions concerning our level of understanding of the sources of nonsea-salt sulfate. For example, the assumption that DMS oxidation accounts for the bulk of the observed nss-SO_4^{2-} may be poorly founded in some cases. Several airborne field studies conducted over low- and mid-latitude oceanic regions have produced data showing an increase in the SO_2 mixing ratio with increasing altitude above the MBL. These observations suggest that the source of MBL nonsea-salt sulfate could have resulted from entrainment of SO_2 air from the FT rather than oxidation of DMS (e.g., Table 8.3, references 12, 60, and 74; Bandy et al., 1992). Possible sources of high-altitude sulfate include long-range transport of anthropogenic SO_2 and/or SO_4^{2-}, volcanic emissions, and convective pumping of precursor species (e.g., DMS, H_2S, CS_2) into the FT (Chatfield and Crutzen, 1984). As discussed earlier in the text, a third factor that may have influenced the reported trend in the MSA/nss-SO_4^{2-} ratio data is this ratio's dependence on the levels of photochemical free radicals, the latter being a strong function of latitude and/or season of the year. This point will be discussed in greater detail later in the text.

The second type of field study for which considerable data have been generated has involved the investigation of DMS diurnal trends under MBL conditions. Under stable meteorological conditions, simple photochemical models predict that for a species such as DMS, having an ocean source and an OH controlled lifetime of approximately 1 day , maximum levels in the DMS mixing ratio should be observed in the early morning hours and minimum levels during the late afternoon. Assuming a fairly constant DMS flux from the ocean surface and no major loss of DMS at night due to NO_3, both the maximum and minimum DMS levels (i.e., points of inflection) should correspond to those times when the DMS flux is exactly balanced by the rate of photochemical removal of DMS. In fact, the results from several field studies designed to look for this diurnal variation have been moderately encouraging, typically showing diurnal amplitude ratios of 1.4–2.3 (Table 8.3) (e.g., Saltzman et al., 1995; Saltzman and Cooper, 1989; Thompson et al., 1993). However, disagreements between observations and predictions have also surfaced, indicating that the uncertainties in these comparison studies are still quite substantial. The uncertainties include (a) the use of sea-to-air fluxes of DMS calculated from oceanic DMS concentration measurements and thin film mass transport approximations (see Section 8.3.1.3), (b) the absence of MBL height data over the time period of DMS measurements, (c) the absence of detailed information on the dynamic exchange of air masses between the MBL and FT, (d) the lack of a detailed description of the daytime photochemical environment, and (e) the lack of a detailed description of the night-time chemical environment as driven by NO_3 chemistry.

In still other diurnal studies the focus of investigation has been on correlations between DMS and DMS oxidation products. For example, during recent shipboard measurements off the Washington coast, Bandy et al. (1992) observed that in daylight periods when DMS was highest (e.g., 180 ppt), SO_2 levels remained statistically nearly the same (e.g., 25 ppt) as on days when DMS levels were low. Bandy et al. estimate that for the season and meteorological conditions encountered and with all OH-oxidized DMS being converted to SO_2, the level of SO_2 observed should have exceeded 60 ppt. These authors further argue that the low SO_2 values cannot be explained by the action of a strong non-photochemical sink, since this would have necessarily resulted in the total depletion of SO_2 at night. Average daytime and night-time SO_2 levels were 30 and 24 ppt, respectively. Bandy et al. suggest that their data appear to be more consistent with the idea that one of the major sources of SO_2 was entrainment from the FT and that SO_2 was basically kept in steady state due to surface deposition. Thus these authors hypothesize that the oxidation of DMS does not include the major formation of SO_2 but rather the production of the sulfate forming species SO_3.

During another shipboard study in the equatorial Pacific, Yvon et al. (reference 75, Table 8.3) recorded fast time resolution measurements of SO_2 which, like the data of Bandy et al., also showed no diurnal variation. By contrast, both DMS and H_2S measurements recorded at the same time revealed very significant variations in their mixing ratios as a function of time of day. From still other measurements reported in the equatorial Pacific by Huebert et al. (1993), very high average DMS/SO_2 mole ratios were recorded (i.e., 18) and again no correlation was observed between DMS and SO_2. Assuming no SO_2 measurement difficulties, the results from the latter two studies, like those reported by Bandy et al., suggest that at each of these sampling sites SO_2 was a minor DMS oxidation product and that the levels of this species were controlled by other sources.

As of this date the only strong correlation reported between DMS and SO_2 in the remote MBL is the seasonal correlation observed at Amsterdam Island ($37°50'$ S, $77°30'$ E) in the Indian Ocean (Putaud et al., 1992). However, as pointed out by Huebert et al. (1993), it is quite possible that even with SO_2 being formed only as a minor product of DMS oxidation, a long-term seasonal coherence between SO_2 and DMS might be found provided that "on average" atmospheric SO_2 levels were controlled by DMS oxidation and not by external sources.

In a still more comprehensive field study by Berresheim et al. (1993a), simultaneous measurements of the gas phase species DMS, DMSO, $H_2SO_4(g)$, MSA(g), and CCN particles were made at a Pacific coastal site in the state of Washington. This study was conducted in parallel with the offshore measurements reported by Bandy et al. (1992). The results from this study indicate that all the measured sulfur gases showed significant diurnal variations, and in most cases these trends were all in phase with each other. Berresheim et al. (1993a) interpreted the observed low mixing ratios for MSA(g) and $H_2SO_4(g)$ relative to DMSO as an indication that DMSO formation was the most important reaction pathway in the oxidation of DMS. Statistical evidence was also obtained which indicated a significant correlation between individual sulfur species and CCN levels.

The photochemical interpretation of the Pacific coast data was not provided in the Berresheim et al. (1993a) publication but has been examined in the text that follows. In this analysis we have used the fact that all gas phase species were measured nearly simultaneously in time. Thus, based on the assumption that the system was in photochemical equilibrium, the DMS and DMSO field data have been used here to assess the relative importance of the addition versus abstraction channels in the oxidation of DMS. We have explored this possibility here by selecting data from the sampling time period of April 24 and 25. During this time period meteorological conditions were reported to be sunny with air advected from the Pacific Ocean, providing optimal conditions for a photochemical study. Taking in this case only those data recorded during the solar window (from 9.30 to 15.00 h, local time), median values for the mixing ratios of DMS, DMSO, $H_2SO_4(g)$, and MSA(g) were reported to be 19.4, 0.95, 0.06, and 0.015 ppt, respectively. For the specific case of DMSO, the atmospheric lifetime is controlled by its reaction with OH and was estimated to be approximately 1 h. The latter estimate was based on an average midday OH level of 3×10^6 molecules cm^{-3} and a value of $k_{OH,DMSO} = 1 \times 10^{-10} cm^3$ molecule^{-1} s^{-1} (Table 8.5(a)). Under these conditions application of the photostationary state approximation allows the rate of formation of DMSO to be equated to its rate of destruction, i.e.

$$k_{OH,DMS}[OH][DMS] \times BR = k_{OH,DMSO}[OH][DMSO] \qquad (8.3)$$

where BR is the branching ratio for the OH addition channel (see Figures 8.5, and 8.6(a)). Given the cited daytime mixing ratios of DMS and DMSO and an average air temperature of 280 K, we calculate a value for the parameter BR of 0.5. For the same temperature the kinetic data shown in Figure 8.5 give a value for the addition channel branching ratio of 0.57. As shown in Figure 8.6(a), the latter channel is the only one leading to the formation of the product DMSO.

The largest uncertainty in the evaluation of the branching ratio is clearly the uncertainty associated with the estimate of the midday average OH concentration. Given that this could be as large as a factor of two, the level of agreement between the field observations and the DMS/OH kinetic data (Figure 8.5) could be viewed as either fortuitous or as reassuring. Taking the more optimistic attitude, if the lifetime of both $H_2SO_4(g)$ and MSA(g) with respect to deposition on pre-existing particles is taken to be 1 h (Berresheim et al., 1993a), and if it is further assumed that all chemical intermediates leading to the formation of MSA(g) and $H_2SO_4(g)$ are at photochemical equilibrium, an estimate can also be made of the rates of production of the latter products relative to the formation of the addition channel product DMSO. For this case, based on the cited ratio of DMSO to MSA, we estimate that the rate of production of MSA from DMS was several factors of 10 lower than that of the addition channel product DMSO. For $H_2SO_4(g)$ the evaluation is not as straightforward in that this species could have been formed either via DMS-generated SO_3 or from gas phase OH-oxidized SO_2. The SO_2 in turn may have been formed from the oxidation of DMS and/or derived from an anthropogenic source. Making the most favorable assumption, i.e., that all $H_2SO_4(g)$ was formed from SO_3 whose origin was DMS oxidation, the observed ratio of $H_2SO_4(g)$ to DMSO would suggest that the SO_3 yield from DMS was very minor relative to that of DMSO. Concerning the yield of SO_2 from DMS, the Pacific coast data are even more difficult to interpret than for SO_3 owing to the strong possibility of rapid heterogeneous processes (Chameides and Stelson, 1992; Hegg, 1985). Thus no definitive statement can be made about the yield of this species from DMS other than it had to be smaller than that for the OH/DMS addition channel, i.e., < 0.5.

In general, the above-cited Pacific coast results do not support Bandy et al.'s (1992) suggestion that for their study SO_3 was the most likely major

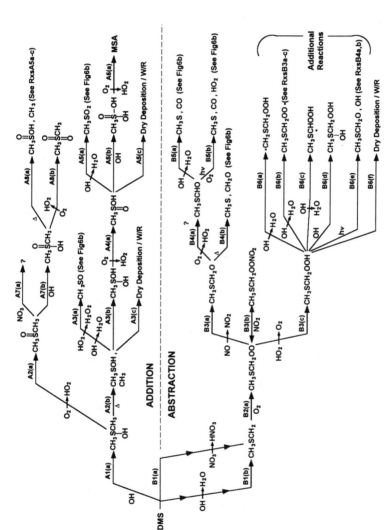

FIGURE 8.6(a). Proposed mechanism for the oxidation of DMS in the remote marine atmosphere: abstraction and addition channels in the OH- and NO₃-initiated oxidation of DMS. W/R = wash-out/rain-out.

"intermediate" product formed in the degradation of DMS. But they do tend to support Bandy et al.'s conclusion that SO_2 was not a major product from DMS oxidation. In both cases, however, it is possible that differences in conditions on the ship versus the coastal study may have influenced the final results.

The Pacific coast results also appear to be in general agreement with recent speculations by Huebert et al. (1993) concerning DMS, SO_2, MSA, and nss-SO_4^{2-} data recorded in the equatorial Pacific. The latter authors, based on the relative lifetimes for the aforementioned species, have speculated that their results were only consistent with the idea that the major products from DMS oxidation were DMSO and/or $DMSO_2$. As noted above, this result would seem to be in agreement with those reported by Berresheim et al. (1993a). On closer

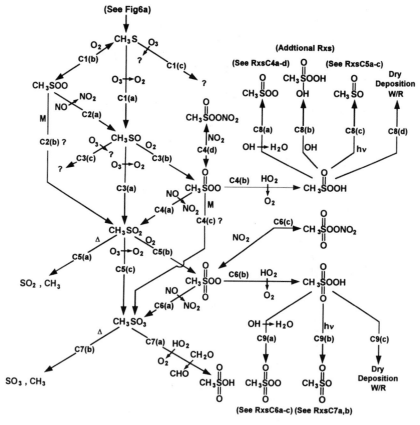

FIGURE 8.6(b). Proposed mechanism for the oxidation of DMS in the remote marine atmosphere: mechanistic scheme for reactions involving CH_3SO_x species. WR = wash-out/rain-out.

inspection, however, one finds that the average MBL temperature reported for this study was quite high, i.e., 300 K (Thompson et al. (1993)). For this case, taking the temperature dependence of the addition versus abstraction pathway in the DMS + OH reaction mechanism as indicated in Figure 8.5, one would predict a corresponding DMSO addition channel rate that was no more than 25% of the total destruction rate of DMS. Thus their data may require a somewhat different explanation as proposed later in this text.

As stated previously, laboratory kinetic studies have been as important as field studies in the elucidation of the atmospheric fate of DMS. In the text that follows, we will discuss in further detail the previous reviews by Turnipseed and Ravishankara (1993), Tyndall and Ravishankara (1991), and Yin et al. (1990). In addition, the present authors will add their own insights to the discussion based on their assessment of more recent field, laboratory kinetic, and modelling studies. Some of the kinetic results with respect to the OH-initiated DMS oxidation have already been mentioned. For example, while it is now certain that hydrogen abstraction from DMS forms the new radical CH_3SCH_2, there is somewhat less certainty about the nature of the product resulting from the OH addition channel. In general, though, the observed O_2 dependence of this reaction suggests that the addition channel yields DMSO as the major product rather than methanesulfenic acid, CH_3SOH (Stickel et al., 1993b). The recent field data reported by Berresheim et al. (1993a), showing a significant correlation between DMS and DMSO mixing ratios in MBL air, also add considerable weight to the argument that DMSO is the major product from the OH addition channel. As shown in Figure 8.6(a), the subsequent reaction of DMSO is somewhat more speculative. Recent kinetic investigations (Hynes et al., reference 4, Table 8.5(a)) as well as sulfur field studies in the Antarctic in which both DMSO and $DMSO_2$ were measured (Eisle and Berresheim, unpublished results) suggest that methanesulfinic acid, $CH_3S(O)OH$, rather than $DMSO_2$ is the more likely dominant stable end-product from the DMS + OH addition channel.

A common mechanistic scheme employed to describe the further oxidation of the CH_3SCH_2 radical involves the sequence of reactions shown in Figure 8.6(a) labelled B2(a), B3(a), and B4(b). This reaction sequence includes the addition of O_2 to form a peroxy radical followed by reaction with NO and the subsequent thermal decomposition of the radical species CH_3SCH_2O to form the CH_3S radical and formaldehyde. As shown in Figure 8.6(b), the chemistry of the CH_3S radical is speculated to be the principal source of the products SO_2, SO_3, MSA(g), and H_2SO_4. As a consequence, the ensuing chemistry from this species has received the greatest attention by laboratory kineticists. However, as shown in Figure 8.6(a), the relative importance of CH_3S chemistry versus other possible DMS oxidation pathways will depend very strongly on the exact conditions encountered in the remote marine atmosphere.

For conditions defined by the tropical and subtropical remote MBL, where NO levels are typically 2–8 ppt (Davis et al., 1987), it appears that the dominant reaction of the peroxy radical, CH_3SCH_2OO, is not with NO but rather with HO_2 radicals. The resulting product should be a stable peroxide species, CH_3SCH_2OOH. For example, assuming an average MBL mixing ratio for NO of 4 ppt and an average daytime value for HO_2 of 4×10^8 molecules cm^{-3} (Davis et al., 1993), the rate for the HO_2 reaction channel is nearly twice that of the NO reaction. Thus, in analogy with the well-studied CH_3OOH species, one might speculate that CH_3SCH_2OOH would undergo a manifold of reactions, i.e., B6(a), B6(b), B6(c), B6(d), B6(e), and B6(f). Interestingly, many of the processes listed are unlikely to lead to the formation of CH_3S radicals, e.g., B6(a), B6(b), B6(d), and B6(f).

Quite significant in the above list of reactions is process B6(f) involving the removal of CH_3SCH_2OOH by surface deposition and/or by wash-out and rain-out processes. In this case, surface deposition velocities of 0.5–1.5 cm s^{-1} would appear to be realistic and thus would define a pathway by which modified DMS was returned directly to the ocean. As shown in Figures 8.6(a) and 8.6(b), it is quite likely that many of the stable peroxides and intermediate acid species CH_3SOH and $CH_3S(O)OH$ would also have significant deposition rates to the ocean surface. If so, they potentially define a short circuit for DMS, preventing it from reaching the oxidized forms SO_2, SO_3, or MSA.

Concerning the chemistry of the CH_3S radical, recent kinetic studies of this species as well as the closely coupled CH_3SO_x species have produced several significant results but have also left many remaining unanswered questions. These have been summarized in the text below.

(a) The most likely reaction of CH_3S under remote MBL conditions is with O_3 to form CH_3SO and still other products not yet identified. Some of these "unidentified products" could in the final analysis turn out to be quite important. The reaction with O_2 to form equilibrated levels of CH_3SOO could also prove to be important. Of particular significance would be the yet unmeasured isomerization reaction involving the CH_3SOO species to give the non-peroxy form of this radical, CH_3SO_2.

(b) Considerable uncertainty still remains regarding the atmospheric fate of CH_3SO, although reactions with O_3 and O_2 are the most likely possibilities. In the first case the product formed under actual atmospheric conditions is unclear, but CH_3SO_2 is considered one of the more likely candidates. Other possibilities reported by Domine et al. (1992), such as CH_2SO, could also significantly alter the final product distribution. The product from the O_2 reaction should be $CH_3S(O)OO$, which, if reasonably stable, could totally dominate the chemistry of CH_3SO owing to the availability of high concentration levels of O_2.

(c) No kinetic studies have been reported involving the $CH_3S(O)OO$ radical, but the competitive processes C4(a) and C4(b) would seem to be the most likely ones. The first reaction produces the radical intermediate CH_3SO_2, whereas, the second gives the potentially stable peroxide $CH_3S(O)OOH$. As in the case of the peroxy radical CH_3SOO, yet a third important reactive channel of the species $CH_3S(O)OO$ might involve its isomerization to form the radical intermediate CH_3SO_3, reaction C4(c).

(d) It has been speculated that the CH_3SO_2 radical is one of the single most important species leading to the formation of SO_2 via the thermal decomposition step C5(a). The most important competing step would likely be C5(b) involving reaction with O_2, giving the new peroxy radical $CH_3S(O)_2OO$. To the extent that reaction C5(b) dominates the chemistry of the sulfonyl radical (CH_3SO_2), very little SO_2 should be seen as a product. It is estimated that the C—S bond strength in CH_3SO_2 is approximately $20 \, kcal \, mol^{-1}$ (Turnipseed and Ravishankara, 1993), which means that reaction C5(a) has a very strong temperature dependence, shifting by nearly a factor of 2 for every 5 K change in temperature in the vicinity of 298 K.

(e) The kinetics of the peroxy radical $CH_3S(O)_2OO$ has not been studied, but it is expected to react in a fashion similar to the lower-oxidation-state peroxy species, except for reaction with O_2. In the latter case the sulfur in $CH_3S(O)_2OO$ is already hexavalent and therefore is unlikely to react further. The two major competing reactions, C6(a) and C6(b), lead to the CH_3SO_3 radical and the peroxide $CH_3S(O)_2OOH$, respectively.

(f) The CH_3SO_3 radical has been speculated to be a critical intermediate in the formation of the DMS oxidation products MSA(g) and H_2SO_4(g) (Bandy et al., 1992; Yin et al., 1990). The product MSA(g) would most likely form via reaction with HO_2 radicals, C7(a), whereas H_2SO_4(g) would be the product resulting from a two-step process, i.e., a thermal decomposition step, C7(b), to form SO_3 followed by the latter species' reaction with H_2O(g). As stated earlier in the text, a key aspect of the latter two-step process is that the H_2SO_4(g) formed by this reaction has the potential to form CCN in the atmosphere. As in the case of the sulfonyl radical, reaction C7(b) involves the breaking of an estimated $22 \, kcal \, mol^{-1}$ bond, meaning that the activation energy would be of at least this magnitude. Thus temperature changes of 5 K would again cause a potential shift in the product yield by a factor of 2. But, unlike the chemistry of the sulfonyl radical (CH_3SO_2), where it is speculated that reaction with O_2 is a major competing step to thermal decomposition, the sulfur in CH_3SO_3 is hexavalent and therefore should have a very low probability of reacting with O_2.

Conclusions: (1) To understand the product distribution from the oxidation of DMS will require a much more exhaustive examination of products that have been identified in studies to date. The most important of these are likely to be peroxide compounds. These range from the simple peroxide species generated from the parent compound, CH_3SCH_2OOH, to the more complex forms generated from the intermediate sulfur peroxy radicals, e.g., $CH_3S(O)_2OOH$. One of the important characteristics of these peroxide products, as well as the intermediate acidic products CH_3SOH and $CH_3S(O)OH$, would be their tendency to undergo direct removal at the ocean surface via dry deposition or to be removed by wash-out or rain-out processes. In all three cases a direct short circuit would be provided for the return of DMS to the ocean surface without the formation of SO_2 or some form of sulfate. (2) The product distribution from DMS is most likely dependent on at least four critical factors: temperature, NO_x levels, HO_x levels, and surface deposition/wash-out rates. This suggests that the product distribution observed in field studies should be a strong function of geographical location, altitude, NO_x sources, MBL mixing times, time of day, and season of the year.

In general, temperature changes should influence the DMS product distribution in two ways: (a) in defining the relative branching in the reaction of OH with DMS into addition and abstraction channels, the abstraction channel being favored at temperatures above 285 K, and (b) in defining the relative importance of the unimolecular thermal decomposition steps C5(a) and C7(b) versus the competing bimolecular processes, i.e., C5(b), C5(c), and C7(a). Thus higher temperatures would seem to promote SO_2 and SO_3 formation.

Higher HO_x levels, on the other hand, would tend to increase the relative importance of OH versus NO_3 with respect to the DMS initiation reaction; and, within the abstraction pathway, higher HO_x levels would also have a strong influence on the relative yields of peroxide species versus SO_2, SO_3, and MSA, i.e., the higher the HO_x level, the greater would be the yield of peroxides.

Higher levels of NO_x would influence the same processes as HO_x (e.g., the nature of the initiation reaction and the branching ratios for the reactions of several peroxy radical species), but these high levels would tend to have the opposite effect. For example, the importance of NO_3 versus OH as the initiating reactant would be promoted by elevated levels of NO_x; and, within the abstraction pathway, the production of SO_2, SO_3, and MSA would also be promoted relative to the peroxides.

Considering the strong interplay of the above variables, to assess the effect of temperature alone on the product distribution (a frequent topic of discussion in many reported field studies) must be considered very problematic. For example, within the MBL in the tropics the temperature tends to be high, HO_x levels are high, NO_x levels are very low, and MBL mixing is fast. At high latitudes, on average, temperatures are much lower, HO_x levels also tend to be lower, but NO_x levels are higher and mixing times tend to be somewhat longer. Thus one would speculate (based on Figures 8.6(a) and 8.6(b) and existing kinetic data) that at high latitudes the product distribution would reflect high yields of DMSO and possibly methanesulfinic acid and relatively lower yields of peroxides, SO_2, SO_3, and MSA. We note that both of the hypothesized major products are typically not measured in field studies. Still a further complicating factor for the high-latitude analysis would be the possible importance of the NO_3 abstraction channel under night-time conditions. In the latter situation SO_2, SO_3, and MSA could be important products. In the tropics, on the other hand, one would speculate that the OH abstraction channel should be dominant, but high HO_x and low NO_x levels together with efficient mixing would tend to promote peroxide formation followed by surface loss. The oxidation products SO_3 and SO_2 as well as MSA might therefore be formed in somewhat lower yields. Again the only products typically measured in this type of setting have been SO_2 and aerosol phase MSA(p) and nss-SO_4^{2-}. Given such a scenario, any comparison of "measured" products between high latitudes and the tropics would appear to be meaningless in terms of the effects of only temperature as a controlling factor.

The above exercise, although highly speculative and qualitative in nature, does serve to point out that future progress in the elucidation of the DMS oxidation mechanism will require both further advances in field studies as well as detailed laboratory kinetic studies. Future field studies in particular need to move toward high-time-resolution measurements involving a much broader cross-section of both stable products and reactive intermediates.

CS_2 and OCS

The dominant sink for atmospheric CS_2 is reaction with OH in the troposphere (Chin and Davis, 1993). The reaction proceeds via reversible adduct formation,

TABLE 8.5(b)   Tropospheric CS_2 Oxidation and Stratospheric Photochemistry of OCS: Important Mechanisms and Rate Constants

Reaction	Rate Constant[a,b]	Reference(s)
OH + CS_2 + M $\leftrightarrow$ CS_2OH + M CS_2OH + O_2 → products	$k_{eff} = \dfrac{1.25 \times 10^{-16}\exp(4550/T)}{T + 1.81 \times 10^{-3}\exp(3400/T)}$	1
OCS + O → CO + SO	$1.6 \times 10^{-11}\exp(-2150/T)$	2
OCS + hv → CO + S		
S + O_2 → SO + O	$2.1 \times 10^{-12\,c}$	2
SO + O_2 → SO_2 + O	$1.4 \times 10^{-13}\exp(-2280/T)$	3, 4
SO_2 + hv → SO + O		
SO_2 + OH $\xrightarrow{\text{multisteps}}$ H_2SO_4 (see Table 8.5(c))		

[a] Units are cm^3 molecule^{-1} s^{-1}.

[b] k_{eff} is the effective rate constant (at 760 Torr), defined as $k_{eff} = \dfrac{1}{[OH][CS_2]}\dfrac{d[product]}{dt}$.

[c] Independent of temperature.

1 Hynes et al. (1988), *J. Phys. Chem.*, **92**, 3846; 2 Atkinson et al. (1992), *J. Phys. Chem. Ref. Data*, **21**, 1125; 3 Black et al. (1982), *J. Chem. Phys. Lett.*, **93**, 598; 4 Goede and Schurath (1983), *Bull. Soc. Chim. Belg.*, **92**, 661.

probably of the SCS–OH isomer (Stickel et al., 1993a, and references therein), and further reaction of the adduct with oxygen. The effective rate constant, k_{eff}, for this mechanism is shown in Table 8.5(b). OCS, CO, and SO_2 have been observed as the corresponding products with yields of 0.83, 0.165, and 1.15, respectively (Stickel et al., 1993b). Based on these results, Chin and Davis (1993) estimated that the CS_2 + OH reaction generates about 30% of the total OCS in the atmosphere.

As mentioned earlier in the text (see Section 8.2 and Figure 8.4), OCS has a relatively long lifetime in the troposphere (4 years) where the reaction with OH represents the only important chemical sink, removing about 22% of OCS (Chin and Davis, 1993). Because it is sufficiently long-lived, it mixes into the stratosphere where the dominant loss processes are OCS photolysis and reaction with $O(^3P)$. Corresponding rate constants are given in Table 8.5(b). The atmospheric cycling of OCS and its potential relevance to global climate change are discussed in more detail in Section 8.4.3.

H_2S and SO_2

Figure 8.7 presents a schematic diagram of the mechanisms of H_2S and gas phase SO_2 oxidation in the atmosphere. Also included are estimates of characteristic times for the individual reactions based on room temperature and typical atmospheric concentrations of the reactants: $[OH] = 1 \times 10^6$, $[O_3] = 1 \times 10^{12}$, $[NO_2] = 2.5 \times 10^{10}$, and $[O_2] = 5 \times 10^{18}$, all in molecule cm^{-3} units. Corresponding reaction rate constants are summarized in Table 8.5(c). The kinetic scheme for oxidation of H_2S to SO_2 is reasonably well understood, primarily as a result of the work of Howard and co-workers (references 2 and 5–7, Table 8.5(c)). It is generally believed that the gas phase oxidation of SO_2 to H_2SO_4 involves four elementary steps. Key uncertainties remain in the rate of the SO_3 + H_2O association reaction and in the rate of isomerization of $SO_3 \cdot H_2O$ to H_2SO_4; hence the generally accepted view that SO_3 is converted to H_2SO_4 with unit efficiency, while probably correct, should be looked upon with at least some degree of skepticism.

The mechanisms of aqueous phase absorption and oxidation of SO_2 have been extensively studied and reviewed (e.g., Martin, 1984). The processes involved represent the principal source of acidity in hydrosols in the remote atmosphere (and also in most of the polluted atmosphere). SO_2 dissolves in water according to the following equilibria:

$$SO_2(g) + H_2O \leftrightarrow SO_2 \cdot H_2O \tag{8.4}$$

$$SO_2 \cdot H_2O \leftrightarrow HSO_3^- + H^+ \tag{8.5}$$

$$HSO_3^- \leftrightarrow SO_3^{2-} + H^+ \tag{8.6}$$

The corresponding equilibrium constants are listed in Table 8.2. Simultaneous absorption of NH_3 significantly enhances SO_2 absorption due to its stabilizing (alkaline) effect on the pH of the hydrosol. Laboratory and field studies suggest that $SO_2(aq)$ ($= SO_2 \cdot H_2O + HSO_3^- + SO_3^{2-}$) is primarily oxidized by aqueous phase O_3 and H_2O_2. The reactions are first-order with respect to both $SO_2(aq)$ and $O_3(aq)$, and both $SO_2(aq)$ and $H_2O_2(aq)$ concentrations. The ozone oxidation reaction is strongly pH-dependent and dominates above approximately pH 5. In contrast, the reaction with H_2O_2 shows only a slight dependence on pH and dominates at low pH levels. Limited availability of oxidants or mass transport limitations in the gas/hydrosol system may reduce the rate of aqueous phase sulfuric acid formation.

In addition to the above processes, a major heterogeneous removal pathway for SO_2 in the MBL may be absorption and aqueous phase oxidation in alkaline sea salt particles. In a recent model study Chameides and Stelson (1992) have shown that the oxidation of $SO_2(aq)$ by $O_3(aq)$ in sea-salt at a typical pH of 8 proceeds sufficiently rapidly to explain nss-SO_4^{2-} levels measured in sea-salt

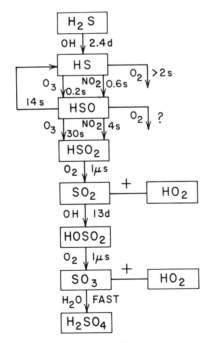

FIGURE 8.7. Major known steps in the atmospheric H_2S oxidation mechanism. Reactants and typical reaction lifetimes at room temperature are indicated next to the arrows. (Reproduced from Wang and Howard, 1990, reference 7, Table 8.5(c), with permission.)

TABLE 8.5(c)   Rate Constants for the Atmospheric Photocemical Oxidation of H_2S and SO_2

Reaction	A^a	E/R^b	$k(298\,K)^a$	Reference(s)
$H_2S \rightarrow SO_2$				
$OH + H_2S \rightarrow SH + H_2O$	630	80	480	1
$NO_3 + H_2S \rightarrow SH + HNO_3$			<0.08	2
$Cl + H_2S \rightarrow SH + HCl$	3690	−210	7470	3
$SH + O_2 \rightarrow$ products			<0.00004	4
$SH + O_3 \rightarrow HSO + O_2$	950	280	370	1
$SH + NO_2 \rightarrow HSO + NO$	2600	−240	5800	4, 5
$HSO + O_2 \rightarrow$ products			<0.002	6
$HSO + O_3 \rightarrow SH + 2O_2$			7	7
$HSO + O_3 \rightarrow HSO_2 + O_2$			4	7
$HSO + NO_2 \rightarrow HSO_2 + NO$			960	6
$HSO_2 + O_2 \rightarrow HO_2 + SO_2$			30	6
$SO_2 \rightarrow H_2SO_4$				
$OH + SO_2 + M \rightarrow HOSO_2 + M$	40^c	-260^c	96^c	1, 8
	10^d	-470^d	48^d	1, 8
$HOSO_2 + O_2 \rightarrow HO_2 + SO_3$	1300	330	40	9
$SO_3 + H_2O + M \rightarrow SO_3 \cdot H_2O + M$			<0.6	10
$SO_3 \cdot H_2O + M \rightarrow H_2SO_4 + M$			e	

[a]Units are $10^{-14}\,cm^3\,molecule^{-1}\,s^{-1}$ except where otherwise indicated.
[b]Units are degrees Kelvin.
[c]$P = 760\,Torr$.
[d]$P = 200\,Torr$.
[e]Generally assumed to be fast even though *ab initio* calculation (ref. 11) suggests a large barrier to isomerization.

1 Atkinson et al. (1992), *J. Phys. Chem. Ref. Data.*, **21**, 1125; 2 Dlugokencky and Howard (1988), *J. Phys. Chem.*, **92**, 1188; 3 Nicovich, Wang, and Wine (1994), *Int. J. Chem. Kinet.*, in press; 4 Stachnik and Molina (1987), *J. Phys. Chem.*, **91**, 4603; 5 Wang et al. (1987), *J. Phys. Chem.*, **91**, 5743; 6 Lovejoy et al. (1987), *J. Phys. Chem.*, **91**, 5749; 7 Wang and Howard (1990), *J. Phys. Chem.*, **94**, 8787; 8 Wine et al. (1984), *J. Phys. Chem.*, **88**, 2095; 9 Gleason and Howard (1988), *J. Phys. Chem.*, **92**, 3414; 10 Wang et al. (1989), *J. Chem. Phys.*, **89**, 4853; 11 Chen and Moore-Plummer (1985), *J. Phys. Chem.*, **89**, 3689.

particles over the North Atlantic. Their calculations further suggest that the bicarbonate/carbonate buffer system in sea-water and the flux of this alkalinity into the atmosphere largely control the removal of sulfur in the MBL $(6-26\,Tg\,SO_2\,year^{-1})$. They estimate that this sink is comparable to SO_2 removal by in-cloud oxidation and dry deposition in the MBL and is roughly comparable to the rate of DMS emission into the MBL. Therefore much of the SO_2 produced from DMS oxidation may be removed by sea-salt scavenging before having a chance to form new CCN particles in the atmosphere. Increasing anthropogenic pollution of the MBL may, however, reduce the sulfur removal capacity of this natural cycle.

The relative importance, on a global scale, of gas and aqueous phase oxidation of SO_2 is still uncertain at this time. Some model calculations suggest that about 80–90% of the transformation of SO_2 takes place in the aqueous phase (e.g., Lelieveld and Heintzenberg, 1992). Corresponding lifetimes of SO_2 in the gas and aqueous phases are on the order of days to weeks and minutes to hours, respectively. However, the major aqueous phase oxidants, O_3 and H_2O_2, are predominantly supplied from the gas phase. Therefore oxidant availability can substantially control the overall efficiency of aqueous phase SO_2 oxidation.

8.3.2.2. The Polluted Atmosphere

In polluted air, metal-catalyzed oxidation by oxygen constitutes a significant additional pathway for the chemical removal of SO_2 in the aqueous phase. The transition metals $Fe(+3)$, $Mn(+2)$, and $Cu(+2)$ are known to be strong catalysts, with $Fe(+3)$ and $Mn(+2)$ showing a synergistic effect on $SO_2(aq)$ oxidation (Martin, 1984). The individual reaction rates are highly pH-dependent, and the reaction orders change with $SO_2(aq)$ and metal concentrations. Carbon (soot) has also been found to catalyze the auto-oxidation of SO_2 adsorbed on the surface of soot particles, but this process appears to be limited by the available reactive surface area.

Aqueous phase absorption of SO_2 can be significantly hindered by a film-like accumulation of organic compounds in the surface layer of hydrosols. Furthermore, as mentioned in Section 8.2.2, the formation of relatively inert bisulfite–aldehyde addition compounds such as HMSA may efficiently reduce the production rate of H_2SO_4 in the aqueous phase. In addition, other organic compounds in solution, e.g., toluene, have been found to inhibit the catalytic oxidation of $SO_2(aq)$ (Martin, 1984).

8.3.3. Deposition

Model studies generally assume that, on a global scale, atmospheric sulfur is cycled back to the Earth's surface mainly in the form of SO_2 and sulfate. Recent estimates of the corresponding dry and wet deposition fluxes for both compounds are summarized in Table 8.6. All studies listed in Table 8.6 have reported an approximate balance between total emission and deposition fluxes in the global atmospheric sulfur cycle. Assuming a balanced atmospheric sulfur budget, the global emission estimate given by the present authors (Table 8.4) – not including sea salt and soil dust contributions – is in relatively good agreement with the total sulfur deposition flux calculated by Langner and Rodhe (1991). Among the four studies listed in Table 8.6, the latter is the only one assuming a relatively low

TABLE 8.6 Global Sulfur Deposition Fluxes Estimated by Different Authors (Tg S year^{-1})

Region	Warneck (1988)	Bribmlecombe et al. (1989)	Langner and Rodhe (1991)[a]	Andreae and Jaeschke (1992)
Continents	71 (SO_2)[b]	33 (dry; SO_2: 17,		28 (SO_2; dry)
	42 (nss-SO_4)[c]	total SO_4: 16)		5 (total SO_4; dry)
	15 (sea-salt)	51 (wet; total S)		70 (total SO_4; wet)
			33 (SO_2; dry)	15 (sea-salt; dry + wet)
			19 (SO_2; wet)	
Oceans	15 (SO_2)	28 (dry; SO_2: 11,	7 (nss-SO_4; dry)	5 (SO_2; dry)
	28 (nss-SO_4)	total SO_4: 17)	38 (nss-SO_4; wet)	13 (nss-SO_4; dry)
	135 (sea-salt)	230 (wet; total S)		65 (nss-SO_4; wet)
				68 (sea-salt; dry)
				110 (sea-salt; wet)
Total	306	342	97[d]	379
	156[e]	178[d]		186[e]

"Dry" and "wet" refer to dry and wet deposition fluxes.

[a]Total (continental + oceanic) deposition fluxes.

[b]Dry deposition of SO_2 removes 50% of total S in urban and regionally polluted air.

[c]Includes 3 Tg S year^{-1} from oceans.

[d]Both sea-salt and soil dust fluxes excluded.

[e]Sea-salt flux excluded.

oceanic DMS emission rate (16 Tg S year^{-1}) which falls into the lower range of our corresponding estimate in Table 8.4. The Langner and Rodhe model also differs in its assumption of a relatively low anthropogenic SO_2 emission rate (67 Tg S year^{-1}). In summary, the major discrepancies among the individual studies listed in Table 8.4 arise from different estimates of oceanic DMS emissions, sulfur emissions from soils and plants, and anthropogenic sulfur emissions.

8.4. POSSIBLE IMPLICATIONS FOR GLOBAL CLIMATE CHANGE

8.4.1. DMS, Sulfate Aerosol, and Cloud Albedo

The role of clouds in the atmospheric radiation budget represents one of the largest uncertainties in predicting future climate change. As pointed out earlier (Sections 8.2.1 and 8.3.2.1), atmospheric sulfur chemistry is believed to be strongly linked to cloud formation via the production of CCN particles from condensable S($+6$) species, predominantly H_2SO_4. Previous measurements suggest that CCN particles consist mainly of sulfate and have a typical radius between 0.05 and 1.0 μm (e.g., Twomey,1971). Binary nucleation of sulfuric acid and water vapor (and possible co-nucleation of NH_3) initially yields low-nanometer-size sulfate particles (condensation nuclei; CN) which may grow into CCN by further condensation of these gases. The number density of CCN is closely related to the droplet size distribution of clouds which in turn has a significant effect on cloud albedo and therefore on global climate.

As our preceding discussion has shown, DMS emission by phytoplankton is believed to be the dominant source of sulfuric acid in the remote MBL. Recognizing the chemical link between DMS emissions and CCN production, Shaw (1983), followed by Charlson et al. (1987), proposed a hypothetical feedback mechanism involving an active thermostatic regulation of marine climate by DMS-producing algae. Specifically, it is assumed that warming of the atmosphere and surface oceans due to solar heat input or greenhouse gases would enhance the biological production of DMS and thereby its emission into the atmosphere. As a result, more CCN would be formed, increasing the cloud albedo and thereby producing a cooling effect which would counteract the assumed initial warming. For example, a 30% increase in CCN population over the oceans affecting only marine stratiform clouds would yield a change in the global heat balance of about $-1\,W\,m^{-2}$ (Charlson et al., 1987). This would significantly reduce the present global climate forcing from atmospheric CO_2 ($+1.5\,W\,m^{-2}$) and other greenhouse gases ($+0.95\,W\,m^{-2}$). Because about one-third of the Earth's surface is covered by oceanic stratus clouds, Charlson et al. (1987) proposed that this cloud-mediated effect of marine DMS emissions may have exerted a major forcing on global climate in pre-industrial times and that it still largely controls the present climate over remote marine regions.

The above hypothesis has been the subject of intense debate in the literature and, to this date, remains highly controversial. Notably, Schwartz (1988) posed a substantial test to the hypothesis based on the argument that anthropogenic SO_2, like DMS, also results in CCN formation in the atmosphere. He argued that despite the fact that sulfate levels in the northern hemisphere have been strongly enhanced due to man-made sulfur emissions, climatological records from the last 100 years have shown no significant differences in cloud albedo or mean surface temperature between the two hemispheres. Based on these data, he concluded that anthropogenic sulfur emissions and, by extension, marine DMS emissions have no significant effect on global cloud albedo.

However, in a lively response to Schwartz's paper, several authors have pointed out that such a clear conclusion cannot yet be drawn. One major argument is that the relationship between anthropogenic sulfur emissions and CCN number density is highly non-linear. Only a small fraction (*ca* 6%) of anthropogenic SO_2 contributes to form new CCN particles in the atmosphere, while typically 50% is removed via dry deposition (Table 8.6), mainly over the continents. Most of the remainder becomes associated with pre-existing particles through in-cloud oxidation processes (Langner et al., 1992). On the other hand, oceanic DMS emissions represent a widely distributed source, and atmospheric DMS oxidation may significantly contribute to CCN formation in remote marine regions where the number of pre-existing CCN is low. Recent model calculations (Lin et al.,1992; Lin and Chameides, 1993) suggest that the degree of coupling between DMS emissions and CCN formation strongly depends on the mechanism by which sulfuric acid is produced from DMS oxidation. The results show that the production of new CCN should be negligible if SO_2 is the dominant intermediate in the $DMS-H_2SO_4$ pathway because of the efficient removal of SO_2 by alkaline sea-salt particles. However, if instead SO_3 is the major intermediate, as discussed earlier, the model predicts a significant CCN production rate of about $24 \, cm^{-3} \, day^{-1}$.

The strongest evidence obtained to date for a DMS–CCN linkage are the results from long-term measurements of aerosol MSA (as tracer for DMS) and CCN concentrations at Cape Grim, Tasmania (Ayers and Gras, 1991). Both MSA(p) and CCN levels showed a strong annual variation with maximum values occurring during austral summer. A statistical evaluation of the data suggests a significant non-linear relationship between atmospheric DMS and CCN levels. In view of the complex chemistry of atmospheric DMS oxidation, it should be expected that the DMS–CCN linkage is non-linear. All the questions addressed in Section 8.3.2.1 must be answered before the role of DMS in global climate regulation can be quantitatively assessed. As per the discussion in Section 8.3.2.1, this requires detailed knowledge not only of reaction rate constants pertinent to atmospheric DMS oxidation but also of the relative yields of intermediate products and end-products resulting from this oxidation. In addition, the physical processes involved in CCN formation also contribute to this non-linearity. As mentioned earlier, the yield of new particles due to homogeneous nucleation of H_2SO_4 (and possibly MSA(g)) is inversely related to the number of already produced or pre-existing particles based on the total surface area available for heterogeneous condensation. Additional processes contributing to the above non-linearity may involve cloud-mediated CCN formation, i.e., the presence of clouds may have a "self-promoting" effect by enhancing the production of CCN (Saxena and Grovenstein, 1993; Hegg et al., 1990). It has been suggested that aqueous phase oxidation of SO_2 in cloud-water followed by evaporation of cloud droplets modifies the aerosol size distribution and chemical composition in a given air mass such that the resulting particles are ideally suited for activation as CCN. Thus relatively more particles may be activated as CCN at a given supersaturation in air which has been processed through clouds. In addition, the supersaturation field itself may be altered in favor of CCN formation. Such mechanisms could explain previous observations of elevated CCN levels in the vicinity of clouds.

The data obtained at Cape Grim are consistent with a climate-stabilizing feedback loop as proposed in the DMS–cloud–climate hypothesis. However, measurements of MSA and nss-SO_4^{2-} in the Vostok ice core (Antarctica) covering the last glacial/interglacial cycle back to 160,000 years have provided evidence apparently in direct conflict with this view (Legrand et al., 1991). Elevated levels of both compounds were found for the period near the end of the last ice age suggesting an increase in oceanic DMS emissions and thus an amplifying rather than stabilizing effect on climate change. It is possible that the observed increase resulted from a shift in favor of DMS-producing biota in the surrounding ocean. On the other hand, changes in aerosol-scavenging processes and/or snow accumulation rates may have also contributed to the observed pattern.

As the previous discussion has shown, large uncertainties still remain in estimating the effects of marine DMS and anthropogenic SO_2 emissions on cloud albedo and global climate. At this time, not even the sign of the proposed DMS–cloud–climate feedback is known (amplification, stabilization, or no regulation of climate). Basically, the response of the DMS-producing system in seawater to changes in climate factors such as temperature and UV radiation are unknown. Overall, the relative importance of DMS versus anthropogenic SO_2 emissions with respect to cloud albedo over the oceans is difficult to estimate because of the many non-linear processes involved. However, recent satellite measurements over the central North Atlantic showed a correlation between chlorophyll concentration, sea surface temperature, and cloud albedo suggesting a predominantly marine source of CCN (Falkowski et al., 1992). By comparison, effects of anthropogenic sulfur emissions from the eastern United States on marine cloud albedo were only detectable in the immediate vicinity of the US coast. In future field programs such measurements should be combined, e.g., with studies of sulfur isotope ratios in marine aerosol to obtain a more quantitative estimate of the source strength and climatic role of biogenic sulfur over the oceans.

More studies are also needed to understand the impact of anthropogenic sulfur emissions on cloud albedo over land. Recently Leaitch et al. [1992] reported measurements made over eastern North America since 1982 indicating a significant enhancement of cloud droplet number concentrations by anthropogenic sulfur pollution. They estimated a corresponding climate forcing of -2 to $-3\,W\,m^{-2}$ for this region. On a global average the cloud albedo forcing by anthropogenic sulfate may be on the order of $-1\,W\,m^{-2}$ (Charlson et al., 1992), most of which is confined to the NH, with large geographical and seasonal variations. However, it should be kept in mind that such estimates are rather crude, mainly because the mass concentration of atmospheric sulfate does not linearly translate into cloud droplet number concentration, which in turn may not be linearly related to cloud albedo.

8.4.2. Radiative Backscattering by Tropospheric Sulfate Aerosol

Much of the subject of the following two sections has already been discussed in Chapter 5, so only a brief overview will be given here. Detailed models have been developed in recent years to estimate the direct climate forcing of sulfur emissions, i.e., the effect of backscattering of short-wave solar radiation by sulfate aerosol. Based on three-dimensional numerical calculations, Charlson et al. (1991) established a global distribution map of the aerosol sulfate mass concentration in the troposphere. The source terms in the model included both anthropogenic and natural sulfur emissions. Using this map and an empirically derived expression for the optical depth of sulfate aerosol, they inferred a global distribution of the backscatter flux of solar radiation from tropospheric sulfate. The results suggest a global average radiative forcing of about $-1\,W\,m^{-2}$, similar but opposite in sign to the present-day greenhouse effect of CO_2 ($+1.5\,W\,m^{-2}$). A subsequent study by Lelieveld and Heintzenberg (1992) showed that much of the direct aerosol forcing may be due to cloud processing of air effecting a shift in the size distribution of sulfate particles (see previous section). The sulfate aerosol resulting from evaporating clouds occurs predominantly in a size range which scatters solar radiation much more efficiently than particles produced through gas phase oxidation of SO_2. More recently Kiehl and Briegleb (1993) recalculated the global climate forcing of sulfate aerosols using more refined assumptions for the optical parameters in their model compared with the model by Charlson et al. (1991). Their results suggest a direct forcing of $-0.54\,W\,m^{-2}$ on a global average and mean values of -0.72 and $-0.38\,W\,m^{-2}$ for the NH and SH, respectively. The relative contribution of anthropogenic versus natural sulfate to this forcing is estimated to be 52% globally, 60% in the NH, and 34% in the SH.

In summary, it appears that the cooling effect of tropospheric sulfate aerosol is dominated by anthropogenic sources over the continents and by natural sources over the oceans. The sum of the direct and indirect (cloud albedo) cooling by sulfate seems to be large enough to significantly offset the total warming from greenhouse gases ($+2.1$ to $+2.5\,W\,m^{-2}$), at least over the highly industrialized regions of the NH. Net greenhouse warming may therefore only occur as a patchy phenomenon, with large geographical and seasonal variations.

8.4.3. The Stratospheric Sulfate (or "Junge") Layer

Relatively little is known at the present time about the potential climate forcing due to changes in the stratospheric sulfate burden. The stratospheric sulfate layer, first discovered by C. Junge in the late 1950s (Junge et al.,1961), envelopes the entire globe and its maximum density occurs at roughly the same altitude (15–23 km) as the stratospheric ozone density maximum. During volcanic quiescent periods and under normal mid-latitude stratospheric conditions (say, 20 km altitude and 220 K temperature) the sulfate aerosol has a number density of some 10 particles cm^{-3}. The particles have an average radius of 0.07 μm and a composition of about 75% H_2SO_4 and 25% H_2O. Crutzen (1976) suggested that the presence of the stratospheric "background" aerosol may be mainly due to OCS photolysis. Many authors have since then discussed the potential for a global climate cooling trend due to increasing OCS levels (see the review by Chin and Davis, 1993, and references therein). However, as discussed earlier in Section 8.2.2, there is no clear evidence to this date that atmospheric OCS levels have actually increased since their earliest measurements in the 1970s. As mentioned earlier, a recent model study by Chin and Davis (1995) has shown that the lifetime of OCS in the stratosphere is about 10 years, or more than two times longer than its total atmospheric lifetime. The model further predicts that most OCS transported into the stratosphere (91%) returns to the troposphere. About 71% of the tropospheric OCS is removed by vegetation and 22% by reaction with OH. Thus previous studies may have largely overestimated the production rate of stratospheric sulfate from OCS and may have underestimated the influence of volcanic eruptions on the stratospheric "background" aerosol.

Powerful volcanic eruptions such as that of Mount Pinatubo in June 1991 can significantly enhance the sulfate particle density in the stratosphere over several years. Since these particles more effectively reflect short-wave solar radiation than they attenuate terrestrial long-wave radiation, the net result of such an event is cooling of the Earth's surface in affected regions. Recent satellite measurements of the radiative effects of the Pinatubo plume (Minnis et al., 1993) unambiguously showed a strong negative forcing (cooling) with an initial latitudinal maximum of $-10\,W\,m^{-2}$ near the equator. Such eruptions may also significantly perturb stratospheric ozone chemistry due to the enhanced surface area supplied by the resulting sulfate aerosols, potentially inducing a higher catalytic destruction rate of ozone. Other linkages between stratospheric ozone chemistry and the atmospheric sulfur cycle may also be important. For example, an enhancement of UV-B radiation at the Earth's surface due to lower stratospheric ozone levels may reduce the population of DMS-producing algae and at the same time increase oceanic emissions of OCS produced by photochemical processes in surface seawater. As we have already pointed out at the beginning of this chapter, scientists are only now beginning to understand the coupling of the atmospheric sulfur cycle to cycles of other elements in the atmosphere and to other earth reservoirs. The need for broad-based interdisciplinary research is obvious in order to find some answers to the many urgent questions outlined above.

Acknowledgments

We would like to thank R. Charlson, W. Chameides, and A. Ravishankara for stimulating discussions contributing to this chapter. We are also grateful to many colleagues for providing us with preprints of their unpublished work, to J. Dignon and C. Howard for providing Figures 8.3 and 8.7, respectively, and to W. Pos for preparing Table 8.1. This research has been supported by the National Science Foundation through grants ATM-9113681 (HB), OPP-9218952 (HB), and ATM-9104807 (PHW), and the National Aeronautics and Space Administration, grants NCC-1-148 (DDD) and NAG-1-1438 (DDD).

References

Andreae, M.O. (1991). Biomass burning: its history, use and distribution, and its impact on environmental quality and global climate. In Levine, J.S., Ed., *Global Biomass Burning: Atmospheric, Climatic, and Biospheric Implications*, MIT Press, Cambridge, MA, pp. 3–21.

Andreae, M.O. and Jaeschke, W.A. (1992). Exchange of sulphur between biosphere and atmosphere over temperate and tropical regions. In Howarth, R.W., Stewart, J.W.B. and Ivanov, M.V., Eds, *Sulphur Cycling on the Continents: Wetlands, Terrestrial Ecosystems, and Associated Water Bodies, SCOPE 48*, Wiley, Chichester, pp. 27–61.

Aneja, V.P. (1990). Natural sulfur emissions into the atmosphere. *J. Air Waste Manage. Assoc.*, **40**, 469–476.

Ayers, G.P. and Gras, J.L. (1991). Seasonal relationship between cloud condensation nuclei and aerosol methanesulfonate in marine air. *Nature*, **353**, 834–835.

Ayers, G.P., Ivey, J.P. and Gillett, R.W. (1991). Coherence between seasonal cycles of dimethyl sulphide, methanesulphonate, and sulphate in marine air. *Nature*, **349**, 404–406.

Bandy, A.R., Scott, D.L., Blomquist, B.W., Chen, S.M. and Thornton, D.C. (1992). Low yields of SO_2 from dimethyl sulfide oxidation in the marine boundary layer. *Geophys. Res. Lett.*, **19**, 1125–1127.

Banwart, W.L. and Bremner, J.M. (1975). Formation of volatile sulfur compounds by microbial decomposition of sulfur-containing amino acids in soils. *Soil Biol. Biochem.*, **7**, 359–364.

Bates, T.S., Calhoun, J.A. and Quinn, P.K. (1992a). Variations in the methanesulfonate to sulfate molar ratio in submicrometer marine aerosol particles over the South Pacific Ocean. *J. Geophys. Res.*, **97**, 9859–9865.

Bates, T.S., Lamb, B.K., Guenther, A., Dignon, J. and Stoiber, R.E. (1992b). Sulfur emissions to the atmosphere from natural sources. *J. Atmos. Chem.*, **14**, 315–337.

Berresheim, H. (1987). Biogenic sulfur emissions from the Subantarctic and Antarctic Oceans. *J. Geophys. Res.*, **92**, 13,245–13,262.

Berresheim, H. and Jaeschke, W. (1983). The contribution of volcanoes to the global atmospheric sulfur budget. *J. Geophys. Res.*, **88**, 3732–3740.

Berresheim, H., Eisele, F.L., Tanner, D.J., Covert, D.S., McInnes, L. and Ramsey-Bell, D.C. (1993a). Atmospheric sulfur chemistry and cloud condensation nuclei (CCN) concentrations over the northeastern Pacific coast. *J. Geophys. Res.*, **98**, 12,701–12,711.

Berresheim, H., Tanner, D.J. and Eisele, F.L. (1993b). Real-time measurement of dimethylsulfoxide in ambient air. *Anal. Chem.*, **65**, 84–86.

Bolin, B., Björkström, A., Keeling, C.D., Bacastow, R. and Siegenthaler, U. (1981). Carbon cycle modelling. In Bolin, B., Ed., *Carbon Cycle Modelling, SCOPE 16*, Wiley, Chichester, pp. 1–28.

Brimblecombe, P., Hammer, C., Rodhe, H., Ryaboshapko, A. and Boutron, C.F. (1989). Human influence on the sulphur cycle. In Brimblecombe P. and Lein, A.Y., Eds, *Evolution of the Global Biogeochemical Sulphur Cycle, SCOPE 39*, Wiley, Chichester, pp. 77–121.

Chameides, W.L. and Stelson, A.W. (1992). Aqueous-phase chemical processes in deliquescent sea-salt aerosols: a mechanism that couples the atmospheric cycles of S and sea salt. *J.Geophys. Res.*, **97**, 20,565–20,580.

Chapman, E.G. (1986). Evidence for S(IV) compounds other than dissolved SO_2 in precipitation. *Geophys. Res. Lett.*, **13**, 1411–1414.

Charlson, R.J., Langner, J., Rodhe, H., Levy, C.B. and Warren, S.G.(1991). Perturbation of the northern hemisphere radiative balance by backscattering from anthropogenic sulfate aerosols. *Tellus*, **43**, 152–163.

Charlson, R.J., Lovelock, J.E., Andreae, M.O. and Warren, S.G. (1987). Oceanic phytoplankton, atmospheric sulphur, cloud albedo and climate. *Nature*, **326**, 655–661.

Charlson, R.J., Schwartz, S.E., Hales, J.M., Cess, R.D., Coakley, J.A., Hansen, J.E. and Hofmann, D.J. (1992). Climate forcing by anthropogenic aerosols. *Science*, **255**, 423–430.

Chatfield, R.B. and Crutzen, P.J. (1984). Sulfur dioxide in remote oceanic air: cloud transport of reactive precursors. *J. Geophys. Res.*, **89**, 7111–7132.

Chin, M. and Davis, D.D. (1993). Global sources and sinks of OCS and CS_2 and their distributions. *Global Biogeochem. Cycles*, **7**, 321–337.

Chin, M. and Davis, D.D. (1995). A re-analysis of carbonyl sulfide as a source of stratospheric background sulfur aerosol. *J. Geophys. Res.*, in press.

Cox, R.A. and Sandalls, F.J. (1974). The photo-oxidation of hydrogen sulphide and dimethyl sulphide in air. *Atmos. Environ.*, **8**, 1269–1281.

Crutzen, P.J. (1976). The possible importance of CSO for the sulfate layer of the stratosphere. *Geophys. Res. Lett.*, **3**, 73–76.

Davis, D.D., Bradshaw, J.D., Rodgers, M.O., Sandholm, S.T. and KeSheng, S. (1987). Free tropospheric and boundary layer measurements of NO over the central and eastern North Pacific Ocean. *J. Geophys. Res.*, **92**, 2049–2070.

Davis, D.D. et al. (1993). A photostationary state analysis of the NO_2–NO system based on airborne observations from the subtropical/tropical North and South Atlantic. *J. Geophys. Res.*, **98**, 23,501–23,523.

Dignon, J. and Hameed, S. (1989). Global emissions of nitrogen and sulfur oxides from 1860 to 1980. *J. Air Pollut. Contr. Assoc.*, **39**, 180–186.

Domine, F., Ravishankara, A.R. and Carlton, C.J. (1992). Kinetics and mechanisms of the reactions of CH_3S, CH_3SO and CH_3SS with O_3 at 300 K and low pressures. *J. Phys. Chem.*, **96**, 2171–2178.

Eisele, F.L. and Tanner, D.J. (1993). Measurement of the gas phase concentration of H_2SO_4 and MSA and estimates of H_2SO_4 production and loss in the atmosphere. *J. Geophys. Res.*, **98**, 9001–9010.

Erickson, D.J., III, Ghan, S.J. and Penner, J.E. (1990). Global ocean-to-atmosphere dimethyl sulfide flux. *J. Geophys. Res.*, **95**, 7543–7552.

Falkowski, P.G., Kim, Y., Kolber, Z., Wilson, C., Wirick, C. and Cess, R. (1992). Natural versus anthropogenic factors affecting low-level cloud albedo over the North Atlantic. *Science*, **256**, 1311–1313.

Fall, R., Albritton, D.L., Fehsenfeld, F.C., Kuster, W.C. and Goldan, P.D. (1988). Laboratory studies of some environmental variables controlling sulfur emissions from plants. *J. Atmos. Chem.*, **6**, 341–362.

Gibson, J.A.E., Garrick, R.C., Burton, H.R. and McTaggart, A.R. (1990). Dimethylsulfide and the alga Phaeocystis pouchetii in Antarctic coastal waters. *Marine Biology*, **104**, 339–346.

Grosjean, D. (1984). Photooxidation of methyl sulfide, ethyl sulfide, and methanethiol. *Environ. Sci. Technol.*, **18**, 460–468.

Guenther, A., Lamb, B. and Westberg, H. (1989). U.S. national biogenic sulfur emissions inventory. In Saltzman, E.S. and Cooper, W.J., Eds, *ACS Symposium Series*, Vol. 393, *Biogenic Sulfur in the Environment*, American Chemical Society, Washington, DC, pp. 14–30.

Hameed, S. and Dignon, J. (1992). Global emissions of nitrogen and sulfur oxides in fossil fuel combustion: 1970–1986. *J. Air Waste Manag. Assoc.*, **42**, 159–163.

Hansen, L.D. and Eatough, D.J. (1991). Organic oxysulfur compounds in the atmosphere. In Hansen, L.D. and Eatough, D.J., Eds, *Organic Chemistry of the Atmosphere*, CRC Press, Boca Raton, FL, pp. 199–232.

Harvey, G.R. and Lang, R.F. (1986). Dimethyl sulfoxide and dimethyl sulfone in the marine atmosphere. *Geophys. Res. Lett.*, **13**, 49–51.

Hatakeyama, S., Izumi, K. and Akimoto, H. (1985). Yield of SO_2 and formation of aerosol in the photo-oxidation of DMS under atmospheric conditions. *Atmos. Environ.*, **19**, 135–141.

Hegg, D.A. (1985). The importance of liquid-phase oxidation of SO_2 in the troposphere. *J. Geophys. Res.*, **90**, 3773–3779.

Hegg, D.A., Radke, L.F. and Hobbs, P.V. (1990). Particle production associated with marine clouds. *J. Geophys. Res.*, **95**, 13,917–13,926.

Huebert, B.J., et al. (1993). Observations of the atmospheric sulfur cycle on SAGA-3. *J. Geophys. Res.*, **98**, 16,985–16,996.

Hynes, A.J., Wine, P.H. and Semmes, D.H. (1986). Kinetics and mechanism of OH reactions with organic sulfides. *J. Phys. Chem.*, **90**, 4148–4156.

Junge, C.E., Chagnon, C.W. and Manson, J.E. (1961). Stratospheric aerosols. *J. Meteorol.*, **18**, 81–108.

Kiehl, J.T. and Briegleb, B.P. (1993). The relative roles of sulfate aerosols and greenhouse gases in climate forcing. *Science*, **260**, 311–314.

Kreidenweis, S.M. and Seinfeld, J.H. (1988). Nucleation of sulfuric acid–water and methanesulfonic acid–water solution particles: implications for the atmospheric chemistry of organosulfur species. *Atmos. Environ.*, **22**, 283–296.

Krouse, H.R. and McCready, R.G.L. (1979). Reductive reactions in the sulfur cycle. In

Trudinger, P.A. and Swaine, D.J., Eds, *Biogeochemical Cycling of Mineral-Forming Elements*. Elsevier, Amsterdam, pp. 315–368.

Langner, J. and Rodhe, H. (1991). A global three-dimensional model of the tropospheric sulfur cycle. *J. Atmos. Chem.*, **13**, 225–263.

Langner, J., Rodhe, H., Crutzen, P.J. and Zimmermann, P. (1992). Anthropogenic influence on the distribution of tropospheric sulphate aerosol. *Nature*, **359**, 712–716.

Leaitch, W.R., Isaac, G.A., Strapp, J.W., Banic, C.M. and Wiebe, H.A. (1992). The relationship between cloud droplet number concentrations and anthropogenic pollution: observations and climatic implications. *J. Geophys. Res.*, **97**, 2463–2474.

Legrand, M., Fenist-Saigne, C., Saltzman, E.S., Germain, C., Barkov, N.I. and Petrov, V.N. (1991). Ice core record of oceanic emissions of dimethylsulfide during the last climate cycle. *Nature*, **350**, 144–146.

Lelieveld, J. and Heintzenberg, J. (1992). Sulfate cooling effect on climate through in-cloud oxidation of anthropogenic SO_2. *Science*, **258**, 117–120.

Lin, X. and Chameides, W.L. (1993). CCN formation from DMS oxidation without SO_2 acting as an intermediate. *Geophys. Res. Lett.*, **20**, 579–582.

Lin, X., Chameides, W.L., Kiang, C.S., Stelson, A.W. and Berresheim, H. (1992). A model study of the formation of cloud condensation nuclei in remote marine areas. *J. Geophys. Res.*, **97**, 18,161–18,172.

Liss, P.S. and Merlivat, L. (1986). Air-sea gas exchange rates: introduction and synthesis. In Buat-Menard, P., Ed., *The Role of Air-Sea Exchange in Geochemical Cycling*, Reidel, Dordrecht, pp. 113–127.

Lovelock, J.E., Maggs, R.J. and Rasmussen, R.A. (1972). Atmospheric dimethyl sulphide and the natural sulphur cycle. *Nature*, **237**, 462–463.

Martin, L.R. (1984). Kinetic studies of sulfite oxidation in aqueous solution. In Calvert, J.G., Ed., *SO_2, NO and NO_2 Oxidation Mechanisms: Atmospheric Considerations*, Butterworths, Boston, MA, pp. 63–100.

Mayewski, P.A., Lyons, W.B., Spencer, M.J., Twickler, M.S., Buck, C.F. and Whitlow, S. (1990). An ice-core record of atmospheric response to anthropogenic sulphate and nitrate. *Nature*, **346**, 554–556.

McFarland, M.D., Kley, D., Drummond, J.W., Schmeltekopf, A.L. and Winkler, R.J. (1979). Nitric oxide measurements in the equatorial Pacific region. *Geophys. Res. Lett.*, **6**, 605–608.

McTaggart, A.R. and Burton, H. (1992). Dimethyl sulfide concentrations in the surface waters of the Australasian Antarctic and Subantarctic Oceans during an austral summer. *J. Geophys. Res.*, **97**, 14,407–14,412.

Minnis, P., Harrison, E.F., Stowe, L.L., Gibson, G.G., Denn, F.M., Doelling, D.R. and Smith, W.L. Jr. (1993). Radiative climate forcing by the Mount Pinatubo eruption. *Science*, **259**, 1411–1415.

Munger, J.W., Tiller, C. and Hoffmann, M.R. (1986). Identification of hydroxy-methanesulfonate in fog water. *Science*, **231**, 247–249.

NAPAP (1991). *National Acid Precipitation Assessment Program, 1990 Integrated Assessment Report*, Washington, DC.

Neftel, A., Beer, J., Oeschger, H., Zürcher, F. and Finkel, R.C. (1985). Sulphate and nitrate concentrations in snow from South Greenland 1895–1978. *Nature*, **314**, 611–613.

Niki, H., Maker, P.D., Savage, C.M. and Breitenbach, L.P. (1983). An FTIR study of the mechanism for the reaction HO + CH_3SCH_3. *Int. J. Chem. Kinet.*, **15**, 647–654.

Orr, W.L. and White, C.M., Eds. (1990). *ACS Symposium Series*, Vol. 429, *Geochemistry of Sulfur in Fossil Fuels*, American Chemical Society, Washington, DC.

Panter, R. and Penzhorn, R.D. (1980). Alkyl sulfonic acids in the atmosphere. *Atmos. Environ.*, **14**, 149–151.

Penner, J.E., Dickinson, R.E. and O'Neill, C.A. (1992). Effects of aerosol from biomass burning on the global radiation budget. *Science*, **256**, 1432–1434.

Pszenny, A.P., Harvey, G.R., Brown, C.J., Lang, R.F., Keene, W.C., Galloway, J.N. and Merrill, J.T. (1990). Measurements of dimethyl sulfide oxidation products in the summertime North Atlantic marine boundary layer. *Global Biogeochem. Cycles*, **4**, 367–379.

Putaud, J.P., Mihalopoulos, N., Nguyen, B.C., Campin, J.M. and Belviso, S. (1992). Seasonal variations of atmospheric sulfur dioxide and dimethylsulfide concentrations at Amsterdam Island in the Southern Indian Ocean. *J. Atmos. Chem.*, **15**, 117–131.

Rennenberg, H. (1991). The significance of higher plants in the emission of sulfur compounds from terrestrial ecosystems. In Sharkey, T.D., Holland, E.A. and Mooney, H.A., Eds. *Trace Gas Emissions by Plants*, Academic Press, San Diego, CA, pp. 217–260.

Robinson, J.M. (1989). On uncertainty in the computation of global emissions from biomass burning. *Climatic Change*, **14**, 243–262.

Saltzman, E.S. and Cooper, D.J. (1989). Dimethyl sulfide and hydrogen sulfide in marine air. In Saltzman, E.S. and Cooper, W.J. Eds, *ACS Symposium Series*, Vol. 393, *Biogenic Sulfur in the Environment*, American Chemical Society, Washington, DC, pp. 330–351.

Saltzman, E.S., et al. (1995). Diurnal variations in atmospheric sulfur gases over the western equatorial Atlantic Ocean. *J. Geophys. Res.*, in press.

Saltzman, E.S., Savoie, D.L., Prospero, J.M. and Zika, R.G. (1986). Methanesulfonic acid and non-sea-salt sulfate in Pacific air: regional and seasonal variations. *J. Atmos. Chem.*, **4**, 227–240.

Saltzman, E.S., Savoie, D.L., Zika, R.G. and Prospero, J.M. (1983). Methane sulfonic acid in the marine atmosphere. *J. Geophys. Res.*, **88**, 10,897–10,902.

Savoie, D.L. and Prospero, J.M. (1989). Comparison of oceanic and continental sources of non-sea-salt sulphate over the Pacific Ocean. *Nature*, **339**, 685–687.

Saxena, V.K. and Grovenstein, J.D. (1993). The role of clouds in the enhancement of cloud condensation nuclei concentrations. *Atmos. Res.*, in press.

Schwartz, S.E. (1988). Are global cloud albedo and climate controlled by marine phytoplankton? *Nature*, **336**, 441–445.

Shaw, G.E. (1983). Bio-controlled thermostasis involving the sulfur cycle. *Climatic Change*, **5**, 297–303.

Spiro, P.A., Jacob, D.J. and Logan, J.A. (1992). Global inventory of sulfur emissions with $1° \times 1°$ resolution. *J. Geophys. Res.*, **97**, 6023–6036.

Stickel, R.L., Zhao, Z. and Wine, P.H. (1993a). Branching ratios for hydrogen transfer in the reactions of OD radicals with CH_3SCH_3 and $CH_3SC_2H_5$. *Chem. Phys. Lett.*, **222**, 312–318.

Stickel, R.L., Chin, M., Daykin, E.P., Hynes, A.J. and Wine, P.H. (1993b). Mechanistic studies of the OH-initiated oxidation of CS_2 in the presence of O_2. *J. Phys. Chem.*, **97**, 13,653–13,661.

Stoiber, R.E., Williams, S.N., and Huebert, B. (1987). Annual contribution of sulfur dioxide to the atmosphere by volcanoes. *J. Volcanol. Geotherm. Res.*, **33**, 1–8.

Thode, H.G. (1991). Sulphur isotopes in nature and the environment: an overview. In Krouse, H.R. and Grinenko, V.A., Eds, *Stable Isotopes. Natural and Anthropogenic Sulphur in the Environment*, Wiley, Chichester, pp. 1–26.

Thompson, A.M. et al. (1993). Ozone observations and a model of marine boundary layer photochemistry during SAGA-3. *J. Geophys. Res.*, **98**, 16,955–16,968.

Turnipseed, A.A. and Ravishankara, A.R. (1993). The atmospheric oxidation of dimethyl sulfide: elementary steps in a complex mechanism. In Restelli, G. and Angeletti, G., Eds., *Dimethylsulphide: Oceans, Atmosphere and Climate*, Kluwer, Dordrecht, pp. 185–196.

Twomey, S. (1971). The composition of cloud nuclei. *J. Atmos. Sci.*, **28**, 377–381.

Tyndall, G.S. and Ravishankara, A.R. (1991). Atmospheric oxidation of reduced sulfur species. *Int. J. Chem. Kinet.*, **23**, 483–527.

Warneck, P. (1988). *Chemistry of the Natural Atmosphere*, Academic Press, San Diego, CA, pp. 484–542.

Watts, S.F., Brimblecombe, P. and Watson, A.J. (1990). Methanesulphonic acid, dimethyl sulphoxide and dimethyl sulphone in aerosols. *Atmos. Environ.*, **24A**, 353–359.

Yin, F., Grosjean, D. and Seinfeld, J.H. (1986). Analysis of atmospheric photooxidation mechanisms for organo-sulfur compounds. *J. Geophys. Res.*, **91**, 14417–14438.

Yin, F., Grosjean, D. and Seinfeld, J.H. (1990). Photooxidation of dimethyl sulfide and dimethyl disulfide. *J. Atmos. Chem.*, **11**, 309–399.

9

Photochemical air pollution

Harvey E. Jeffries

9.1. INTRODUCTION

Photochemical air pollution has been a recognized problem in the US since the 1950s when researchers in Los Angeles connected plant injury symptoms with gases produced during the irradiation of hydrocarbons and oxides of nitrogen (Haagen-Smit et al., 1951). In the 1960s and 1970s limited air quality monitoring began in many US cities and the "LA smog" problem was identified as a widespread phenomenon. By 1971 atmospheric scientists had a general explanation for the creation of *secondary* air pollutants. Secondary air pollutants such as ozone, O_3, are formed in the air by chemical reactions among primary air pollutants (e.g., oxides of nitrogen, NO_x, and volatile organic compounds, VOCs, including hydrocarbons) that are emitted directly into the urban atmospheric environment. Primary pollutants are mostly caused by anthropogenic activities associated with energy production and utilization. Since the 1970s photochemical air pollution has been observed in virtually all major cities in the world.

The Clean Air Act of 1970 created the US Environmental Protection Agency (EPA) and required the Administrator of the EPA to identify ubiquitous pollutants that may reasonably be anticipated to endanger public health or welfare and to issue air quality criteria for them. In 1971 the EPA promulgated the National Ambient Air Quality Standard (NAAQS) for ozone, and standards for other air pollutants soon followed. The primary (based on public health) and secondary (based on welfare) standards in effect in 1994 are given in Table 9.1. In spite of efforts to enforce these standards in the US, in 1991 the EPA estimated that 140 million people lived in areas that violated the NAAQS for ozone. Figure 9.1 shows the 10 year trend in ozone monitored in US cities. While there has been a very slight downward trend, it is not clear how much of this is due to effective

TABLE 9.1 National Ambient Air Quality Standards in Effect in 1994

Pollutant	Ave. Time	Primary Standard	Primary Level	Secondary Standard	Secondary Level
PM-10	Annual arithmetic mean	$50\,\mu g\,m^{-3}$		Same	
	24 h	$150\,\mu g\,m^{-3}$		Same	
SO_2	Annual arithmetic mean	$80\,\mu g\,m^{-3}$	0.03 ppm	$1300\,\mu g\,m^{-3}$	0.50 ppm
	24 h	$365\,\mu g\,m^{-3}$	0.14 ppm		
CO	8 h	$10\,\mu g\,m^{-3}$	9 ppm		
	1 h	$40\,\mu g\,m^{-3}$	35 ppm		
NO_2	Annual arithmetic mean	$100\,\mu g\,m^{-3}$	0.053 ppm	Same	
O_3	Maximum daily 1 h average	$235\,\mu g\,m^{-3}$	0.12 ppm	Same	
Pb	Maximum quarterly average	$1.5\,\mu g\,m^{-3}$		Same	

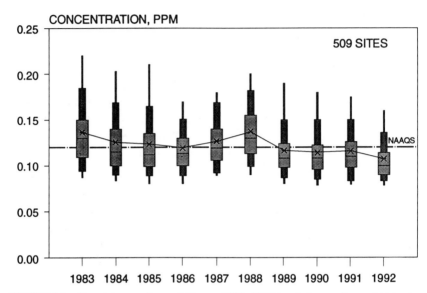

FIGURE 9.1. Box plot comparisons of 10 year (1983–1992) trends in annual second-highest daily maximum 1 h ozone concentrations. Upper box and stems are 95th, 90th, and 75th percentiles; the X is the composite average; the line is the median; lower box and stems are 25th, 10th, and 5th percentiles.

control programs and how much is due to natural variation in meteorological conditions.

Ozone is a natural component in the Earth's atmosphere with its major reservoir in the stratosphere (Warneck, 1988). Owing to the downward transport of this ozone, there is at the surface of the Earth a natural global background concentration of ozone (between 10 and 60 ppb monthly mean 1 h values). The research of the 1970s and 1980s has shown, however, that the unhealthy concentrations occurring in urban areas are mostly produced by atmospheric chemistry of locally emitted precursors. While a general understanding of the processes creating such secondary air pollutants as ozone was developed in the 1960s and 1970s, a complete mechanistic understanding still escapes scientists and policy makers in the 1990s (NRC, 1991).

Our understanding of urban atmospheric composition and chemistry is derived from three kinds of studies: field monitoring campaigns, laboratory kinetics and reaction pathways investigations, and mathematical modeling exercises. The field-monitoring programs have revealed the complexity of the urban environment and have provided observations for comparison with mathematical model predictions. The laboratory studies have yielded detailed understanding of important reaction processes and have provided many of the necessary parameters for the mathematical models. Mathematical modeling studies have attempted to mimic the urban chemical environment by simulating the combined effects of emissions, transport, and reaction of urban chemical compounds responsible for air pollution effects. Difficulties in achieving agreements between ambient urban observations and model predictions have led to improved design of the field campaigns and to the recognition that certain additional laboratory data are required, and have confirmed that formulations of the mathematical models needs to be improved. Full turns of this cycle require 10–15 years.

The most challenging problem in photochemical air pollution science has been the accurate description and prediction of the secondary air pollutants and in particular ozone. To describe fully the origins of urban O_3, not only do the emissions of the precursor compounds have to be described accurately, but their subsequent transport, dispersion, and intermixing must also be described with equal exactness. During their transport the precursors absorb solar radiation and this initiates a complex series of photochemical reactions that transform the primary pollutants into secondary pollutants. Compounding the problem of accurately describing this situation is the mixing of older, partially reacted, emissions with fresh emissions. These processes can occur within a single day between upwind and downwind emission sites in the same urban area, or they can occur across days during stagnant meteorological episodes via "carry-over" of pollutants from one day to the next. Further, they can also occur in the form of partially reacted material from an upwind urban area impacting another downwind urban area, usually after overnight transport.

The challenge for the field scientist has been to make sufficient measurements that observed concentrations can be explained. One difficulty is the variability and complexity of the urban environment. Most measurement methods characterize the point of sampling but conditions a few blocks away may be quite different. Some of the difficult problems have been associated with the vertical dimension. Buildings have boundary layers that may prevent air near them from being representative of the atmosphere at a particular altitude. Aircraft that can be used to sample in the vertical dimension are every expensive to operate routinely. A second field problem has been the creation of accurate emission inventories (see Chapter 4). Mobile sources are most significant in urban areas, and some tunnel measurement studies (Pierson et al., 1990) have suggested that emission models for mobile sources may be inaccurate by a factor of 2 or 3. Reconciliation between ambient measurements and predictions of models based on emission inventories has been difficult in part because of the transport and mixing processes that make it unclear as to which emissions might have contributed to the measurements and in part because the emissions at the time of the measurement were poorly characterized.

The challenge for the laboratory kineticist has been the determination of complete product mass balances for two important classes of organic compounds, the aromatic (e.g., toluene and xylenes) and the biogenic (e.g., isoprene and the terpenes) hydrocarbons. The most difficult problem has been the creation of analytical methods for detecting, identifying, and quantifying the reaction intermediates after initial oxidative attack on these compounds. These intermediates are probably multifunctional, containing double bonds, carbonyl groups, hydroxyl groups, and perhaps even nitrate groups on the same species. Commercial sources for such compounds are not available for use as standards or for experiments and this has made progress difficult. At present the identity of more than half of the reacted carbon for aromatic species is unknown, yet these species are most likely the origin of the significant reactivity observed when aromatic compounds and NO_x are irradiated in smog chamber experiments.

The challenge for the mathematical modeler has been striking a balance between the level of generalization, distortion, and deletion in his mathematical formulation of the model and the computer resources needed to solve the resulting equations numerically. The urban and regional photochemical simulation problem has been classified as one of the "grand challenges" in high-performance computing, requiring gigaflops of processing power, gigabytes of memory, and terabytes of storage and display (NSF, 1990). In present models, solving the chemical transformations occupies 80–90% of the computing time and therefore most economies of formulation occur in the representation and solution of the chemistry. Typically the complex mixture of urban hydrocarbons (more than 200 species routinely identified) are represented by only 8–12 model species.

Smog chamber and modeling studies have shown that the atmospheric chemistry of NO_x and VOCs leading to O_3 production exhibits significant non-linearity (Dodge, 1977). That is, a decrease in precursor concentrations may or may not lead to a decrease in O_3, and in some circumstances such decreases may lead to *increases* in O_3. It is this non-linearity that caused the regulatory community to depend upon mathematical models to describe the probable response to control measures (Freas et al., 1978). Thus the Clean Air Act Amendments of 1990 (CAA90, Public Law 101–549) require that states with cities in violation of the National Ambient Air Quality Standard for O_3 use mathematical air quality models to demonstrate that proposed control measures would, in the future, be effective in preventing the re-occurrence of violations. In the United States, therefore, scientific models have been transformed into legal instruments. This has created a tension between advancing scientific understanding and the need for regulatory consistency (Jeffries and Arnold, 1987).

The following sections will emphasize the composition of urban air and the chemical processes among these species that lead to urban ozone formation. Results from air quality model simulations will be used to provide examples for analysis, but little emphasis will be placed upon the emission and transport processes that are necessary components of these models.

9.2. REACTANTS, INTERMEDIATES, AND PRODUCTS

9.2.1. Inorganic Compounds

Table 9.2 lists important inorganic compounds in urban areas, gives the chemical formulas, and briefly describes their origins and sinks. A free radical is a species with an unshared or non-bonded electron in its structure. In reactions, this unshared electron can essentially be "passed" among different species. The symbols for free radicals often have a "dot" to represent the unshared electron, e.g., $^{\cdot}OH$ or $HO_2^{\cdot}$ radicals.

9.2.2. Volatile Organic Compounds

Table 9.3 gives the average composition of 1 ppmC of urban VOC mixture. This average urban composition was based upon data, extensively quality-assured, obtained in a US EPA ambient 06.00–09.00 EST urban hydrocarbon study (Lonneman, 1987). This EPA study used canisters collected in 29 US cities over the period 1984–1988. Additional data were obtained from an EPA five-site ambient aldehyde data set (Richter, 1989). More than 1000 canister compositions were analyzed by Jeffries et al. (1989) to produce an average urban composition of *ca* 200 different organic species.

9.2.3. Model Representations

Chemical reaction mechanisms used in air quality models (AQMs) generally have between 32 and 55 model species. The inorganic species are unique and no generalization of their chemistries is possible. Thus current AQM reaction mechanisms use at least 14 model species to represent explicitly the inorganic compounds listed in Table 9.2. In an explicit representation the model species is intended to have exactly the same properties as the real species and to undergo the same reactions in the model as those of the real species. The organic species are included in AQM mechanisms via a mixture of explicit (but compressed) representations and generalized representations. The first members of homologous series generally tend to have unique chemistries compared with the higher members, and thus these species usually appear in most model mechanisms as themselves (e.g., formaldehyde, ethene, and toluene). Other hydrocarbons appearing in Table 9.3 are represented via some type of generalization or distortion principles or are deleted from the representation. The three most common generalization schemes are the carbon bond method (Whitten et al., 1980), the surrogate species method (Atkinson et al., 1982; Lurmann et al., 1987), and lumped species methods (Stockwell, 1986; Stockwell et al., 1990; Carter, 1990).

TABLE 9.2   Important Inorganic Compounds in Urban Air

Species	Relevance
$NO_x = NO + NO_2$	Oxides of nitrogen; nitric oxide and nitrogen dioxide. Sources are any combustion process using air. These species play a major role in chain propagation and in production of ozone. Mixing ratios in urban areas are tens of ppb.
$NO_3^{\bullet}$, N_2O_5	Nitrate radical and dinitrogen pentoxide. $NO_3^{\bullet}$ is a transient intermediate formed by reaction of O_3 with NO_2. It can react with NO_2 to form the stable species N_2O_5. Mixing ratios in urban areas are <5–500 ppt for NO_3 and <15 ppb for N_2O_5.
HNO_3	Nitric acid. A stable final oxidation product of NO_2 formed by reaction with $^{\bullet}OH$. Removed from air by dry deposition and rain-out. Mixing ratios in urban areas are tens of ppb.
$HOONO_2$	Peroxy nitric acid (PNA). An *unstable* oxidation product of NO_2 formed by reaction with $HO_2^{\bullet}$. Rapidly decomposes to reactants at a highly temperature-dependent rate. Mixing ratios in urban areas are less than 2 ppb.
$HONO$	Nitrous acid. An *unstable* product of OH + NO. Removed from air by rapid photolysis to produce $^{\bullet}OH$ and NO again. Mixing ratios in urban areas are 0.03–8 ppb.
$O(^3P)$, $O(^1D)$	Ground state atomic oxygen and excited atomic oxygen. Formed in the troposphere by photolysis of NO_2 or O_3. $O(^3P)$ reacts with O_2 to produce O_3. $O(^1D)$ is collisonally deactivated by other gas molecules (M) to give $O(^3P)$, or it reacts with H_2O to give two OH radicals. Mixing ratios are 10^5–10^7 atoms cm^{-3}.
O_3	Ozone. Formed from $O(^3P) + O_2$. Its photolysis is an important source of $^{\bullet}OH$ in the troposphere. Major health hazard in urban air pollution. Mixing ratios are 20–450 ppb.
$^{\bullet}OH$	Hydroxyl radical. The key oxidizer of carbon and nitrogen compounds in the troposphere. Sources are photolysis processes. Loss is via termination reactions, mostly with NO_2. Peak mixing ratios in urban areas are about 10^{-4} ppb.
$HO_2^{\bullet}$	Hydroperoxy radical. Oxidizes NO and makes $^{\bullet}OH$ or reacts with self to make H_2O_2. It is the major source of "re-created" $^{\bullet}OH$. Peak mixing ratios in urban areas are about 10^{-2} ppb.
H_2O_2	Hydrogen peroxide. A major termination product in the absence of NO_x. removed from air by dry deposition, wash-out and rain-out. Mixing ratios in urban areas are 1–2 ppb.
CO	Carbon monoxide. A major chain propagation species. Produced by photolysis of formaldehyde and during combustion. Mixing ratios in urban areas are 200–800 ppb during the summer months.

TABLE 9.3   Average Distribution of Volatile Organic Compounds in Urban Air

Compound	Amount[a]	Compound	Amount[a]
Acyclic alkanes		1-Pentene	3.9
Ethane	38.9	1-Octene	0.9
Propane	45.5	1-Nonene	3.0
n-Butane	72.1	2-Methyl-propene	8.6
n-Butane	7.1	2-Methyl-1-butene	4.2
n-Pentane	30.5	3-Methyl-1-butene	1.4
n-Hexane	13.0	2-Methyl-1-pentene	3.5
n-Heptane	7.3	4-Methyl-1-Pentene	1.1
n-Octane	5.4	2,3,3-Trimethyl-1-butene	7.6
n-Nonane	7.4	2,3,3-Trimethyl-1-butene	2.3
n-Decane	18.6	*c*-2-Butene	3.5
2-Methyl-propane	32.4	*t*-2-Butene	4.2
2-Methyl-butane	78.0	*c*-2-Pentene	8.3
2-Methyl-pentane	20.7	*t*-2-Pentene	4.2
3-Methyl-pentane	15.0	*t*-2-Hexene	1.1
2-Methyl-hexane	13.0	*c*-2-Heptene	0.8
3-Methyl-hexane	8.2	*t*-2-Heptene	1.1
3-Methyl-heptane	5.3	*c*-2-Octene	0.5
2-Methyl-heptane	4.4	2-Methyl-2-butene	0.3
4-Methyl-octane	3.9	2-Methyl-2-pentene	2.1
3-Methyl-octane	3.0	*t*-4-Methyl-2-pentene	1.8
4-Methyl-nonane	21.2	*c*-4-Methyl-2-pentene	0.6
2,2-Dimethyl-butane	2.6	*t*-3-Methyl-2-pentene	1.2
2,3-Dimethyl-butane	5.9	*Dialkenes*	
2,3-Dimethyl-pentane	7.7	Isoprene	2.4
2,4-Dimethyl-pentane	4.1	1,3-Butadiene	2.1
3,3-Dimethyl-pentane	1.4	*Cycloalkenes*	
2,5-Dimethyl-hexane	2.7	Cyclohexene	1.2
2,4-Dimethyl-hexane	2.7	*Terpenes*	
2,5-Dimethyl-hexane	1.5	A-Pinene	5.4
2,4-Dimethyl-heptane	2.0	δ-3-Carene	1.7
2,3-Dimethyl-heptane	1.5	*Aromatics*	
2,5-Dimethyl-heptane	0.9	Benzene	20.1
2,2,4-Trimethyl-pentane	12.4	Toluene	63.7
2,3,4-Trimethyl-pentane	4.7	Ethylbenzene	10.0
2,2,5-Trimethyl-hexane	3.8	Isoamylbenzene	4.5
Cycloalkanes		*n*-Propylbenzene	3.3
Cyclopentane	3.9	*sec*-Butylbenzene	3.0
Cyclohexane	4.7	Isopropylbenzene	2.4
Cycloheptane	0.9	*n*-Amylbenzene	1.8
Methyl-cyclopentane	10.0	*n*-Hexylbenzene	0.6
Methyl-cyclohexane	5.3	*m,p*-xylene	34.2
Ethyl-cyclopentane	0.0	*o*-Xylene	14.7
Ethyl-cyclohexane	1.5	*m*-Ehtyltoluene	9.4
Acyclic alkenes		*p*-Ethyltoluene	7.0
Ethene	27.2	*o*-Ethyltoluene	6.4
Propene	9.1	1,2-Diethylbenzene	2.1

TABLE 9.3 (*Continued*)

Compound	Amount[a]	Compound	Amount[a]
Aromatics (Continued)		*p,m,o*-Methylstyrene	3.2
1,3-Diethylbenzene	7.0	2,6-Dimethylstyrene	3.0
1,4-Diethylbenzene	4.7	*Alkynes*	
1-Methyl-4-*i*-propylbenzene	5.4	Acetylene	15.3
1-Methyl-4-*i*-butylbenzene	0.0	*Aldehydes*	
1,2,4-Trimethylbenzene	23.1	Formaldehyde	10.0
1,3,5-Trimethylbenzene	6.8	Acetaldehyde	10.0
1,2,3-Trimethylbenzene	5.6	*Unspeciated*	
1,2-Dimethyl-3-ethylbenzene	7.3	Alkanes	23.1
1,2-Dimethyl-4-ethylbenzene	2.1	Alkenes	21.6
1,2,4,5-Tetramethylbenzene	1.5	Aromatics	10.6
1,2,3,5-Tetramethylbenzene	1.4	Unknown	7.9
1,2,3,4-Tetramethylbenzene	1.2		
Class totals			
Alkanes	530.9		
Alkenes	115.2		
Aromatics	270.7		
Aldehydes	20.0		
Unspeciated	63.2		

[a]Units are ppbC in 1 ppmC total.
To obtain percentage of each species or class of species on a carbon basis, divide numbers by 10.

The carbon bond method is based upon a complete representation of all the carbon in a reacting system by the use of mechanism species that represent carbon–carbon bond types. For example, in the Carbon Bond Four (CB4) mechanism (Gery et al., 1989) used in the EPA OZone Isopleth Package (OZIPR) (Gery and Crouse, 1990), in the EPA Urban Airshed Model (EPA, 1990), and in the EPA Regional Oxidant Model (EPA, 1991) nine organic species are used to represent all the compounds appearing in Table 9.3: (1) single-bonded carbons, including those occurring as side-chains on olefins and aromatic species, are represented as a one-carbon PAR; (2) ethene is represented as itself, i.e., a two-carbon ETH; (3) all other terminal double-bonded carbons are represented as a two-carbon OLE with additional PARs included in the model's mixture composition as needed to account for the rest of the compound's single-bonded carbon; (4) all internal double-bonded carbons are represented as two two-carbon ALDs with additional PARs as needed; (5) toluene is represented as a seven-carbon TOL; (6) other mono-alkylbenzenes are represented as a seven-carbon TOL with additional PARs as needed; (7) xylenes are represented as an eight-carbon XYL; (8) other di- and tri-alkylbenzenes are represented as an eight-carbon XYL with additional PARs as needed; (9) formaldehyde is represented as itself, HCHO; (10) all other carbonyl bonds are represented as a two-carbon ALD with additional PARs as needed; and (11) some species are explicitly represented as non-reactive (e.g., ethane) by use of as many one-carbon NR species as needed to represent each carbon in the non-reactive molecule. When the necessary reaction product species are added to the primary organic species, the CB4 mechanism uses a total of 18 organic species for all the organic chemistry in the model.

The surrogate species approach uses the chemistry of an explicit species to represent the chemistry of all other species of the same class, e.g., *n*-butane to represent all alkanes. In the case of using *n*-butane as the surrogate, a molecule of propane or a molecule of hexane would be treated as if it were an *n*-butane molecule, and the nearly explicit chemistry of *n*-butane would be included in the model mechanism. In the propane case the model representation gains carbon mass and in the hexane case loses carbon mass relative to the actual carbon concentration, and thus this system is not carbon conservative. Some mechanism formulations use a "missing carbon" species to account for the unrepresented or overrepresented carbon, but unlike the PAR species in CB4, this species is merely used for bookkeeping, i.e., the missing carbon has no chemistry (but the overrepresented carbon would). In a manner similar to the *n*-butane example, propene and all higher terminal olefins would be represented by the explicit chemistry of propene and toluene, all higher mono-substituted benzenes would be represented as toluene, and so forth. This method tends to use more organic species and the total mechanism tends to be larger than with the carbon bond approach. The Lurmann 1987 mechanism, for example, had two alkanes, three olefins, three aromatics, three aldehydes, and an unreactive species to represent the organic mix. When the necessary product species are added to the primary organic species, the complete mechanism has 37 organic species.

The lumped species approach is similar to the surrogate species approach, but the model species characteristics are determined before each simulation by a mole fraction weighting of the characteristics of the individual explicit mechanisms for the species that are being lumped together. Thus the reaction rate constant of the lumped species included in the model is a weighted value based upon the actual species rate constants and the actual molecular distribution of species that are included in the lumped species. Likewise, the product stoichiometries are mole-fraction-weighted values derived from the product stoichiometries of the explicit species mechanisms for each species to be included as part of the lumped model species. These are also weighted by the mole fraction of each explicit species or by the estimation of the amount of each species that will react during the simulation. Thus these types of mechanisms change with every application depending upon the distribution of the organic species to be represented. These types of mechanisms are more difficult to apply, because the mechanism parameters must be computed for each application. This method tends to have fewer model species than the surrogate species approach but more than the number in the carbon bond approach.

9.3. CHEMICAL PROCESSES

9.3.1. Overview

The chemical phenomenon the urban photochemist seeks to explain is the production of O_3 in and downwind of urban areas. It was recognized early in the study of urban photochemistry (Blacet, 1952) that the photolysis of NO_2 could produce O_3 via the reactions

$$NO_2 + h\nu \ (\lambda \leqslant 430 \, nm) \rightarrow O(^3P) + NO \tag{9.1}$$

$$O(^3P) + O_2 + M \rightarrow O_3 + M \tag{9.2}$$

However, owing to the rapid back reaction

$$O_3 + NO \rightarrow NO_2 + O_2 \tag{9.3}$$

O_3 production by reactions (9.1) and (9.2) would be very limited unless there were a chemical process for oxidizing NO to NO_2 which was competitive with reaction (9.3).

While free radicals had been a well-known phenomenon in combustion chemistry since the early 1930s and Phillip Leighton's 1961 book *Photochemistry of Air Pollution* had suggested that $^{\bullet}OH$ radicals could have a role in urban air pollution, it was not until Greiner (1967) made the first measurements of the reactions of $^{\bullet}OH$ with alkanes that $^{\bullet}OH$ radicals became the center of attention for those seeking to explain photochemical smog. Two groups (Heicklen et al., 1969; Stedman et al., 1970) put forth the supposition that it was a chain process involving $^{\bullet}OH$ and HO_2 radicals that was responsible for both the oxidation of the organics and the oxidation of NO to NO_2. The basic process they suggested can be illustrated most easily as

$$^{\bullet}OH + CO \rightarrow CO_2 + H^{\bullet} \tag{9.4}$$

$$H^{\bullet} + O_2 \rightarrow HO_2^{\bullet} \tag{9.5}$$

$$HO_2^{\bullet} + NO \rightarrow NO_2 + {}^{\bullet}OH \tag{9.6}$$

Greiner had shown that $^{\bullet}OH$ could abstract hydrogens from alkanes (RH, where R— is chemical shorthand for any carbon chain, e.g., R— stands for CH_3—, CH_3CH_2—, $CH_3CH(CH_3)$—, and so forth). It was subsequently reasoned that in the presence of oxygen the alkyl radicals would produce alkyl peroxy radicals

($RO_2^{\bullet}$). In the presence of NO these peroxy radicals were expected to produce NO_2 and alkyoxy radicals ($RO^{\bullet}$). The fate of the $RO^{\bullet}$ radicals depends upon the structure of the alkyl chain, but one reaction possible for those radicals that contain a hydrogen on the carbon atom adjacent to the oxygen radical carrier is to lose this hydrogen via abstraction by an oxygen molecule and to form an aldehyde or ketone. The abstracted $H^{\bullet}$ forms an $HO_2^{\bullet}$ radical. Put together, this reaction sequence forms a chain, i.e.,

$$^{\bullet}OH + RH \rightarrow R^{\bullet} + H_2O \tag{9.7}$$

$$R^{\bullet} + O_2 \rightarrow RO_2^{\bullet} \tag{9.8}$$

$$RO_2^{\bullet} + NO \rightarrow NO_2 + RO^{\bullet} \tag{9.9}$$

$$RO^{\bullet} + O_2 \rightarrow R'CHO + HO_2^{\bullet} \tag{9.10}$$

$$HO_2^{\bullet} + NO \rightarrow NO_2 + {}^{\bullet}OH \tag{9.6}$$

where the last reaction re-creates an $^{\bullet}OH$ radical to perhaps begin another chain.

Because of the recognition of the central importance of $^{\bullet}OH$ radical reactions with organic compounds, many laboratory kinetics studies have been performed to determine the rates of $^{\bullet}OH$ reactions and to determine the particular peroxy radicals formed and their subsequent rates of reaction with NO (see discussion in Atkinson, 1985, 1990; Wallington et al., 1992).

A complete description of the process of NO oxidation by radicals includes explaining the origins of the $^{\bullet}OH$ radical and its ultimate fate. In addition, characteristics of the chain process, especially the process of photochemical radical branching, need further explanation here. The remainder of this section will provide formal definitions and explanations of the chemical processes leading to O_3 formation.

9.3.2. Photolysis

In a real sense, atmospheric photochemistry begins with a molecule absorbing a photon of light. Each photon represents an energy quantum $\mathscr{E} = h\nu$, where h is Planck's constant, ν is the frequency of radiation, and $\nu = c/\lambda$, with c the speed of light and λ the wavelength in nanometers. Spectral radiant flux density is radiant energy per unit area per unit time, or $\mathfrak{D}_\lambda = d\mathscr{E}_\lambda/(da\,dt)$, and its units are $W\,m^{-2}\,nm^{-1}$. $\mathfrak{D}_\lambda$ is also called spectral irradiance. Because molecules can respond to a photon regardless of its arriving direction, it is *spherical* spectral irradiance that must be considered for photochemistry, and thus upwelling radiation at a point in space as well as the downwelling radiation must be included (i.e., a full 4π steradians of solid angle). Thus, unlike radiant flux density arriving at a horizontal surface, spherical irradiance does not exhibit a cosine law dependence. In the atmospheric photochemistry community, spherical spectral radiant flux density is called *actinic flux* (meaning photochemically effective flux) and its values are given in units of photons $\text{cm}^{-2}\,\text{s}^{-1}\,\text{nm}^{-1}$ over a series of narrow wavelength intervals.

9.3.2.1. Actinic Flux

There are no instruments that measure actinic flux. Spectroradiometers are available for measuring spectral irradiance on a solid horizontal receiving surface, but the absence of spherical receiving surfaces to measure over 4π or even 2π steradians in the ultraviolet region prevents these instruments from making actinic flux measurements. Chemical actinometry, the measurement of broadband actinic flux by the measurement of the loss of an absorbing gas in a quartz tube or bulb exposed to sunlight, has had limited success (e.g., Parrish et al., 1983). Therefore in models the actinic flux is computed by a radiation transfer submodel, or, if not internal to the model, species photolysis rates are external inputs to the model. These external rates have to be computed using some type of actinic flux before such models can be run.

Outside the Earth's atmosphere the solar beam can be represented as a uniform parallel flux of photons called the extraterrestrial solar spectral irradiance $\mathcal{H}_{0\lambda}$, or just solar flux. Fortunately, high accuracy 2 nm bandwidth data (see Figure 9.2) are readily available (ASTM, 1973; Iqbal, 1983). Actinic flux ($\mathcal{A}_\lambda$) arriving at a point in the air above an urban surface is made up of three components,

$$\mathcal{A}_\lambda = \mathcal{N}_\lambda + \widetilde{\mathcal{S}}_\lambda + \widetilde{\mathcal{R}}_\lambda$$

where $\mathcal{N}_\lambda$ is the direct beam or photon flux normal to the Sun (this is shown in Figure 9.2), $\widetilde{\mathcal{S}}_\lambda$ is the diffuse spherical irradiance, and $\widetilde{\mathcal{R}}_\lambda$ is the reflected spherical irradiance. $\mathcal{N}_\lambda$ is easily related to $\mathcal{H}_{0\lambda}$ by the application of a multicomponent Beer–Lambert law

$$\mathcal{N}_\lambda = \mathcal{H}_{0\lambda} \prod_{i=1}^{n} \exp\left(-k_\lambda^i \int C_z^i \, ds\right) \tag{9.11}$$

where k_λ^i is the ith species absorption or scattering coefficient per unit path length per unit concentration, C_z^i is the concentration of the ith absorber or scatterer as a function of altitude z, s is the slant path length, and n is the number of absorbing or scattering species. For example, stratospheric O_3 and O_2 absorb nearly all radiation below 300 nm and stratospheric O_3 has a significant impact on the amount of radiation between 300 and 320 nm (see the O_3 transmission curve in Figure 9.2). Thus, to compute the radiation reaching the Earth's surface, the total O_3 column must be known. Likewise, atmospheric water vapor absorbs above 750 nm and again the total H_2O column is needed. Figure 9.2 shows examples of

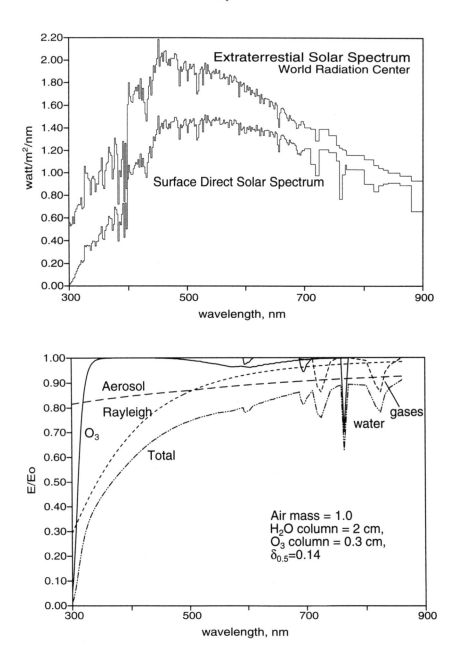

FIGURE 9.2. Transmission of extraterrestrial solar flux through the atmosphere as a function of various absorbers and scattering bodies.

each of the exponential terms in equation (9.11) and also shows the total product. Gases in the atmosphere absorb (water vapor, methane, and ozone) and scatter (nitrogen and oxygen molecules) the solar beam as it passes through the atmosphere. In addition, aerosols, both in the stratosphere and in the troposphere, scatter and absorb radiation. Figure 9.2 shows how each of these processes varies with wavelength for a particular set of conditions.

The computation of $\widetilde{\mathcal{S}}_\lambda$ is much more complex than that of $\mathcal{N}_\lambda$ because multiple scattering is highly dependent upon complex relationships among the diameters of the scattering centers and the wavelength of the light and involves the application of both Rayleigh and Mie scattering theories. Other factors such as the single-scattering albedo of particles affect the ratio between absorption and scattering, further complicating the calculations. $\widetilde{\mathcal{R}}_\lambda$ is related not only to the reflectivity or albedo of urban surfaces but also to the downward reflectivity of the atmosphere for upwelling radiation, which is in turn related to aerosol and molecular scattering.

The computation of actinic flux in the urban environment is a complex process. Because the Sun's position relative to the zenith at a given location varies, the length of the solar beam's path through the atmosphere changes as a function of time of day and year. This gives rise to large changes in the spectral distribution of the transmitted energy. In addition, seasonal variations in factors such as the total O_3 column can cause large variations in actinic flux for different episodes.

9.3.2.2. Absorption Cross-Sections and Quantum Yields

Each molecule has a unique radiation absorption spectrum related to the electronic structure of the atoms and bonds that make up the molecule. These absorption spectra must be measured in laboratory experiments. The units most often used to express these spectra are $cm^2\,molecule^{-1}$, called a "cross-section" and the quantities are symbolized by σ_λ. Some molecular absorption cross-sections have sharp spectral features that require subnanometer resolution to capture. When used with 1 nm actinic flux data, such σ_λ values have to be *interval averaged* to be commensurate. Such interval-averaged absorption cross-sections for formaldehyde (HCHO) are shown in Figure 9.3 (Rogers, 1990).

After a molecule has absorbed a photon, it may undergo a number of primary photochemical processes. These include giving off excess energy as radiation (fluorescence), dissipating energy by collisions (quenching), undergoing chemical transformation (isomerization, dissociation, ionization, etc.), transferring all or some of the energy to other molecules that then react further (sensitization), or reacting with another compound. These are all called *primary* photochemical processes, because the excited molecule formed by the initial absorption of the photon is a reactant. Secondary photochemical processes include the subsequent thermal reactions.

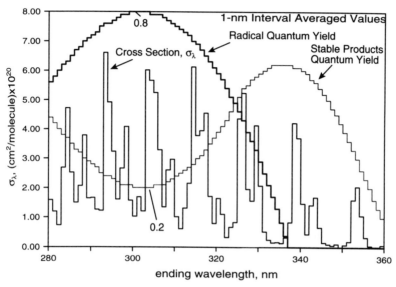

FIGURE 9.3. Absorption cross-sections (Rogers, 1990) and quantum yields (Moortgat et al., 1983) for HCHO photolysis to radicals and to stable products.

Some molecules undergo different dissociation reactions depending upon the energy of the absorbed photon. For example, formaldehyde has two dissociation pathways, one leading to radical products and one leading to only stable products:

$$HCHO \xrightarrow{h\nu} H^{\bullet} + H\overset{\displaystyle O}{\overset{\displaystyle \|}{C}}^{\bullet} \tag{9.12}$$

$$HCHO \xrightarrow{h\nu} H_2 + CO \tag{9.13}$$

The term quantum yield (Φ_λ) is used to describe the extent to which each of these occurs as a function of wavelength; it is defined as

$$\Phi_\lambda = \frac{\text{product molecules formed per cm}^3\,\text{s}^{-1}\,\text{nm}^{-1}}{\text{photons absorbed per cm}^3\,\text{s}^{-1}\,\text{nm}^{-1}}$$

Figure 9.3 shows the quantum yields for reactions (9.12) and (9.13) as a function of wavelength (Moortgat et al., 1983).

9.3.2.3. Photolysis Rates

The spectral photolysis rate j_λ of a compound such as NO_2 is given by

$$j_\lambda(NO_2) = \mathcal{A}_\lambda \sigma_\lambda(NO_2) \Phi_\lambda(NO_2) \qquad (9.14)$$

where λ is the wavelength, $\mathcal{A}_\lambda$ is the actinic flux at λ at the site of photolysis and is a function of the Sun's position, $\sigma_\lambda(NO_2)$ is the absorption cross-section for NO_2 at λ, and $\Phi_\lambda(NO_2)$ is the quantum yield for disassociation of NO_2 at λ. The photolysis rate $j(NO_2)$ is the integral of the spectral rate over the range where $j_\lambda > 0$:

$$j(NO_2) = \int_{\lambda_m}^{\lambda_n} j_\lambda (NO_2) \, d\lambda \qquad (9.15)$$

In practice, the terms of equations (9.14) and (9.15) are replaced with interval-averaged values and a summation is applied:

$$j(NO_2) \approx \sum_{k=m}^{n} \overline{\mathcal{A}_k} \, \overline{(\sigma\Phi)_k} \, (\lambda_k - \lambda_{k-1}) \qquad (9.16)$$

where the overbar represents a value averaged over each $(\lambda_k - \lambda_{k-1})$ interval. The units of j are min^{-1} or s^{-1}.

Figure 9.4 shows an example calculation of the NO_2 photolysis rate, $j(NO_2)$, at 13.16 EDT (solar noon) at the University of North Carolina outdoor chamber site just outside of Pittsboro, NC (latitude 35.708°, longitude 79.109°).

In chemical models of urban atmospheric chemistry, such calculations must be performed about every 10 min for tens of chemical species that are significant photoabsorbers. Depending upon the model, these include NO_2, NO_3^-, O_3, HONO, HNO_3, various organic nitrites and nitrates, H_2O_2, HCHO, CH_3CHO, CH_3OOH, glyoxal (HC(O)C(O)H) and methyl glyoxal, acetone, and methyl ethyl ketone.

Photolytic reactions introduce a strong environmental-dependent factor into urban chemistry and one that is presently poorly treated in most air quality models.

9.3.3. Radical Initiation, Propagation, and Termination

9.3.3.1. Radical Initiation Reactions

Radical initiation is almost always a photolysis process or the reaction of a species which was previously formed by a photolysis process (e.g., O_3 + olefins). Because radical initiation produces radicals where there were none before, we refer to these radicals as "new radicals". Present photochemical mechanisms produce new ˙OH radicals directly in only two ways:

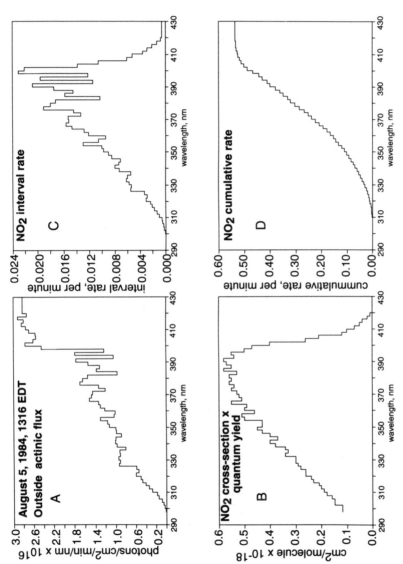

FIGURE 9.4. Spectral terms in the calculation of the photolysis rate of NO$_2$ at a given time and date.

$$O_3 \xrightarrow{hv} O(^1D) + O_2 \qquad (9.17)$$

$$O(^1D) + H_2O \longrightarrow 2\,{}^{\bullet}OH \qquad (9.18)$$

and

$$HONO \xrightarrow{hv} {}^{\bullet}OH + NO \qquad (9.19)$$

There are, however, several photolysis reactions that directly produce new $HO_2^{\bullet}$ radicals, including

$$HCHO \xrightarrow[O_2]{hv} 2HO_2^{\bullet} + CO \qquad (9.20)$$

$$CH_3CHO \xrightarrow[O_2]{hv} CH_3O_2^{\bullet} + HO_2^{\bullet} + CO \qquad (9.21)$$

$$H_2O_2 \xrightarrow{hv} 2HO_2^{\bullet} \qquad (9.22)$$

In the presence of NO these reactions become indirect sources of new ${}^{\bullet}OH$ via

$$HO_2^{\bullet} + NO \longrightarrow {}^{\bullet}OH + NO_2 \qquad (9.6)$$

and via the sequence

$$CH_3O_2^{\bullet} + NO \longrightarrow CH_3O^{\bullet} + NO_2 \qquad (9.23)$$

$$CH_3O^{\bullet} + O_2 \longrightarrow HO_2^{\bullet} + HCHO \qquad (9.24)$$

$$HO_2^{\bullet} + NO \longrightarrow {}^{\bullet}OH + NO_2 \qquad (9.6)$$

Therefore it is possible to convert each new radical of any origin into an equivalent amount of new ${}^{\bullet}OH$ by developing a set of radical-to-radical yields based upon the time-dependent mass balance analysis of a reacting system. For example, where $RO_2^{\bullet}$ is some organic peroxy radical produced by photolysis,

$$d[{}^{\bullet}OH] = \frac{d[{}^{\bullet}OH]}{d[HO_2^{\bullet}]} \frac{d[HO_2^{\bullet}]}{d[RO_2^{\bullet}]} d[RO_2^{\bullet}] \qquad (9.25)$$

and $d[{}^{\bullet}OH]/d[HO_2^{\bullet}]$ and $d[HO_2^{\bullet}]/d[RO_2^{\bullet}]$ are determined from the net reactions of $HO_2^{\bullet}$ and $RO_2^{\bullet}$. The net reactions are computed by summing the mass throughput of all the processes that produce and consume each radical; the amount of $HO_2^{\bullet}$ produced divided by the amount of $RO_2^{\bullet}$ consumed is the yield: $d[HO_2^{\bullet}]/d[RO_2^{\bullet}]$. Hence we can describe the effective radical input to a system by describing the total magnitude of the new ${}^{\bullet}OH$ radicals and by displaying the relative contributions of each VOC or other source to this total magnitude. A significant fraction of new $^{\bullet}OH$ comes from $O_3 + h\nu$, which is usually treated as an inorganic source, although clearly organic reactions were mostly responsible for creating the O_3 in the first place. The magnitude of new $^{\bullet}OH$ and its distribution among the sources demonstrate how the model reaction mechanisms are starting and maintaining their oxidation processes.

9.3.3.2. Radical Propagation Reactions

Radical propagation steps are where VOCs are oxidized and NO-to-NO$_2$ conversions occur. For a reaction step to be a propagation step, there must be as many radicals produced as were consumed in the step. The reaction sequence described by reactions (9.7)–(9.9), followed by (9.6) is a propagation chain in which a VOC molecule is converted to one or more carbonyl molecules and in which two molecules of NO are converted to two molecules of NO$_2$; finally, the free radical emerges as $^{\bullet}OH$ again at the end of the chain. This $^{\bullet}OH$ is described as "re-created $^{\bullet}OH$" in contrast with the new $^{\bullet}OH$ from initiation processes. Note that in these chains the initial $^{\bullet}OH$ could be either a new $^{\bullet}OH$ or a re-created $^{\bullet}OH$, but only re-created $^{\bullet}OH$ emerges.

The propagation chain illustrated by reactions (9.4)–(9.6) only oxidizes one NO molecule to NO$_2$ per cycle instead of two molecules when an RO$_2^{\bullet}$ radical is involved. Another case where only one NO molecule would be oxidized is

$$^{\bullet}OH + HCHO \longrightarrow HC^{\bullet}(=O) + H_2O \tag{9.26}$$

$$HC^{\bullet}(=O) + O_2 \longrightarrow HO_2^{\bullet} + CO \tag{9.27}$$

$$HO_2^{\bullet} + NO \longrightarrow NO_2 + {}^{\bullet}OH \tag{9.6}$$

For C$_{5+}$ alkanes the RO$^{\bullet}$ radical formed after the first oxidation of NO to NO$_2$ can undergo isomerization via a six-member ring to create a hydroxyl alkyl radical (HOR$^{\bullet}$) via internal hydrogen abstraction. This radical subsequently reacts with O$_2$ to form an HORO$_2^{\bullet}$ radical capable of again oxidizing NO to NO$_2$. Eventually the hydroxy alkoxy radical does decompose to produce hydroxy carbonyls and an HO$_2^{\bullet}$ radical. Thus some VOC oxidation chains are capable of oxidizing *three* or more NO molecules in each cycle.

9.3.3.3. Radical Termination Reactions

Radical termination stops propagation, because the reacting radical is incorporated into stable products. The most ubiquitous termination reaction is

$$\cdot OH + NO_2 \longrightarrow HNO_3$$

The reason why this reaction dominates termination is that radical propagation steps produce NO_2, thus making NO_2 available during the time that a large flux of radicals is also available. Other termination reactions are

$$HO_2^{\cdot} + HO_2^{\cdot} \longrightarrow H_2O_2$$

$$RO_2^{\cdot} + HO_2^{\cdot} \longrightarrow ROOH$$

$$RO_2^{\cdot} + NO \longrightarrow RONO_2$$

$$CH_3C(O)O_2^{\cdot} + NO_2 \longrightarrow CH_3C(O)O_2NO_2 \text{ (i.e., PAN)}$$

$$R'C(O)O_2^{\cdot} + NO_2 \longrightarrow R'C(O)O_2NO_2$$

where ROOH is an organic hydro peroxide, $RONO_2$ is an organic nitrate, and the last product above is an organic peroxy nitrate. The peroxy nitrates are highly temperature-labile and the net termination from these reactions depends upon the condition of the equilibrium. Radical–radical reactions such as the first two shown in this example only occur at significant rates after most of the NO_x has been removed, otherwise the radicals would react with NO in propagation steps.

9.3.4. Photochemical Cycles

In this section the detailed steps described above will be assembled into an interacting system with emphasis upon the feedback paths that regulate the various processes. Ozone production will be seen to be an autocatalytic process in which the production of O_3 leads to more production of NO_2 and O_3, yet, the greater production of NO_2 and O_3 directly leads to a loss of NO_2 and this eventually stops the production of O_3. Thus, the O_3 time profile in non-diluted, long-irradiation experiments shows the characteristic "S-shaped" curve of an autocatalytic process.

Figure 9.5 shows two interconnected cycles, each driven by sunlight and each needing the other to function. The top boxes include mostly organic processes and the bottom boxes include mostly inorganic processes. The interaction of these cycles leads to ozone production. The consumption of VOCs and NO_x occurs by a free radical cycle in which the principal processes are radical initiation, propagation, and termination. Closely coupled to this radical chain is another cycle in which new (i.e., emitted) NO is first oxidized to NO_2. The newly created NO_2 then photolyzes to produce NO again, which can once again be oxidized to NO_2. Competing with NO_2 photolysis are radical reactions that convert NO_2 into inorganic and organic nitrogen products and thereby terminate

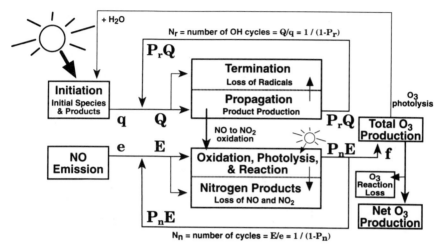

FIGURE 9.5.   Schematic of ˙OH chain, VOC, and NO$_x$ oxidation cycle.

the cycling of that particular nitrogen atom. Figure 9.6 gives details about the inside of the top boxes and a later figure will give details about the inside of the bottom boxes in Figure 9.5.

In radical initiation, ˙OH and other radicals are created (see the top right side of Figure 9.6). New RO$_2$ and new HO$_2$ radical production, however, actually enters the ˙OH cycle part of the way through the cycle in Figure 9.6, but by application of equation (9.25), these initiation processes can be expressed as just new ˙OH radicals (see the right side cycle in Figure 9.6). During radical propagation, either new (just initiated) or old (re-created) radicals attack VOCs to produce peroxy radicals (RO$_2$ and HO$_2$). NO is oxidized to NO$_2$ by these peroxy radicals, leading to the production of aldehydes (RHCO), ketones (RR′CO), and smaller organic peroxy radicals or HO$_2$ radicals. The actual frequencies of the different pathways through the top of Figure 9.6 are dependent upon the specific reaction rate constant for ˙OH + VOC and upon the molecular concentration of each VOC, which change with time.

Also note that as primary VOCs are oxidized to secondary aldehydes and ketones, the aldehydes and ketones join the primary VOCs in the list of organic compounds that can be attacked by ˙OH (see the right side of the list in Figure 9.6). The secondary aldehydes also photolyze to create even more new ˙OH – this is a type of photochemical radical branching, because two new radicals are produced at each aldehyde photolysis while the original radical that led to the production of the aldehydes may still be cycling. Without this photochemical branching, the entire process would very rapidly stop owing to termination.

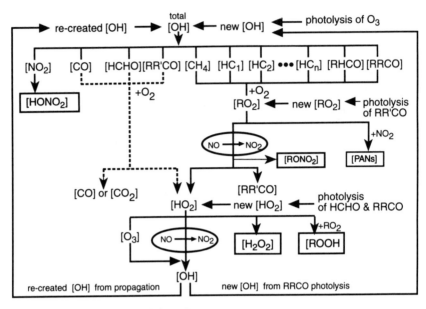

FIGURE 9.6. Details of ˙OH chain cycle.

Some of the $RO_2^•$ radicals, usually less than 10%, react with NO to produce $RONO_2$ instead of propagating. These are termination reactions that remove both a radical and nitrogen oxide. The usual fate of $HO_2^•$ radicals is to react with NO and re-create ˙OH radicals, which is the left-side cycle in Figure 9.6. However, as shown at the bottom of Figure 9.6, when NO concentrations are very low, $HO_2^•$ can react with O_3 to make ˙OH, or with itself to make H_2O_2, or it could react with an $RO_2^•$ to make ROOH, the first reaction being a type of propagation reaction and the latter two being radical termination reactions. In the latter case, termination stops this particular radical cycle.

In the NO oxidation cycle, either new (fresh emissions) or old (re-created) NO is oxidized to NO_2 by the radicals in the organic cycle. The newly produced NO_2 either photolyzes to continue the cycle or reacts with radicals to terminate the nitrogen cycle (which also terminates a radical cycle). When NO_2 photolyzes, it produces an $O(^3P)$ atom that reacts with O_2 to make O_3. If there is still new NO in the system, this newly made O_3 will immediately react with the new NO to make back NO_2 (or even with the old NO, but at this point $[NO]_{new} > [NO]_{old}$ and so the O_3 is much more likely to react with a new NO). That is, the peroxy radical oxidation of NO is stored in the form of NO_2 until all of the new NO is oxidized. Only after $[NO]_{new}$ has fallen significantly below $[NO_2]$ would $[NO]_{re-created} > [NO]_{new}$ and would the $HO_2^•$ and $RO_2^•$

radicals become competitive with O_3 for the NO, thus allowing an accumulation of O_3 to occur. *Thus, O_3 cannot appear in the system if there is any new NO.*

Once the emitted NO has been oxidized, any re-created NO can be cycled several times to produce a new O_3 each time the cycle runs. In this way, one NO molecule may make three or four O_3 molecules. Some of the accumulated O_3 is lost in other nitrogen cycles, resulting in a yield of O_3 per NO_2-photolyzed ("f" in Figure 9.5) of less than 1.0. Further, some O_3 reacts with organic species and is lost, and some ozone photolyzes to produce $O(^1D)$ which leads to new ˙OH.

Finally, the reaction of NO_2 with radicals forms nitrogen products such as nitric acid (HNO_3) and organic peroxy nitrates and brings a halt to the production of O_3 from this particular nitrogen atom. When all nitrogen has reacted this way, all O_3 production stops and the system is said to be "NO_x-limited".

Return to Figure 9.6 and examine the top row where ˙OH reactions are shown. The termination reaction ˙OH + NO_2 is in competition with ˙OH-propagation reactions. This is also true inside the cycle for $RO_2^˙$ radicals. An important parameter of any NO_x/VOC system would be the probability that the free radical associated with each ˙OH that enters the cycle would make it all the way through and emerge as a re-created ˙OH, and thus have the potential to start another oxidation chain. This probability is called the *radical propagation factor* and it describes the fraction of re-created ˙OH per total ˙OH reacting in a process. At the top of Figure 9.5, q = new ˙OH, Q = total ˙OH reacting, and P_r = radical propagation factor. The relationship between these is $P_r = 1 - q/Q$, where Q/q is defined as the average number of times each new ˙OH cycles before being lost to termination. Typical values of P_r are about 0.75 and each new ˙OH reacts approximately four times in propagation before being terminated. Note that eventually *all* the new ˙OH must be lost in termination, some by organic and some by inorganic reactions.

In a similar manner a *nitrogen propagation factor* P_n, gives the probability that an NO_2 molecule will be photolyzed rather than reacted to nitrogen products. P_n describes the fraction of re-created NO per total NO reacting. At the bottom of Figure 9.5, e = new NO, E = total NO reacting, and P_n = nitrogen propagation factor. As with the radicals, the relationship between these is $P_n = 1 - e/E$, where E/e is defined as the average number of times each new NO cycles before being lost to termination. Typical values of P_n are 0.75–0.8 and each new NO reacts four to five times in propagation before being lost. If the reacting system has a low hydrocarbon-to-NO_x starting ratio, not all the NO_x will be removed by termination before the day is over and NO_2 will remain. These systems are said to be "VOC-limited", although "radical-limited" would be a better descriptor.

In tracking the formation of O_3, it is helpful to create another lumped species, odd-oxygen or O_x, which is defined as

$$O_x = O(^3P) + O_3 + NO_2 + 2NO_3^{\bullet} + 3N_2O_5 + HNO_3$$
$$+ HO_2NO_2 + PAN \tag{9.28}$$

O_x contains one-half of all the O_2 that was "split" when $HO_2^{\bullet}$ and $RO_2^{\bullet}$ reacted with NO, less any O_x lost by chemical reaction. Such losses are, for example,

$$O_3 \xrightarrow[H_2O]{hv} 2{}^{\bullet}OH + O_2 \text{ via } O(^1D)$$

$$O_3 + HO_2^{\bullet} \longrightarrow {}^{\bullet}OH + O_2 + O_2$$

$$O_3 + {}^{\bullet}OH \longrightarrow H_2O + O_2$$

$$O_3 + \text{olefins} \longrightarrow \text{aldehydes} + \text{others}$$

$$O(^3P) + HCHO \longrightarrow {}^{\bullet}OH + HO_2^{\bullet} + CO$$

$$O(^3P) + \text{olefins} \longrightarrow \text{radical products}$$

$$NO_3^{\bullet} \xrightarrow{hv} NO + O_2$$

$$NO_3^{\bullet} + HCHO \longrightarrow HNO_3 + HO_2^{\bullet} + CO$$

$$NO_3^{\bullet} + \text{olefins} \longrightarrow NO_2 + \text{radical products}$$

These generally account for only about 10% of the O_x production. Thus, about 90% of the O_x produced should be present as NO_2, the oxidized forms of NO_2, or O_3. The oxidized forms of NO_2 are related to the number of new ${}^{\bullet}OH$ radicals produced, as most of these terminate via reaction with NO_2. The amount of O_3 that accumulates is equal to the net O_x produced minus all the new NO that had to be oxidized to NO_2; this is just the sum of the species in equation (9.28) that contain nitrogen.

9.4. EXAMPLES OF OZONE FORMATION

9.4.1. Simple Trajectory Model Simulation

9.4.1.1. Simulation Conditions

Figure 9.7 shows the time profiles of NO_x and O_3 predicted by the CB4 mechanism (Gery et al., 1989) in an OZIPR (Gery and Crouse, 1990) simple trajectory model simulation of a particular day in Atlanta, Georgia. In this model a well-mixed box is moved at the average wind speed along a trajectory through the urban area. As the box moves, its height (H) increases owing to the mixing height rise from the Sun's heating. This rise results in a decrease in the concentrations of the primary emitted species in the box. In addition, aged materials (e.g., O_3 and partially reacted VOCs) that were in the air above H are mixed down into the box as H increases. At the same time, fresh emissions are

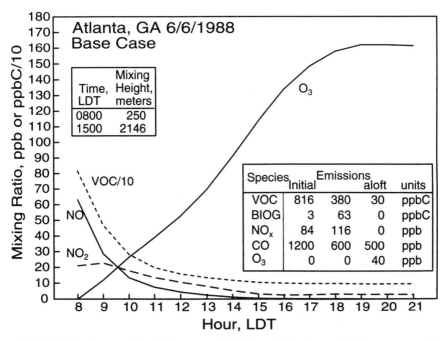

FIGURE 9.7.   Ozone, oxides of nitrogen, and VOC time profiles for an OZIPR trajectory model simulation of a "design day" in Atalanta, Georgia.

added through the bottom of the box, increasing the concentrations in the box for some species. In this example the box starts at 08.00 LDT with H = 250 m and at 15.00 LDT the final H = 2146 m. The increase represents nearly a factor-of-9 dilution of the initial box contents. This large dilution is evident in Figure 9.7 and is mostly responsible for the large decreases in NO and VOCs from 08.00 to 12.00 LDT. Emissions vary from hour to hour depending upon where the box is in the urban area. The table in Figure 9.7 gives the mixing ratios of the species initially included in the 250 m box. These were based on observed center city values (Baugues, 1989). The VOC composition was based on the data in Table 9.3. The table in Figure 9.7 also gives the amounts by which the initial values would have increased owing to emissions into the 250 m box if there were no dilution or reactions. These values depend upon the trajectory and the county-wide emission inventories. Also given in this table are the mixing ratios of the species assumed to be aloft, i.e., above H. These were based on typical background values. Note that this simulation also included emissions of "biogenic" hydrocarbons (the species BIOG in the table in Figure 9.7). These are based on the EPA Biogenic Emissions Inventory System (EPA, 1987). Without

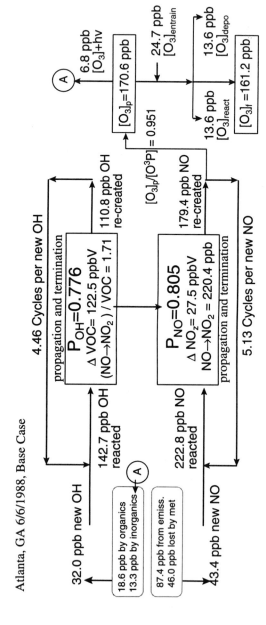

FIGURE 9.8. Process analysis for the Atlanta simulation.

reaction and dilution, these natural emissions would have produced 63 ppbC of biogenic hydrocarbons in the 250 m box by the end of the simulation.

Figure 9.7 shows that even in the presence of a large dilution rate the NO_2 mixing ratio increased in the first hour. This was caused mostly by a titration reaction between the initial NO and the entrained O_3 from aloft as H increased, but there was radical oxidation of the NO as well. After 09.00 LDT, all primary species mixing ratios decreased owing to the dilution. The mixing ratio for O_3, however, increased throughout the entire simulation until the Sun set.

9.4.1.2. Process Analysis

Figure 9.8 shows a "process analysis" based on the concepts illustrated in Figure 9.5. The numerical values in Figure 9.8 were obtained by integrating the rates of the mechanism's reactions throughout the course of the simulation and then summing the integrated reactions to obtain net reactions for the entire simulation. This technique is called *integrated reaction rate analysis* (Crouse, 1990).

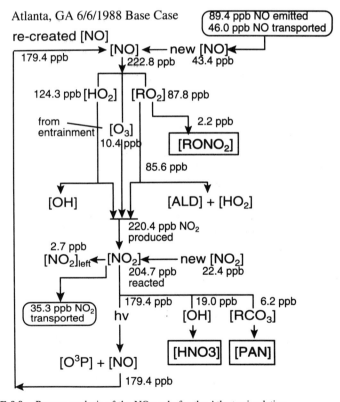

FIGURE 9.9.   Process analysis of the NO cycle for the Atlanta simulation.

The figure shows that 32 ppb of new $^\bullet$OH was responsible for the gross production of 171 ppb of O_3 by its cycling an average of 4.5 times to react with a total of 122.5 ppb of VOCs. For each reaction of an $^\bullet$OH with a VOC, 1.71 molecules of NO were oxidized to NO_2, yielding 210 ppb of NO-to-NO_2 conversion.

Details on this conversion are given in Figure 9.9, which shows that $RO_2^\bullet$, $HO_2^\bullet$, and O_3 entrained from aloft converted 222 ppb of new NO to 220 ppb NO_2, and 2.2 ppb of organic nitrates, $RONO_2$, were formed. Altogether, each ppb of 43 ppb of new NO was oxidized an average of 5.13 times. After the first oxidation the other four NO oxidations resulted in the gross production of 179.4 ppb of $O(^3P)$ via the photolysis of the NO_2 produced. About 5% of this $O(^3P)$ was consumed in reactions with organics or was lost in reactions with NO_2, resulting in about 95% of it forming O_3 or about 171 ppb of O_3. Photolysis reactions, other loss reactions, vertical transport, and surface deposition consumed about 4% of the gross O_3 production, leaving 161 ppb of O_3 in the model box at the end of the simulation.

Another way to look at the data in Figure 9.8 is that for every new $^\bullet$OH radical added to the system, about five O_3 molecules were formed.

9.4.1.3. VOC Reactivities

Figure 9.10 examines some VOC reactivity measures in the CB4 reaction mechanism for this simulation. The one-carbon PAR molecule in CB4 makes up about 55% of the total carbon in urban VOCs (bottom of Table 9.3), so it is not surprising that more of it reacted with $^\bullet$OH than any other species (upper left pie chart in Figure 9.10). One-third of the $^\bullet$OH reactions, however, were with CO. CO has a *lower* reaction rate constant than any of the species in the VOC mixture. In this simulation CO had a larger initial mixing ratio than the VOCs and, because of the background levels of CO assumed to be in the air above the mixing height, CO was diluted much less than were the VOCs. Therefore CO had significantly higher mixing ratios relative to the VOCs throughout the simulation and this compensated for the lower $^\bullet$OH + CO rate constant. The primary reason why all the rest of the urban VOC species had relatively small amounts reacting with $^\bullet$OH is that their molecular concentrations were low, i.e., they comprise less than one-half the mix carbon and they have carbon numbers from two to eight, resulting in small molecular concentrations relative to PAR, CO, and even methane. The species with the highest $^\bullet$OH reaction rate constants were completely consumed, some very early in the simulation. Nevertheless, these species only accounted for less than one-fourth of the total $^\bullet$OH + organic reactions.

The pie chart in the upper right of Figure 9.10 shows the origin of the $HO_2^\bullet$ and $RO_2^\bullet$ radicals that produced the total O_x. In computing these, the radical production of daughter product reactions was assigned back to the parent

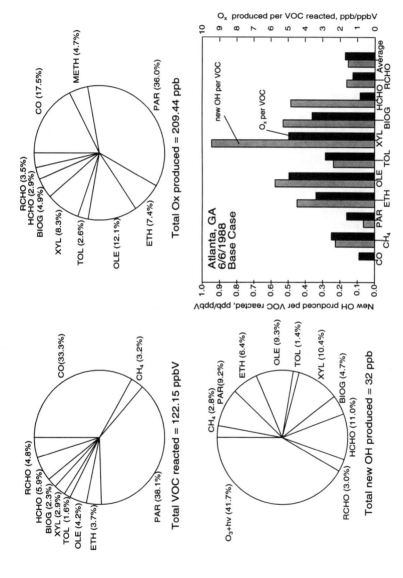

FIGURE 9.10. Species distributions and yields for the production of new radicals and O_x in the Atlanta simulation.

compound, i.e., O_x production from HCHO produced during the reaction of ETH was assigned to ETH and not to HCHO. The O_x production and yields for HCHO reported in the chart are just for primary HCHO included in the initial concentrations and in the emissions. *In examining the chart, it is clear that more than one-half of the odd-oxygen production came from peroxy radicals produced by the species usually considered the least reactive: CO, methane, and the alkanes.* This result is primarily due to the abundance of these three relative to the other species in the urban environment. At present, EPA excludes methane from any controls relative to O_3 issues, yet, as shown in this simulation, methane contributed nearly 5% of the gross O_x production, i.e., about the same amount as contributed by the biogenic or natural hydrocarbons from vegetation which have received so much recent attention (Chameides et al., 1988; Cardelino and Chameides, 1990).

The bar chart in Figure 9.10 shows the species specific O_x yields (i.e., O_x per VOC species consumed) and new $^\bullet$OH yields for this simulation. As with the pie chart, in this chart O_x and radicals originating in reactions of daughter products have been assigned back to the parent species. The OLE (or $\geq C_3$ alkenes) and XYL (or xylenes) molecules produced 5 O_x molecules when they reacted, the BIOG VOCs produced 3.5 O_x molecules, and CO produced less than 1 O_x molecule when it reacted with $^\bullet$OH. The average O_x yield of just the VOC species was 2.1, but when the large amount of CO with its low O_x yield was included, the total average O_x yield per molecule of (VOC + CO + CH_4) mix reacted was only 1.7.

The pie chart in the lower left of Figure 9.10 shows the origin of the new $^\bullet$OH radicals in this simulation. Slightly less than one-half of the total new $^\bullet$OH was produced by O_3 photolysis. The next largest amounts were produced by HCHO, then XYL, then OLE. These three actually produced $HO_2^\bullet$ radicals via photolysis reactions, and then the $HO_2^\bullet$ radicals were converted to $^\bullet$OH radicals by reaction (9.6). Because NO_x was present throughout the entire simulation (Figure 9.9 shows that the final NO_2 mixing ratio was 2.7 ppb), essentially all the $HO_2^\bullet$ produced reacted with NO, and reactions leading to H_2O_2 and ROOH were less than 1% of the total $HO_2^\bullet$ reactions. The bar chart shows the $^\bullet$OH yields by VOC species. XYL is a powerful radical source in this mechanism, producing nearly 0.9 new $^\bullet$OH radical per XYL molecule reacting. HCHO, BIOG, OLE, and ETH all produce about 0.5 new $^\bullet$OH radical per molecule.

9.4.2. Ozone Isopleth Diagrams

A useful tool for understanding the effects of precursor changes on secondary pollutants is an isopleth diagram. Figure 9.11 shows an O_3 isopleth diagram based on the meteorological scenario for June 6, 1988 for Atlanta, GA. The abscissa of the diagram is the initial VOCs and the ordinate is the initial NO_x that

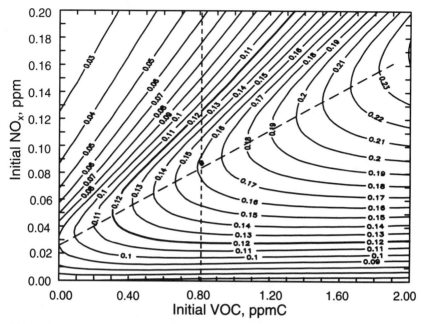

FIGURE 9.11.   Ozone isopleth diagram for Atlanta, Georgia. The diagonal line is the so-called "ridge line" and the black dot represents the base-case simulation shown in Figure 9.6. The vertical dashed line shows the constant hydrocarbon conditions studied in this section.

were used in a series of trajectory simulations. The interior of the diagram is made up of isopleths, lines of constant value, for O_3. Together, these form a response surface for precursor change. Although the axes are labeled as initial precursor, they represent both initial and emission precursor levels which are scaled together. That is, if the initial VOCs were doubled, the VOC emissions were also doubled. The "dot" in Figure 9.11 represents the conditions shown in Figure 9.7, and such simulations are called the "base-case" conditions. The diagonal dashed line approximates the locus of the so-called "ridge-line" which is the maximum O_3 that can be formed for each level of VOC. It is readily seen that the base-case condition in Figure 9.7 has slightly more initial NO_x than would result in forming the maximum possible O_3 that could be formed for the base-case VOC precursor levels. At 0.0 ppmC of initial VOCs, the ridge line intercepts the NO_x axis and the O_3 isopleth is not zero at this point. This is because the simulations all include 0.04 ppm of O_3 and 0.03 ppmC of VOCs aloft which are entrained into the well-mixed box to provide some reactivity even when the initial and emitted VOCs are zero. The area of the diagram below the ridge line is often described as being "NO_x-limited" and the area above the ridge line as being "hydrocarbon-limited."

It is apparent from this O_3 isopleth diagram that O_3 is non-linearly related to its precursor levels and that there are initial and emission conditions for which a reduction in precursors would cause an increase in maximum O_3, e.g., 0.14 ppm of NO_x and 0.80 ppmC of VOCs. It is also apparent that there are precursor conditions for which a very large reduction in VOC emissions may have almost no effect, e.g., 0.04 ppm of NO_x and 1.60 ppmC of VOCs.

To understand why the NO_x-limited region of the isopleth diagram has the shape it does, one only has to consider that at the ridge line, each trajectory simulation completely converted all the NO_x into nitrogen products just at the end of the simulation and thus there was no NO to participate in propagation reactions nor NO_2 to photolyze after that point. As more VOCs were added to the initial precursor level, this total consumption of the NO_x just happened sooner in the simulation. There is not, however, a constant relationship between the initial NO_x level and the final O_3 level, e.g., 0.02 ppm of initial NO_x yields 0.1 ppm of O_3, or 5 O_3 molecules per NO_x molecule, 0.04 ppm of initial NO_x yields 0.135 ppm of O_3 or 3.3 O_3 molecules per NO_x molecule, and 0.08 ppm of initial NO_x yields 0.19 ppm of O_3 or 2.3 O_3 molecules per NO_x molecule. This change in "efficiency" is related to the NO_3^*/N_2O_5 cycle which includes the reactions

$$O_3 + NO_2 \longrightarrow NO_3^* + O_2$$

$$NO_3^* \xrightarrow{h\nu} NO + O_2 \tag{9.29}$$

$$NO_3^* + \text{organics} \longrightarrow HNO_3 + R^* \tag{9.30}$$

Reaction (9.29) is a loss of two O_x and (9.30) is a loss of one. The higher the O_3 and NO_2 levels, the more mass runs through these reactions and the greater is the O_x loss. Thus the net efficiency for making O_x decreases with increasing NO_x and VOCs.

To understand why the VOC-limited region of the isopleth diagram has the shape it does requires additional analysis. Figure 9.12 shows the principal process analysis parameters for conditions falling on the vertical dashed line of Figure 9.11, i.e., at a constant initial VOC concentration of 0.816 ppmC, but with initial NO_x varying from 0.01 to 0.20 ppm.

In plot A of Figure 9.12 the inorganic (i.e., $O_3 + h\nu$) and organic (e.g., HCHO + $h\nu$) sources of new $\cdot$OH are plotted as a function of initial NO_x at constant VOC concentration. Their sum, the total new $\cdot$OH, is also plotted. In Figure 9.11 the maximum O_3 at this initial VOC occurred at 81 ppb of initial NO_x. The O_3 radical source peaked at about 50 ppb of initial NO_x and fell sharply thereafter. The organic sources peaked at somewhere between 100 and 150 ppb of initial NO_x. The total new $\cdot$OH therefore peaked at about 60 ppb of initial NO_x.

In plot B of Figure 9.12 the total new $\cdot$OH from plot A is replotted on a scale that also accommodates the total $\cdot$OH reacted. On the right-side scale is the

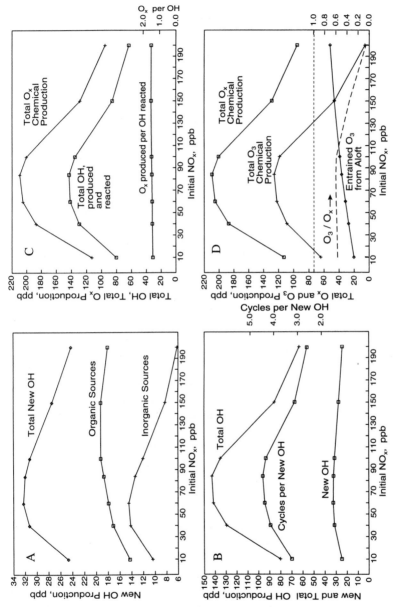

FIGURE 9.12.   Significant process parameters as a function of constant VOC and variable NO_x conditions for the Atlanta scenario.

number of times each $^{\bullet}$OH reacted before being terminated. At low initial NO_x mixing ratios the $^{\bullet}$OH cycles were low because there was insufficient NO to convert all the $HO_2^{\bullet}$ into $^{\bullet}$OH, i.e., there was too little NO for radical propagation, and radical–radical reactions caused termination. At NO_x above 80 ppb, the $^{\bullet}$OH cycles decreased because of increased termination due to $^{\bullet}$OH + NO_2 rather than $^{\bullet}$OH propagation by $^{\bullet}$OH + VOC. The peak $^{\bullet}$OH cycles occurred at about 85 ppb of initial NO_x. The total $^{\bullet}$OH reacted is the product of new $^{\bullet}$OH and the number of $^{\bullet}$OH cycles; therefore total $^{\bullet}$OH peaked at 85 ppb of initial NO_x.

In plot C of Figure 9.12 the total $^{\bullet}$OH produced from plot B is replotted along with the total O_x produced. On the right-side axis is the average number of O_x produced per $^{\bullet}$OH reacting with VOCs. This number increased monotonically from 1.41 to 1.51 as initial NO_x increased. Time profiles of O_x in each simulation reveal that the cause of this increase was that the higher NO mixing ratios were more competitive for the $RO_2^{\bullet}$ radicals, thus they produced more O_x in the form of NO_2. The peak of the total O_x produced was at about 85 ppb of initial NO_x.

In plot D of Figure 9.12, the total O_x produced, the total O_3 chemical production, and the O_3 entrained from aloft are plotted as symbols and solid lines. On the right-side axis is the ratio O_3/O_x, i.e., the fraction of O_x that is O_3, plotted as a dashed line. The fraction of O_x that is O_3 remains essentially constant at initial NO_x below the maximum O_x production, but decreases monotonically as initial NO_x increases above the O_x maximum. This is because, below the maximum O_x, all NO_x has been converted to nitrogen products and no further $O(^3P)$ production can occur after this happens. Above the maximum O_x point, however, some of the initial NO_x remained at the end of the simulation, and as initial NO_x increased, the system was increasingly unable to convert NO_2 into products and the remaining NO_2 made up an increasingly larger fraction of the O_x. Basically, the system could not go to completion because there were insufficient radicals to both recycle the NO and remove the NO_2. Above 150 ppb of initial NO_x, more of the final O_3 was due to entrainment from aloft than was due to chemical production and at 200 ppb, essentially all the O_3 in the simulation was from aloft whereas all the O_x was in the form of NO_2 and its products.

In summary, increasing initial NO_x mixing ratios inhibited O_3 production in three ways: (1) a larger fraction of the total O_x production was required to oxidize the larger mass of NO; (2) new $^{\bullet}$OH initiation from O_3 photolysis was reduced because the O_3 mixing ratios increased later and later in the simulations owing to (1) and therefore further and further past the solar peak; and (3) $^{\bullet}$OH propagation was reduced because increased NO_2 mixing ratios caused increased termination of $^{\bullet}$OH radicals. The reduction in new $^{\bullet}$OH initiation and the increased $^{\bullet}$OH termination as initial NO_x increased above the point where all NO_x was converted to products resulted in less production of O_x which resulted in less O_3. This created a negative feedback via decreased new $^{\bullet}$OH from O_3 which further reduced the O_3 mixing ratio and accelerated the decrease in reactivity.

This phenomenon is almost exclusively one associated with the inorganic chemistry which is believed to be nearly completely and fully understood and therefore, unless there is an extraordinary external radical source, O_3 formation should always have the characteristics described here.

9.5. CONTROL STRATEGIES

Knowledge that O_3 is a secondary pollutant formed by reactions of NO_x and VOC leads immediately to the conclusion that control of O_3 must be based on control of its precursors, and the issue for regulators was not what to do, but how much each of the precursors should be reduced. As the science of atmospheric photochemistry evolved, it became clear that complex, non-linear, feedback-controled processes relate NO_x and VOC precursors mixing ratios to O_3 mixing ratios. After some initial attempts to use linear or observational relationships (EPA, 1971), increasing scientific knowledge about O_3 formation led regulators to adopt mathematical models that included the best representations of the non-linear chemistry as the basis of needed control computations (Freas et al., 1978). Because of the health-basis of National Ambient Air Quality Standards, the legal basis of control was focused on O_3, however, and not the underlying chemically important O_x. This has caused the first of a number of complexities in the control decisions: the need to predict accurately not just the level of O_x, but also its partitioning, which is strongly related to emitted NO.

The first truly scientific understanding of ozone's relationship to its precursors that was adopted by regulatory decision makers came when O_3 isopleth diagrams were created. While it was possible to create such diagrams in 1976, the lack of accurate inputs for the trajectory model (i.e., mixing height versus time, trajectories, initial conditions, emissions) limited the accuracy of the model predictions. For these reasons, EPA developed a method to "scale" the model predictions to observations of O_3, and this method became known as the Empirical Kinetics Modeling Approach or EKMA (Dodge, 1977). The basic idea was to "calibrate" the isopleth diagram axes by using the observed VOC-to-NO_x morning ratio to locate a unique point on the isopleth line corresponding to the observed maximum O_3. The changes in calibrated precursor mixing ratios needed to move this "design" point to the locus of the 0.120 ppm O_3 isopleth line defined all the possible control scenarios for the set of meteorological conditions used to produce the O_3 isopleth diagram.

In choosing among this set of all possible reductions, the decision makers initially preferred those paths that reduced VOCs rather than those that reduced NO_x. There are at least three reasons for this: (1) reductions of VOCs *never* lead to an increase in O_3, but, for the upper half of the diagram, reductions in NO_x can; (2) in a join reduction strategy, reductions of NO_x result in needing larger reductions in VOCs to reach the standard than if no NO_x reduction occurred; and

(3) technologies for VOC reductions were more readily available and were cheaper than technologies for NO_x reductions. Many of the center city VOC-to-NO_x ratios were thought to be at or above the ridge line ratios on the O_3-isopleth diagrams that were produced in the late 1970s and thus VOC reduction appeared effective. Since 1980 the number of locations violating the NAAQS for O_3 has increased and many are in the southern portion of the US. These tend to have VOC-to-NO_x ratios well below the ridge line where large reductions of VOCs lead to almost no change in O_3, but where reductions of NO_x would be beneficial. It has also been argued that large natural sources of VOCs bias the southern urban locations even more toward "NO_x-limited" conditions (Chameides et al., 1988) and thus would require draconian reductions of anthropogenic VOCs to limit net O_3 production to the NAAQS.

Since 1987 it has been recognized that O_3 and its precursors are being transported between urban areas and that local controls alone are incapable of attaining the NAAQS for O_3 (EPA, 1991). This has led to abandonment of the simple trajectory model, with its inherent morning–afternoon, single-day cause-and-effect structure, as the method of choice to make control calculations. Instead, the preferred tools are urban-scale, three-dimensional Eulerian models (e.g., the UAM) that operate with time varying boundary conditions obtained from larger, regional-scale Eulerian models (e.g., the ROM). In the Clean Air Act these are called "photochemical grid models", but only the UAM and the ROM are being used. Both these models use the same photochemical reaction mechanism as was used in OZIPR to make the plots in this chapter. In these multilayered grid models, however, meteorological factors are much more influential on predicted O_3 concentrations than in the OZIPR example shown here, and the treatment of transported O_3 can have an overwhelming effect on the predicted controls needed to achieved the NAAQS for O_3.

9.6. CONCLUSIONS

It has been easier to determine that there *are* several ubiquitous air pollutants that may reasonably be anticipated to endanger public health or welfare than it has been to determine how to minimize exposures to these substances without sacrificing our "standard of living". Some difficulties in achieving our clean air goals are associated with a lack of complete science and thus a failure to recognize the true complexity of the secondary pollutant problems. Because chemistry plays such a central role for secondary air pollutants such as ozone, this chapter has emphasized the composition of urban air and the chemical processes, as we currently know them, among the species that lead to urban ozone formation.

The creation of urban ozone is recognized as a non-linear, feedback-regulated process among a large number (> 200) of compounds. This means that it is possible to reduce emissions of a precursor and to increase rather than decrease the concentration of ozone, or it may mean that large decreases in emissions of a precursor may be ineffective in reducing the concentration of ozone. The energy for the chemical transformations is derived from ambient solar radiation, especially the short-wavelength or ultraviolet portion of the spectrum. There are no routinely monitored data to characterize city radiation fields, and the intensity of the driving force must be inferred by the use of radiation models. The central actors in the ozone creation process are free radicals whose atmospheric concentrations are extremely low, e.g., 10^{-6}–10^{-4} ppm, and whose lifetimes are very short, e.g., milliseconds. Measurements of these radicals have only occurred in a few research laboratories. Routine ambient and laboratory measurements can only quantify the effects of the radical reactions. Therefore one must infer the actual magnitudes of the sources and fates of these important atmospheric entities via mathematical modeling. These models may suffer from internal compensating errors and thus, while they may explain a particular situation, may fail to forecast accurately for a significantly changed situation as might occur in the future.

Examples of models of the ozone formation process show the complex relationships described above and explain the origins of VOC "reactivities". One counter-intuitive result is that more than one-half of the formation of urban O_3 occurs by reaction of the so-called "least reactive" species: CO, methane, and the paraffins. This is because the large concentrations of these compounds make them very competitive for the hydroxyl radical compared with the much smaller concentration of the so-called "most reactive" HCs. Thus the real role of the latter is to supply new radicals to the system, while the former use these radicals to create most of the ozone. One concern this raises is the extent to which cities have radical sources that are not well represented in the models, e.g., heterogeneous formation of nitrous acid, a powerful radical source, from NO_x emissions. Finally, O_3 is itself a powerful radical source, supplying nearly half of all the new radicals in urban situations. The transport of O_3 both from the stratosphere and from upwind urban areas may have a large effect on the "reactivity" of a downwind area, thus such mixing of older and newer emissions must be careful modeled if we expect to make accurate forecasts of the effects of emission reductions. This requires very complex three-dimensional models that will simulate the vertical component of the atmosphere much more accurately than do present regulatory models.

Acknowledgments

Much of the material presented here was developed under various US Environmental Protection Agency cooperative agreements with the Atmospheric Research and Environmental Assessment Laboratory under the directions of Dr Marcia Dodge and Dr Joseph Sickles. Other support was from the

Western States Petroleum Association under the direction of Dr Allan Hirita. I am grateful to Shawn Tonnesen for performing the Atlanta simulations and producing the plots used in that discussion, and to Dr Kenneth Sexton for producing the data in Table 9.2.

References

ASTM (1973). Standard solar constant and air mass zero solar spectral irradiance tables. In *Annual Book of ASTM Standards, Part 41*, ASTM, Philadelphia, PA, ASTM Standard E490–73a.

Atkinson, R. (1985). Kinetics and mechanisms of the gas-phase reactions of the hydroxyl radical with organic compounds under atmospheric conditions. *Chem. Rev.*, **85**(1), 69–201.

Atkinson, R. (1990). Gas-phase tropospheric chemistry of organic compounds: a review. *Atmos. Environ.*, **24A**(1), 1–41.

Atkinson, R., Lloyd, A.C. and Winges, L. (1982). An updated chemical mechanism for hydrocarbon/NO_x/SO_2 photooxidations suitable for inclusion in atmospheric simulation models. *Atmos. Environ.*, 16, 1341.

Baugues, K. (1989). *Initial Conditions for RIA/OZIP Simulations*, US Environmental Protection Agency, Research Triangle Park, NC.

Blacet, F.E. (1952). Photochemistry in the lower atmosphere. *Ind. Eng. Chem.*, **44**, 1339.

Cardelino, C.A. and Chameides, W.L. (1990). Natural hydrocarbons, urbanization, and urban ozone. *J. Geophys. Res.*, **95**, 13,971–13,979.

Carter, W.P.L. (1990). A detailed mechanism for the gas-phase atmospheric reactions of organic compounds. *Atmos. Environ.*, **24A**(3), 481–518.

Chameides, W.L., Lindsay, R.W., Richardson, J. and Kiang, C.S. (1988). The role of biogenic hydrocarbons in urban photochemical smog: Atlanta as a case study. *Science*, **241**, 1473–1474.

Crouse, R.R. (1990). Integrated reaction rate analysis of ozone production in a photochemical oxidant model. *Master's Thesis, Department of Environmental Science and Engineering*, University of North Carolina, Chapel Hill, NC.

Dodge, M.C. (1977). Combined use of modeling techniques and smog chamber data to derive ozone precursor relationships. In Dimtriades, B., Ed., *Proc. Int. Conf. on Photochemical Oxidant Pollution and Its Control, EPA-600/3-77-001b*, Environmental Protection Agency, Research Triangle Park, NC, pp. 881–889.

EPA (1971). Federal Register, **36**, August 14, 15,489.

EPA (1987) Criteria pollutant emission factors for the 1985 NAPAP emission inventory. *Report EPA-60017-87-015*, Office of Research and Development US Environmental Protection Agency, Washington, DC.

EPA (1990). *User's Guide for the Urban Airshed Model, Volume I: User's Manual for UAM (CB4), EPA-450/4-90-007a*. Office of Air Quality Planning and Standards, US Environmental Protection Agency, Research Triangle Park, NC.

EPA (1991). *Regional Modeling for Northeast Transport (ROMNET), EPA-450/4-91-002a*, US Environmental Protection Agency, Research Triangle Park, NC.

Freas, W.P., Martinez, E.L., Meyer, E.L., Sennet, D.H. and Summerhays, J.E. (1978).

Procedures for Quantifying Relationships Between Photochemical Oxidants and Precursors: Supporting Documentation, EPA-450/2-77-021b. US Environmental Protection Agency, Research Triangle Park, NC.

Gery, M.W. and Crouse, R.R. (1990). *User's Guide for Executing OZIPR, 9D2196NASA,* Atmospheric Research and Exposure Laboratory, US Environmental Protection Agency, Research Triangle Park, NC.

Gery, M.W., Whitten, G.Z., Killus, J.P. and Dodge, M.C. (1989). A photochemical kinetics mechanism for urban and regional scale computer modeling. *J. Geophys. Res.*, **94**(D10), 12,925–12,956.

Greiner, N.R. (1967). *J. Chem. Phys.*, **46**, 3389.

Haagen-Smit, A.J., Darley, E.F., Zaitlin, M., Hull, H. and Nobel, W. (1951). Investigation on injury to plants from air pollution in the Los Angeles area. *Plant Physiol.*, **27**, 18.

Heicklen, J., Westberg, K. and Cohen, N. (1969). Conversion of NO to NO_2 in polluted atmospheres. Center for Air Environment Studies Publication No. 115–69. University Park: Pennsylvania State University, 5 pp.

Iqbal, M. (1983). *An Introduction to Solar Radiation*, Academic Press, New York, NY.

Jeffries, H.E. and Arnold, J.R. (1987). The science of photochemical reaction mechanism development and evaluation. In Atkinson, R., Jeffries, H.E., Whitten, G.Z. and Lurmann, F.W., Eds, *Proc. EPA Workshop on Evaluation and Documentation of Chemical Mechanisms, EPA/600/9-87/024*, US Environmental Protection Agency, Research Triangle Park, NC.

Jeffries, H.E., Sexton, K.G. and Arnold, J.R. (1989). *Validation Testing of New Mechanisms with Outdoor Chamber Data Volume 2: Analysis of VOC Data for the CB4 and CAL Mechanisms*, EPA-600/3-89-010b (NTIS PB 89-159-040/AS), US Environmental Protection Agency, Research Triangle Park, NC.

Leighton, P.A. (1961). *Photochemistry of Air Pollution*, Academic Press, New York, NY.

Lonneman, W.A. (1987). Comparison of 0600–0900 AM hydrocarbon compositions obtained from 29 cities. *APCA/EPA Symp. on Measurements of Toxic Air Pollutants, Raleigh, NC*, APCA, Pittsburgh, PA.

Lurmann, F.W., Carter, W.P.L. and Coyner, L.A. (1987). *A Surrogate Species Chemical Reaction Mechanism for Urban Scale Air Quality Simulation Models Volume I: Adaptation of the Mechanism*, EPA-600/3-87/014a, US Environmental Protection Agency, Research Triangle Park, NC.

Moortgat, G.K., Seiler, W. and Warneck, P. (1983). Photodissociation of HCHO in air: CO and H_2 quantum yields at 220 and 300 K. *J. Chem. Phys.*, **78**(3), 1185–1190.

NRC (1991). *Rethinking the Ozone Problem in Urban and Regional Air Pollution*, National Academy Press, Washington, DC.

NSF (1991). Committee on Physical, Mathematical, and Engineering Sciences of the Federal Coordinating Council for Science, Engineering, and Technology, "Grand Challenges: High Performance Computing and Communications, The FY1991 US Research and Development Program", c/o National Science Foundation, Computer and Information Science and Engineering, Washington, DC, February.

Parrish, D., Murphy, P.C., Albritton, D.L. and Fehsenfeld, F.C. (1983). The measurement of the photodisociation rate of NO_2 in the atmosphere. *Atmos. Environ.*, **17**, 1365.

Pierson, W.R., Gertler, A.W. and Bradow, R.L. (1990). Comparison of the SCAQS tunnel study with other on-road vehicle emissions data. *J. Air Waste Manag. Assoc.*, **40**, 1495.

Richter, H. (1989). *Aldehyde Data for EPA Nonmethane Organic Compound Monitoring Program.* US Environmental Protection Agency, Research Triangle Park, NC.

Rogers, J.D. (1990). Ultraviolet absorption cross sections and atmospheric photodissociation rate constants of formaldehyde. *J. Phys. Chem.*, **90**, 4011–4015.

Stedman, D.H., Morris, E.D., Daby, E.E., Niki, H. and Weinstock, B. (1970). The role of ˙OH radicals in photochemical smog reactions. *160th National Meeting of the American Chemical Society, Chicago, IL*, ACS, Washington, DC.

Stockwell, W.R. (1986). A homogeneous gas-phase mechanism for use in a regional acid deposition model. *Atmos. Environ.*, **20**, 1615–1632.

Stockwell, W.R., Middleton, P. and Chang, J.S. (1990). The second generation regional acid deposition model chemical mechanism for regional air quality modeling. *J. Geophys. Res.*, **95**(D10), 16,343–16,367.

Wallington, T.J., Dagaut, P. and Kurylo, M.J. (1992). Ultraviolet absorption cross sections and reaction kinetics and mechanisms for peroxy radicals in the gas phase. *Chem. Rev.*, **92**(4), 667–710.

Warneck, P. (1988). *Chemistry of the Natural Atmosphere*, Academic Press,, New York, NY.

Whitten, G.Z., Hogo, H. and Killus, J.P. (1980). The carbon bond mechanism: a condensed kinetic mechanism for photochemical smog. *Environ. Sci. Technol.*, **14**, 690–700.

10

Ozone in the troposphere

Paul J. Crutzen

10.1. INTRODUCTION

Although ozone (O_3) makes up less than one-millionth of the volume of the atmosphere, the importance of this gas in atmospheric chemistry is great and manifold. About 90% of all ozone is located in the stratosphere (10–50 km), where it provides a critical service to the biosphere by absorbing most of the harmful solar ultraviolet radiation (UV-B and UV-C) at wavelengths less than about 315 nm. The absorbed radiation is the main energy source of the stratosphere and establishes much of its temperature structure and dynamics. Because of this and the strong absorption of terrestrial radiation in the infrared

349

"window" near 9.6 μm ozone also plays a significant role in the Earth's radiation balance. Changes in the atmospheric concentrations of tropospheric and stratospheric ozone can thus play a role in anthropogenic climate change (Fishman et al., 1979a). Molecule for molecule, tropospheric and lower stratospheric ozone is about 2000 times more effective than CO_2 as a greenhouse gas (IPCC, 1990; Lacis et al., 1990; Schwarzkopf and Ramaswamy, 1993; Wang et al., 1991; see also Chapter 13). Human activities, in particular emissions of chlorofluorocarbons have caused substantial loss of stratospheric ozone (Farman et al., 1985; Gleason et al., 1993; see also Chapter 11) with indirect consequences for climate and, as we will discuss, for tropospheric chemistry (Schnell et al., 1991; Madronich and Granier, 1992).

The troposphere contains only about 10% of all atmospheric ozone. For a long time it was commonly accepted that most ozone in the troposphere emanated from the stratosphere and that it did not play any role in tropospheric chemistry. However, at the beginning of the 1970s it began to become clear that this view had to be revised. First, Levy (1971) drew attention to the fact that the photolysis of ozone at wavelengths (λ) shorter than about 315 nm, largely coinciding with the biologically harmful UV-B and UV-C wavelength regions, gives rise to electronically excited $O(^1D)$ atoms according to

$$O_3 + h\nu \rightarrow O(^1D) + O_2 \quad (\lambda < 315\,\text{nm}) \tag{10.1}$$

which possess enough energy to react with water vapor to produce highly reactive hydroxyl (OH) radicals:

$$O(^1D) + H_2O \rightarrow 2\,OH \tag{10.2}$$

The importance of OH radicals in atmospheric chemistry lies in the fact that, in contrast with about 10^{13} times more abundant molecular oxygen, they react with almost all gases that are emitted by natural processes and anthropogenic activities into the atmosphere. Examples of such gases are hydrocarbons, carbon monoxide, most sulfur- and halogen-containing compounds, as well as NO and NO_2. Owing to their very low solubility in water, only a few of these gases can be significantly removed from the atmosphere by precipitation processes. Best estimates are (e.g., Prinn et al., 1992; Thompson, 1992) that the average, global volume mixing ratio of OH in the troposphere is about 3×10^{-2} ppt. Thus, despite the fact that too much tropospheric ozone and UV-B radiation can be damaging to the biosphere (e.g., SCOPE, 1992; UNEP, 1992; Adams et al., 1985; Reich and Amundson, 1985; Lippmann, 1988), causing health effects, forest decline, and an estimated cost to agriculture in the US alone of several billion dollars each year, they are also of critical importance for keeping the atmosphere "clean". As a consequence, the average lifetimes of most atmospheric gases are determined by their reactivity with OH, varying from a few hours for highly reactive gases such as isoprene (C_5H_8) or NO_x (= NO + NO_2), to 1–2 months for CO and almost 10 years for CH_4. Whenever a compound does not react with OH in the troposphere, it will reach the stratosphere, where it can interfere with stratospheric ozone chemistry, such as is the case for the chlorofluorocarbon gases ($CFCl_3$, CF_2Cl_2) and N_2O. The concentrations of OH therefore determine the oxidation efficiency (also termed oxidation capacity or oxidation power) of the atmosphere, the distance over which gases can be transported in the atmosphere, and their temporal variability. Isoprene will thus only be found above or in the close vicinity of forests, its main emission source, while methane, despite discrete local to regional sources such as wetlands, rice fields, coal mines, etc., is rather evenly distributed around the globe, with just a little more in the northern hemisphere (NH) than in the southern hemisphere (SH). This also implies that CH_4 and CO are the main reaction partners of OH in the background troposphere, roughly 40% reacting with CH_4 and oxidation products, and 60% with CO, other gases having been largely removed nearer the source.

Having shown the fundamental role of OH in the chemistry of the atmosphere, the same holds true for tropospheric ozone. Contrary to the belief of only about 20 years ago, it is now clear, as we will again demonstrate in this study, that ozone in the troposphere stems only partly from the stratosphere (Crutzen, 1973). To start with, it is clear that reactions (10.1) and (10.2) together destroy ozone. However, ozone can also be produced in the troposphere as a result of the catalytic action of the NO_x gases in the oxidation chains of CO, CH_4 and other hydrocarbons (Crutzen, 1973), e.g., via

$$CO + OH\ (+ O_2) \rightarrow CO_2 + HO_2 \tag{10.3}$$

$$HO_2 + NO \rightarrow OH + NO_2 \tag{10.4}$$

$$NO_2 + h\nu \rightarrow NO + O \quad (\lambda < 420\,nm) \tag{10.5}$$

$$O + O_2 + M \rightarrow O_3 + M \tag{10.6}$$

net C1: $CO + 2\,O_2 \rightarrow CO_2 + O_3$

Note that in this oxidation chain NO, NO_2, OH, and HO_2 serve as catalysts; e.g., the OH radical which is lost by reaction (10.3) reappears through reaction (10.4). The reaction chain therefore does not influence the sum of the concentrations of OH + HO_2 (= HO_x) and of NO + NO_2 (= NO_x), but it can affect the ratios of the concentrations of NO to NO_2, and of OH to HO_2. Addition of NO_x to the troposphere favors O_3 production, e.g., via the above reaction cycle C1, and, via reactions (10.1) + (10.2) and (10.4), also higher OH concentrations. As an appreciable fraction of the atmospheric input of NO_x now occurs by a variety of anthropogenic activities (see Table 10.1), one should expect ozone concentra-

TABLE 10.1   Tropospheric NO Sources
($Tg\,N\,year^{-1}$)

Natural	
Soils	5–20
Lightning	2–20
From stratosphere	0.5
Anthropogenic	
Fossil fuel combustion	24
Biomass burning	2.5–13
Aircraft emissions	0.6

Source: IPCC (1992).

tions to have increased in the NH. This is indeed the case (e.g., Claude et al., 1992; Staehelin and Schmid, 1991; Volz and Kley, 1988; Crutzen, 1988; Logan, 1989; Oltmans and Levy, 1993). Because of this and ozone loss in the stratosphere (Gleason et al., 1993), which allows more photochemically active radiation to penetrate into the troposphere, OH formation in the troposphere via reactions (10.1) and (10.2) was enhanced by about 4% during the 1980s (Madronich and Granier, 1992). Owing to several chemical feedbacks, however, this does not automatically imply that OH concentrations have increased by a similar ratio. The issue is further complicated by the fact that the concentrations of CH_4 and CO have been increasing as well: CH_4 by $0.5–0.8\%\,year^{-1}$ worldwide and CO by $0.7–1.1\%\,year^{-1}$ at mid-latitudes in the NH until the mid-1980s (IPCC, 1992; Zander et al., 1989; see also Chapter 3).

This chapter provides an overview of the important role of ozone in the photochemistry of the troposphere, emphasizing global processes, the main factors which determine the ozone budget and distribution, and the changes which have occurred or may be expected to occur owing to anthropogenic activities.

10.2. TROPOSPHERIC OZONE: ITS SOURCES AND SINKS

10.2.1. Supply from the Stratosphere

Downward transfer of ozone from the stratosphere to the troposphere occurs mainly at extratropical latitudes, often in connection with tropopause folding events (Danielsen, 1968). Despite the fact that the importance of these events for the transfer of stratospheric constituents, in particular ozone, has been known for a long time, there still exists considerable quantitative uncertainty about the fluxes involved. Estimates of these have mostly been derived by combining observations of ozone with those of other chemical and meteorological tracers. The most useful, approximately conservative, meteorological tracer is potential vorticity (PV). Measurements or model calculations of this property around tropopause folds, together with those of O_3, both of which increase rapidly in the transition region between the troposphere and the stratosphere, have been used by several authors to derive the downward flux of ozone into the troposphere. Likewise, data on the decay of the stratospheric content of radioactive nuclear bomb test materials such as ^{90}Sr have been used for this purpose. In this way, Danielsen (see Danielsen and Mohnen, 1977) estimated an extratropical downward flux of stratospheric air of 4.3×10^{20} g year^{-1} in the NH, approximately equal to the NH stratospheric mass. A quasi-sinusoidal, seasonal dependence was deduced, with the maximum flux in mid-April being about three times larger than the minimum in October. By combining correlations between PV and ^{90}Sr and between ^{90}Sr and O_3, obtained in a number of field experiments and through meteorological analyses of ozone sonde measurements, Danielsen and Mohnen (1977) calculated the downward flux of ozone to be equal to 7.8×10^{10} molecules cm^{-2} s^{-1} averaged over the year. They also concluded that this flux was maximum in spring, with a value five times larger than in the fall, the large seasonal effect reflecting seasonal variability in both downward air mass fluxes and ozone mixing ratios in the lower stratosphere, which likewise maximize in spring.

As discussed by Vaughan (1988), there are still considerable uncertainties with this method, especially regarding the small-scale structure of the mentioned properties in the lower stratosphere, which are highly variable in space and time, differing from fold to fold. Thus there is no acceptable agreement even on the mass flux from the stratosphere to the troposphere. A more recent theoretical/observational study by Holton (1990) based on the "downward control" principle, in which the average meridional circulation in the stratosphere is derived from the continuity equation and stratospheric temperature deviations from radiative equilibrium, caused by energy dissipation due to large-scale dynamic perturbations, gives an annual downward air mass flux of 1.3×10^{20} g from the stratosphere in the NH, only 30% of that estimated by Danielsen. For the SH the downward mass flux was estimated to be half of that in the NH.

An alternative, novel method to derive the stratosphere to troposphere transfer of ozone was recently introduced by Murphy et al. (1992). From a large set of simultaneous measurements of the concentrations of NO_y (= all reactive nitrogen except N_2O, i.e., NO_x + HNO_3 + PAN + . . .) and O_3 in the lower stratosphere, and knowledge of the stratospheric production of NO_y via the reaction,

$$N_2O + O(^1D) \rightarrow 2 NO \qquad (10.7)$$

which can be estimated from photochemical calculations using measured distributions of N_2O and O_3 in the stratosphere, these authors derived an average global downward flux of O_3 in the range of $(1.9\text{–}6.4) \times 10^{10}$ molecules cm^{-2} s^{-1}, which is smaller than the estimate of Danielsen and Mohnen (1977) for the NH and which may be partially explained by the fact that the downward air mass flux in the NH is larger than in the SH (Holton, 1990). The relatively large range of ozone flux estimates by Murphy et al. (1993) is mainly caused by the variability of the measured NO$_y$ to O_3 concentration ratios, which may not be conservative properties owing to possible loss of NO$_y$ by particle settling and photochemical ozone loss in the lower stratosphere during winter and spring, especially in the SH. Furthermore, there are also remaining uncertainties in the production of NO by reaction (10.7) which may largely be resolved by the current global observations of stratospheric N_2O and O_3 on the Upper Atmosphere Research Satellite (UARS). The method of Murphy et al. (1993) assumes, largely supported by measurements of NO$_y$ and N_2O, that most NO$_y$ in the lower stratosphere emanates from reaction (10.7). Contributions from other potential NO$_y$ sources, e.g., downward flux from the mesosphere, stratospheric aircraft emissions, and upward transport of lightning-produced NO$_y$ in the tropics, would lead to higher O_3 fluxes. From the available information, however, it appears, that these are less important than reaction (10.7).

The above estimates of downward ozone fluxes fall generally rather well in line with most results obtained with general circulation models: $(3.8\text{–}6.6) \times 10^{10}$ molecules cm^{-2} s^{-1} in the NH and $(2.5\text{–}3.6) \times 10^{10}$ molecules cm^{-2} s^{-1} in the SH (Mahlman et al., 1980; Gidel and Shapiro, 1980; Allam and Tuck, 1984; Levy et al., 1985). However, because of the coarse grid resolution of the models that were used, these results must be regarded as quite preliminary. Ebel et al. (1992), using a more detailed, higher-resolution mesoscale model and statistics on tropopause folding and "cut-off-low" events, calculated a much higher downward O_3 flux in the NH of $(12\text{–}15) \times 10^{10}$ molecules cm^{-2} s^{-1}, two to eight times larger than the range of estimates of Murphy et al. (1992). This large discrepancy cannot be explained by additional NO$_y$ sources. The estimate by Ebel et al. (1992), based on model simulations for April, may only be correct for the springtime period in the NH. As indicated by Danielsen and Mohnen (1977), ozone fluxes maximize in spring, being five times larger than in the fall according to these authors. In the statistical analysis of Ebel et al. (1992) all tropopause-folding events were weighted about equally, giving a relatively small seasonal dependence of downward ozone fluxes.

In summary, the extratropical downward flux of ozone from the stratosphere to the troposphere in the NH seems to be in the range of $3\text{–}8 \times 10^{10}$ molecules cm^{-2} s^{-1}. The SH flux may be half as large. Much more work is clearly still needed to establish more reliable estimates of the stratospheric source of tropospheric ozone, including its seasonal and spatial dependences. It is interesting to compare these stratosphere-to-troposphere fluxes with the average production of ozone by direct photolysis of O_2 and reaction (10.6), which is equal to 5×10^{13} molecules cm^{-2}s^{-1}. Therefore only about 0.1% of all ozone produced in the stratosphere leaks down into the troposphere. This fraction may be a sensitive function of the general circulation of the atmosphere, in particular the rapidity of transport out of the main ozone production region in the 25–35 km height region in the tropics. For the future, it may be asked whether increasing greenhouse warming in the tropics and accompanying latent heat release may stimulate the upward branch of the Brewer–Dobson/Hadley circulation, with the return flow at extratropical latitudes being enforced by reduced solar and infrared heating as a consequence of ozone loss in the lower stratosphere and enhanced radiative cooling due to increasing concentrations of other radiatively active greenhouse gases, CO_2, CH_4, and N_2O.

10.2.2. Destruction at the Earth's Surface

This process was for a long time believed to be the main sink for ozone in the troposphere (e.g., Junge, 1962) with early estimates (e.g., Fabian, 1973; Fabian and Junge, 1970) giving values in line with estimated fluxes from the stratosphere. A study by Fishman (1985) of destruction fluxes at the Earth's surface, based on a methodology developed by Galbally and Roy (1980), gave average fluxes of 17×10^{10} molecules cm^{-2}s^{-1} for the NH and 5.6×10^{10} molecules cm^{-2}s^{-1} in the SH, the marked difference reflecting much more efficient destruction over land than ocean. In fact, some recent aircraft observations using the eddy correlation method (Kawa and Pearson, 1989) indicate that ozone destruction at the ocean surface may even be about four times less efficient than estimated by Galbally and Roy (1980) and Fishman (1985). Although much more information is now available, no updates of these estimates have yet been made.

10.2.3. Photochemical Sinks and Sources Derived from CO and CH$_4$ Oxidation

The fundamental reactions (10.1) and (10.2), leading to the formation of OH radicals also cause ozone loss. Quantitative estimates of this can be made with available information on water vapor and ozone distributions, the latter allowing calculations of photochemically active UV fluxes. According to our own estimates, which were obtained with the global tropospheric photochemical model MOGUNTIA (Crutzen and Zimmermann, 1991), the global average tropospheric ozone loss by these reactions is 14×10^{10} molecules cm^{-2}s^{-1}. Note that this ozone loss plus destruction flux to the Earth's surface is already larger than the estimated influx of ozone from the stratosphere.

Further ozone loss occurs through the reaction of O_3 with HO_2 radicals, e.g., as a consequence of one of the branches of the CO oxidation cycle

$$CO + OH\ (+O_2) \rightarrow CO_2 + HO_2 \tag{10.3}$$

$$HO_2 + O_3 \rightarrow OH + 2O_2 \tag{10.8}$$

net C2: $CO + O_3 \rightarrow CO_2 + O_2$

If this set of reactions were the only one causing CO oxidation to CO_2, a loss of almost 35×10^{10} molecules cm^{-2} s^{-1} of O_3 would be the result, estimated on the basis of the observed global distribution of CO and OH concentrations that were calculated with the MOGUNTIA model (see Table 10.3(a)). Clearly the estimated flux of ozone from the stratosphere ($\approx 5 \times 10$ molecules cm^{-2} s^{-1}) is insufficient to balance ozone loss by destruction at the Earth's surface ($\approx 10 \times 10^{10}$ molecules cm^{-2} s^{-1}) + destruction by reactions (10.1) + (10.2) (14 $\times 10^{10}$ molecules cm^{-2} s^{-1}) + maximum loss by the reaction sequence C2 (35 $\times 10^{10}$ molecules cm^{-2} s^{-1}), implying that ozone must also be produced within the troposphere. Because solar radiation capable of direct O_3 production by photolysis of O_2 and reaction (10.6) is absorbed in the stratosphere, tropospheric ozone production needs catalysis by NO_x, e.g., via the CO oxidation branch C1 already mentioned and via CH_4 oxidation:

$$CH_4 + OH\ (+ O_2) \rightarrow CH_3O_2 + H_2O \tag{10.9}$$

$$CH_3O_2 + NO \rightarrow CH_3O + NO_2 \tag{10.10}$$

$$CH_3O + O_2 \rightarrow CH_2O + HO_2 \tag{10.11}$$

$$HO_2 + NO \rightarrow OH + NO_2 \tag{10.4}$$

$$NO_2 + h\nu \rightarrow NO + O\ (2\times)\quad (\lambda < 420\,nm) \tag{10.5}$$

$$O + O_2 + M \rightarrow O_3 + M\ (2\times) \tag{10.6}$$

net C3: $CH_4 + 4 O_2 \rightarrow CH_2O + H_2O + 2 O_3$

Further oxidation of CH_2O to CO, e.g., via

$$CH_2O + h\nu \rightarrow H + HCO\quad (\lambda < 350\,nm) \tag{10.12}$$

$$H + O_2 + M \rightarrow HO_2 + M \tag{10.13}$$

$$HCO + O_2 \rightarrow CO + HO_2 \tag{10.14}$$

$$HO_2 + NO \rightarrow OH + NO_2\ (2\times) \tag{10.4}$$

$$NO_2 + h\nu \rightarrow NO + O\ (2\times)\quad (\lambda < 420\,nm) \tag{10.5}$$

$$O + O_2 + M \rightarrow O_3 + M\ (2\times) \tag{10.6}$$

net C4: $CH_2O + 4 O_2 \rightarrow CO + 2 OH + 2 O_3$

constitutes another significant source for O_3, as well as for HO_x radicals. Detailed analysis of the CO and CH_4 oxidation cycles in NO_x-rich environments indicates the potential production of almost three O_3 molecules per CH_4 molecule that is oxidized to CO. As the average tropospheric global column destruction rate of CO by reaction with OH, calculated with MOGUNTIA, is equal to 3.5×10^{11} molecules $cm^{-2} s^{-1}$ and that of CH_4 about 10^{11} molecules $cm^{-2} s^{-1}$, the global average, maximum total column tropospheric O_3 production rate would add up to almost 6.5×10^{11} molecules $cm^{-2} s^{-1}$ if the entire troposphere were in the NO_x-rich state, far too much to balance the loss of ozone by reactions (10.1) + (10.2) (1.4×10^{11} molecules $cm^{-2} s^{-1}$) and surface deposition ($\approx 10 \times 10^{11}$ molecules $cm^{-2} s^{-1}$). Not yet considered in this simple analysis is the additional production of ozone by the interactions of anthropogenic NO_x and non-methane hydrocarbon (NMHC) emissions. In polluted atmospheres photochemical smog is a manifestation of such reactions (see Chapter 9). Clearly we must conclude that O_3 is being both produced and destroyed by chemical reactions in the troposphere, more than is supplied from the stratosphere, with O_3 production occurring in NO_x-rich and destruction in NO_x-poor environments. As anthropogenic activities contribute substantially to NO_x emissions (see Table 10.1), O_3 production is most pronounced in the NH. This may explain the observations of higher O_3 concentrations in most of the NH background troposphere, as summarized in Figure 10.1 and discussed further in Section 10.3.1. Most importantly, however, reviews of ozone observations prior to the Second World War, some of them going back to the last century and the 1930s (Volz and Kley, 1988; Bojkov, 1988; Crutzen, 1988) clearly indicate much lower ozone concentrations during the earlier times. More recent records indicate continued ozone increases during the past decades at various monitoring stations in the NH (Logan, 1985; Staehelin and Schmid, 1991; Bojkov, 1988; WMO, 1992). A long-term trend analysis for some "background" stations based on the measurements of Oltmans and Levy (1993) is given in Table 10.2. They indicate substantial upward trends for the NH stations Barrow (Alaska) and Mauna Loa (Hawaii). Especially during summer at Barrow, the upward ozone trend has been very large, $(1.73 \pm 0.58)\%$ year^{-1}. Besides the possibility of an underlying large-scale NH trend, this finding probably also reflects a substantial influence from regional pollution due to oil production activities. In contrast, at the South Pole there is a substantial downward trend which may reflect the development of the ozone hole in the lower stratosphere, allowing less transport of ozone from the stratosphere and more ozone destruction by reactions (10.1) + (10.2) and (10.8) owing to higher UV fluxes (Schnell et al., 1991).

10.2.4. Production Involving Interactions between NO_x and Reactive Hydrocarbons

The discovery of high ozone concentrations in southern California during photochemical smog episodes led to the proposal by Haagen-Smit and Fox (1956) and Leighton (1961) that the ozone was produced by photochemical

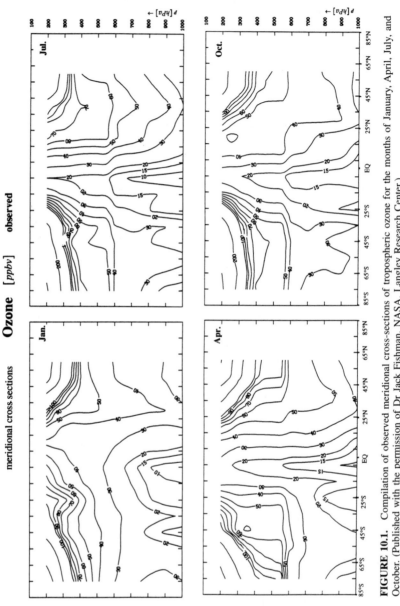

FIGURE 10.1. Compilation of observed meridional cross-sections of tropospheric ozone for the months of January, April, July, and October. (Published with the permission of Dr Jack Fishman, NASA, Langley Research Center.)

TABLE 10.2 Trends in De-seasonalized Surface Ozone Mixing Ratio (% year^{-1})

Station	$z/\varphi/\lambda$	Period	Annual	Winter	Spring	Summer	Autumn
Barrow	11/71° N/157° W	3/73–2/92	0.67 ± 0.30	−0.07 ± 0.81	0.85 ± 1.26	1.73 ± 0.58	0.50 ± 0.61
Mauna Loa	3397/20° N/156° W	10/73–9/92	0.37 ± 0.26	0.56 ± 0.67	0.55 ± 0.87	0.34 ± 0.78	0.04 ± 0.63
Samoa	82/14° S/171° W	1/76–12/91	0.03 ± 0.44	0.22 ± 0.86	0.00 ± 0.94	−0.82 ± 1.36	0.22 ± 1.35
South Pole	2835/90° S	1/75–2/92	−0.68 ± 0.23	−0.22 ± 0.56	−0.66 ± 1.03	−1.42 ± 0.72	0.66 ± 0.73

Including 95% confidence interval, based on Student's t-test (reproduced from Oltmans and Levy, 1993), and including information on station elevation z (m), latitude φ, longitude λ, and observation period.

reactions involving the oxidation of hydrocarbons (RH) in the presence of NO_x, in which NO is oxidized to NO_2 not by ozone but by reactions with HO_2 and RO_2 radicals, followed by rapid photolysis of NO_2 into NO and O and recombination of O with O_2 to from O_3:

$$RH + OH (+ O_2) \rightarrow RO_2 + H_2O \qquad (10.15)$$

$$RO_2 + NO \rightarrow RO + NO_2 \qquad (10.16)$$

$$RO + O_2 \rightarrow Carb + HO_2 \qquad (10.17)$$

$$HO_2 + NO \rightarrow OH + NO_2 \qquad (10.4)$$

$$NO_2 + h\nu \rightarrow NO + O \ (2\times) \quad (\lambda < 420\,nm) \qquad (10.5)$$

$$O + O_2 + M \rightarrow O_3 + M \ (2\times) \qquad (10.6)$$

$$\text{net C5:} \quad RH + 4O_2 \rightarrow Carb + H_2O + 2O_3$$

yielding two ozone molecules for each hydrocarbon molecule that is oxidized. Further degradation of the carbonyl compounds (Carb) leads to additional RO_2 and HO_2 formation with ensuing O_3 production. Reaction cycle C3 is thus a special case of C5 involving CH_4 (i.e., $R = CH_3$). Subsequent studies have shown that high regional surface ozone concentrations in polluted regions have become ubiquitous, involving the same kind of reactions that were shown in cycle C5 (e.g., Hov et al., 1978; Isaksen et al., 1978; see also Chapter 9).

For a long time it was believed that the hydrocarbons involved in ozone production were mainly of anthropogenic origin, and regulatory measures to limit the build-up of ozone therefore generally concentrated on limiting hydrocarbon emissions. In many situations the results were disappointing. It is now clear that the main reason for this failure is that, especially during summertime, large quantities of reactive hydrocarbons are also produced by trees, to such an extent that even in some metropolitan areas, e.g., Atlanta in the US, the natural emissions can match the anthropogenic sources of reactive hydrocarbons. Furthermore, as the natural hydrocarbon emissions peak during summer and in the early afternoon hours owing to temperature and light dependences, they may be especially effective in producing high concentrations of ozone during these periods (Chameides et al., 1988). Depending on regional natural hydrocarbon production, very large reductions in anthropogenic hydrocarbon emissions may be needed to effect a significant reduction of ozone levels. The little success in lowering ozone concentrations by hydrocarbon emission abatement efforts, especially in rural areas, is thus readily explained (Chameides et al., 1988; Trainer et al., 1987).

Efforts are therefore increasingly directed toward reducing NO emissions. The dependence of ozone concentrations on NO_x levels has also been clearly documented, e.g., by observations at a rural background station in the Colorado mountains which indicate a daily increase in ozone volume mixing ratios equal to 14–20 times the NO_x level during summer (June 1–September 1) for atmospheric mixing ratios of NO_x below 1 ppb (Parrish et al., 1986). No significant increase was detected in the winter season (December 1–March 1), indicating the combined effects of much less photochemically active solar radiation and lower emissions of natural hydrocarbon precursors for ozone formation.

Since the pioneering work by Sanadze (1957) and Rasmussen and Went (1965), great advances have been made in estimating the emissions of natural and anthropogenic hydrocarbons, as summarized in two review papers by Singh and Zimmerman (1992) and Fehsenfeld et al. (1992). The natural emissions of isoprene and monoterpenes plus other reactive hydrocarbons dominate with contributions of 350–450 and 400–500 Tg, respectively (Singh and Zimmerman, 1992), close to the early global estimates by Zimmerman et al. (1978). These are mostly very reactive hydrocarbons which are mainly emitted during the warm summertime period over forested regions and which are thus rapidly lost, mostly by reaction with OH, on time scales of only a few hours. In comparison, global, anthropogenic hydrocarbon emissions are estimated to be close to 100 Tg, 30% consisting of ethane (C_2H_6) and propane (C_3H_8) which have lifetimes of 1 month or more and are thus widely transported. The anthropogenic hydrocarbon emissions in the US are estimated to be near $20 \, Tg \, year^{-1}$, compared with natural emissions of $33–71 \, Tg \, year^{-1}$ (Singh and Zimmerman, 1992; see also Lamb et al., 1987). Extrapolation of the results of modeling studies for parts of the eastern US by Fishman et al. (1985) would lead to an average ozone production flux of about 5×10^{10} molecules $cm^{-2} \, s^{-1}$ in the NH during summer time. At other times of the year the ozone production will be lower owing to reduced photochemical activity and comparatively larger loss of catalytic NO_x through reactions of N_2O_5, and possibly NO_3, on sulfate and other aerosol particles, leading to HNO_3 which is readily removed by precipitation (Dentener and Crutzen, 1993).

An important result of the chemical interactions between reactive hydrocarbons and NO_x is also the production of organic nitrates, in particular peroxyacetylnitrate (PAN = $CH_3C(O)O_2NO_2$), which in contrast with HNO_3, is little water-soluble (see also Chapter 6). Thus, although the formation of PAN implies a sink for NO_x, this compound merely serves as a reservoir species which, owing to its greater chemical stability at the colder temperatures prevailing in the middle and upper troposphere, can be transported over long distances, releasing its NO_x content in the warmer regions of the lower troposphere (Crutzen, 1979; Singh and Hanst, 1981). PAN can be formed both in polluted boundary layers containing large quantities of reactive hydrocarbons and NO_x and from the oxidation of ethane and propane in the free troposphere. Surface measurements of PAN at various rural sites in eastern North America (Parrish et al., 1993a) during late summer and early fall showed that PAN constituted 12–25% of the NO_y compounds, 20–30% being made up of HNO_3. In agreement with what may be expected from its greater thermal stability at lower temperatures, the relative contribution of PAN to NO_y increases with height and toward higher latitudes, where PAN becomes more abundant than other forms of NO_y (Singh et al., 1985, 1992; Hübler et al., 1992). Frequently PAN concentrations are positively correlated with those of O_3 (Singh et al., 1990a, b), indicating the importance of upward transport from polluted boundary layers in agreement with the conclusions reached by Kanakidou et al. (1991), whose model study substantially underestimated PAN concentrations when only C_2H_6 and C_3H_8 oxidation reactions were taken into account. Thus transfer of PAN from polluted boundary layers into the middle and upper troposphere, especially by fast vertical transport processes, and formation from more reactive hydrocarbons than C_2H_6 and C_3H_8, probably play an important role in the PAN and NO_x concentration distributions and consequently in the O_3 chemistry of the troposphere.

NO_x often being the controling factor for tropospheric ozone formation, Liu et al. (1987) introduced an NO_x-based ozone production efficiency given by the number of O_3 molecules that are produced per NO molecule emitted into the atmosphere. This factor is, however, not a unique number and depends non-linearly on the prevailing NO_x and hydrocarbon concentrations, being higher at lower NO_x concentrations (Liu et al., 1987). For the eastern US with large NO emissions the ozone production efficiency is about 10 for the summertime period. Given an anthropogenic NO_x production of $6\,Tg\,N\,year^{-1}$ in the US, Liu et al. (1987) derived a production of $5 \times 10^{13}\,g$ of O_3 ($\approx 10^{12}\,mol$) for the three summer months in the US due to industrial NO emissions. Extrapolation to current global conditions, which may be permissible as most anthropogenic NO_x emissions take place at temperate latitudes, would yield an ozone production of about $3 \times 10^{12}\,mol$ for the summer months, or an equivalent production flux of about $10^{11}\,molecules\,cm^{-2}\,s^{-1}$ for the NH. At least for the summertime period, therefore the anthropogenic ozone production involving the oxidation of reactive hydrocarbons may be quite significant in comparison with the influx of ozone from the stratosphere into the troposphere. It should be noted, however, that the above estimates were based on analysis of high-pressure photochemical smog episodes. Under these conditions, a large fraction of the ozone that is produced can again be destroyed by chemical reactions in the boundary layer and uptake at the Earth's surface, so that it may not become available for export to other regions of the troposphere. Recognizing this, Parrish et al. (1993b) used measurements of boundary layer CO and O_3 in the outflow of eastern North American air masses along the Canadian Atlantic coast to derive estimates of the net export of ozone from the eastern US to the Atlantic. Obtaining a ratio of about 0.3 between the O_3 and CO concentration enhancements over background values and given a production of about $3.3 \times 10^{11}\,mol$ of CO from anthropogenic sources during the summer months in the eastern US, they deduce an export of only 10^{11} mol of ozone to the Atlantic. This may, however, be an underestimate, as CO can also be transported to the free troposphere and along directions other than the main outflow pathway. Furthermore, CO can also be produced by the oxidation of the natural hydrocarbons that are emitted by vegetation. Nevertheless, the above outflow of 10^{11} mol during summer would supply about the same input to the northern Atlantic as the stratospheric influx over this area (Parrish et al., 1993b). In the most recent model study on this subject, Jacob et al. (1993a, b) calculate a summertime export of boundary-layer-produced ozone of 4×10^{11} mol to the global atmosphere, corresponding to a mean hemispheric production rate of 2×10^{10} molecules $cm^{-2} s^{-1}$ from US emissions alone. They also estimate that a similar amount of ozone may be produced by reactions in the free troposphere from the NO_x and hydrocarbons that escape from the boundary layer. Extrapolation of these findings to northern hemispheric summertime conditions on the basis of anthropogenic NO emissions (see Table 10.1) would yield an average NH ozone production of about 10^{11} molecules $cm^{-2} s^{-1}$, agreeing with the extrapolation of the results of Liu et al. (1987) to NH conditions as shown before. In Section 10.3.4 we will compare these estimates of photochemical ozone formation in polluted regions with those derived from CH_4 and CO oxidation using our global photochemical model.

We conclude this section by emphasizing that many studies of the impact of NMHC oxidation on ozone formation were performed for photochemical smog situations under anticyclonic, fair weather conditions. Little analysis has been made of the possible ozone formation resulting from NO_x-catalyzed hydrocarbon oxidation under meteorological conditions with stronger vertical mixing, in particular involving convective and frontal activity, which may be accompanied by NO production due to lightning, setting the stage for more efficient net ozone production than in the boundary layer (Pickering et al., 1992a, b; Luke et al., 1992) by a combination of the following factors:

- NO_x dilution leading to a greater efficiency of O_3 production per NO molecule emitted.
- Lack of surface deposition of NO_x and O_3.
- Slower photochemical loss of O_3 and NO_x.

The effect may be particularly strong in the tropics. While Pickering et al. (1992a) calculated an enhancement in ozone production by a factor of 4 following entrainment of the Oklahoma City plume in a convective system, for Manaus, Amazonia this figure was estimated to be as high as 35.

Provided there were enough NO_x in the atmosphere, the global potential for tropospheric O_3 production from NMHC oxidation would be enormous. With a maximum yield of about 2.5 O_3 per emitted C (Singh et al., 1981) and a total NMHC emission of about 10^{15} g $year^{-1}$ (Singh and Zimmerman, 1992), the theoretically maximum worldwide O_3 production would be 2×10^{14} mol year^{-1}, corresponding to a global flux average of almost 10^{12} molecules cm^{-2} s^{-1}. Such a high production of ozone is not reached owing to low natural emissions of NO_x, especially over the oceans, as well as rapid conversion of NO_x to nitric acid and organic nitrates (Madronich and Calvert, 1990; Singh and Hanst, 1981; Crutzen, 1979), and loss of organic material into compounds that are removed by precipitation or that attach to aerosol. In order to understand the ozone budget of the troposphere, it is clearly necessary to enhance knowledge about the transfer and chemistry of NO_x, NMHC, and derived oxygenates to the free troposphere. As most NMHC emissions take place in the tropics, future increases in NO emissions due to population growth and industrial development, which are inevitable, are likely to cause substantial increases in ozone concentrations in these regions. This is clearly an issue of large significance. As shown by the observations of high tropospheric ozone concentrations during the dry season due to biomass burning (see Section 10.2.5), ozone production efficiency in the tropics may be very high.

10.2.5. Production due to Biomass Burning

Between 1.8×10^{15} and 4.7×10^{15} gC of biomass are burned each year in the tropics, mostly as a result of a variety of land clearing and agricultural activities, producing about 50 Tg of NMHC and 5 Tg N of NO each year (Crutzen and Andreae, 1990; Crutzen et al., 1985; Crutzen and Goldammer, 1993). The composition of the NMHC is on the average 20% alkanes, 44% alkenes, 10% alkynes, 13% aromatics, and 10% oxygenates, some 60–70% thus consisting of reactive compounds. As a consequence of these emissions and the plentiful availability of photochemically active UV radiation, high levels of ozone can build up during the dry season, which coincides with the winter months in the hemispheres. The occurrence of widespread photochemical ozone formation in the tropics during the dry season has been documented by many surface and aircraft measurements both in Brazil during the NCAR (Delany et al., 1985; Crutzen et al., 1985) and ABLE 2A field expeditions (Browell et al., 1988; Gregory et al., 1988; Kirchhoff, 1988) and in Africa during the DECAFE expedition in northern Congo during February 1988 (Fontan et al., 1992; Andreae et al., 1992). Figure 10.2 shows the composite of the ozone volume mixing ratio measurements over northern Congo during 1988, while Figure 10.3 summarizes the average profiles of ozone that were obtained during the NCAR expedition during August–September 1980 in the tropical humid forest and savanna (cerrado) regions of Brazil. The Congo measurements were restricted to the atmosphere below 3.5 km and were mostly taken over forested regions. Comparing Figures 10.2 and 10.3, qualitative agreement in ozone volume mixing

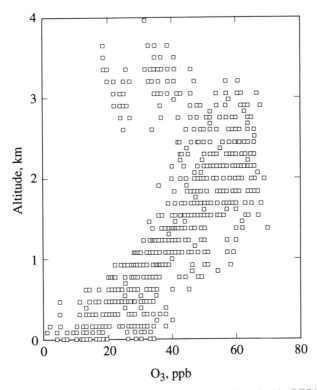

FIGURE 10.2.   Composite of ozone profile measurements made during the DECAFE 88 expedition in northern Congo. (From Andreae et al., 1992.)

ratio exists between the Congo and Amazon forest (selva) profiles, both starting at about 20 ppb at the forest canopy level and increasing to about 50 and 40 ppb, respectively, at 3.5 km altitude. Above 3.5 km the Brazil measurements show only a relatively small increase in ozone volume mixing ratios by 5 ppb towards 9 km (250 hPa), the ceiling of the aircraft. Similar measurements of ozone over the Brazilian savanna (cerrado) regions showed on average an increase from about 45 ppb at low altitudes, to a maximum of 55 ppb at the top of the boundary layer near 3 km altitude. In the free troposphere above 3 km, rather similar mixing ratios were observed as over the forested regions.

Figure 10.4 shows some general features of the ozone volume mixing ratio distribution for different seasons and regions in the tropics (compiled by Crutzen and Andreae, 1990, based on Browell et al., 1988, 1990; Gregory et al., 1988, 1990; Kirchhoff, 1988, 1990; Delany et al., 1985). A clear increase from lowest values in the equatorial Pacific to higher values over the continents is shown,

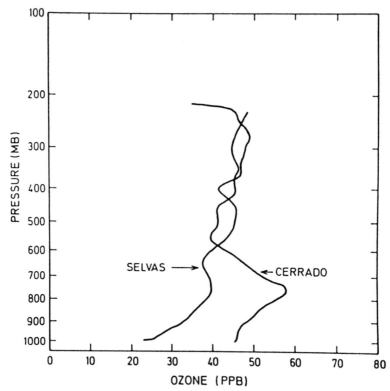

FIGURE 10.3. Ozone profiles measured over the cerrado (savanna) and over fire-free areas above the selva (tropical rain forest) of Brazil. (From Delany et al., 1985.)

peaking during the dry season, thus indicating more O_3 production over the continents during all seasons, most likely owing to larger emissions of hydrocarbons and NO_x, e.g., by lightning (Torres and Buchan, 1988; Kaplan et al., 1988). Dry season mixing ratio maxima in the boundary layer at 2–3 km altitude over northern Congo and over the cerrado regions of Brazil clearly indicate intense O_3 production triggered by the injection of NO from biomass fires. Additional significant emissions of NO from savanna soils during the dry season can, however, also play a significant role. As shown by airborne measurements of Harris et al. (1993) in South Africa, strongly enhanced emissions of NO can temporarily occur over a period of a few days after rainfall. The frequency of such events will increase toward the end of the dry season. Although NO emissions from soils during the wet season in the savanna regions may well be larger, this may not stimulate ozone production, as a substantial fraction of the NO_x that is produced may be taken up by the lush vegetation, in

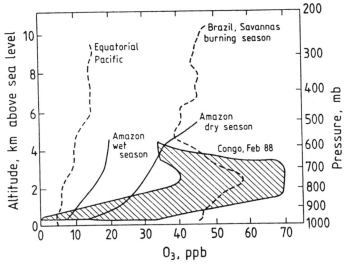

FIGURE 10.4.   Compilation of ozone vertical profiles over various regions and during the dry and wet seasons in the tropics. (From Crutzen and Andreae, 1990.)

contrast with conditions during the dry season when most above-ground herbaceous vegetation is dead.

Analysis of satellite observations by Fishman et al. (1990), who derived the tropospheric ozone column amount from the difference between the total ozone measured by TOMS (Total Ozone Mapping Spectrometer) and the stratospheric component measured by SAGE (Stratospheric Aerosol and Gas Experiment), indicates a marked effect of biomass burning on tropospheric ozone, affecting a large geographical region in the SH, stretching from the east coast of Brazil, over the southern Atlantic, southern Africa, the Indian Ocean, all the way down to southern Australia. Over this region, which covers about 10% of the Earth's surface ($= 5 \times 10^{17} \, cm^2$), or a total of about $(3-8) \times 10^{12} \, mol$, an increase in tropospheric ozone by 5–10 Dobson Units ($= 1-2.5) \times 10^{17} \, molecules \, cm^{-2}$) was determined to have occurred from June–August to September–November. This may have been due to both enhanced transport from the stratosphere and tropospheric ozone production. With a free tropospheric lifetime of ozone of approximately 1 month, the observed increase corresponds to a net column rate of increase of $(4-10) \times 10^{10} \, molecules \, cm^{-2} \, s^{-1}$ over the affected areas. How much of this net increase is due to stratospheric influx, how much to the export of photochemically produced ozone in the polluted boundary layer over the Brazilian and southern African continents, and how much to ozone formation in the free troposphere from NO_x and hydrocarbons that escaped from the boundary layer is a matter of research. If all the increase in ozone were due to photochemical production, this would indicate an ozone production efficiency of 15–40, up to two times higher than found for the US (Liu et al., 1987; Jacob et al., 1993a), but two to five times smaller than that calculated for the free troposphere as a result of CH_4 and CO oxidation (see Section 10.3.3).

Although it is clear that substantial ozone formation takes place during the dry season over the continents of Africa and South America, and some first attempts have been made to model these (Crutzen et al., 1985; Crutzen and Carmichael, 1993; Jacob and Wofsy, 1988; Pickering et al., 1992b), there are several critical questions that must be addressed by future studies.

- Can the increase in tropospheric ozone abundance during the dry season in the southern hemisphere be explained entirely by photochemical reactions among the biomass burning effluents?
- Alternatively, could ozone intrusion from the stratosphere in the tropics and subtropics contribute to this phenomenon?
- What are the effects of NO_x contributions from sources other than biomass burning, such as lightning and thermal decomposition of long-range transported PAN produced in the NH from industrial NO_x and natural or anthropogenic hydrocarbons (Singh et al., 1990a, b). In this context we note that during June–October the ITCZ (intertropical convergence zone) is significantly displaced northward into the NH, so that, e.g., the effluents of the Indian subcontinent can affect the chemistry of the SH.
- Why is the impact of biomass burning on ozone during the dry season so much more noticeable in the NH than in the SH? Is this due to lower NO_x emission rates and a lower efficiency for ozone production per unit of NO_x addition as a consequence of the larger industrial NO_x background in the NH?
- Although there are indications that NO emissions from the tropical forests to the atmosphere during the wet season are less than during the dry season (Bakwin et al., 1990), the input of NO by lightning should maximize in the most convective regions during the wet season. The convection also rapidly transports large quantities of reactive hydrocarbons that are emitted by the forests (Zimmerman et al., 1988; Rasmussen and Khalil, 1988), as well as partially oxidized hydrocarbons that result from photochemical reactions in the boundary layer (see, e.g., Madronich and Calvert, 1990) to the upper troposphere, where they are mixed with lightning-produced NO_x. Under these circumstances one would expect intense photochemical ozone production in the middle and upper troposphere. The observations, however, show higher upper tropospheric ozone concentrations during the dry season. It is hard to see why this is so, except if destruction of ozone is likewise strongly enhanced by aqueous phase photochemical loss reactions in clouds (Lelieveld and Crutzen, 1990, 1991) or on the surface of ice particles, as briefly discussed in the following section.

10.2.6. Effects of Heterogeneous Reactions

As shown by Lelieveld and Crutzen (1990, 1991), the formation of cloud droplets generally leads to a photochemical loss of ozone compared with the cloud-free case owing to a combination of factors.

- Prevention of ozone formation due to the separation of especially HO_2 and RO_2 radicals from NO by the uptake of the peroxy radicals in the liquid phase, NO remaining in the gas phase owing to its low solubility in water. Consequently, reactions (10.4) and (10.16) which are critical for tropospheric ozone formation can much less take place, even in NO_x-rich environments. (Note that this will most likely also prevent ozone formation in fogs.)
- Ozone loss in the aqueous phase by uptake and dissociation of HO_2 in the aqueous phase:

$$HO_2 = H^+ + O_2^- \qquad (10.18)$$

immediately followed by the reactions

$$O_3 + O_2^- \rightarrow O_3^- + O_2 \qquad (10.19)$$

$$H^+ + O_3^- \rightarrow OH + O_2 \qquad (10.20)$$

Reactions (10.18)–(10.20) thus have the same net result as reaction (10.8). However, while reaction (10.8) in the gas phase occurs only with a probability of about 10^{-5} per collision, reactions (10.19) and (10.20) occur at every collision in the aqueous phase, so that they can become very significant, especially under low-acidity conditions.

- Conversion of photochemically active NO_x to HNO_3 via the set of reactions

$$NO + O_3 \rightarrow NO_2 + O_2 \qquad (10.21)$$

$$NO_2 + O_3 \rightarrow NO_3 + O_2 \qquad (10.22)$$

$$NO_3 + NO_2 = N_2O_5 \qquad (10.23)$$

$$N_2O_5 + H_2O \rightarrow HNO_3 \quad \text{(on droplet, sea-salt, ice,} \qquad (10.24)$$
$$\text{or aerosol surfaces)}$$

which is activated during nighttime, as NO_3 is very efficiently photolyzed during daytime.

From statistics of the frequencies of occurrence of various cloud types, their average height range, and estimates of typical upward velocities (i.e., mass flows) through these cloud types (Lelieveld et al., 1989), Lelieveld and Crutzen (1990) derived spatial and temporal estimates of typical time periods which air parcels spend in and outside of clouds. By letting the air parcels undergo successive sequences of gas phase photochemistry during cloud-free periods and aqueous phase photochemistry during cloudy periods, Lelieveld and Crutzen (1990, 1991) estimated the overall influence of clouds on the O_3 and CO budgets as well on HO_x radical concentrations for various locations and seasons. Very significant effects were derived, resulting in substantially enhanced ozone destruction in NO_x-poor environments and reduced ozone production in NO_x-rich environments. A recent study by Dentener (1993) using the MOGUNTIA model, however, showed that the introduction of aqueous phase chemistry, on top of the aerosol surface reactions (10.21)–(10.24), causes a drop in tropospheric ozone concentrations by only 5–15%. This response appears rather small compared with what may be expected from the study by Lelieveld and Crutzen (1990, 1991). The issue deserves further study.

Lelieveld and Crutzen (1990) in their analysis only considered the influence of aqueous phase reactions. However, it is quite possible that reactions (10.18)–(10.20) also take place on ice particle surfaces, where they can lead to particularly efficient ozone loss (C. Wang and P.J. Crutzen, work in progress). Furthermore, efficient daytime conversion of NO_x to HNO_3 may also occur on the surface of ice particles via

$$NO + O_2^- \rightarrow O_2NO \rightarrow ?NO_3^-$$ (10.25)

As the concentrations of HO_2 are two orders of magnitude larger than those of OH, in the presence of a sufficient surface area of ice particles reaction (10.25) can be more important than the gas phase formation reaction

$$OH + NO_2 (+ M) \rightarrow HNO_3 (+ M)$$ (10.26)

Laboratory simulation of the behavior of HO_2 radicals on the surface of ice particles is therefore a matter of great interest. In general, except for the stratosphere, where they play a critical role in ozone depletion, the potential significance of reactions taking place on ice particles has so far not been investigated, despite the fact that the surface area and volume of ice particles in the troposphere are orders of magnitude larger than in the stratosphere.

Another set of heterogeneous reactions of great significance involves chemical reactions taking place on aerosol particles. Of particular interest may be the reaction (10.24) of N_2O_5 with H_2O on sulfate and other aerosol particles, leading to the production of HNO_3, and thereby loss of reactive NO_x. As shown by Dentener and Crutzen (1993), the loss of NO_x by these heterogeneous reactions on aerosol particles is of major importance especially during winter at mid- and high-latitudes (up to 80% reduction). As a consequence, O_3 and OH concentrations are reduced as well, by 5–25% and by up to 30%, respectively, in the NH.

10.3. GLOBAL TROPOSPHERIC OZONE DISTRIBUTIONS AND BUDGETS

10.3.1. Observations

Despite the demonstrated great importance of tropospheric ozone in atmospheric chemistry, there is still a dearth of ozone observations in many regions of the troposphere outside industrial regions, except for the occasional ozone measurements made during research campaigns, e.g., the GAMETAG aircraft missions over the Pacific in 1977 and 1978 (Routhier et al., 1980), the NCAR 1979 and 1980 dry season expeditions in Brazil (Delany et al., 1985; Crutzen et al., 1985), the GTE/CITE 1 aircraft flights over the northern Pacific (Fishman et al., 1987), the ABLE field programs in the Amazon basin during the dry season of 1985 (Browell et al., 1988; Gregory et al., 1988; Kirchhoff, 1988) and the wet season of 1987 (Browell et al., 1990; Gregory et al., 1990; Kirchhoff et al., 1990), and the DECAFE field project in northern Congo during February 1988 (Fontan et al., 1992; Andreae et al., 1992). Aircraft observations by Seiler in the 1970s (Seiler and Fishman, 1981) and by Marenco and Said (1989) over the Atlantic between Europe and North America and along the coasts of North and South America, as well as East Africa have also significantly contributed to the ozone database. Extensive ozone measurements on research ships were carried out in the marine boundary layer over the Pacific and Indian Ocean in July 1986, May–August 1987 and April–May 1988 (Johnson et al., 1990) and over the Pacific by Liu et al. (1983) and Piotrowicz et al. (1991). Characteristic for most of the observations over the Pacific were the frequent occurrences of very low volume mixing ratios of surface ($\approx 10\,\mathrm{ppb}$ or less) and free tropospheric ozone, indicative of significant photochemical ozone destruction. Although less drastic than for the Pacific, the same pattern of low ozone concentrations in the troposphere was also reported by Smit et al. (1991) over the Atlantic between about $10°$ of latitude north and south of the ITCZ.

Outside the most industrialized regions, where plentiful ground-based ozone measurements are made on a regular basis, mainly for public health and environmental protection reasons, only few ground-based ozone measurement stations exist. Among these we mention the networks run by US scientists as part of the CMDL (Climate Monitoring and Diagnostics Laboratory) activities of the National Oceanic and Atmospheric Administration at Barrow/Alaska, Mauna Loa/Hawaii, American Samoa and South Pole, and as part of the AEROCE Program over the North Atlantic at Bermuda, Reykjavik, the Westman Islands and Mace Head/Ireland, Niwot Ridge/Colorado, and Izana/Canary Islands. Additional stations making regular surface ozone measurements at coastal sites are run by the Japanese at Syowa/Antarctica, by the South Africans at Cape Point, and by the Australians at Cape Grim/Tasmania. These observations have recently been summarized by Oltmans and Levy (1993). In Brazil, surface ozone measurements are regularly carried out by Kirchhoff and co-workers (e.g. Kirchhoff et al., 1991), showing the importance of biomass burning in the production of elevated ozone concentrations. A network of rural surface ozone stations has existed for the last decade in Japan, showing the effects of stratospheric intrusions, marine tropical clean air advection during summer with low ozone concentrations, and regional photochemical ozone production (Murao et al., 1990; Ogawa and Miyata, 1985; Chang et al., 1989). Regular ozone measurements have also been made since the beginning of the 1980s in Malaysia by Ilyas (1987) and since late 1986 in Indonesia (Komala and Ogawa, 1991; Pardede, 1991). Vertical profiles of tropospheric ozone over India have recently been summarized by Srivastav et al. (1993). The observations show especially over New Delhi the build-up of an ozone maximum below a rather persistent temperature inversion located between about 440 and 560 hPa during summer, indicative of tropospheric ozone production.

Despite the paucity of data over remote areas, the main characteristics of the global distributions of tropospheric ozone are sufficiently known to allow a first global compilation of the data. Figure 10.1 shows zonal average meridional cross-sections of O_3 in the background atmosphere as derived by Dr Jack Fishman of NASA, Langley Research Center (unpublished material presented with his permission). These data show the following most striking features, which were also discussed by Fehsenfeld and Liu (1993).

- A general increase in ozone volume mixing ratios with altitude, indicative of both significant downward flux from the stratosphere and more efficient ozone production in the upper troposphere. Clearly discernible also is a correlation with the height of the tropopause.
- Higher ozone concentrations in the NH than in the SH, consistent with both a larger influx of ozone from the stratosphere and a higher photochemical ozone production, mainly due to much higher anthropogenic NO_x emissions in the NH, as first hypothesized by Fishman et al. (1979a, b). Recent research, summarized in Section 10.2.5, has shown, however, that there may also be a considerable effect of anthropogenic activities on ozone in the SH due to biomass burning.
- The occurrence of highest ozone concentrations during the spring months in both the NH and SH, most likely reflecting periods of largest stratospheric inputs, although it has been proposed by Penkett and Brice (1986) that an accumulation of pollutants at higher latitudes in the NH during winter may stimulate strong ozone production during the spring.
- Minimum ozone concentrations in the tropics due to a low supply of ozone from the stratosphere and maximum ozone destruction by reactions (10.1) + (10.2) and (10.8). It should also be noted that, as shown in Figure 10.4 and discussed in Section 10.2.5, there exist substantial seasonal and zonal

variations in tropical ozone concentrations, with highest concentrations during the dry season over the continents and lowest concentrations in the tropical oceanic regions, in particular over the Pacific.

10.3.2. Comparison with Model Calculations

At several research institutions, regional and global photochemical models aimed at describing the tropospheric ozone distributions are being developed. Here we present some of the first results obtained with the three-dimensional, global transport–photochemistry model MOGUNTIA which has been developed at the author's research division at the Max Planck Institute for Chemistry in Mainz, Germany (Zimmermann, 1988; Zimmermann et al., 1989; Crutzen and Zimmermann, 1991). Figure 10.5, which shows the seasonal and meridional ozone volume mixing ratio distribution for the middle of the past decade, indicates some of the main features that are also present in the observations (see Figure 10.1), the main difference being that, although yielding lowest concentrations in the tropics, the model nevertheless calculates too high ozone concentrations in that region. This discrepancy may have several causes.

- A too large influx of ozone from the stratosphere to the upper troposphere at extratropical latitudes and from there too large horizontal transport to the tropical upper troposphere.
- Too large NO production from lightning, leading to too large ozone production.
- Insufficient consideration of the effects of aqueous and heterogeneous reactions (Lelieveld and Crutzen, 1990, 1991) or loss of ozone on ice particles due to reactions (10.18)–(10.20) (see Section 10.2.6).

10.3.3. Ozone and Hydroxyl Concentrations: Industrial versus Pre-industrial Periods

With the large growth in tropospheric NO emissions due mainly to industrial and automotive fossil fuel combustion, and accompanying enhancements in the atmospheric CH_4 concentrations and CO emissions, there should have been a significant increase in the tropospheric ozone concentrations, especially in the northern hemisphere, since pre-industrial times. Theoretical estimates of the pre-industrial ozone distributions have been made with the same global model MOGUNTIA used for the simmulation of the present atmospheric situation, leading to the results presented in Figure 10.6. Comparing Figures 10.5 and 10.6, we note up to a doubling of surface ozone concentrations in the NH. In the SH and the tropics the calculated ozone concentration increased during the same period by up to about 50%. In Figures 10.7 and 10.8 we also present for the

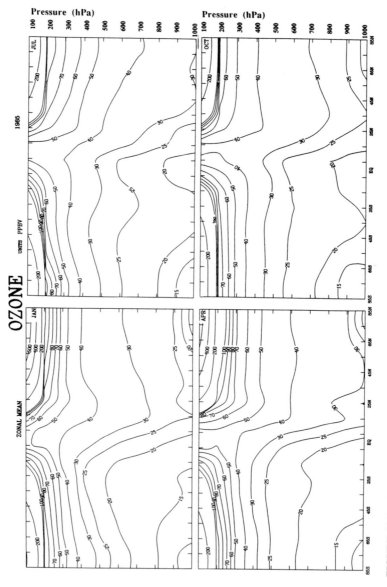

FIGURE 10.5. As Figure 10.1, but calculated with the MOGUNTIA model for the mid-1980s.

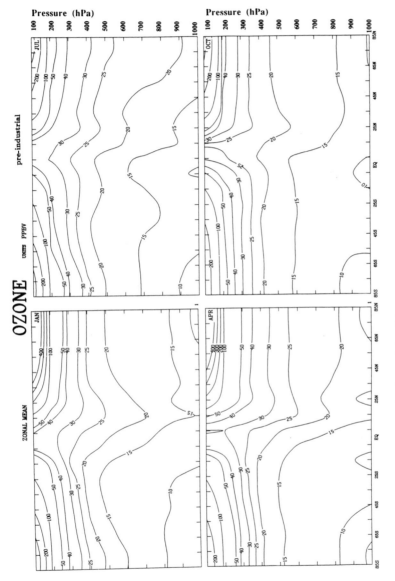

FIGURE 10.6.   As Figure 10.5, but for the pre-industrial era.

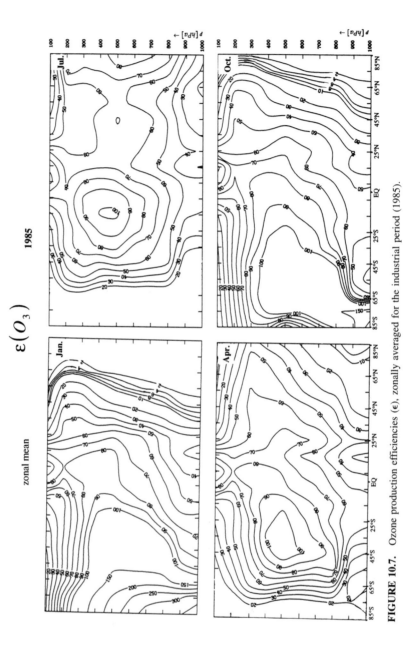

FIGURE 10.7. Ozone production efficiencies (ε), zonally averaged for the industrial period (1985).

$\varepsilon(O_3)$

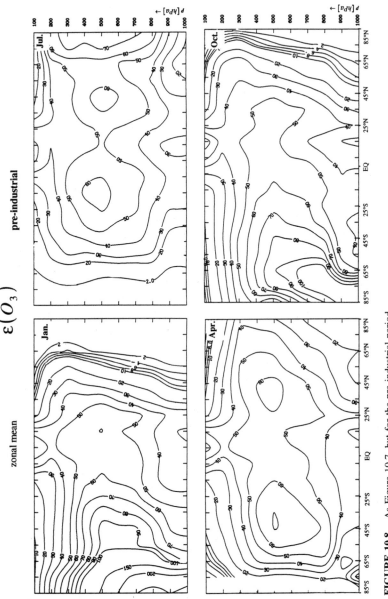

FIGURE 10.8.   As Figure 10.7, but for the pre-industrial period.

industrial and pre-industrial cases, respectively, for the months of January, April, July and October, the calculated zonal average ozone production efficiencies (ϵ), defined by Liu et al. (1987) as the number of ozone molecules that are produced for each NO_x molecule that is removed by chemical reaction from the atmosphere, so that for this study

$$\epsilon = \frac{[NO]\{k_4\,[HO_2] + k_{10}\,[CH_3O_2]\}}{k_{26}\,[NO_2]\,[OH] + \text{``}2k_{25}\,[N_2O_5]\,[\text{Aerosol}]\text{''}}$$

The calculated numbers reflect both the inverse dependence on NO_x concentrations (increasing efficiencies with altitude and substantially higher values for each of the seasons in the SH than in the NH) as well as the effect of the heterogeneous NO_x-removing reactions (10.21)–(10.24) on aerosol particles (low values at middle and high latitudes during winter). In the tropics, for the industrial case, ozone production efficiencies average about 70%. This value is an order of magnitude larger than the ozone production efficiencies calculated for the summertime boundary layer of the US (Jacob et al., 1993a, b), emphasizing the potentially large role of CH_4 and CO oxidation reactions and NO_x transport in the global ozone distribution and budget. The calculated ozone production efficiencies for the industrial era are about 40% larger than for the pre-industrial era. Although the opposite might be expected, based on the inverse dependence of the ozone production efficiencies on NO_x concentrations, this effect is overcompensated by that following from increased background CH_4 and CO concentrations. Note again that these results only reflect increases in the concentrations and emissions of NO_x, CH_4 and CO, neglecting the effects of anthropogenic and natural hydrocarbon emissions.

A substantial growth in tropospheric ozone concentrations in the NH has indeed been established by observations. According to Volz and Kley (1988), reliable surface ozone measurements were already made by A. Levy and co-workers at Montsouris on the outskirts of Paris, which showed year-round ozone volume mixing ratios of 10 ppb or less for air coming from the rural sector, about four times lower than presently observed background observations at mid-latitudes in the NH. The more extensive measurements made in the 1920s to 1940s in Western Europe, including many optical observations, indicate up to a doubling toward present-day summertime tropospheric ozone concentrations at 3–4 km altitude (Crutzen, 1988). At the northern hemispheric ozone measurement sites of Barrow, Mauna Loa, and Fichtelberg and Hohenpeissenberg/ Germany on average close to a 1% $year^{-1}$ increase in surface ozone concentrations has occurred between 1966 and 1986 (Bojkov, 1988). Interestingly, measured ozone increases at 850 hPa were almost two times higher. The recent analyses of background ozone measurements by Oltmans and Levy (1993), as shown in Table 10.2, and by Oltmans et al. (1989) show slower upward trends of ozone at the Mauna Loa station in Hawaii and indicate a slowing down of the upward trend during the past decade, probably due to stagnating NO_x emissions and enhanced ozone destruction by reactions (10.2) and (10.8) as a consequence of loss of stratospheric ozone.

We also present in Figures 10.9 and 10.10 annual mean OH concentrations calculated for the industrial and pre-industrial eras. As a test for the validity of the calculated OH fields, we give in Figure 10.11 the calculated and observed methylchloroform (CH_3CCl_3) volume mixing ratios at the five stations at which such measurements are regularly carried out (Prinn et al., 1992; see also Chapter 7). The agreement between the two data sets appears very good. The greatest deviations are noted for the Irish station. However, even there the correspondence between theoretical and observed mixing ratios was very good until 1985, indicating the possibility of a calibration problem in the new (or old) set of measurements. At the most southerly station, Tasmania, the model slightly overpredicts the CH_3CCl_3 concentrations. However, the model does not yet take into account the relatively small loss of CH_3CCl_3 due to hydrolysis in oceanic waters (Butler et al., 1991). Consideration of this would bring the theoretical values into better agreement with the observations made since 1985. Comparing the results presented in Figures 10.9 and 10.10, we calculate for the tropics and the SH overall decreases by about 25% in zonally and annually averaged OH concentrations from pre-industrial to industrial times due to increases in CH_4 and CO abundances. The OH reductions are somewhat larger than those presented

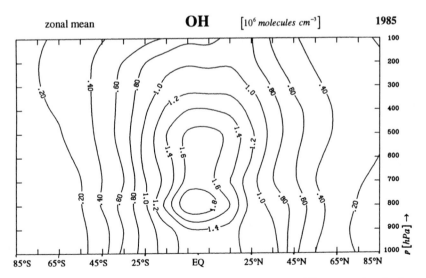

FIGURE 10.9. Meridional cross-section of the annual, 24 h average OH concentrations for the industrial case, (1985).

zonal mean   **OH**   [10^6 *molecules* cm^{-3}]   **pre-industrial**

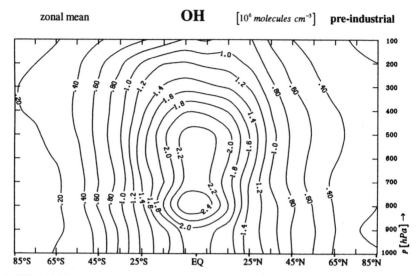

FIGURE 10.10.   Same as Figure 10.9, but for the pre-industrial case.

earlier by Crutzen and Zimmermann (1991), mainly owing to the inclusion of heterogeneous reactions of N_2O_5 on aerosol (Dentener and Crutzen, 1993). The calculations do not, however, include the opposing effect of stratospheric ozone loss since the beginning of the 1980s (Prinn et al., 1992).

The high OH concentrations in the tropics lead to pronounced maxima in the atmospheric destruction rates of CH_4 and CO for both the industrial and pre-industrial periods (see Figure 10.12), showing the great importance of the tropics in atmospheric chemistry, both as an atmospheric sink as well as a source region through emissions at the Earth's surface. Altogether we calculate for the industrial period atmospheric destruction rates for CO and CH_4 of 2532 and 421 Tg year^{-1}, respectively. For the pre-industrial period the corresponding numbers are about two times lower, 1244 Tg CO year^{-1} and 207 Tg CH_4 year^{-1}. The differences mainly reflect increasing sources due to human activities. For the tropics these involve especially rice production, biomass burning, cattle holdings, and organic waste disposal.

10.3.4. Tropospheric Ozone Budgets

In Tables 10.3(a) and 10.3(b) we present for both the NH and SH the various terms which enter into the ozone budget of the troposphere, as obtained with or assumed in MOGUNTIA, including transfer from the stratosphere to the troposphere, photochemical production by reactions of HO_2 and CH_3O_2 with

METHYLCHLOROFORM

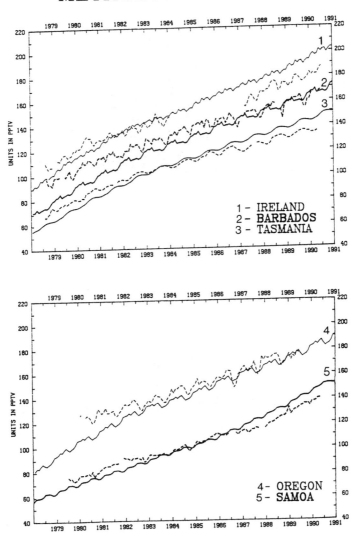

FIGURE 10.11. Model calculated (full curves) and measured (dashed curves) volume mixing ratios of CH_3CCl_3 at five background stations.

CH_4 and CO - Oxidation by OH

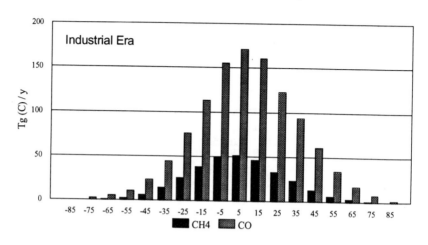

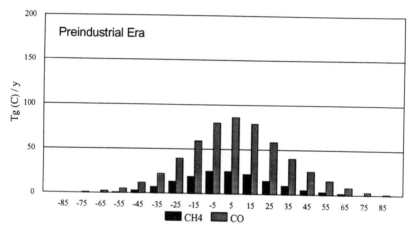

FIGURE 10.12.   Latitudinal distributions in 10° belts of the model-calculated atmospheric sinks of CH_4 and CO in the present and pre-industrial atmospheres.

NO, photochemical destruction by reactions (10.1) + (10.2) and (10.8), as well as uptake at the Earth's surface. The results clearly show the great importance of *in-situ* photochemical processes involving CH_4 and CO oxidation in the ozone budget, which are characterized by high ozone production efficiencies (see Figures 10.7 and 10.8). The strong ozone production is significantly due to anthropogenic activities, as can be seen by comparison of the budget analyses for the industrial and pre-industrial atmospheres (Tables 10.3(a) and 10.3(b)). Nevertheless, even in the pre-industrial era, photochemical reactions in the troposphere were strong contributors to tropospheric ozone, mainly owing to reactions in the tropics. It should again be noted that the budget analyses presented in Tables 10.3(a) and 10.3(b) only consider CH_4 and CO as "fuels" for the ozone formation processes. The contributions coming from the oxidation of more reactive hydrocarbons may be significant but are very difficult to estimate, in particular for that fraction which may be produced in the free troposphere following the escape of NO_x and reactive hydrocarbons from the boundary layer. Extrapolation of theoretical estimates of O_3 production during the summer season over the US (Liu et al., 1987; Jacob et al., 1993a) to the entire NH would indicate the possibility of an average ozone production flux of 10^{11} molecules $cm^{-2} s^{-1}$ (see Section 10.2.4), about one third of the global, annual mean production flux which is calculated to come from CH_4 and CO oxidation. We might thus reach the conclusion that the contribution of "photochemical smog ozone" to the free troposphere is substantially smaller than that coming from CH_4 and CO oxidation, especially if we accept the analysis of Parrish et al. (1993b)

TABLE 10.3(a)   Tropospheric Ozone Budgets, Globally and for the Northern and Southern Hemispheres (10^{13} mol year^{-1}, Upper Numbers), and Hemispheric and Global Averages (10^{10} molecules $cm^{-2} s^{-1}$, Lower Numbers) for the Industrial Case

	Global	NH	SH
Sources			
HO_2 + NO	6.5 26.0	4.1 32.8	2.4 19.2
CH_3O_2 + NO	1.7 6.8	1.0 8.0	0.7 5.7
Transport from stratosphere	1.0 4.0	0.7 5.5	0.3 2.6
Sinks			
$O(^1D)$ + H_2O	3.8 14.4	2.2 17.6	1.6 12.8
HO_2 + O_3 and OH + O_3	2.8 11.2	1.8 14.4	1.0 8.0
Deposition on surface	2.7 10.8	1.8 14.4	0.9 7.2
Net chemical source	1.6 6.4	1.1 8.8	0.5 4.0

Only CH_4 and CO oxidation cycles considered.

TABLE 10.3(b)   As Table 10.3(a), but for the Pre-industrial Case

	Global	NH	SH
HO_2 + NO	2.9	1.6	1.3
	11.4	12.8	10.4
CH_3O_2 + NO	0.8	0.4	0.4
	3.2	3.3	3.2
Transport from stratosphere	1.0	0.7	0.3
	4.0	5.5	2.6
Sinks			
$O(^1D)$ + H_2O	2.2	1.2	1.0
	8.8	9.5	8.0
HO_2 + O_3 and OH + O_3	1.2	0.7	0.5
	4.8	5.7	3.8
Deposition of surface	1.4	0.9	0.5
	5.6	7.2	4.1
Net chemical source	0.3	0.1	0.2
	1.4	0.9	2.0

of the net ozone export from the eastern US to the background troposphere. However, the calculated ozone production from reaction cycles C1 and C3 may be overestimates, because the coarse grid spacing (10° longitude by 10° latitude) that is used in the MOGUNTIA model artificially enhances the dispersion of short-lived NO_x, causing higher ozone production efficiency (e.g. Chatfield and Delany, 1990; Jacob et al., 1993a, b; Kanakidou and Crutzen, 1993). Consequently, at least for the summer months, the contribution of ozone produced in the polluted boundary layers of the US and other industrial regions may well be significant for the ozone budget of the NH. This issue requires substantially more scientific investigation.

10.4. CONCLUSIONS AND RECOMMENDATIONS

An overview has been given of the important role of ozone in tropospheric chemistry and the processes which determine its production, destruction, and distribution. It is likely that *in-situ* photochemical reactions play a more important role in the tropospheric ozone budget than flux from the stratosphere. However, although we have presented a model-derived tropospheric ozone budget which indicates this, the model needs substantial further development, especially with regard to its spatial resolution, treatment of transport and chemistry of NO_x and treatment of heterogeneous reactions in clouds and on particulate matter. The significance of anthropogenic and natural, reactive hydrocarbon oxidation for the global tropospheric ozone budget is likewise not well known. Especially the ozone-producing potential of those fractions of the hydrocarbons and NO_x that escape from the boundary layer into the free troposphere has been insufficiently treated in past studies. We note the great importance of the organic nitrates as well as the potential significance of reactions taking place on surfaces, including ice particles, for tropospheric NO_x and O_3. However, there exists a large gap of knowledge in this area. To improve this situation, more atmospheric measurements are needed of O_3, NO_x and NO_y, CO, and natural and anthropogenic organics, including their partially oxidized products. Laboratory simulations of reactions taking place on surfaces are likewise much needed.

Our model calculations show large increases in tropospheric ozone concentrations due to anthropogenic activities, leading to higher emissions of NO_x, CH_4 and CO. The calculated growth in ozone is substantiated by observations during the past century at several NH stations. It appears, however, that ozone increases have been leveling off during the past decade, probably owing to a smaller growth in the emissions of industrially produced NO_x and enhanced ozone destruction by reactions (10.1) + (10.2) and (10.8), caused by stratospheric ozone loss. Our model likewise calculates a drop in global average OH concentrations by about 25% in the tropics and SH from the pre-industrial to the "pre-ozone hole" years. The trend over the past decade may, however, to some degree have gone in the opposite direction due to the depletion of stratospheric ozone, which allows more OH-producing ultraviolet radiation to penetrate into the troposphere (reactions (10.1) and (10.2)).

In the future we foresee substantial increases in ozone concentrations in the populous and rapidly developing countries of the tropics and subtropics, which, owing to low total ozone, also happen to be the sites of the most intensive photochemical activity in the troposphere. It is thus highly important to strongly expand the network of observational sites in this part of the world in order to enhance our knowledge of the role of the tropics in tropospheric chemistry and to detect and interpret future changes in atmospheric chemical composition in these regions.

Acknowledgments

Many thanks to Dr Peter Zimmermann for his help in the preparation of this chapter and to Dr Jos Lelieveld for comments. Very special thanks go to Dr Jack Fishman for his permission to publish his compilation of the global ozone measurements presented in Figure 10.1.

References

Adams, R.M., Hamilton, S.A. and Carl, B.A. (1985). An assessment of the economic effects of ozone on U.S. agriculture. *J. Air Pollut. Control Assoc.*, **35**, 938–943.

Allam, R.B. and Tuck, A.F. (1984). Transport of water vapour in a stratosphere-troposphere general circulation model. I. Fluxes. *Q. J. R. Meteorol. Soc.*, **110**, 321–356.

Andreae, M.O., Chapuis, A., Cros, B., Fontan, J., Helas, G., Justice, C., Kaufman, Y.J., Minga, A. and Nganga, D. (1992). Ozone and Aitken nuclei over Equatorial Africa: airborne observations during DECAFE 88. *J. Geophys. Res.*, **97**(D7), 6137–6148.

Bakwin, P.S., Wofsy, S.C., Fan, S.-M., Keller, M., Trumbore, S.E. and DaCosta, J.M. (1990). Emission of nitric oxide (NO) from tropical forest soils and exchange of NO between the forest canopy and atmospheric boundary layers. *J. Geophys. Res.*, **95**(D10), 16755–16764.

Bojkov, R.D. (1988). *Ozone changes at the surface and in the free troposphere.* In Isaksen, I.S.A. Ed., *Tropospheric Ozone*, D. Reidel, Dordrecht, pp. 83–96.

Browell, E.V., Gregory, G.L., Harriss, R.C. and Kirchhoff, V.W.J.H. (1988). Tropospheric ozone and aerosol distributions across the Amazon. *J. Geophys. Res.*, **93**(D2) 1431–1451.

Browell, E.V., Gregory, G.L., Harriss, R.C. and Kirchhoff, V.W.J.H. (1990). Ozone and aerosol distributions over the Amazon basin during the wet season. *J. Geophys. Res.*, **95**(D10), 16,887–16,901.

Butler, J., Elkins, J., Thompson, T., Hall, B., Swanson, T. and Koropalov, V. (1991). Oceanic consumption of CH_3CCl_3: implications for tropospheric OH. *J. Geophys. Res.*, **96**, 22,347–22,355.

Chameides, W.L., Lindsay, R.W., Richardson, J. and Kiang, C.S. (1988). The role of biogenic hydrocarbons in urban photochemical smog, *Science*, **241**, 1473–1474.

Chang, Y.S., Carmichael, G.R., Kurita, H. and Ueda, H. (1989). The transport and formation of photochemical oxidants in central Japan. *Atmos. Environ.*, **23**, 363–393.

Chatfield, R. and Delany, A.C. (1990). Convection links biomass burning to increased ozone: however, model will tend to overpredict O_3. *J. Geophys. Res.*, **95**, 18473–18488.

Claude, H., Vandersee, W. and Wege, K. (1992). On long term ozone trends at Hohenpeissenberg. *Proc. Quadrennial Ozone Symp. Charlottesville, VA, June 4–13, 1992*, in press.

Crutzen, P.J. (1973). A discussion of the chemistry of some minor constituents in the stratosphere and troposphere. *Pure Appl. Geophys.*, **106–108**, 1385–1399.

Crutzen, P.J. (1979). The role of NO and NO_2 in the chemistry of the troposphere and stratosphere. *Ann. Rev. Earth Planet. Sci.*, **7**, 443–472.

Crutzen, P.J. (1988). Tropospheric ozone: a review. In Isaksen, I.S.A., Ed., *Tropospheric Ozone*, Reidel, Dordrecht, pp 3–32.

Crutzen, P.J. and Andreae, M.O. (1990). Biomass burning in the tropics: impact on atmospheric chemistry and biogeochemical cycles. *Science*, **250**, 1669–1678.

Crutzen, P.J. and Carmichael, G.R. (1993). Modelling the influence of fires on atmospheric chemistry. In Crutzen, P.J. and Goldammer, J.G., Eds., *Fire in the Environment. The Ecological, Atmospheric, and Climatic Importance of Vegetation Fires. Dahlem Workshop Report*, Wiley, Chichester.

Crutzen, P.J., Delany, A.C., Greenberg, J., Haagenson, P., Heidt, L., Lueb, R., Pollock, W.,

Seiler, W., Wartburg, A. and Zimmerman, P. (1985). Tropospheric chemical composition measurements in Brazil during the dry season. *J. Atmos. Chem.*, **2**, 233–256.

Crutzen, P.J. and Goldammer, J.G., Eds (1993). *Fire in the Environment. The Ecological, Atmospheric, and Climatic Importance of Vegetation Fires. Dahlem Workshop Report*, Wiley, Chichester.

Crutzen, P.J. and Zimmermann, P.H. (1991). The changing photochemistry of the troposphere, *Tellus*, **43AB**, 136–151.

Danielsen, E.F. (1968). Stratospheric–tropospheric exchange based on radioactivity, ozone and potential vorticity. *J. Atmos. Sci.*, **25**, 502–518.

Danielsen, E.F. and Mohnen, V.A. (1977). Project Dustorm report: ozone transport, in situ measurements and meteorological analyses of tropopause folding. *J. Geophys. Res.*, **82**, 5867–5877.

Delany, A.C., Haagenson, P., Walters, S., Wartburg, A.F. and Crutzen, P.J. (1985). Photochemically produced ozone in the emission from large-scale tropical vegetation fires. *J. Geophys. Res.*, **90**, 2425–2429.

Dentener, F. J. (1993). Heterogeneous chemistry in the troposphere. *Ph.D. Thesis*, University of Utrecht.

Dentener, F.J. and Crutzen, P.J. (1993). Reaction of N_2O_5 on tropospheric aerosol: impact on the global distributions of NO_x, O_3, and OH. *J. Geophys. Res.*, **98**(D4), 7149–7164.

Ebel, A., Elbern, H. and Oberreuter, A. (1992). Stratosphere–troposphere air mass exchange and cross-tropopause fluxes of ozone. In Thrane, E.V., Ed. *Coupling Processes in the Lower and Middle Atmosphere*, Kluwer, Dordrecht, pp. 49–65.

Fabian, P. (1973). A theoretical investigation of tropospheric ozone and stratospheric–tropospheric exchange processes. *Pure Appl. Geophys.*, **106–108**, 1044–1057.

Fabian, P. and Junge, C.E. (1970). Global rate of ozone destruction at the Earth's surface. *Arch. Meteorol. Geophys. Bioklim.*, **A19**, 161–172.

Farman, J.C., Gardiner, B.G. and Shanklin, J.D. (1985). Large losses of total ozone in Antarctica reveal seasonal ClO_x/NO_x interaction. *Nature*, **315**, 207–210.

Fehsenfeld, F.C., Calvert, J., Fall, R., Goldan, P., Guenther, A.B., Hewett, C.N., Lamb, B., Liu, S., Trainer, M., Westberg, H. and Zimmerman, P. (1992). Emissions of volatile organic compounds from vegetation and the implications for atmospheric chemistry. *Global Biogeochem. Cycles*, 6, 389–430.

Fehsenfeld, F.C. and Liu, S.C. (1993). Tropospheric ozone: distribution and sources. In Hewitt, C.N. and Sturgess, W.T., Eds., *Global Atmospheric Chemical Change*. Elsevier, New York, pp 169–231.

Fishman, J. (1985). Ozone in the troposphere. In Whitten, R.C. and Prasad, S.S., Eds., *Ozone in the Free Atmosphere*, Van Nostrand Reinhold, New York, pp. 161–194.

Fishman, J., Gregory, G.L., Sachse, G.W., Beck, S.M. and Hill, G.F. (1987). Vertical profiles of ozone, carbon monoxide, and dew-point temperature obtained during GTE/CITE 1, October–November 1983, *J. Geophys. Res.*, **92**(D2), 2083–2094.

Fishman, J., Ramanathan, V., Crutzen, P.J. and Liu, S.C. (1979a). Tropospheric ozone and climate. *Nature*, **282**, 818–820.

Fishman, J., Solomon, S. and Crutzen, P.J. (1979b). Observational and theoretical evidence in support of a significant in-situ photochemical source of tropospheric ozone. *Tellus*, **31**, 432–446.

Fishman, J., Vukovich, F.M. and Browell, E.V. (1985). The photochemistry of synoptic-scale ozone synthesis: implications for the global tropospheric ozone budget. *J. Atmos. Chem.*, **3**, 299–320.

Fishman, J., Watson, C.E., Larsen, J.C. and Logan, J.A. (1990). Distribution of tropospheric ozone determined from satellite data. *J. Geophys. Res.*, **95**(D4), 3599–3617.

Fontan, J., Druilhet, A., Benech, B., Lyra, R. and Cros, B. (1992). The DECAFE experiments: overview and meteorology. *J. Geophys. Res.*, **97**(D6), 6123–6136.

Galbally, I.E. and Roy, C.R. (1980). Destruction of ozone at the Earth's surface. *Q. J. R. Meteorol. Soc.*, **106**, 599–620.

Gidel, L.T. and Shapiro, M.A. (1980). General circulation model estimates of the net vertical flux of ozone in the lower stratosphere and the implications for the tropospheric ozone budget. *J. Geophys. Res.*, **85**, 4049–4058.

Gleason, J.F., Bhartia, P.K., Herman, J.R., McPeters, R., Newman, P., Stolarski, R.S., Flynn, L., Labow, G., Larko, D., Seftor, C., Wellemeyer, C., Komhyr, W.D., Miller, A.J. and Planet, W. (1993). Record low global ozone in 1992. *Science*, **260**, 523–526.

Gregory, G.L., Browell, E.V. and Warren, L.S. (1988). Boundary layer ozone: An airborne survey over the Amazon basin. *J. Geophys. Res.*, **93**(D2), 1452–1468.

Gregory, G.L., Browell, E.V., Warren, L.S. and Hudgins, C.H. (1990). Amazon basin ozone and aerosol: wet season observations. *J. Geophys. Res.*, **95**(D10), 16903–16912.

Haagen-Smit, A.J. and Fox, M.M. (1956). Ozone formation in photochemical oxidation of organic substances. *Ind. Eng. Chem.*, **48**, 1484–1501.

Harris, G.W., Wienhold, F.G. and Zenker, T. (1993). Airborne observations of strong biogenic NO_x emissions from the Namibian savanna at the end of the dry season. *Nature*, in press.

Holton, J.R. (1990). On the global exchange of mass between the stratosphere and troposphere . J. Atmos. Sci., **47**, 392–395.

Hov, O., Hesstvedt, E. and Isaksen, I.S.A. (1978). Long-range transport of tropospheric ozone. *Nature*, **273**, 341–344.

Hübler, G., Fahey, D.W., Ridley, B.A., Gregory, G.L. and Fehsenfeld, F.C. (1992). Airborne measurements of total reactive odd nitrogen (NO_y). *J. Geophys. Res.*, **97**, 9833–9850.

Ilyas, M. (1987). Equatorial measurements of surface ozone. *Atmos. Environ.*, **21**, 1799–1803.

IPCC (Intergovernmental Panel on Climate Change) (1990). *Climate Change, The IPCC Scientific Assessment, WMO/UNEP*, Cambridge University Press, Cambridge.

IPCC (1992). *The Supplementary Report to the IPCC Scientific Assessment*, Cambridge University Press, Cambridge.

Isaksen, I.S.A., Hov, O. and Hesstvedt, E. (1978). Ozone generation over rural areas. *Environ. Sci. Technol.*, **12**, 1279–1284.

Jacob, D.J., Logan, J.A., Gardner, G.M., Yevich, R.M., Spivakovsky, C.M., Wofsy, S.C., Sillman, S. and Prather, M.J. (1993a). Factors regulating ozone over the United States and its export to the global atmosphere. *J. Geophys. Res.*, **98**(D8), 14817–14826.

Jacob, D.J., Logan, J.A., Yevich, R.M., Gardner, G.M., Spivakovsky, C.M., Wofsy, S.C., Munger, J.W., Sillman, S., Prather, M.J., Rodgers, M.O., Westberg, H. and Zimmerman,

P.R. (1993b). Simulation of summertime ozone over North America. *J. Geophys. Res.*, **98**(D8), 14,797–14,816.

Jacob, D.J. and Wofsy, S.C. (1988). Photochemistry of biogenic emissions over the Amazon forest. *J. Geophys. Res.*, **93**(D2), 1477–1486.

Johnson, J.E., Gammon, R.H., Larsen, J., Bates, T.S., Oltmans, S.J. and Farmer, J.C. (1990). Ozone in the marine boundary layer over the Pacific and Indian Oceans: latitudinal gradients and diurnal cycles. *J. Geophys. Res.*, **95**(D8), 11,847–11,856.

Junge, C.E. (1962). Global ozone budget and exchange between stratosphere and troposphere. *Tellus*, **XIV**, 363–377.

Kanakidou, M. and Crutzen, P.J. (1993). Scale problems in global tropospheric chemistry modeling: comparison of results obtained with a three-dimensional model, adopting longitudinally uniform and varying emissions of NO_x and NMHC. *Chemosphere*, **26**, 787–801.

Kanakidou, M., Singh, H.B., Valentin, K.M. and Crutzen, P.J. (1991). A two-dimensional study of ethane and propane oxidation in the troposphere. *J. Geophys. Res.*, **96**(D8), 15,395–15,413.

Kaplan, W.A., Wofsy, S.C., Keller, M. and DaCosta, J.M. (1988). Emission of NO and deposition of O_3 in a tropical forest system. *J. Geophys. Res.*, **93**(D2), 1389–1395.

Kawa, S.R. and Pearson, R., Jr. (1989). Ozone budgets from the dynamics and chemistry of marine stratocumulus experiment. *J. Geophys. Res.*, **94**(D7), 9809–9817.

Kirchhoff, V.W.J.H. (1988). Surface ozone measurements in Amazonia. *J. Geophys. Res.*, **93**(D2), 1469–1476.

Kirchhoff, V.W.J.H., Barnes, R.A. and Torres, A.L. (1991). Ozone climatology at Natal, Brazil, from in-situ ozonesonde data. *J. Geophys. Res.*, **96**(D6), 10,889–10,910.

Kirchhoff, V.W.J.H., Da Silva, I.M.O. and Browell, E.V. (1990). Ozone measurements in Amazonia. Dry season versus wet season. *J. Geophys. Res.*, **95**(D10), 16913–16926.

Komala, N. and Ogawa, T. (1991). Diurnal and seasonal variations of the tropospheric ozone in Indonesia. In Ilyas, M., Ed., *Ozone Depletion*, University of Science, Penang, and United Nations Environmental Programme, Nairobi, pp. 178–188.

Lacis, A. A., Wuebbles, D.J. and Logan, J.A. (1990). Radiative forcing by changes in the vertical distribution of ozone. *J. Geophys. Res.*, **95**, 9971–9981.

Lamb, B., Guenther, A., Gay, D. and Westberg, H. (1987). A national inventory of biogenic hydrocarbon emissions. *Atmos. Environ.*, **21**, 1695–1705.

Leighton, P.A. (1961). *Photochemistry of Air Pollution*, Academic Press, San Diego, CA.

Lelieveld, J. and Crutzen, P.J. (1990). Influences of cloud photochemical processes on tropospheric ozone. *Nature*, **343**, 227–233.

Lelieveld, J. and Crutzen, P.J. (1991). The role of clouds in tropospheric photochemistry. *J. Atmos. Chem.*, **12**, 229–267.

Lelieveld, J., Crutzen, P.J. and Rodhe, H. (1989). Zonal average cloud statistics for global atmospheric chemistry modeling. *GLOMAC Report CM-74*, International Meteorological Institute, University of Stockholm.

Levy, H. II (1971). Normal atmosphere: large radical and formaldehyde concentrations predicted. *Science*, **173**, 141–143.

Levy, H. II, Mahlman, J.D., Moxim, W.J. and Liu, S.C. (1985). Tropospheric ozone: the role of transport. *J. Geophys. Res.*, **90**(D2), 3753–3772.

Lippmann, M. (1988). Health effects of ozone: a critical review. *J. Air Pollut. Control Assoc.*, **39**, 672–695.

Liu, S.C., McFarland, M., Kley, D., Zafiriou, O. and Huebert, B. (1983). Tropospheric NO_x and O_3 budgets in the equatorial Pacific. *J. Geophys. Res.*, **88**, 1360–1368.

Liu, S.C., Trainer, M., Fehsenfeld, F.C., Parrish, D.D., Williams, E.J., Fahey, D.W., Hübler, G. and Murphy, P.C. (1987). Ozone production in the rural troposphere and the implications for regional and global ozone distributions. *J. Geophys. Res.*, **92**(D4), 4191–4207.

Logan, J.A. (1985). Tropospheric ozone: seasonal behaviour, trends and anthropogenic influence. *J. Geophys. Res.*, **90**(D6), 10,463–10,482.

Logan, J.A. (1989). Ozone in rural areas of the United States. *J. Geophys. Res.*, **94**, 8511–8532.

Luke, W.T., Dickerson, R.R., Ryan, W.F., Pickering, K.E. and Nunnermacker, L.J. (1992). Tropospheric chemistry over the lower Great Plains of the United States. 2. Trace gas profiles and distributions. *J. Geophys. Res.*, **97**(D18), 20,647–20,670.

Madronich, S. and Calvert, J.G. (1990). Permutation reactions of organic peroxy radicals in the troposphere. *J. Geophys. Res.*, **95**(D5), 5697–5715.

Madronich, S. and Granier, C. (1992). Impact of recent total ozone changes on tropospheric ozone photodissociation, hydroxyl radicals, and methane trends. *Geophys. Res. Lett.*, **19**, 37–40.

Mahlman, J.D., Levy, H., II. and Moxim, W.J. (1980). Three-dimensional tracer structure and behavior as simulated in two ozone precursor experiments. *J. Atmos. Sci.*, **37**, 655–685.

Marenco, A. and Said, F. (1989). Meridional and vertical ozone distribution in the background troposphere (70° N–60° S; 0–12 km altitude) from scientific aircraft measurements during the STRATOZ III experiment (June 1984). *Atmos. Environ.*, **23**, 201–214.

Murao, N., Ohta, S., Furuhashi, N. and Mizoguchi, I. (1990). The causes of elevated concentrations of ozone in Sapporo. *Atmos. Environ.*, **24A**, 1501–1507.

Murphy, D.M., Fahey, D.W., Proffitt, M.H., Liu, S.C., Chan, K.R., Eubank, C.S., Kawa, S.R. and Kelly, K.K. (1992). Reactive nitrogen and its correlation with ozone in the lower stratosphere and upper troposphere. *J. Geophys. Res.*, **98**, 8751–8773.

Ogawa, T. and Miyata, A. (1985). Seasonal variation of the tropospheric ozone: a summer minimum in Japan. *J. Meteorol. Soc. Jpn.*, **63**, 937–946.

Oltmans, J., Komhyr, W.D., Franchois, P.R. and Matthews, W.A. (1989). Tropospheric ozone: variations from surface and ozonesonde observations. In Bojkov, R. and Fabian, P., Eds., *Ozone in the Atmosphere*, A. Deepak Publ. Hampton, VA,, pp 539–543.

Oltmans, S.J. and Levy, II, H. (1993). Surface ozone measurements from a global network. *Atmos. Environ.*, in press.

Pardede, L.S. (1991). Surface ozone measurements at Bandung. In Ilyas, M., Ed., *Ozone Depletion*, University of Science, Penang and United Nations Environmental Programme, Nairobi, pp. 189–195.

Parrish, D.D., Buhr, M.P., Trainer, M., Norton, R.B., Shimshock, J.P., Fehsenfeld, F.C., Anlauf, K.G., Bottenheim, J.W., Tang, Y.Z., Wiebe, H.A., Roberts, J.M., Tanner, R.L., Newman, L., Bowersox, V.C., Olszyna, K.J., Bailey, E.M., Rodgers, M.O., Wang, T., Berresheim, H., Roychowdhury, U.K. and Demerjian, K.L. (1993a). The total reactive oxidized nitrogen levels and the partitioning between the individual species at six rural sites in eastern North America. *J. Geophys. Res.*, **98**(D2), 2927–2939.

Parrish, D. D., Fahey, D.W., Williams, E.J., Liu, S.C., Trainer, M., Murphy, P.C., Albritton, D.L. and Fehsenfeld, F.C. (1986). Background ozone and anthropogenic ozone enhancement at Niwot Ridge, Colorado. *J. Atmos. Chem.*, **4**, 63–80.

Parrish, D.D., Holloway, J.S., Trainer, M., Murphy, P.C., Forbes, G.L. and Fehsenfeld, F.C. (1993). Export of North American ozone pollution to the North Atlantic Ocean. *Science*, **259**, 1436–1439.

Penkett, S.A. and Brice, K.A. (1986). The spring maximum in photo-oxidants in the northern hemisphere troposphere. *Nature*, **319**, 655–658.

Pickering, K.E., Thompson, A.M., Scala, J.R., Tao, W.-K., Dickerson, R.R. and Simpson, J. (1992a). Free tropospheric ozone production following entrainment of urban plumes into deep convection. *J. Geophys. Res.*, **97**(D16), 17,985–18,000.

Pickering, K.E., Thompson, A.M., Scala, J.R., Tao, W.-K. and Simpson, J. (1992b). Ozone production potential following convective redistribution of biomass burning emissions. *J. Atmos. Chem.*, **14**, 297–313.

Piotrowicz, S.R., Bezdek, H.F., Harvey, G.R., Springer-Young, M. and Hanson, K.J. (1991). On the ozone minimum over the equatorial Pacific Ocean. *J. Geophys. Res.*, **96**(D10), 18,679–18,687.

Prinn, R., Cunnold, D., Simmonds, P., Alyea, F., Boldi, R., Grawford, A., Fraser, P., Gutzler, D., Hartley, D., Rosen, R. and Rasmussen, R. (1992). Global average concentration and trend for hydroxyl radicals deduced from ALE/GAGE trichloroethane (methyl chloroform) data from 1978–1990. *J. Geophys. Res.*, **97**, 2445–2462.

Rasmussen, R. and Went, F. (1965). Volatile organic material of plant origin in the atmosphere. *Proc. Natl. Acad. Sci.*, **53**, 215–220.

Rasmussen, R.A. and Khalil, M.A.K. (1988). Isoprene over the Amazon basin. *J. Geophys. Res.*, **93**(D2), 1417–1421.

Reich, P.B. and Amundson, R.G. (1985). Ambient levels of ozone reduce net photosynthesis in tree and crop species. *Science*, **230**, 566–570.

Routhier, F., Dennett, R., Davis, D.D., Wartburg, A., Haagenson, P. and Delany, A.C., (1980). Free tropospheric and boundary-layer airborne measurements of ozone over the latitude range of 58° S to 70° N. *J. Geophys. Res.*, **85**(C12), 7307–7312.

Sanadze, G.A. (1957). The nature of gaseous substances emitted by leaves of Robina pseudoacacia. *Akad. Nauk Gruz*, **19**, 83–86.

Schnell, R.C., Liu, S.C., Oltmans, S.J., Stone, R.S., Hofmann, D.J., Dutton, E.G., Deshler, T., Sturges, W.T., Harder, J.W., Sewell, S.D., Trainer, M. and Harris, J.M. (1991). Decrease of summer tropospheric ozone concentrations in Antarctica, *Nature*, **351**, 726–729.

Schwarzkopf, M.D. and Ramaswamy, V. (1993). Radiative forcing due to ozone in the 1980's: dependence on altitude of ozone change. *Geophys. Res. Lett.*, **20**, 205–208.

SCOPE (1992). *Effects of Increased Ultraviolet Radiation on Biological Systems, SCOPE/ UNEP*, Paris.

Seiler, W. and Fishman, J. (1981) The distribution of carbon monoxide and ozone in the free troposphere. *J. Geophys. Res.*, **86**, 7255–7266.

Singh, H.B., Condon, E., Vedder, J., O'Hara, D., Ridley, B.A., Gandrud, B.W., Shetter, J.D., Salas, L.J., Huebert, B., Hübler, G., Carroll, M.A., Albritton, D.L., Davis, D.D.,

Bradshaw, J.D., Sandholm, S.T., Rodgers, M.O., Beck, S.M., Gregory, G.L. and Le Bel, P.J. (1990). Peroxyacetyl nitrate measurements during CITE 2: Atmospheric distribution and precursor relationships. *J. Geophys. Res.*, **95**(D7), 10,163–10,178.

Singh, H.B. and Hanst, P.L. (1981). Peroxyacetyl nitrate (PAN) in the unpolluted atmosphere: an important reservoir for nitrogen oxides. *Geophys. Res. Lett.*, **8**, 941–944.

Singh, H.B., Herlth, D., O'Hara, D., Salas, L., Torres, A.L., Gregory, G.L., Sachse, G.W. and Kasting, J.F. (1990b). Atmospheric peroxyacetyl nitrate measurements over the Brazilian Amazon basin during the wet season: Relationships with nitrogen oxides and ozone. *J. Geophys. Res.*, **95**(D10), 16,945–16,954.

Singh, H.B., Martinez, J.R., Hendry, D.G., Jaffe, R.J. and Johnson, W.B. (1981). Assessment of the oxidant-forming potential of light saturated hydrocarbons. *Environ. Sci. Tech.*, **15**, 113–119.

Singh, H.B., O'Hara, D., Herlth, D., Bradshaw, J.D., Sandholm, S.T., Gregory, G.L., Sachse, G.W., Blake, D.R., Crutzen, P.J. and Kanakidou, M.A. (1992). Atmospheric measurements of peroxyacetyl nitrate and other organic nitrates at high latitudes: possible sources and sinks. *J. Geophys. Res.*, **97**(D15), 16,511–16,522.

Singh, H.B., Salas, L.J., Ridley, B.A., Shetter, J.D., Donahue, N.M., Fehsenfeld, F.C., Fahey, D.W., Parrish, D.D., Williams, E.J., Liu, S.C., Hübler, G. and Murphy, P.C. (1985). Relationship between peroxyactetyl nitrate and nitrogen oxides in the clean troposphere. *Nature*, **318**, 347–349.

Singh, H.B. and Zimmerman, P.B. (1992). Atmospheric distribution and sources of nonmethane hydrocarbons. In Nriagu, J.O., Ed., *Gaseous Pollutants: Characterization and Cycling*, Wiley, Chichester, pp. 177–235.

Smit, H.G.J., Gilge, S. and Kley, D. (1991). Ozone profiles over the Atlantic Ocean between 36° S and 52° N in March/April 1987 and September/October 1988. *Report Jül.-2567*, ISSN 0366–0885, Forschungszentrum Jülich GmbH, Jülich.

Srivastav, S.K., Ali H. and Peshin, S.K. (1993). Summer maxima of lower tropospheric ozone over tropics. *J. Climate*, in press.

Staehelin, J. and Schmid, W. (1991). Trend analysis of tropospheric ozone concentrations utilizing the 20-year data set of balloon soundings over Payerne (Switzerland). *Atmos. Environ.*, **25A**, 1739–1740.

Thompson, A.M. (1992). The oxidizing capacity of the Earth's atmosphere: probable past and future changes. *Science*, **256**, 1157–1165.

Torres, A.L. and Buchan, H. (1988). Tropospheric nitric oxide measurements over the Amazon basin. *J. Geophys. Res.*, **93**(D2), 1396–1406.

Trainer, M., Williams, E.J., Parrish, D.D., Buhr, M.P., Allwine, E.J., Westberg, H.H., Fehsenfeld, F.C. and Liu, S.C. (1987). Models and observations of the impact of natural hydrocarbons on rural ozone. *Nature*, **329**, 705–707.

UNEP (1992). Madronisch, S., Björn, L.O., Ilyas, M. and Caldwell, M.M. *Environmental Effects Panel Report. Chapter 1: Changes in Biologically Active Ultraviolet Radiation Reaching the Earth's Surface.* WMO, Geneva.

Vaughan, G. (1988). Stratosphere-troposphere exchange of ozone. In Isaksen, I.S.A., Ed., *Tropospheric Ozone*, 125–135.

Volz, A. and Kley, D. (1988). Evaluation of the Montsouris series of ozone measurements made in the nineteenth century. *Nature*, **332**, 240–242.

Wang, W.C., Dudek, M.P., Liang, X.-Z. and Kiehl, J.T. (1991). Inadequacy of effective CO_2 as a proxy in simulating the greenhouse effect of other radiatively active gases. *Nature*, **350**, 573–577.

WMO (1992). Scientific assessment of ozone depletion: 1991. *Global Ozone Research and Monitoring Project–Report No. 25*, WBO, Geneva.

Zander, R., Demoulin, Ph., Ehhalt, D.H., Schmidt, U. and Rinsland, C.P. (1989). Secular increase of the total vertical column abundance of carbon monoxide above Central Europe since 1950. *J. Geophys. Res.*, **94**, 11,021–11,028.

Zimmerman, P.R., Chatfield, R.B., Fishman, J., Crutzen, P.J. and Hanst, P.L. (1978). Estimates on the production of CO and H_2 from the oxidation of hydrocarbons emissions from vegetation. *Geophys. Res. Lett.*, **5**, 679–682.

Zimmerman, P.R., Greenberg, J.P. and Westberg, C.E. (1988). Measurements of atmospheric hydrocarbons and biogenic emission fluxes in the Amazon boundary layer. *J. Geophys. Res.*, **93**(D2), 1407–1416.

Zimmermann, P. (1988). MOGUNTIA: a handy global tracer model. In Van Dop, H., Ed., *Air Pollution Modeling and its Applications VI*, NATO/CCMS, Plenum, New York, pp 593–608.

Zimmermann, P., Feichter, J., Rath, H., Crutzen, P. and Weiss, W. (1989). A global three-dimensional source–receptor model investigation using 85Kr. *Atmos. Environ.*, **23**, 25–35.

11

Stratospheric pollution and ozone depletion

Tun-Li Shen, Paul J. Wooldridge, and Mario J. Molina

11.1. INTRODUCTION

The lowest layers of the atmosphere, as defined by temperature structure, are the troposphere (0–10 km), the stratosphere ($\approx$ 10–50 km), and the mesosphere (50–150 km). An important feature of the stratosphere is that it contains about 90% of the total amount of ozone in the atmosphere, the significance of which is its strong absorption of ultraviolet radiation. The absorption is essentially complete between 200 and 290 nm and less strong in the 290–330 nm region. The heat from this absorption of solar radiation by ozone causes the temperature to increase with altitude, as illustrated in Figure 2.2 of Chapter 2. This "inverted" temperature profile is largely responsible for the dynamic stability of the stratosphere toward vertical mixing, in contrast with the rapid mixing of the troposphere. This chapter focuses on the link between the amount of ozone in the stratosphere and changes in its chemical composition due to anthropogenic sources, i.e., stratospheric pollution.

11.2. FORMATION AND DESTRUCTION OF OZONE IN THE STRATOSPHERE

11.2.1. Chapman's Mechanism

The formation of ozone in the stratosphere is initiated by photodissociation of molecular oxygen by solar radiation at wavelengths shorter than 242 nm, within the Herzberg continuum (200–220 nm) and the Schumann–Runge band (185–200 nm) of oxygen's absorption spectrum:

$$O_2 + h\nu \ (< 242\,\text{nm}) \rightarrow O + O \tag{11.1}$$

The oxygen atoms released from reaction (11.1) rapidly combine with oxygen molecules to form ozone:

$$O + O_2 + M \rightarrow O_3 + M \tag{11.2}$$

where M is N_2 or O_2. Photodissociation generates atomic oxygen with unit quantum yield:

$$O_3 + h\nu \ (<300\,nm) \rightarrow O + O_2 \qquad (11.3)$$

This is the primary source of atomic oxygen in the stratosphere. The net result of reactions (11.2) and (11.3) is the conversion of solar energy to heat; ozone is not destroyed in this process. However, ozone is very reactive and can be destroyed by various other processes such as by reaction with oxygen atoms:

$$O + O_3 \rightarrow 2O_2 \qquad (11.4)$$

which converts "odd"-oxygen (defined as the sum of ozone and atomic oxygen) back to "even"-oxygen, O_2. Normal ozone abundances peak in the 6–8 ppm range at an altitude around 20–25 km. Column amounts, i.e. vertical integrals, typically vary from 290 to 310 Dobson units (1 Dobson unit = 10^{-3} atm cm $\approx 2.7 \times 10^{16}$ molecules cm^{-2}) on a globally averaged basis.

The above four steps form the model proposed by Sidney Chapman in 1930. For 20 years this simple model, involving only oxygen species, appeared sufficient to explain the balance between the production and destruction of ozone in the stratosphere. It is of interest to note that in the dark, such as during the polar night, there should be no production or destruction according to this mechanism.

Refinements in measurements revealed that ozone abundances were noticeably smaller than those predicted by Chapman's reactions. In the 1950s and 1960s other ozone destruction pathways were proposed, based on the photochemistry of atmospheric water and the influence of the reactive radicals on the distribution of odd-oxygen in the atmosphere. More recently the importance of nitrogen oxides (Crutzen, 1970; Johnston, 1971) and chlorine compounds (Stolarski and Cicerone, 1974; Molina and Rowland, 1974; Rowland and Molina, 1975) has become apparent, and of great concern owing to the anthropogenic perturbations in the concentrations of these chemicals in the stratosphere. The main ozone destruction processes to be added to Chapman's reactions can be considered as catalytic cycles of the form

$$X + O_3 \rightarrow XO + O_2$$
$$\underline{XO + O \rightarrow X + O_2}$$
$$\text{net:} \quad O + O_3 \rightarrow 2O_2$$

where the free radical X can be H, OH, NO, Cl, or Br. Note that X is not consumed by these two reactions and that each cycle leads to the destruction of two odd-oxygen species. Other free radical species are less important for ozone destruction owing to either low abundances, endothermic reactions, or rapid transformation to non-reactive forms (e.g., fluorine species to the strongly bound HF). The relative importance of these catalytic cycles is determined by the concentrations of the active radicals and the reaction kinetics.

Discussed next are the atmospheric photo-oxidation pathways of source gases, such as H_2O, CH_4, N_2O and the CFCs to yield the radicals involved in the catalytic cycles, together with the interconversion of these active radicals to "reservoir" species, which are those that do not participate directly in ozone destruction reactions.

11.2.2. HO_x Chemistry

HO_x refers to the family of water-based free radicals H, OH, and HO_2. Water in the stratosphere is scarce: at the temperature of the tropopause (see Figure 11.1), its low vapor pressure results in very little being transported up from below, a "freeze-drying" or "cold trap" effect. The water mixing ratio is at most $\approx 5-6$ ppm (of the order of 1000 times less than in the troposphere), with roughly half coming from the multistep oxidation of CH_4:

$$CH_4 + 2O_2 \rightarrow \rightarrow 2H_2O + CO_2$$

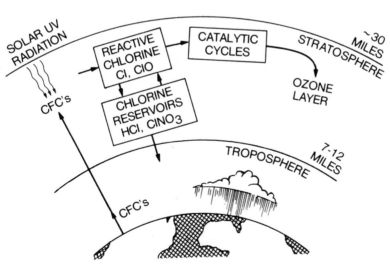

FIGURE 11.1.   Simplified diagram of the atmospheric behavior of chlorofluorocarbons.

HO_x is produced predominantly by reactions between water or methane and electronically excited oxygen atoms generated by the photodissociation of ozone, reaction (11.3'):

$$O_3 + h\nu \ (<300\,\text{nm}) \rightarrow O(^1D) + O_2 \qquad (11.3')$$

$$O(^1D) + H_2O \rightarrow 2OH \qquad (11.5)$$

$$O(^1D) + CH_4 \rightarrow OH + CH_3$$

The main HO_x catalytic cycle responsible for the removal of ozone is

$$OH + O_3 \rightarrow HO_2 + O_2 \qquad (11.6)$$

$$\underline{HO_2 + O \rightarrow OH + O_2} \qquad (11.7)$$

$$\text{net:} \quad O + O_3 \rightarrow 2O_2$$

The relative importance of this and the several other HO_x catalytic cycles which can be thought of depends on the altitude under consideration. For example, in the upper stratosphere, where the abundances of O and H are relatively high, the following cycle destroys odd-oxygen:

$$OH + O \rightarrow H + O_2 \qquad (11.8)$$

$$H + O_2 + M \rightarrow HO_2 + M \qquad (11.9)$$

$$\underline{HO_2 + O \rightarrow OH + O_2} \qquad (11.7)$$

$$\text{net:} \quad 2O \rightarrow O_2$$

On the other hand, in the lower stratosphere the following catalytic cycle, in which O atoms do not participate, becomes important:

$$OH + O_3 \rightarrow HO_2 + O_2 \qquad (11.6)$$

$$\underline{HO_2 + O_3 \rightarrow OH + 2O_2} \qquad (11.10)$$

$$\text{net: } 2O_3 \rightarrow 3O_2$$

The efficiencies of the above catalytic cycles are strongly affected by the OH/HO_2 ratio. In addition, both OH and HO_2 also play critical roles by interacting with species in the NO_x and ClO_x families (see Section 11.2.5).

An additional HO_x reservoir species, besides H_2O, is hydrogen peroxide; it affects the concentrations of OH and HO_2 through the following reactions:

$$HO_2 + HO_2 \rightarrow H_2O_2 + O_2$$

$$H_2O_2 + hv \rightarrow OH + OH$$

$$H_2O_2 + OH \rightarrow H_2O + HO_2$$

Hydrogen peroxide concentrations in the stratosphere are quite small, as it is easily photolyzed.

11.2.3. NO_x Chemistry

The sum of NO and NO_2 is referred to as NO_x. Transport of NO_x to the stratosphere from below is negligible owing to the short residence time for these species in the troposphere (*ca* 1 day). The chief natural source of NO_x in the stratosphere is N_2O, which is produced by biological processes in soil and is essentially inert in the troposphere. Minor sources of stratospheric NO_x include galactic cosmic rays and solar proton events (Crutzen, 1979). In addition, as discussed in Section 11.5.3, direct injection of NO_x into the stratosphere by proposed fleets of high-altitude aircraft is a potential future source. About 95% of N_2O in the stratosphere is destroyed by photolysis:

$$N_2O + hv \rightarrow N_2 + O(^1D)$$

and the remainder reacts with $O(^1D)$:

$$O(^1D) + N_2O \rightarrow 2NO$$

$$O(^1D) + N_2O \rightarrow N_2 + O_2$$

In the upper stratosphere the most important cycle controlling ozone levels is the following:

$$NO + O_3 \rightarrow NO_2 + O_2 \tag{11.11}$$

$$\underline{NO_2 + O \rightarrow NO + O_2} \tag{11.12}$$

$$\text{net:} \quad O_3 + O \rightarrow 2O_2$$

An interesting aspect of nitrogen oxide chemistry is the diurnal, seasonal, and latitudinal behavior of the interconversion between NO_x and the reservoir species N_2O_5, which is determined by reactions whose relative importance depends on the available sunlight:

● night-time reactions

$$NO + O_3 \rightarrow NO_2 + O_2 \tag{11.11}$$

$$NO_2 + O_3 \rightarrow NO_3 + O_2 \tag{11.13}$$

$$NO_2 + NO_3 + M \rightarrow N_2O_5 + M \tag{11.14}$$

● daytime reactions

$$NO_2 + hv \; (<390\,nm) \rightarrow NO + O \tag{11.15}$$

$$N_2O_5 + hv \rightarrow NO_2 + NO_3 \tag{11.16}$$

$$NO_3 + hv \rightarrow NO + O_2 \tag{11.17}$$

$$NO_3 + hv \rightarrow NO_2 + O \tag{11.18}$$

Thus NO increases at sunrise and decreases following sunset. Observations of the interconversion of the reactive nitrogen species provide useful checks for model simulations of atmospheric chemistry.

11.2.4. ClO_x Chemistry

The following catalytic cycle involving chlorine species is very efficient in destroying ozone:

$$Cl + O_3 \rightarrow ClO + O_2 \tag{11.19}$$

$$\underline{ClO + O \rightarrow Cl + O_2} \tag{11.20}$$

$$net: \quad O_3 + O \rightarrow 2O_2$$

The ClO_x cycle is initiated when Cl atoms are released by photodissociation of chlorinated compounds, and can be interrupted by the formation of reservoir species, mainly HCl, $ClONO_2$, and HOCl. Atomic chlorine may abstract hydrogen from CH_4, H_2, etc., to form HCl:

$$Cl + CH_4 \rightarrow HCl + CH_3$$

The recombination of ClO and NO_2 radicals forms $ClONO_2$, and the reaction of ClO and HO_2 forms HOCl (see Section 11.2.5). As a reservoir species, HOCl, though, is generally less important owing to its rapid photolysis.

HCl from the oceans or from volcanic eruptions is a very minor source of chlorine in the stratosphere, because the water that is always present with these sources very effectively dissolves the HCl and returns it to the surface in rain (see, e.g., Tabazadeh and Turco, 1993). Rather, surface emissions of CH_3Cl comprise the chief natural source of chlorine to the stratosphere. Anthropogenic sources, namely manufactured chlorofluorocarbons (CFCs), provide a larger source, which poses a threat to the ozone layer (Molina and Rowland, 1974).

CFCs are used as refrigerants, solvents, aerosol propellants, and blowing agents for plastic foams (see Chapter 7 for more details). These chemicals are very inert: they have no troposphere sinks and are not water-soluble, and thus are mixed rapidly throughout the troposphere and gradually through the stratosphere, where eventually they photodissociate, as depicted in Figure 11.1. CFCs need not rise above most of the atmospheric O_2 and O_3, because they can be photodissociated by wavelengths which penetrate to lower altitudes, in the 185–210 nm spectral window (between strong absorptions of O_2 to shorter wavelengths, and of O_3 to longer). As examples consider the two most prominent CFCs that reach the stratosphere, CFC-11 ($CFCl_3$) and CFC-12 (CF_2Cl_2):

$$CFCl_3 + h\nu \rightarrow CFCl_2 + Cl$$

$$CF_2Cl_2 + h\nu \rightarrow CF_2Cl + Cl$$

Subsequent reactions of the $CFCl_2$ and CF_2Cl radicals lead to the rapid release of the remaining chlorine atoms, which then initiate the catalytic cycles.

A bromine cycle also exists, analogous to that involving reactions (11.19) and (11.20). As with chlorine, there are natural sources of stratospheric bromine (mainly methyl bromide, CH_3Br) as well as man-made: the "halons" CF_3Br, CF_2ClBr, etc., used in fire extinguishers, and CH_3Br, used as a fumigant. Bromine species are present in the stratosphere in small amounts relative to chlorine species, but molecule for molecule they are much more effective in destroying ozone, because the bromine reservoirs are significantly less stable (HBr forms much more slowly than HCl; also, $BrONO_2$ photolyzes more readily than $ClONO_2$; etc.).

11.2.5. Coupling of $HO_x/NO_x/ClO_x$ Reactions

The importance of a family of species and its cycles depends on the abundance of the active radicals that initiate the chain reactions, which in turn depends on the amount held in reservoir form and the likelihood of radical regeneration. The segregation of reactions of various species into individual catalytic cycles is a useful tool for understanding the nature of the ozone destruction processes. Note, however, that the choice of the cycles is arbitrary; the concentration of any given species is dependent on all the reactions in which it participates. An important set of reactions is that which couples the different radical families, examples of which are discussed below.

The most important reactions coupling HO_x and NO_x are

$$HO_2 + NO \rightarrow OH + NO_2 \qquad (11.21)$$

$$OH + NO_2 + M \rightarrow HNO_3 + M \qquad (11.22)$$

$$HNO_3 + h\nu \rightarrow OH + NO_2 \qquad (11.23)$$

$$HO_2 + NO_2 + M \rightarrow HO_2NO_2 + M \qquad (11.24)$$

$$HO_2NO_2 + h\nu \rightarrow HO_2 + NO_2 \qquad (11.25)$$

$$HO_2NO_2 + OH \rightarrow NO_2 + H_2O + O_2 \qquad (11.26)$$

Reaction (11.21) strongly affects the partitioning of OH and HO_2, and hence the relative contributions of the various HO_x catalytic cycles.

The effectiveness of the NO_x cycle is reduced when NO_2 is tied-up through reaction (11.22), a key reaction coupling the HO_x and NO_x cycles. Reaction (11.24) also ties-up NO_2; the reverse of these two reactions is the release of NO_2 by photolysis, i.e., reactions (11.23) and (11.25). Reactions that couple HO_x and ClO_x are

$$OH + HCl \rightarrow H_2O + Cl \qquad (11.27)$$

$$HO_2 + ClO \rightarrow HOCl + O_2 \qquad (11.28)$$

As can be seen, an increase in OH has opposite effects on the NO_x and ClO_x catalytic cycles. As OH increases, HCl is converted to Cl by reaction (11.27), and the impact of ClO_x is enhanced. On the other hand, OH transforms NO_2 into its reservoir species HNO_3 by reaction (11.22), decreasing the effect of NO_x on ozone depletion.

Reactions that play key roles in the interaction of NO_x and ClO_x cycles are

$$ClO + NO \rightarrow NO_2 + Cl \qquad (11.29)$$

$$ClO + NO_2 + M \rightarrow ClONO_2 + M \qquad (11.30)$$

$$ClONO_2 + h\nu \rightarrow Cl + NO_3 \qquad (11.31)$$

The ClO_x catalytic cycle is influenced most by reaction (11.29) in the upper stratosphere, and by reaction (11.30) in the lower stratosphere. As mentioned above, the formation of $ClONO_2$ via reaction (11.30) provides a temporary reservoir for ClO.

11.2.6. SO_x Chemistry

OCS and SO_2 are believed to be the primary sources of stratospheric sulfur. Being continuously released mainly by various biological processes, OCS, which is very long-lived in the troposphere, provides a continuous source of sulfur in the stratosphere, where it is photolysed. The second sulfur source, SO_2, is important when injected directly into the stratosphere by major volcanic eruptions. The role of sulfur photochemistry has been investigated in some detail following the eruptions of El Chichon in 1982 and Mount Pinatubo in 1991, and the increases in the stratospheric sulfur loading (more than an order of magnitude) following those major eruptions have been closely monitored along with the various chemical perturbations and ozone changes.

In the stratosphere SO_2 first reacts with OH to begin its oxidation to H_2SO_4, which proceeds via the following scheme (Stockwell and Calvert, 1983):

$$SO_2 + OH + M \rightarrow HSO_3 + M \qquad (11.32)$$

$$HSO_3 + O_2 \rightarrow HO_2 + SO_3 \qquad (11.33)$$

$$SO_3 + H_2O \rightarrow \ \rightarrow H_2SO_4 \qquad (11.34)$$

Note that no HO_x radicals are consumed in this process. The detailed mechanism of reaction (11.34) is not yet known, but it is clear that the extremely hygroscopic product, H_2SO_4, combines with water vapor to form sulfate aerosol particles, which play a major role in stratospheric chemistry by providing surfaces for heterogeneous reactions as well as by being involved in the formation of polar stratosphere clouds. These play a crucial role in the near complete seasonal destruction of ozone which occurs over Antarctica.

11.3. HIGH-LATITUDE OZONE LOSS AND HETEROGENEOUS CHEMISTRY

Before 1985 there was scant evidence of a long-term decline in stratospheric ozone levels. However, the discovery of the Antarctic "ozone hole" in 1985 changed this: ozone column measurements showed a dramatic decline of the monthly mean value in October at Halley Bay from 300 to 350 DU (Dobson units) in the mid-1970s to values lower than 200 DU (Farman et al., 1985). Furthermore, ozone vertical profiles from balloon measurements indicated that the ozone depletion was occurring at altitudes from about 10 to 20 km (Hofmann et al., 1987), see Figure 11.2. Based on the steady increase in tropospheric and stratospheric halocarbons, Farman et al. suggested a possible link between the growth of active chlorine in the stratosphere (released by CFCs) and the ozone losses. The discovery of the ozone hole was surprising not only because of its magnitude but also its location: based only on gas phase chemical models of the stratosphere, it was anticipated that chlorine-initiated ozone depletion would occur predominantly at middle and lower latitudes and at altitudes between 35 and 45 km (see, e.g., NRC, 1982), not in the lower polar stratosphere.

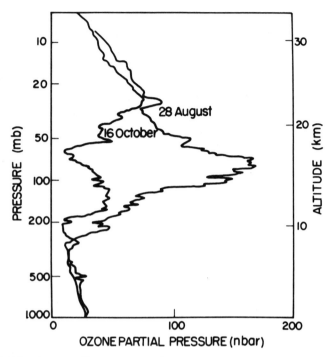

FIGURE 11.2.   Ozone profiles measured in 1987 over McMurdo Station, Antarctica (From Hofmann et al., 1987.)

An initial question was whether the Antarctic "ozone hole" is merely a natural phenomenon, only never before noticed. Overwhelming evidence, however, has since pointed to anthropogenic emissions, specifically CFCs, as the cause (see, e.g., Solomon, 1990; Anderson et al., 1991; WMO, 1992). Early theories put forth to explain the origin of the Antarctic ozone hole included as principal causes solar cycles, atmospheric dynamics, and chemistry. Callis and Natarajan (1986) suggested that the decline in ozone could be related to NO_x generated in the upper stratosphere by solar UV radiation in connection with the 11 year solar cycle. This turned out to be inconsistent with the observations that ozone depletion takes place in the lower rather than upper polar stratosphere; and that the concentration of NO_x there is remarkably low. The central idea of the dynamics theory (e.g., Tung et al., 1986) was that, upon first sunrise, warming of the Antarctic stratosphere leads to a net upward lifting of ozone-poor air from the troposphere or lower stratosphere, leading to the springtime decline of ozone. Verification of this theory would have come from the observations of upward air flow; however, tracer (e.g., CFCs, CH_4, and N_2O) studies revealed a strong downward flux within the polar vortex, where ozone depletion is most severe. The chemical theory, as discussed below, is an extension of the known radical-initiated catalytic cycles, but with consideration of the unique conditions of the polar stratosphere.

By 1988 it also became clear that significant ozone depletion was taking place at high northern latitudes—e.g., more than 20% between $53°\,N$ and $64°\,N$—in the winter months. The depletion is not as severe and localized as over Antarctica, but is certainly significant (WMO, 1990a).

11.3.1. Free Radical Chemistry of the Polar Stratosphere

In the polar stratosphere very little ozone is produced, as the large solar zenith angle (low Sun elevation) results in essentially no photodissociation of oxygen. This also means that catalytic cycles involving atomic oxygen are ineffective. Several catalytic cycles which result in the net reaction

$$2O_3 \rightarrow 3O_2$$

and which do not require the presence of O atoms have been proposed since the discovery of the ozone hole.

Solomon et al. (1986) suggested a cycle based on the coupling of the HO_x and ClO_x families:

$$OH + O_3 \rightarrow HO_2 + O_2 \tag{11.6}$$

$$Cl + O_3 \rightarrow ClO + O_2 \tag{11.19}$$

$$ClO + HO_2 \rightarrow HOCl + O_2 \tag{11.28}$$

$$\underline{HOCl + h\nu \rightarrow OH + Cl}$$

$$\text{net:} \quad 2O_3 + h\nu \rightarrow 3O_2$$

The importance of this cycle can be estimated from measurements of HOCl. Observations of this species in the Antarctic stratosphere, however, indicate that this cycle makes only minor contributions to ozone depletion (see, e.g., Toon and Farmer, 1989).

Another proposal was a catalytic cycle involving bromine (McElroy et al., 1986):

$$Cl + O_3 \rightarrow ClO + O_2 \tag{11.19}$$

$$Br + O_3 \rightarrow BrO + O_2 \tag{11.35}$$

$$\underline{ClO + BrO \rightarrow Cl + Br + O_2} \tag{11.36}$$

$$\text{net:} \quad 2O_3 \rightarrow 3O_2$$

Actually, the reaction between ClO and BrO has three channels:

$$ClO + BrO \rightarrow Cl + Br + O_2 \tag{11.36}$$

$$ClO + BrO \rightarrow BrCl + O_2 \tag{11.37}$$

$$ClO + BrO \rightarrow Br + OClO \tag{11.38}$$

Laboratory studies have shown that the first and third channels are equally fast and that the production of BrCl seems to be a minor channel (see, e.g., Hills et al., 1987; Friedl and Sander 1989; Turnipseed et al., 1991), important only at night—photolysis of this species occurs rapidly during the day:

$$BrCl + h\nu \rightarrow Br + Cl \tag{11.39}$$

A mechanism involving the formation of a ClO dimer was also proposed (Molina and Molina, 1987):

$$ClO + ClO + M \rightarrow Cl_2O_2 + M \tag{11.40}$$

$$Cl_2O_2 + h\nu \rightarrow Cl + Cl + O_2 \tag{11.41}$$

$$\underline{2(Cl + O_3 \rightarrow ClO + O_2)} \tag{11.19}$$

$$\text{net:} \quad 2O_3 + h\nu \rightarrow 3O_2$$

Like the reaction between BrO and ClO, the reaction between two ClOs has three bimolecular channels:

$$ClO + ClO \rightarrow Cl_2 + O_2$$

$$ClO + ClO \rightarrow Cl + ClOO$$

$$ClO + ClO \rightarrow Cl + OClO$$

These bimolecular reactions are very slow and hence of little atmospheric importance. On the other hand, in the lower polar stratosphere the termolecular reaction (11.40) that leads to the formation of Cl_2O_2, can occur efficiently, as it is facilitated by the higher pressures and lower temperatures there.

The structure of Cl_2O_2 formed by this reaction is important. The asymmetric dimer ClOClO would not lead to ozone destruction. In the case of the symmetric dimer (chlorine peroxide) there are two possible channels for photodissociation:

$$ClOOCl + h\nu \rightarrow Cl + ClOO \tag{11.42}$$

$$ClOOCl + h\nu \rightarrow ClO + ClO \tag{11.43}$$

In the dimer cycle, production of chlorine atoms by reaction (11.42) competes with reaction (11.43) and the thermal dissociation of ClOOCl, both of which yield two ClO radicals, which leads to no ozone depletion. However, thermal dissociation is slower than photolysis at temperatures below $\approx 220\,K$; furthermore, experiments (Cox and Hayman, 1988; Molina et al., 1990) indicate that reaction (11.42) is the major photolysis path. In addition, the structure of Cl_2O_2 has been shown to have the symmetric form by both theory and experiment (McGrath et al., 1990; Birk et al., 1989; Cheng and Lee, 1989). The product of reactions (11.36) and (11.42) could be ClOO or $Cl + O_2$; however, the Cl—OO bond strength is only $\approx 5\,kcal\,mol^{-1}$, so that even under polar stratospheric conditions ClOO rapidly decomposes to yield free Cl atoms.

11.3.2. Polar Stratospheric Clouds (PSCs) and Sulfate Aerosols

As purely gas phase reactions do not support large concentrations of ClO, the explanation of the Antarctic ozone hole required a shift in thinking, from just gas phase chemistry to a more comprehensive picture which includes heterogeneous chemistry—linking gas phase and aerosol chemistry.

Compared with the troposphere, the stratosphere is extremely dry and practically cloudless. Under normal conditions there is a sparse layer of aerosol at altitudes of 12–30 km, as first described by Junge in 1961. This background "sulfate" layer, made up of small sulfuric acid droplets (typically 75% H_2SO_4 and 25% H_2O at mid-latitudes) of radius roughly 0.1 μm and number density 1–10 cm^{-3}, is present at all latitudes. The number density and size of particles in the stratosphere are seen to vary widely. Increases by factors of 10–100 in the aerosol mass are observed following the injection of SO_2 directly into the stratosphere by major volcanic eruptions such as El Chichon in 1982 and Mount Pinatubo in 1991, which decay over a period of a few years. A possible role of increased sulfate aerosols in ozone depletion was suggested by Hofmann and Solomon (1989). Recent observations by Gleason et al. (1993) and Deshler et al. (1992) support these expectations; the reason is explained in Section 11.3.3.

In addition to various ground-based observations, long-term global trends of stratospheric particle concentrations are being recorded by satellite instruments (SAGE, SAM), the first of which were launched in the late 1970s. Over the polar regions, stratospheric aerosols and clouds have been observed by the Stratospheric Aerosol Measurement II (SAM II) instrument on board the Nimbus 7 satellite. As the general characteristics of these clouds have previously been reviewed elsewhere (e.g., Turco et al., 1989, Hamill and Toon, 1991), only a brief summary is presented here (see also Table 11.1).

Observations, including satellite, lidar, and *in situ* particle measurements, have indicated that the nature of the PSCs depends mainly on temperature and water

TABLE 11.1   Properties of the Background Sulfate Aerosol and Polar Stratospheric Clouds

Property	Sulfate Aerosol	Type I PSC	Type II PSC
Composition	40–80 wt% H_2SO_4	$HNO_3 \cdot 3H_2O$ crystals	H_2O ice
Size (μm)	0.01–1	0.3–3	1–100
Formation temperature (K)	195–240	< 195	< 187

vapor concentration. PSCs fall into two distinct types. The small (0.5–2 μm) class of particles, type I, are most likely crystals of nitric acid trihydrate (NAT), as their formation and existence correlate with the partial pressures of water and nitric acid over NAT (e.g., 3×10^{-4} Torr H_2O and 4×10^{-7} Torr HNO_3 at 198 K) but are above the frost point of pure ice (Toon et al., 1986; Crutzen and Arnold, 1986; McElroy et al., 1986). Type II PSCs are observed to form only when the temperature drops below the ice frost point (191 K for 3×10^{-4} Torr H_2O), and they grow into significantly larger (up to 100 μm) crystals.

The current understanding is that there are basically three chemical components involved in the formation of PSCs: H_2O, HNO_3, and H_2SO_4. Large amounts of HCl do not condense: inspection of a phase diagram of the HCl/H_2O system (Figure 11.3) reveals that HCl solutions or hydrates are not stable under polar stratospheric conditions (Molina, 1992).

We discuss next the binary systems H_2O/H_2SO_4 and H_2O/HNO_3, and the ternary system $H_2O/H_2SO_4/HNO_3$. The H_2O/H_2SO_4 system is of interest for the lower stratosphere at all latitudes. When the temperature is above the threshold of PSC formation, stratospheric aerosol particles are mainly composed of supercooled aqueous sulfuric acid droplets, for example, in the mid-latitude lower stratosphere at about 16 km the temperature is roughly 220 K and the sulfate aerosol particles in equilibrium with 5 ppm H_2O ($\approx 4 \times 10^{-4}$ Torr) have compositions of 70–75 wt% H_2SO_4 (see Figure 11.4). The equilibrium composition of these droplets changes with temperature: they absorb water to become increasingly more dilute   reaching about 40 wt% H_2SO_4 around 195 K. The thermodynamically stable forms for these aerosols are actually crystalline hydrates ($H_2SO_4 \cdot nH_2O$; n = 1, 2, 3, 4, 6.5, 8); however, sulfuric acid solutions have a strong tendency to supercool: laboratory and field observations indicate that sulfate aerosols remain liquid under most conditions outside the polar vortices (Zhang et al., 1993a; Toon et al., 1993).

As mentioned above, type I PSCs most likely consist of nitric acid trihydrate. Laboratory measurements of vapor pressures over NAT (Hanson and Mauersberger 1988a,b) confirmed that under typical lower stratospheric conditions, i.e., ≈ 3–5 ppm H_2O and ≈ 5–10 ppb HNO_3, NAT is stable at temperatures ≈ 5 K higher than the ice frost point (see Figure 11.5). The slopes from plots of $\log P_{HNO_3}$ versus $\log P_{H_2O}$ over NAT were found to be -3, as required by the

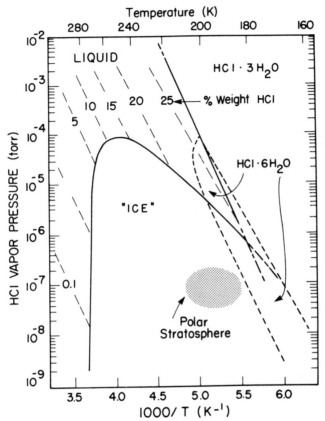

FIGURE 11.3. HCl vapor pressure as a function of temperature for the HCl/H$_2$O system. The long-dashed lines represent the vapor pressures of liquids whose composition are given in wt% HCl, and the solid lines give the coexistence conditions for the two condensed phases. The short-dashed lines enclose the thermodynamic stability region for the HCl hexahydrate, which only nucleates from liquid solutions at temperatures below 170 K and hence is not likely to be formed in the stratosphere.

Duhem–Margules equation, considering that the crystal structure allows only negligible variations from the 3:1 stoichiometry. Worsnop et al. (1993) explored the long-term (days) evolution of the HNO$_3$/H$_2$O system, finding that, in addition to NAT, it is possible for HNO$_3$•2H$_2$O (nitric acid dihydrate, NAD) crystals to exist under stratospheric conditions if nucleated and if the more thermodynamically stable NAT is not present. Infrared absorption spectra of amorphous and crystalline films of H$_2$O/HNO$_3$ mixtures (Ritzhaupt and Devlin, 1991; Tolbert and Middlebrook, 1990; Middlebrook et al., 1992; Smith et al., 1991; Koehler et

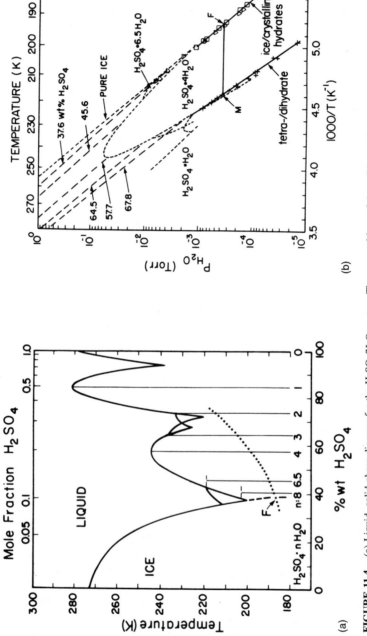

FIGURE 11.4. (a) Liquid–solid phase diagram for the H_2SO_4/H_2O system. The compositions of the solids are given by the vertical lines. (b) H_2O vapor pressure as a function of temperature for the H_2SO_4/H_2O system. Superimposed on (a) is a line representing the equilibrium compositions of supercooled liquid droplets in the stratosphere, assuming 3 ppm of H_2O at 100 mbar, i.e., ≈ 16 km altitude. F is the ice frost point and M is where crystalline $H_2SO_4 \cdot 4H_2O$ would melt upon warming under these conditions.

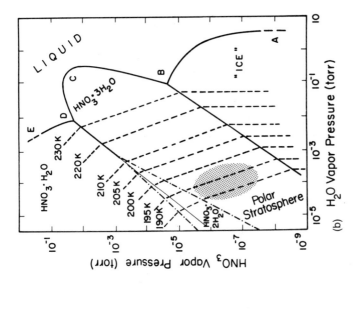

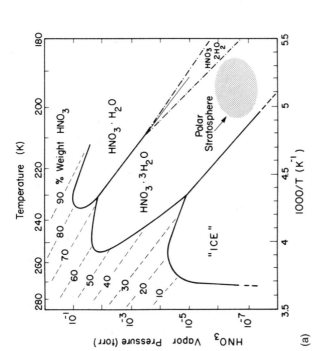

FIGURE 11.5. (a) Nitric acid vapor pressure as a function of temperature for the HNO_3/H_2O system. The dashed lines represent the vapor pressures of liquids whose compositions are given in wt% HNO_3, and the solid lines represent coexistence for two condensed phases. (b) Log of nitric acid partial pressure versus log of water partial pressure for the $HNO_3 \cdot nH_2O$ crystalline hydrates. The dashed lines indicate temperatures, and the solid lines represent coexistence for two condensed phases. Note that the slopes of the dashed lines are $-n$ and that for a given water vapor pressure NAT ($HNO_3 \cdot 3H_2O$) is stable at ≈ 5 K warmer than ice under typical polar stratospheric conditions (shaded region).

al., 1992) as well as for small NAT and NAD aerosol particles have been reported (Barton et al., 1993) and agree with the vapor pressure studies.

A number of researchers have carried out measurements on the uptake of nitric acid by sulfuric acid solutions (see, e.g., Reihs et al., 1990; Van Doren et al., 1991), but only recently over the ranges of composition and temperature expected for the polar stratosphere (Zhang et al., 1993b). Figure 11.6 illustrates the expected composition of $H_2O/H_2SO_4/HNO_3$ liquid droplets in equilibrium with fixed water and nitric acid vapor partial pressures as a function of

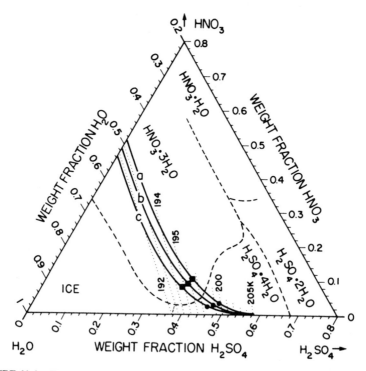

FIGURE 11.6. Ternary diagram for the $H_2SO_4/HNO_3/H_2O$ system. The dashed lines indicate the eutectics between the crystalline phases (Carpenter and Lehrman, 1925). Superimposed upon this are vapor–liquid equilibrium dilution curves (solid lines) for stratospheric aerosol droplets at 100 mbar (≈ 16 km) and at ambient mixing ratios of 5 ppm H_2O, and 10 ppb HNO_3 (a), 5 ppb HNO_3 (b), and 2.5 ppb HNO_3 (c), as estimated from $H_2SO_4/HNO_3/H_2O$ vapor pressure data. Also along the dilution lines are the temperatures (dotted lines), the frost point of crystalline $HNO_3 \cdot 3H_2O$ (i.e., NAT supersaturation, $S_{NAT} = 1$, ●), and the point at which the HNO_3 vapor pressure reaches a supersaturation of 10 with respect to NAT ($S_{NAT} = 10$, ■). Liquid vapor pressures used for the interpolation were those measured over ternary solutions containing 35–70 wt% H_2SO_4 (Zhang et al., 1993b) and for 0% H_2SO_4 extrapolations of those reported for supercooled $HNO_3(aq)$ (Hanson, 1990) to lower temperatures.

temperature. Notice that upon cooling, the equilibrium composition of the liquid changes very rapidly below 195 K to increased water and nitric acid contents. Solutions with these lower H_2SO_4 concentrations were observed to crystallize readily into NAT and H_2SO_4 hydrates (Molina et al., 1993). Hence a likely mechanism for the formation of type I PSCs involves incorporation of HNO_3 vapor into the liquid H_2O/H_2SO_4 droplets, followed by crystallization of NAT in the droplets and by subsequent growth of the NAT crystals by absorption of additional amounts of HNO_3 and H_2O vapors.

11.3.3. Reactions Involving PSCs and Sulfate Aerosols

It is now well-established that heterogeneous reactions occurring on PSCs play a central role in the chlorine activation leading to polar ozone depletion (see, e.g., Molina, 1991). The impact of these heterogeneous reactions is twofold: (1) chlorine is repartitioned from the inert reservoirs HCl and $ClONO_2$ into much more photolabile forms, mainly Cl_2 and HOCl, and (2) NO_x is removed from the gas phase ("denoxification") in the lower polar stratosphere through incorporation of nitric acid into PSCs and is thus unable to interfere with the chlorine cycle by forming $ClONO_2$. Various techniques, e.g., Knudsen cell reactors, wall-coated flow tubes, and droplet train flow tubes, have been used in the last few years to measure the reaction probability per collision, γ, of various chlorine and nitrogen species on surfaces representative of stratospheric particles: ice, NAT, and sulfuric acid solutions. An extensive review of these heterogeneous processes has been presented by Kolb et al. (1993).

The most important chlorine activation reaction in the polar stratosphere is

$$ClONO_2 + HCl \rightarrow Cl_2 + HNO_3 \qquad (11.44)$$

The net result of this reaction taken together with the gas phase reactions (11.19) and (11.30), and chlorine photolysis is the conversion of HCl and NO_x to ClO and HNO_3:

$$ClONO_2 + HCl \rightarrow Cl_2 + HNO_3 \qquad (11.44)$$

$$Cl_2 + h\nu \rightarrow 2Cl \qquad (11.45)$$

$$2(Cl + O_3 \rightarrow ClO + O_2) \qquad (11.19)$$

$$\underline{ClO + NO_2 + M \rightarrow ClONO_2 + M} \qquad (11.30)$$

net: $HCl + NO_2 + 2O_3 \rightarrow ClO + HNO_3 + 2O_2$

As a gas phase reaction, (11.44) is extremely slow, having an upper limit of 10^{-19} cm^3 s^{-1} (Molina et al., 1985; DeMore et al., 1992). It was noted, however, that one must take extreme care to exclude water from the measurement apparatus in order to determine good upper limits to the homogeneous gas phase rate constant, as water adsorbed on surfaces is very effective in promoting the reaction (Molina et al., 1985; Rowland et al., 1986).

To account for ozone depletion over Antarctica, Solomon et al. (1986) suggested that reaction (11.44) is promoted by PSCs. Studies carried out on water-ice and NAT have indicated that it does proceed with high efficiency, having $\gamma > 0.2$ (Molina et al., 1987; Moore et al., 1990; Leu et al., 1991; Hanson and Ravishankara, 1991a; Abbatt and Molina, 1992b). The product Cl_2 rapidly desorbs, but the nitric acid remains at the surface. Although it is clear that the reaction occurs on the surfaces of PSC particles, it is unlikely for both $ClONO_2$ and HCl molecules to simultaneously collide on the same active site. A plausible mechanism involves incorporation of HCl or $ClONO_2$ or both into the surface layers of the particles, followed by reaction. Studies of the HCl uptake by ice crystals under these conditions have shown a much larger surface coverage than could be explained in terms of intact HCl molecules interacting with the surface through hydrogen bonding: instead, the results are consistent with the surface layers relaxing to a liquid-like configuration, where the HCl is ionically solvated (Molina, 1992; Abbatt et al., 1992).

The reaction between $ClONO_2$ and H_2O, which is also very slow in the gas phase, has also been studied on ice and NAT surfaces (Molina et al., 1987; Tolbert et al., 1987; Leu 1988a; Leu, et al., 1991; Moore et al., 1990; Hanson and Ravishankara, 1991a; Abbatt and Molina, 1992b). Here the reactant H_2O is already in the condensed phase:

$$ClONO_2 + H_2O \rightarrow HOCl + HNO_3 \qquad (11.46)$$

and again nitric acid remains at the surface. The HOCl product is easily photolyzed to yield a free chlorine atom. Furthermore, HOCl can participate in a subsequent heterogeneous reaction, which also proceeds rapidly on water-ice and NAT surfaces (Hanson and Ravishankara, 1992; Abbatt and Molina, 1992a):

$$HCl + HOCl \rightarrow Cl_2 + H_2O \qquad (11.47)$$

The combination of reactions (11.46) and (11.47) is equivalent to reaction (11.44)—both convert chlorine reservoirs to gas phase Cl_2. This suggests that the mechanism of reaction (11.44) involves the two steps—reactions (11.46) and (11.47).

If N_2O_5 is abundant, another heterogeneous reaction which may convert HCl to an easily photolyzed gas is

$$N_2O_5 + HCl \rightarrow ClNO_2 + HNO_3 \tag{11.48}$$

This reaction has also been studied on ice and NAT surfaces (Tolbert et al., 1988b; Leu, 1988b; Hanson and Ravishankara, 1991a).

As mentioned in Section 11.2.3, NO_x is converted to the reservoir species N_2O_5 by reactions (11.13) and (11.14), or into the HNO_3 reservoir by the reaction of NO_2 with OH, reaction (11.22). Hydrolysis of N_2O_5 yields the more stable HNO_3:

$$N_2O_5 + H_2O \rightarrow 2HNO_3 \tag{11.49}$$

In the gas phase, though, this reaction is negligibly slow (DeMore et al., 1992). However, it takes places very efficiently, with $\gamma \approx 0.1$, on sulfuric acid solutions throughout the concentration range of the stratospheric aerosols (Mozurkewich and Calvert, 1988; Van Doren et al., 1991; Hanson and Ravishankara, 1991b).

The high reaction probability of reaction (11.49) independent of the sulfuric acid concentration is in marked contrast with the behavior of reactions (11.44) and (11.46) on sulfuric acid solutions. Laboratory studies of reaction (11.46) indicate that γ is strongly dependent on the composition of the aerosols (Tolbert et al., 1988a; Hanson and Ravishankara, 1991b), increasing from 1.9×10^{-4} (75 wt% sulfuric acid solution) at 230 K to 6.4×10^{-2} (40 wt% sulfuric acid solution) at 215 K, (cf. Table 11.2). Since the solubility of HCl in 50–80 wt% H_2SO_4 solutions is very low, reaction (11.44) on sulfate aerosols is not important at mid-latitudes. At higher latitudes, however, where the temperature can drop to ≤ 200 K and the aerosol droplets absorb more water and HCl, reactions (11.44) and (11.46) on liquid sulfate aerosols become very important.

The net effect of reaction (11.49) is to convert catalytically active nitrogen oxides to the inert reservoir HNO_3, and to repartition hydrogen and chlorine species, as a consequence of the strong coupling between the various chemical

TABLE 11.2   Reaction probabilities (γ) for Heterogeneous Reactions Important in the Stratosphere

Reaction	γ_{NAT}	γ_{ice}	$\gamma_{H_2SO_4 \text{ solution}}$
$ClONO_2 + HCl$	0.3 (200–202 K)	0.3 (200–202K)	
$ClONO_2 + H_2O$	0.006 (200–202K)	0.3 (200–202K)	0.064 (40 wt%, 218 K)[a]
$HCl + HOCl$	0.1 (195–200 K)	0.3 (195–200 K)	—
$N_2O_5 + HCl$	0.003 (200 K)	0.03 (190–220 K)	—
$N_2O_5 + H_2O$	0.0006 (200 K)	0.03 (195–200 K)	0.1 (200–230 K)

[a]Log $\gamma = 1.87 - 0.0747W$, for $40 < W < 75$; W is H_2SO_4 wt%.

Source: DeMore et al. (1992).

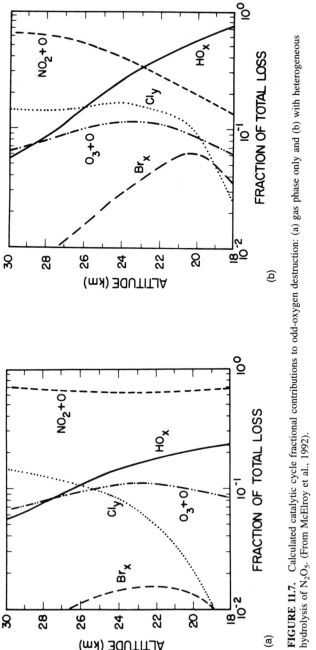

FIGURE 11.7. Calculated catalytic cycle fractional contributions to odd-oxygen destruction: (a) gas phase only and (b) with heterogeneous hydrolysis of N_2O_5. (From McElroy et al., 1992).

families (see Section 11.2.5). The result is to decrease the importance of NO_x in ozone depletion, and to increase it for hydrogen and chlorine species, as shown in Figure 11.7. Analysis of recent field measurements of nitrogen and chlorine species indicates that this heterogeneous reaction does play an important role in the lower stratosphere at mid-latitudes, particularly in the presence of the enhanced sulfate aerosol concentrations which followed the eruption of Mount Pinatubo in 1991 (see, e.g., Fahey et al., 1993); the observations cannot be explained with gas phase chemistry alone. Hence the effect of volcanic eruptions on stratospheric ozone is indirect: increased ozone depletion may occur, but only because chlorine chemistry is enhanced. In the presence of only natural levels of chlorine the effect is expected to be rather small.

11.4. FIELD OBSERVATIONS AND COMPUTER MODELS

Concerns about various anthropogenic emissions leading to global ozone depletion prompted research efforts directed at developing an understanding of stratospheric processes as well as to assess the impact of chemical perturbations to the atmosphere. Research in three areas has significantly refined our understanding of the chemistry of the stratosphere. One area of study involves laboratory measurements of elementary rate constants and absorption cross-sections of numerous reactions important in the stratosphere (DeMore et al., 1992). Another area of research focuses on measuring the concentrations of the important trace gases in the stratosphere. The third area involves the development of model simulations of the stratosphere. Advances in these three areas over the last two decades have been impressive.

11.4.1 Measurement Techniques and Results

A large number of measurements on individual species by *in situ* and remote sensing techniques from ground, aircraft, balloon, and spacecraft have been reported in the last decade. Intercomparison campaigns using different techniques to provide observations of the same species at the same time and place have validated the accuracy and precision of some measurements; see, e.g., the special issue of *J. Atmos. Chem.*, **10**(2) (1990).

11.4.1.1. Ozone

Although stratospheric ozone concentrations are determined predominantly by photochemical and dynamical processes, levels are also influenced by galactic cosmic rays, solar proton events, and volcanic eruptions, in addition to human activity. Thus, to understand the relationship between the variable natural stratospheric ozone levels and anthropogenic perturbations, it is essential to establish long-term ozone trends. A number of techniques have been employed to measure the total ozone column abundances and vertical ozone profiles. The main instruments for measuring column abundances include ground-based Dobson and other filter spectrometers, and satellite-based Backscattered Ultraviolet (BUV), Solar Backscattered Ultraviolet (SBUV), and the Total Ozone Mapping Spectrometer (TOMS). Current profile measurement techniques include balloon-borne ozonesondes and also a wide range of satellite-based instruments, including BUV, SBUV, and the Stratospheric Gas and Aerosol Experiment (SAGE). Figure 11.8 shows representative profile and column measurements of ozone abundances.

In the upper stratosphere (≈ 30–50 km) the dominant factors controlling ozone levels are photochemical processes involving atoms or small molecules, with less important roles being played by reservoir species with short photochemical lifetimes such as $HOCl$, $ClONO_2$, etc. Below 30 km the chemistries of HO_x, NO_x, and ClO_x are closely coupled and the concentration of ozone is controlled by both chemical and transport processes. To assess our understanding of the complex chemistry of ozone and the catalytic cycles, monitoring of the source gases (e.g., H_2O, N_2O, CH_4, CFCs) and the key trace gases in the oxygen family (O, O_3), nitrogen family (NO, NO_2, NO_3, N_2O_5, HNO_3, HO_2NO_2), hydrogen family (OH, HO_2, H_2O_2), and chlorine family (ClO, $HOCl$, $ClONO_2$, HCl, halocarbons) is essential. We discuss next observations of key species responsible for stratosphere ozone depletion.

11.4.1.2. HO_x

The key HO_x species are H_2O, OH, HO_2, and H_2O_2. Since water is the primary source of the HO_x radicals, measurements of water mixing ratios are important to the understanding of HO_x chemistry. Two main *in situ* techniques have been used for the measurement of H_2O vapor. The Lyman-α hygrometer monitors the fluorescence emitted by electronically excited hydroxyl radicals, OH^*, produced from dissociating H_2O with Lyman-α (121.6 nm) light (Kley and Stone, 1978). The frost point hygrometer relies on the equilibrium between water vapor in the atmosphere and an ice surface at the frost point temperature (Masterbrook and Oltmans, 1983). Remote sensing techniques commonly utilize the strong infrared lines; for example, one of the six spectral channels of the Limb Infrared Monitor of the Stratosphere (LIMS) on board the Nimbus 7 satellite is centered on emissions of H_2O (1370–1560 cm^{-1}); (the other five channels are NO_2 (1560–1630 cm^{-1}), O_3 (926–1141 cm^{-1}), HNO_3 (844–917 cm^{-1}), and CO_2 (579–755 and 637–673 cm^{-1}); (see WMO, 1990a).

The most important observations in the HO_x family are on the OH and HO_2 radicals. Fluorescence induced by solar flux, resonance lamps, and lasers has been the principal technique for OH measurement. Laser-induced fluorescence (LIF) has utilized A–X electronic transitions of OH (excitation of the (0–1) band

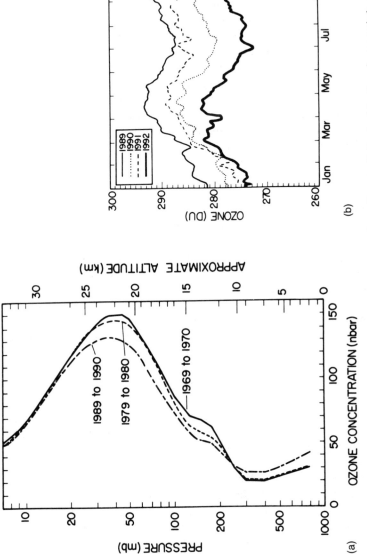

FIGURE 11.8. (a) Average ozone concentration versus altitude measured over Payerne, Switzerland for three 2 year periods, 1969–1970, 1979–1980, and 1989–1990. Periods were chosen to be approximately during solar maximum and 2 years were used to remove most of any quasi-biennial oscillation effect. (From Stolarski et al., 1992.) (b) Daily global ozone amount (area-weighted 65° S to 65° N) from NOAA-11 SBUV/2. The 1992 data are represented by the thick solid line. The 1991 data are represented by the dotted line. The 1990 data are represented by the dashed line. The 1989 data are represented by the solid line. (From Gleason et al., 1993.)

at 282 nm or the (0–0) band at 309 nm) (Anderson, 1987). Two *in situ* balloon-borne techniques have been used to measure HO_2. One is cryogenic trapping of free radicals on a liquid-nitrogen-cooled surface, analyzed subsequently by electron spin resonance spectroscopy in the laboratory (Mihelcic et al., 1978; Helten et al., 1984). Another technique utilizes the chemical conversion of HO_2 to OH by addition of NO, i.e. reaction (11.21), followed by the detection of OH via induced fluorescence (Stimpfle et al., 1990). Ground-based detection of HO_2 above 35 km from a rotational emission line near 265.7 GHz has been reported (de Zafra et al., 1984). Also, vertical HO_2 profiles from about 20 to 50 km have been obtained by Traub et al. (1990) using a far-infrared spectrometer. Simultaneous LIF measurements of OH, HO_2, H_2O, and O_3 have recently been carried out (Stimpfle et al., 1990; Wennberg et al., 1990), providing a most useful comparison between models and observations. H_2O_2 has been observed by microwave limb sounding (Waters et al., 1981), ground-based millimeter wave spectrometry (de Zafra et al., 1985), and far-infrared limb sounding (Chance et al., 1991); the first two observations, however, provide only upper limits. In addition, simultaneous measurements of OH, HO_2, and H_2O_2 been obtained by Park and Carli (1991) using far-infrared spectroscopy.

11.4.1.3. Nitrogen Oxides

NO, NO_2, NO_3, and N_2O_5 are the principal species here. The NO/NO_2 ratio is affected by interconversion within the NO_2 family (discussed in Section 11.2.3) as well as by reactions with O_3, HO_2, or ClO (Sections 11.2.3 and 11.2.5), and thus the comparison between observed ratios of NO/NO_2 and model calculations provides an important test of our understanding of NO_x in the stratosphere. The principal technique for *in situ* detection of NO and NO_2 is based on chemiluminescence: NO is converted to electronically excited (luminescent) NO_2^* via the reaction $NO + O_3 \rightarrow NO_2^* + O_2$. Measurement of NO_2 is based on the same principle, except that NO_2 is first photolyzed to yield NO. Extensive data sets are available from *in situ* balloon-, aircraft- and rocket-borne measurements (Ridley et al., 1987; Kondo et al., 1990). In addition, the sum of NO_x, N_2O_5 and HNO_3 has been measured *in situ* by the catalytic conversion of all nitric oxides to NO, followed by detection using NO/O_3 chemiluminesence (Fahey et al., 1989, 1990a). Other methods, mainly infrared and visible spectrophotometry, have also been used to measure NO, NO_2, and NO_3. NO_2 in the polar stratosphere has been measured by high-resolution infrared (Farmer et al., 1987b; Coffey et al., 1989; Mankin et al., 1990) and ultraviolet spectroscopy (Wahner et al., 1989b, 1990a; Sanders et al., 1989). Ground-based observation of NO_3 has utilized the strong absorption near 662 nm with moon and stars as a light source (Solomon et al., 1989a, and references therein). Further techniques for NO and NO_2 measurement include pressure-modulated radiometry (with a device which selectively modulates the emission from a gas by using the absorption lines of the same gas as an optical filter) (Drummond and Jarnot, 1978; Roscoe et al., 1986), long-path absorption (Louisnard et al., 1983), and the Balloon-Borne Laser In-Situ Sensor (BLISS) instrument (Webster and May, 1987). The BLISS instrument employed long-path tunable diode laser infrared absorption to monitor species including NO ($1854\,\text{cm}^{-1}$), NO_2 ($1598\ \text{cm}^{-1}$), O_3 ($1063\,\text{cm}^{-1}$), HNO_3 ($1333\,\text{cm}^{-1}$), and N_2O ($1525\,\text{cm}^{-1}$). The capability of the BLISS instrument to simultaneously measure chemically coupled nitrogen species provides one of the best checks of our knowledge of NO_x chemistry (Webster et al., 1990). There have been a number of observations of N_2O_5 in the stratosphere, most of which measured absorption or emission of the strong infrared band at $1240\,\text{cm}^{-1}$ (see, e.g., the review by Roscoe, 1991). Combined with information from satellite data for NO_2 (LIMS, SAGE, and SME, the Solar Mesosphere Explorer) and for HNO_3 (LIMS), this provides a basis for estimating NO_y (NO_y = $NO + NO_2 + NO_3 + HNO_3 + 2(N_2O_5) + HO_2NO_2 + ClONO_2$, of which HNO_3 is the major component in the lower stratosphere). Simultaneous measurements of the important nitrogen species have also been carried out by Abbas et al. (1991) as well as by the shuttle-borne ATMOS instrument; the latter provided the most complete simultaneous measurements of the nitrogen family so far, making observations of all the NO_y species including NO, NO_2, N_2O_5, HNO_3, HO_2NO_2, and $ClONO_2$. (Russell et al., 1988; Allen and Delitsky, 1990).

11.4.1.4. Chlorine Species

These include the CFC source gases, the reservoirs HCl, $ClONO_2$, and HOCl, and the ClO radical. The majority of the CFC vertical profiles have been obtained from grab-samples (canisters filled with air at the collection point) from aircraft or balloon flights and later analyzed by gas chromatography (WMO, 1986). Several techniques have been used to measure stratospheric ClO: *in situ* resonance fluorescence, balloon-borne and ground-based millimeter wave and microwave spectroscopy. In the *in situ* resonance fluorescence method ClO is converted to atomic Cl by the reaction $ClO + NO \rightarrow Cl + NO_2$ and a resonance lamp is then used to monitor Cl near 120 nm (Weinstock et al., 1981). On the ER-2 aircraft in the Antarctic and Arctic missions (AAOE and AASE) ClO was converted to Cl which was then detected by atomic resonance fluorescence, while ozone was simultaneously detected by UV absorption (Brune et al., 1989b, 1990). Ground-based millimeter wave spectrometric measurements of a ClO rotational line emission were reported by, e.g., Solomon et al. (1984). Also, de Zafra et al. (1987) used this technique to measure ClO vertical profiles over Antarctica. Waters et al. (1981, 1988) and Stachnik et al. (1992) used a balloon-borne millimeter wavelength heterodyne spectrometer to measure limb emission from a ClO rotational line near 640 GHz (along with nearby lines of O_3, HCl, ClO, and HO_2). In addition, ClO (as well as O_3, SO_2, and H_2O) has been measured globally by the microwave limb sounder aboard the UARS launched in September 1991 (Waters et al., 1993). Observations of BrO in the polar stratosphere have been made using near-ultraviolet absorption spectroscopy (see, e.g., Carroll et al., 1989; Wahner et al., 1990b) and *in situ* by a similar method to that used for ClO (Brune et al., 1989a; Toohey et al., 1990).

Monitoring $ClONO_2$ is important because this species links the stratospheric nitrogen and chlorine cycles. Measurements of $ClONO_2$ utilizing the infrared 1292, 809, and $780\,cm^{-1}$ infrared bands have been made by balloon-borne solar absorption spectroscopy (see, e.g., Massie et al., 1987) as well as the ATMOS (Zander et al., 1990) and UARS (Roche et al., 1993) instruments. The abundances of $ClONO_2$ as well as HCl in the polar stratosphere have also been recorded by techniques including high-resolution infrared spectrophotometry (e.g., Farmer et al., 1987a,b; Coffey et al., 1989; Mankin et al., 1990). As HCl is usually the most abundant stable reservoir species for chlorine radicals, our knowledge of the concentration of this species is important to the understanding of the chlorine budget. The strong infrared absorptions of HCl and HF have been utilized by ground-based, balloon (Farmer and Raper, 1977), aircraft (Mankin and Coffey, 1983), and ATMOS (Raper et al., 1987) and UARS (Reber et al., 1993) measurements.

11.4.1.5. Multi-species Measurements

The rather large variability in the concentrations of many of the key species and the highly coupled nature of the chemical processes make it important to simultaneously monitor as many of the reactive species as possible. Also, measurements of concentration ratios for certain species have also proved to be useful for testing models. For example, the $ClO/(HCl + ClONO_2)$ ratio is a measure of the fraction of the inorganic chlorine that participates in ozone destruction cycles, i.e., the active versus reservoir forms. As another example, simultaneous measurements of HCl and HF provide an indication of the relative contributions of natural and anthropogenic sources to chlorine in the stratosphere. The observed increases in both HCl and HF, as well as the ratio of their concentrations, which has decreased in a 13 year period between 1977 and 1990 from about 7 to about 4 (e.g., Rinsland et al., 1991), point to CFCs as the source of the increased halogen burden in the stratosphere: this ratio has decreased because there are no natural sources of HF, while HCl is formed from the decomposition of biogenic CH_3Cl as well as CFCs.

Several comprehensive experiments have been initiated in the past few years. Examples include the space-shuttle-borne Atmospheric Trace Molecule Spectroscopy (ATMOS) experiment, the Upper Atmosphere Research Satellite (UARS), the Airborne Arctic Stratospheric Expedition (AASE), and the Airborne Antarctic Ozone Experiments (AAOE). Equipped with a high-resolution Fourier transform near- and middle-infrared spectrometer, the ATMOS instrument is capable of monitoring more than 40 major and minor gaseous species from altitudes above ≈ 16 km (WMO, 1986; Farmer et al., 1987a). The primary goals of UARS complement ATMOS: besides monitoring 15 important species in the upper atmosphere belonging to the HO_x, NO_x, and ClO_x families (WMO, 1986), it also measures solar irradiance, particle energy deposition, temperatures, and wind fields. In addition to observations of ozone levels, instruments on board the Nimbus 7 satellite have provided other important measurements: LIMS (HNO_3, NO_2, O_3, H_2O, temperature), SAMS (CH_4, N_2O, temperature), SBUV/TOMS (ozone, solar flux), and SME (ozone, NO_2, aerosols, solar flux). The AAOE— (see *J. Geophys. Res.*, **94**, 11,179–11,737 (part 1) and 16,437–16,857 (part 2)—and AASE—(see *Geophys. Res. Lett.*, **17**, 313–564—campaigns, using two aircraft (ER-2 and DC-8) platforms, were designed to understand the perturbed chemistry and rapid ozone loss in the high-latitude lower stratosphere. The instruments were capable of making *in situ* measurements of important trace gases including ClO, BrO, OClO, $ClONO_2$, NO, NO_2, NO_y, HNO_3, H_2O, O_3, N_2O, CH_4, CO, and CFCs, particulates, as well as meteorological parameters.

11.4.2. Observations of the Polar Stratosphere

The meteorological structure which facilitates rapid ozone loss in the strato-sphere over Antarctica, i.e., the "ozone hole", is the polar vortex, which is a strong circumpolar wind pattern that forms over the pole in the dark winter (see, e.g., Schoeberl and Hartmann, 1991). Over the South Pole it is nearly circular and approximately the size of the Antarctic continent, and the chemically perturbed region is roughly coincident with this vortex. In the northern hemisphere the different land mass distribution and tall mountains lead to a less intense and less symmetric vortex. Field studies of vertical profiles of long-lived tracers (e.g., CFCs and N_2O) provide evidence of a strong downward flux inside the vortex. The vortex structure, coupled with the very cold temperature, sets the scene for the perturbed chemistry in the lower polar stratosphere that leads to severe, localized, ozone depletion.

   Field campaigns, organized soon after the discovery of the Antarctic ozone hole, have monitored concentrations of the key species that participate in the various catalytic cycles over Antarctica and the Arctic. These expeditions included the National Ozone Expeditions (NOZE I and II) in 1986 and 1987, the Airborne Antarctic Ozone Experiment (AAOE) in 1987, and the Airborne Arctic Stratosphere Expeditions (AASE I and II) in 1989 and 1992. Very strong evidence linking ozone loss to active chlorine radicals comes from the field observations along with laboratory measurements and modeling. The observa-tional techniques were chiefly those mentioned above, though in some cases they were adapted to the special conditions of the polar stratosphere and the sampling platforms. Some results are briefly summarized next.

11.4.2.1. Ozone and ClO

As all the important polar ozone-destroying catalytic cycles involve ClO, it is a key species to monitor along with O_3. Measurements of these two species provided striking evidence linking ClO levels to ozone losses, as illustrated in Figure 11.9. Near 20 km, abundances of ClO are $\approx 1-2$ ppb in both the Antarctic and Arctic vortices, which is orders of magnitude higher than concentrations in the unperturbed mid-latitude stratosphere, revealing almost complete conversion of chlorine species to reactive forms within the chemically perturbed regions. The high levels of ClO at these altitudes and the anticorrelation with O_3 are consistent with the ClO dimer catalytic mechanism, since the rate of ClO dimer formation, reaction (11.40), increases quadratically with [ClO] (Anderson et al., 1991). Recent satellite observations of ClO (UARS) have also shown extremely high levels in portions of both the northern and southern polar winter vortices, particularly in regions where the temperature drops below the threshold for PSC formation (Waters et al., 1993).

11.4.2.2. BrO and OClO

Significant enhancements in BrO compared with mid-latitudes have also been observed over both Antarctica and the Arctic. The diurnal behavior shows that

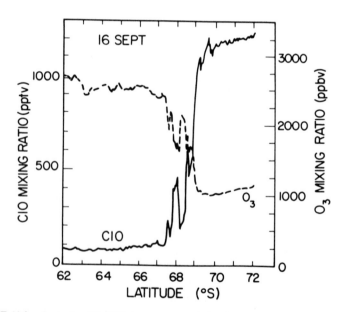

FIGURE 11.9. September 16, 1987 measurements of chlorine monoxide and ozone carried out during the Airborne Antarctic Ozone Experiment. (From Anderson et al., 1989.)

BrO (similarly to ClO) can combine with NO_2 to form the reservoir species $BrONO_2$ during night-time. Since the BrO + ClO reaction is believed to be the only source of OClO (see Section 11.3.1), an enhancement of its abundance as measured by ground-based and airborne UV spectrometers (Solomon et al., 1987, 1989b; Wahner et al., 1989a; Schiller et al., 1990) indicates that the ClO–BrO cycle indeed plays a significant role in the chemistry of the lower polar stratosphere. In addition, Solomon et al. (1993) observed high OClO concentrations over Antarctica in autumn 1992, when PSCs were unlikely to occur, supporting the role of sulfate aerosols in chlorine and bromine activation in the cold stratosphere at high latitudes when the aerosol loading is large, in this case after the Pinatubo eruption.

11.4.2.3. Halogen Reservoirs

Only a few per cent of the inorganic chlorine is in free radical form at mid-latitudes (see Section 11.2.5). In contrast, the sustained high levels of ClO within the polar vortex, along with the significant reductions in the column abundances of HCl, $ClONO_2$, and NO_2, clearly demonstrate the extent of the perturbation to chlorine chemistry: up to 70–80% of the chlorine burden can be in catalytically active form.

Another indicator of perturbed chlorine chemistry in the polar stratosphere is the HCl/HF ratio. Observations indicate that in the chemically perturbed region of the Antarctic vortex the HCl/HF ratio is reduced from the current mid-latitude value of ≈ 4–5 to ≈ 1, corresponding to a conversion of $\approx 80\%$ of the HCl to other chlorine species (Coffey et al., 1989). In the Arctic the ratio is reduced to ≈ 2 within the vortex, corresponding to 60% conversion (Mankin et al., 1990). Gas phase chemistry alone cannot account for this: heterogeneous reactions need to be taken into account.

11.4.2.4. Nitrogen Species

Measurements of N_2O and reactive nitrogen, NO_y, reveal a marked decrease in NO_y inside the polar vortex. Decreases in N_2O levels should be accompanied by increases in NO_y (with a linear negative correlation), as the principal source of NO_y is N_2O (Fahey et al., 1990b,c). Using this relationship, the difference between the expected and measured NO_y concentrations can be used to estimate the extent of denitrification (condensation of nitric acid into particulates, followed by sedimentation) in the polar stratosphere. For example, inside the Antarctic vortex, instead of an NO_y value of 10 ppb calculated from the observed, N_2O value, a much lower value of 1–4 ppb was observed, this denitrification being explainable in terms of the heterogeneous reactions coupled with sedimentation of particles containing HNO_3. This and the liberation of chlorine from its reservoirs are clearly evident from measurements of the various species

SUMMARY OF LATITUDE VARIATIONS
OF STRATOSPHERIC TRACE GASES

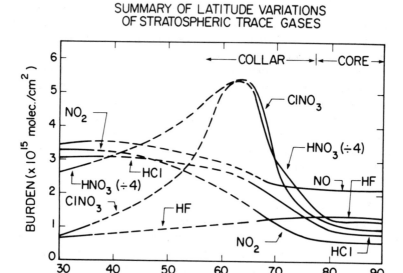

FIGURE 11.10.   Variation in trace gases from mid-latitudes to the core of the Antarctic polar vortex during September 1987. The dashed lines are for the region where there were no measurements. In the "collar" region the large $ClONO_2$ concentrations may be the result of mixing ClO-rich air from inside the vortex with NO_2-rich air from outside (From Toon et al., 1989.)

versus latitude; see Figure 11.10 (Toon et al., 1989). Inert species such as HF show normal behavior, while the concentrations of those species involved in the reactions discussed here are markedly perturbed.

11.4.3. Modeling and Assessment

The simultaneous actions of radiative, dynamical, and chemical processes affect the concentrations of trace species in the stratosphere; therefore a realistic theoretical model must consider all the above. According to the level of sophistication, one-, two-, and three-dimensional models have been employed. One-dimensional (altitude only) models coupling chemical and transport processes have been used extensively in the past as explorative tools for understanding the effects of anthropogenic perturbations to the stratosphere, as they may include very detailed chemistry. Two-dimensional models seek to predict latitudinal and seasonal variations and may include feedback between radiation, dynamics, and photochemistry. Finally, three-dimensional models which aspire to rely less extensively on the parameterization of various transport processes are being further developed as more computer power and observational data become available.

The physical and chemical processes that determine the concentration for any atmospheric chemical species are represented by the continuity equation, in which the number density of a general species i changes with time as

$$\frac{\partial n_i}{\partial t} = P_i - L_i + \nabla \cdot \phi_i$$

where P_i and L_i are the chemical and photochemical production and loss terms, respectively, and $\nabla \cdot \phi_i$ is the flux divergence, which accounts for transport. Typically this system of equations, which may involve 100 or more chemical and photochemical reactions, is integrated numerically using an implicit finite difference representation. Under some circumstances these equations can be simplified: at steady state the time derivative is equal to zero; for short-lived species, chemistry is much faster than transport, so the flux divergence for such species can be neglected; etc. For example, in the winter polar stratosphere, vertical transport within the vortex as well as horizontal transport in and out of the vortex is slower than the characteristic time for ozone depletion reactions. Thus the rate of chemical ozone loss can be approximated by a simplified continuity equation in terms of the two catalytic cycles found to be most important, considering also the fact that the production of O_3 is negligible:

$$d[O_3]/dt \approx -2(k_{40}[ClO][ClO][M] + k_{36}[ClO][BrO])$$

Using this approach with measured values of (ClO), Anderson et al. (1991) showed that the rapid ozone loss could be explained, with the first term being responsible for as much as 75% of the total Antarctic ozone loss. Notice that the loss processes are quadratic in [XO] and thus the ozone destruction rate increases non-linearly with increases in stratospheric loading of chlorine and bromine.

A typical set of mid-latitude model results is shown in Figure 11.11, taken from McElroy and Salawitch (1989). In general, the results are in fairly good agreement with the observations, taking atmospheric variability into account. However, some questions remain. For example, models have underestimated ozone concentrations above 35 km by as much as 30–50% (the "40 km ozone problem"). That problem is particularly striking, since in the chemistry-dominated upper stratosphere the concentration of ozone is controlled by the well-established photochemical processes represented by various catalytic cycles (see, e.g., Toumi et al., 1991; Eluszkiewicz and Allen, 1993).

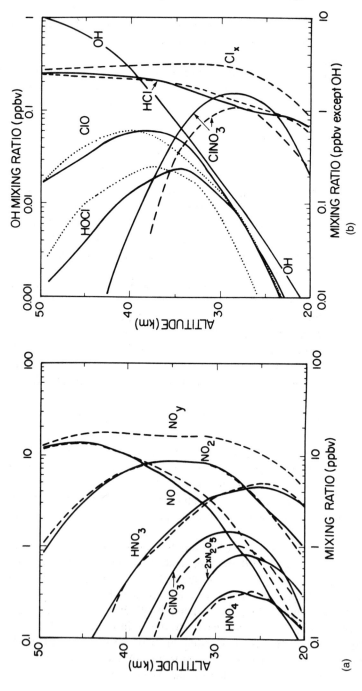

FIGURE 11.11. Calculated and observed concentrations for (a) the odd nitrogen family and (b) the inorganic chlorine family (total inorganic chlorine $Cl_x = HCl + ClONO_2 + ClO + HOCl$). (From McElroy and Salawitch, 1989.) Data from the ATMOS experiment for 30°N on April 30 and May 1, 1985 are shown as dotted lines (but with observed $ClONO_2$ profiles updated with the results of Zander et al., 1990. Model results are shown by solid lines. Also shown in (b) is the calculated profile for OH.

11.5. THE FUTURE OF STRATOSPHERIC OZONE

Understanding the long-term impact of halogenated chemicals on the atmosphere is important because of the very long atmospheric lifetimes of these compounds. Investigations of the effects of the CFCs and potential CFC substitutes such as hydrochlorofluocarbons (HCFCs) on stratospheric ozone have been initiated in the last few years. Here we discuss the atmospheric fates of halocarbons and how relative "ozone depletion potentials" (ODPs) of halocarbons are determined.

11.5.1. Determining the Lifetimes of Halocarbons in the Atmosphere

A halocarbon's atmospheric lifetime, specifically in relation to the time required for diffusion into the stratosphere, determines its influence on stratospheric ozone depletion. Substitute compounds such as HCFCs have been introduced with the intent to reduce the amount of chlorine reaching the stratosphere; examples include HCFC-134a (CF_3CFH_2) as a substitute for CFC-12 (CF_2Cl_2), and HCFC-141b (CH_3CFCl_2), HCFC-123 (CF_3CFCl_2), and HCFC-22 (CHF_2Cl) as substitutes for CFC-11 ($CFCl_3$) (Manzer, 1990). Unlike the CFCs, the HCFCs are destroyed to a large extent in the troposphere by reaction with OH radicals and thus have shorter atmospheric lifetimes. In order to elucidate the atmospheric fate of these new chemicals, knowledge of the reaction rates and of tropospheric OH levels is needed. The chemical lifetime due to reaction with OH, τ_{OH}, can be written as

$$\tau_{OH} = (1/k_{OH\ +\ halcarbon})\ [\overline{OH}]$$

where $k_{OH\ +\ halcarbon}$ is the rate constant for the reaction with OH radicals and $[\overline{OH}]$ is an average tropospheric OH concentration. Increasing efforts are being directed at measuring and predicting the reaction rates between OH and halocarbons and the subsequent reaction products; see Table 11.3.

As one can see, the accuracy of lifetime estimates hinges upon our knowledge not only of the rate constants but also of the average concentration of OH radicals in the troposphere. Direct measurements of tropospheric OH concentrations are extremely difficult; however, in recent years there has been significant progress: techniques using a ^{14}C tracer method, laser-induced fluorescence, as well as long-path absorption spectroscopy and chemical ionization mass spectroscopy have been employed to measure local tropospheric OH concentrations (see, e.g., Mount and Eisele, 1992). An indirect method to determine the global mean OH concentrations compares measured levels of methyl chloroform (CH_3CCl_3), which has no natural sources and whose main removal process is the reaction

TABLE 11.3   OH Rate Constants, Atmospheric Lifetimes, and Ozone Depletion Potentials (OPDs) of Selected Halocarbons

Molecule	k_{OH} (at 280 K)	Estimated Lifetime (years)	ODP
CFCs			
CFC-11 ($CFCl_3$)	–	55	1.0
CFC-12 (CF_2Cl_2)	–	116	≈ 1.0
CFC-113 ($CFCl_2CF_2Cl$)	–	110	1.07
CFC-114 (CF_2ClCF_2Cl)	–	220	≈ 0.8
CFC-115 (CF_2ClCF_3)	–	550	≈ 0.5
HCFCs, etc.			
HCFC-22 (CF_2HCl)	3.3×10^{-15}	15.8	0.055
HCFC-123 (CF_3CHCl_2)	3.1×10^{-14}	1.7	0.02
HCFC-124 (CF_3CHFCl)	7.6×10^{-15}	6.9	0.022
HCFC-141b (CH_3CFCl_2)	4.4×10^{-15}	10.8	0.11
HCFC-142b (CH_3CF_2Cl)	2.3×10^{-15}	22.4	0.065
HCFC-225ca ($CF_3CF_2CHCl_2$)	1.9×10^{-14}	2.8	0.025
HCFC-225cb (CF_2ClCF_2CHFCl)	6.5×10^{-15}	8.0	0.033
CCl_4		47	1.08
CH_3CCl_3	8.1×10^{-15}	6.1	0.12
Brominated compounds			
H-1301 (CF_3Br)	–	66–69	≈ 16
H-1211 (CF_2ClBr)	–	19–20	≈ 4
H-1202 (CF_2Br_2)	–	4	≈ 1.25
H-2402 (CF_2BrCF_2Br)	–	22–30	≈ 7
H-1201 (CF_2HBr)	7.1×10^{-15}	6	≈ 1.4
H-2401 (CF_3CHFBr)	1.6×10^{-14}	2	≈ 0.25
H-2311 ($CF_3CHClBr$)	5.2×10^{-14}	1	≈ 0.14
CH_3Br	3.3×10^{-14}	1–2	≈ 0.6

Source: WMO (1990b, 1992).

with OH, with expected concentrations based on known industrial emissions (see Chapter 7 for detailed discussion). In a recent study that used this method, the estimated average OH concentration was found to be about 8×10^5 molecule cm^{-3} (Prinn et al., 1992).

11.5.2. Ozone Depletion from CFC Substitutes: the Ozone Depletion Potential

The ozone depletion potential (ODP) of a halocarbon x is usually defined as the steady state ozone destruction that results from each mass unit of the particular species x relative to that of CFC-11 (Wuebbles, 1983):

$$ODP (x) = \Delta O_3(x)/\Delta O_3 \text{ (CFC–11)}$$

As a relative measure, the ODP does not predict absolute ozone losses but shows the expected effects of releasing one molecule versus another one into the atmosphere. The primary factor that determines a compound's ODP is its tropospheric lifetime, which, as discussed above, for an HCFC is determined by its reaction rate with hydroxyl. Some example reaction rate constants and the derived atmospheric lifetimes and ODPs for several halocarbons are shown in Table 11.3.

While substitution of hydrogen-containing halocarbons for CFCs will generally result in less chlorine being delivered to the stratosphere, long-term massive use could still have deleterious effects. Thus such compounds are thought of as transitional substances. As steady state models are inappropriate for predicting their impact on ozone in the near future, time dependent ODPs may need to be considered for compounds such as the HCFCs (Solomon and Albritton, 1992).

As discussed earlier, bromine species are one to two orders of magnitude more efficient at destroying ozone than those of chlorine, on a molecule-per-molecule basis, as is reflected by the very large ODPs of the halons (cf. Table 11.3). Similarly, despite the relatively short tropospheric lifetime of methyl bromide (CH_3Br), this species has a sizable ODP of about 0.6 and hence its industrial production might be regulated—or even phased out—in the future (see, e.g., WMO, 1992).

11.5.3. Other Potential Threats: Supersonic Aircraft, etc.

Human activity impacts the ozone layer not only by pollutants emitted at the surface which enter the stratosphere by transport processes (e.g. CFCs), but also by the release of pollutants (e.g., NO_x, HCl, H_2O, etc.) emitted directly into the stratosphere by high-altitude aircraft, rockets, etc. Potential ozone depletion caused by the engine emissions of projected fleets of supersonic aircraft has attracted attention since the early 1970s, focusing mostly on the NO_x combustion by-product. When only gas phase chemistry was considered, models indicated that the release of NO_x near 20 km would deplete ozone through catalytic cycles, reactions (11.11) and (11.12), whereas release around 10–12 km (such as the current subsonic fleet) may lead to a slight increase in ozone, the latter being promoted by the following reactions involving NO_x (the "smog" reactions) which are initiated by the reaction between OH radical and hydrocarbons (RH):

$$OH + RH \rightarrow H_2O + R$$

$$R + O_2 \rightarrow RO_2$$

$$RO_2 + NO \rightarrow RO + NO_2$$

$$NO_2 + h\nu \rightarrow NO + O$$

$$O + O_2 + M \rightarrow O_3 + M$$

Since the fate of NO_2 is altitude dependent, there exists a cross-over point above which increased NO_x leads to ozone depletion (where reaction (11.12), $NO_2 + O \rightarrow NO + O_2$, is favored) and below which ozone is produced (where NO_2 photolysis is favored). However, the ozone changes are sensitive to the balance of competing catalytic cycles as well as to atmospheric circulation. The calculated impact of a fleet of supersonic aircraft, now referred to as "high-speed civil transports") (HSCTs), is currently being re-examined (e.g., Johnston et al., 1989). Recent indications are that the impact of HSCTs is highly dependent on heterogeneous chemistry, especially the $N_2O_5 + H_2O \rightarrow 2HNO_3$ reaction on sulfate aerosols: the predicted effects of NO_x emissions are smaller when this reaction is taken into account. On the other hand, H_2O and sulfur emissions are now also important considerations in the evaluation of the impact of these proposed aircraft (HSRP/AESA, 1993).

Space shuttle and other rocket launches, particularly those using solid fuel boosters with chlorine compounds as fuel components, have been considered as possible sources of ozone depletion, as they emit chlorine directly into the stratosphere. The chlorine burden due to these sources, however, is hundreds of times smaller than that due to the CFCs. On the other hand, the effects due to other emissions—mainly particulates such as alumina, etc.—have not yet been quantified (WMO, 1992).

11.5.4. International Regulations on Ozone-Depleting Compounds

The first international agreement limiting CFCs, the Montreal Protocol on Substances That Deplete the Ozone Layer, was approved on September 16, 1987 while the AAOE mission was only starting to gather the data that would firmly link CFCs to the formation of the "ozone hole". It provided for a staged control of five CFCs and three bromine-containing halocarbons, to ultimately freeze consumption of the bromine compounds at 1986 levels and to cut CFC consumption to 50% of 1986 levels by the year 2000. It was quickly recognized that the restrictions did not go far enough (see Figure 11.12), leading to amendments negotiated in London in 1990 and in Copenhagen in 1992. Even with full compliance with stringent international regulation, the global ozone

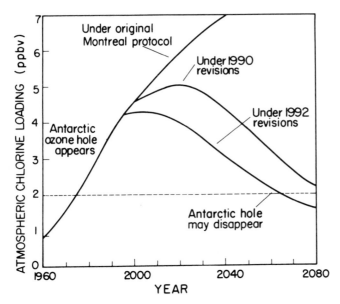

FIGURE 11.12.   Measured (1960–1990) and projected (from 1990 on) atmospheric chlorine loadings with and without international protocols. (From *C&E News,* May 24, 1993, pp. 8–18.)

declines will not be reversed in the near future (see, e.g., Prather and Watson, 1990) and the "ozone hole" is expected to develop annually over Antarctica for several decades to come.

References

Abbas, M.M., Kunde, V.G., Brasunas, J.C., Herman, J.R., and Massie, S.T. (1991). Nighttime reactive nitrogen measurements from stratospheric infrared thermal emission observations. *J. Geophys. Res., 96,* 10,885–10,897.

Abbatt, J.P.D., Beyer, K.D., Fucaloro, A.F., McMahon, J.R., Wooldridge, P.J., Zhang, R. and Molina, M.J. (1992). Interactions of HCl vapor with water ice: implications for the stratosphere. *J. Geophys. Res., 97,* 15,819–15,826.

Abbatt, J.P.D. and Molina, M.J. (1992a). The heterogeneous reaction HOCl + HCl → Cl_2 + H_2O on ice and nitric acid trihydrate: reaction probabilities and stratospheric implications. *Geophys. Res. Lett., 19,* 461–464.

Abbatt, J.P.D. and Molina, M.J. (1992b). Heterogeneous interactions of $ClONO_2$ and HCl on nitric acid trihydrate at 202 K. *J. Phys. Chem., 96,* 7674–7679.

Allen, M. and Delitsky, M.L. (1990). Stratosphere NO, NO_2, and N_2O_5: a comparison of model results with spacelab 3 atmospheric trace molecule spectroscopy (ATMOS) measurements. *J. Geophys. Res., 95,* 14,077–14,082.

Anderson, J.G. (1987). Free radicals in the earth's atmosphere: their measurements and interpretation. *Ann. Rev. Phys. Chem., 38,* 489–520.

Anderson, J.G., Brune, W.H. and Proffitt, M.H. (1989). Ozone destruction by chlorine radicals within the Antarctic vortex: the spatial and temporal evolution of $ClO-O_3$ anticorrelation based on in situ ER-2 data. *J. Geophys. Res.*, **94**, 11,465–11,479.

Anderson, J.G., Toohey, D.W. and Brune, W.H. (1991). Free radicals within the Antarctic vortex: the role of CFCs in Antarctic ozone loss. *Science*, **251**, 39–46.

Barton, N., Rowland, B. and Devlin, J.P. (1993). Infrared spectra of large acid hydrate clusters: formation conditions of submicron particles of $HNO_3 \cdot 2H_2O$ and $HNO_3 \cdot 3H_2O$. *J. Phys. Chem.*, **97**, 5848–5851.

Birk, M., Friedl, R.R., Cohen, E.A., Pickett, H.M. and Sander, S.P. (1989). The rotational spectrum and structure of chlorine peroxide. *J. Chem. Phys.*, **91**, 6588–6597.

Brune, W.H., Anderson, J.G. and Chan, K.R. (1989a). In situ observations of BrO in the Antarctica: ER-2 aircraft results from 54° S to 72° S latitude. *J. Geophys. Res.*, **94**, 16,639–16,647.

Brune, W.H., Anderson, J.G. and Chan, K.R. (1989b). In situ observations of ClO in the Antarctic: ER-2 aircraft results from 54° S to 72° S latitude. *J. Geophys. Res.*, **94**, 16,649–16,663.

Brune, W.H., Toohey, D.W., Anderson, J.G. and Chan, K.R. (1990). In situ observations of ClO in the Arctic stratosphere: ER-2 aircraft results from 59° N to 80° N latitude. *Geophys. Res. Lett.*, **17**, 505–508.

Callis, L.B. and Natarajan, M. (1986). The Antarctic ozone minimum: relationship to odd nitrogen, odd chlorine, the final warming, and the 11-year solar cycle. *J. Geophys. Res.*, **91**, 10,771–10,796.

Carpenter, C.D. and Lehrman, A. (1925). The solid–liquid temperature relation in the ternary system $H_2SO_4-HNO_3-H_2O$ and its relation to the ternary system $N_2O_5-H_2O-SO_3$. *Trans. Am. Inst. Chem. Eng.*, **17**, 35–73.

Carroll, M.A., Sanders, R.W., Solomon, S. and Schmeltekopf, A.L. (1989). Visible and near-ultraviolet spectroscopy at McMurdo station, Antarctica 6. Observations of BrO. *J. Geophys. Res.*, **94**, 16,633–16,638.

Chance, K.V., Johnson, D.G., Traub, W.A. and Jucks, K.W. (1991). Measurements of the stratospheric hydrogen peroxide concentration profile using far infrared thermal emission spectroscopy. *Geophys. Res. Lett.*, **18**, 1003–1006.

Cheng, B.-M. and Lee, Y.P. (1989). Production and trapping of gaseous dimeric ClO: the infrared spectrum of chlorine peroxide (ClOOCl) in solid argon. *J. Chem. Phys.*, **90**, 5930–5935.

Coffey, M.T., Mankin, W.G. and Goldman, A. (1989). Airborne measurements of stratospheric constituents over Antarctica in the austral spring 1987, 2. Halogen and nitrogen trace gases. *J. Geophys. Res.*, **94**, 16,597–16,613.

Cox, R.A. and Hayman, G.D. (1988). The stability and photochemistry of dimers of the ClO radical and implications for Antarctic ozone depletion. *Nature*, **332**, 796–800.

Crutzen, P.J. (1970). The influence of nitrogen oxides on atmospheric ozone content. *Q. J. R. Meteorol. Soc.*, **96**, 320–325.

Crutzen, P.J. (1979). The role of NO and NO_2 in the chemistry of the troposphere and stratosphere. *Ann. Rev. Earth Planet. Sci.*, **7**, 443–472.

Crutzen, P.J. and Arnold, F. (1986). Nitric acid cloud formation in the cold Antarctic stratosphere: a major cause for the springtime ozone hole. *Nature*, **324**, 651–655.

DeMore, W.B., Sander, S.P., Golden, D.M., Hampson, R.F., Kurylo, M.J., Howard, C.J.,

Ravishankara, A.R., Kolb, C.E. and Molina, M.J. (1992). Chemical kinetics and photochemical data for use in stratospheric modeling; Evaluation No.10. *JPL Publ. 92–20.* Jet Propulsion Laboratory, Pasadena, CA.

Deshler, T., Adriani, A., Gobbi, G.P., Hofmann, D.J., Di Donfrancesco, G. and Johnson, B.J. (1992). Volcanic aerosol and ozone depletion within the Antarctic polar vortex during the austral spring of 1991. *Geophys. Res. Lett.,* **19**, 1819–1822.

de Zafra, R.L., Jaramillo, M., Parrish, A., Solomon, P., Connor, B. and Barrett, J. (1987). High concentrations of chlorine monoxide at low altitudes in the Antarctic spring stratosphere: diurnal variation. *Nature,* **328**, 408–413.

de Zafra, R.L., Parrish, A., Barrett, J. and Solomon, P. (1985). An observed upper limit on stratospheric hydrogen peroxide. *J. Geophys. Res.,* **90**, 13,087–13,090.

de Zafra, R.L., Parrish, A., Solomon, P.M. and Barrett, J.W. (1984). A measurement of stratospheric HO_2 by ground-based millimeter-wave spectroscopy. *J. Geophys. Res.,* **89**, 1321–1326.

Drummond, J.R. and Jarnot, R.F. (1978). Infrared measurements of stratospheric composition II. Simultaneous NO and NO_2 measurements. *Proc. R. Soc. Lond. A.,* **364**, 237–254.

Eluszkiewicz, J. and Allen, M. (1993). A global analysis of the ozone deficit in the upper stratosphere and lower mesosphere. *J. Geophys. Res.* **98**, 1069–1082.

Fahey, D.W., Kawa, S.R. and Chan, K.R. (1990a). Nitric oxide measurements in the Arctic winter stratosphere. *Geophys. Res. Lett.,* **17**, 489–492.

Fahey, D.W., Kawa, S.R., Woodbridge, E.L., Tin, P., Wilson, J.C., Jonsson, H.H., Dye, J.E., Baumgardner, D., Borrmann, S., Toohey, D.W., Avallone, L.M., Proffitt, M.H., Margitan, J., Loewenstein, M., Podolske, J.R., Salawitch, R.J., Wofsy, S.C., Ko, M.K.W., Anderson, D.E., Schoeberl, M.R. and Chan, K.R. (1993). In situ measurements constraining the role of sulfate aerosols in mid-latitude ozone depletion. *Nature,* **363**, 509–514.

Fahey, D.W., Kelly, K.K., Kawa, S.R., Tuck, A.F., Loewenstein, M., Chan, K.R., and Heidt, L.E. (1990b). Observations of denitrification and dehydration in the winter polar stratospheres. *Nature,* **344**, 321–324.

Fahey, D.W., Murphy, D.M., Kelly, K.K., Ko, M.K.W., Proffitt, M.H., Eubank, C.S., Ferry, G.C., Loewenstein, M. and Chan, K.R. (1989). Measurements of nitric oxide and total reactive nitrogen in the Antarctic stratosphere: observations and chemical implications. *J. Geophys. Res.,* **94**, 16,665–16,681.

Fahey, D.W., Solomon, S., Kawa, S.R., Loewenstein, M., Podolske, J.R., Strahan, S.E. and Chan, K.R. (1990c). A diagnostic for denitrification in the winter polar stratospheres. *Nature,* **345**, 698–702.

Farman, J.C., Gardiner, B.G. and Shanklin, J.D. (1985). Large losses of total ozone in Antarctica reveal seasonal ClO_x/NO_x interaction. *Nature,* **315**, 207–210.

Farmer, C.B. and Raper, O.F. (1977). The HF:HCl ratio in the 14–38 km region of the stratosphere. *Geophys. Res. Lett.,* **4**, 527–529.

Farmer, C.B., Raper, O.F. and O'Callaghan, F.G. (1987a). Final report on the first flight of the ATMOS instrument during the spacelab 3 mission, April 29 through May 6, 1985. *JPL Publ. 87–32,* Jet Propulsion Laboratory, Pasadena, CA.

Farmer, C.B., Toon, G.C., Schaper, P.W., Blavier, J.F. and Lowes, L.L. (1987b). Stratospheric trace gases in the spring 1986 Antarctic atmosphere. *Nature,* **329**,

126–130.

Friedl, R.R. and Sander, S.P. (1989). Kinetics and product studies of the reaction ClO + BrO using discharge-flow mass spectrometry. *J. Phys. Chem.,* **93**, 4756–4764.

Gleason, J.F., Bhartia, P.K., Herman, J.R., McPeters, R., Newman, P., Stolarski, R.S., Flynn, L., Labow, G., Larko, D., Seftor, C., Wellemeyer, C., Komhyr, W.D., Miller, A.J. and Planet, W. (1993). Record low global ozone in 1992. *Science,* **260**, 523–526.

Hamill, P. and Toon, O.B. (1991). Polar stratospheric clouds and the ozone hole. *Phys. Today,* **44**, 34–42.

Hanson, D. (1990). The vapor pressures of supercooled HNO_3/H_2O solutions. *Geophys. Res. Lett.,* **17**, 421–423.

Hanson, D. and Mauersberger, K. (1988a). Vapor pressures of HNO_3/H_2O solutions at low temperatures. *J. Phys. Chem.,* **92**, 6167–6170.

Hanson, D. and Mauersberger, K. (1988b). Laboratory studies of the nitric acid trihydrate: implications for the south polar stratosphere. *Geophys. Res. Lett.,* **15**, 855–858.

Hanson, D.R. and Ravishankara, A.R. (1991a). The reaction probabilities of $ClONO_2$ and N_2O_5 on polar stratosphere cloud materials. *J. Geophys. Res.,* **96**, 5081–5090.

Hanson, D.R. and Ravishankara, A.R. (1991b). The reaction probabilities of $ClONO_2$ and N_2O_5 on 40 to 75% sulfuric acid solutions. *J. Geophys. Res.* **96**, 17,307–17,314.

Hanson, D.R. and Ravishankara, A.R. (1992). Investigation of the reactive and nonreactive processes involving $ClONO_2$ and HCl on water and nitric acid doped ice. *J. Phys. Chem.,* **96**, 2682–2691.

Helten, M., Patz, W., Trainer, M., Fark, H., Klein, E. and Ehhalt, D.H. (1984). Measurements of stratospheric HO_2 and NO_2 by matrix isolation and E.S.R. spectroscopy. *J. Atmos. Chem.,* **2**, 191–202.

Hills, A.J., Cicerone, R.J., Calvert, J.G. and Birks, J.W. (1987) Kinetics of the BrO + ClO reactions and implications for stratospheric ozone. *Nature,* **328**, 405–408.

Hofmann, D.J., Harder, J.W., Rolf, S.R. and Rosen, J.M. (1987). Balloon-borne observations of the development and vertical structure of the Antarctic ozone hole in 1986. *Nature,* **326**, 59–62.

Hofmann, D.J. and Solomon, S. (1989). Ozone destruction through heterogeneous chemistry following the eruption of El Chichon. *J. Geophys. Res.,* **94**, 5029–5041.

HSRP/AESA (1993). 1993 HSRP/AESA interim assessment. *NASA Ref. Publ. 1333*, National Aeronautics and Space Administration, Washington, DC, in press.

Johnston, H.S. (1971). Reduction of stratospheric ozone by nitrogen oxide catalysts from supersonic transport exhaust. *Science,* **173**, 517–522.

Johnston, H.S., Kinnison, D.E. and Wuebbles, D.J. (1989). Nitrogen oxides from high-altitude aircraft: an update of potential effects on ozone. *J. Geophys. Res.,* **94**, 16351–16363.

Kley, D. and Stone, E.J. (1978). A measurement of water vapor in the stratosphere by photodissociation with Ly(α) (1216 Å) light. *Rev. Sci. Instrum.,* **49**, 691–697.

Koehler, B.G., Middlebrook, A.M. and Tolbert, M. (1992). Characterization of model polar stratospheric cloud films using Fourier transform infrared spectroscopy and temperature programmed desorption. *J. Geophys. Res.,* **97**, 8065–8074.

Kolb, C.E., Worsnop, D.R., Zahniser, M.S., Davidovits, P., Keyser, L.F., Leu, M.-T., Molina, M.J., Hanson, D.R., Ravishankara, A.R., Williams, L.R. and Tolbert, M.R. (1993). Laboratory studies of atmospheric heterogeneous chemistry. In Barker, J.R.,

Ed., *Current Problems in Atmospheric Chemistry*, in press.

Kondo, Y., Aimedieu, P., Pirre, M., Matthews, W.A., Ramaroson, R., Sheldon, W.R., Benbrook, J.R. and Iwata, A. (1990). Diurnal variation of nitric oxide in the upper stratosphere. *J. Geophys. Res., 95*, 22,513–22,522.

Leu, M.T. (1988a). Laboratory studies of sticking coefficients and heterogeneous reactions important in the Antarctic stratosphere. *Geophys. Res. Lett., 15*, 17–20.

Leu, M.T. (1988b). Heterogeneous reactions of N_2O_5 with H_2O and HCl on ice surfaces: implications for Antarctic ozone depletion. *Geophys. Res. Lett., 15*, 851–854.

Leu, M.-T., Moore, S.B. and Keyser, L.F. (1991). Heterogeneous reactions of chlorine nitrate and hydrogen chloride on type I polar stratospheric clouds. *J. Phys. Chem., 95*, 7763–7771.

Louisnard, N., Fergant, G., Girard, A., Gramont, L., Lado-Bordowsky, O., Laurent, J., LeBoiteux, S. and Lemaitre, M.P. (1983). Infrared absorption spectroscopy applied to stratospheric profiles of minor constituents. *J. Geophys. Res., 88*, 5365–5376.

McElroy, M.B. and Salawitch, R.J. (1989). Changing composition of the global stratosphere. *Science, 243*, 763–770.

McElroy, M.B., Salawitch, R.J. and Minschwaner, K. (1992). The changing stratosphere. *Planet. Space Sci., 40*, 373–401.

McElroy, M.B., Salawitch, R.J., Wofsy, S.C. and Logan, J.A. (1986) Reduction of Antarctic ozone due to synergistic interactions of chlorine and bromine. *Nature, 321*, 759–762.

McGrath, M.P., Clemitshaw, K.C., Rowland, F.S. and Hehre, W.J. (1990). Structures, relative stabilities, and vibrational spectra of isomers of Cl_2O_2: the role of chlorine oxide dimer in Antarctic ozone depleting mechanisms. *J. Phys. Chem., 94*, 6126–6132.

Mankin, W.G. and Coffey, M.T. (1983). Latitudinal distributions and temporal changes of stratospheric HCl and HF. *J. Geophys. Res., 88*, 10776–10784.

Mankin, W.G., Coffey, M.T., Goldman, A., Schoeberl, M.R., Lait, L.R. and Newman, P.A. (1990). Airborne measurements of stratospheric constituents over the Arctic in the winter of 1989. *Geophys. Res. Lett., 17*, 473–476.

Manzer, L.E. (1990). The CFC–ozone issue: progress on the development of alternatives to CFCs. *Science, 249*, 31–35.

Massie, S.T., Davidson, J.A., Cantrell, C.A., McDaniel, A.H., Gille, J.C., Kunde, V.G., Brasunas, J.C., Conrath, B.J., Maguire, W.C., Goldman, A. and Abbas, M.M. (1987). Atmospheric infrared emission of $ClONO_2$ observed by a balloon-borne Fourier spectrometer. *J. Geophys. Res., 92*, 14,806–14,814.

Masterbrook, H.J. and Oltmans, S.J. (1983). Stratospheric water vapor variability for Washington, DC/Boulder, CO: 1964–82. *J. Atmos. Sci. 40*, 2157–2165.

Middlebrook, A.M., Koehler, B.G., McNeill, L.S. and Tolbert, M.A. (1992). Formation of model polar stratospheric cloud films. *Geophys. Res. Lett., 24*, 2417–2420.

Mihelcic, D., Ehhalt, D.H., Kulessa, G.F. Klomfass, J., Trainer, M., Schmidt, U. and Rohrs, H. (1978). Measurements of free radicals in the atmosphere by matrix isolation and electron paramagnetic resonance. *Pure Appl. Geophys., 116*, 530–536.

Molina, L.T. and Molina, M.J. (1987). Production of Cl_2O_2 from the self-reaction of the ClO radical. *J. Phys. Chem., 91*, 433–436.

Molina, L.T., Molina, M.J., Stachnik, R.A. and Tom, R.D. (1985). An upper limit to the rate of the HCl + ClONO$_2$ reaction. *J. Phys. Chem.,* **89**, 3779–3781.

Molina, M.J. (1991) Heterogeneous chemistry on polar stratospheric clouds. *Atmos. Environ.,* **25A**, 2535–2537.

Molina, M.J. (1992). In Calvert, J.G., Ed., *The Chemistry of the Atmosphere: Its Impact on Global Change.* Blackwell Sci. Publ., Oxford, 1994.

Molina, M.J., Colussi, A.J., Molina, L.T., Schindler, R.N. and Tso, T.-L. (1990) . Quantum yield of chlorine-atom formation in the photodissociation of chlorine peroxide (ClOOCl) at 308 nm. *Chem. Phys. Lett.,* **173**, 310–315.

Molina, M.J. and Rowland, F.S. (1974). Stratospheric sink for chlorofluoromethanes: chlorine atom-catalyzed destruction of ozone. *Nature,* **249**, 810–812.

Molina, M.J., Tso, T.-L., Molina, L.T. and Wang, F.C.-Y. (1987) Antarctic stratospheric chemistry of chlorine nitrate, hydrogen chloride, and ice: Release of active chlorine. *Science,* **238**, 1253–1257.

Molina, M.J., Zhang, R., Wooldridge, P.J., McMahon, J.R., Kim., J.E., Chang, H.Y. and Beyer, K. (1993). Physical chemistry of the H$_2$SO$_4$/HNO$_3$/H$_2$O system: implication for the formation of polar stratospheric clouds and heterogeneous chemistry. *Science,* **261**, 1418–1423.

Moore, S.B., Keyser, L.F., Leu, M.T., Turco, R.P. and Smith, R.H. (1990). Heterogeneous reactions on nitric acid trihydrate. *Nature,* **345**, 333–335.

Mount, G.H. and Eisele, F.L. (1992). An intercomparison of tropospheric OH measurements at Fritz Peak Observatory, Colorado. *Science,* **256**, 1187–1190.

Mozurkewich, M. and Calvert, J.G. (1988). Reaction probability of N$_2$O$_5$ on aqueous aerosols. *J. Geophys. Res.,* **93**, 15,889–15,896.

NRC (National Research Council) (1982). *Causes and Effects of Stratospheric Ozone Reduction: An Update.* National Academy Press, Washington, DC.

Park, J.H. and Carli, B. (1991). Spectroscopic measurements of HO$_2$, H$_2$O$_2$, and OH in the stratosphere. *J. Geophys. Res.,* **96**, 22,535–22,541.

Prather, M.J. and Watson, R.T. (1990). Stratospheric ozone depletion and future levels of atmospheric chlorine and bromine. *Nature,* **344**, 729–734.

Prinn, R., Cunnold, D., Simmonds, P., Alyea, F., Boldi, R., Crawford, A., Fraser, P., Gutzler, D., Hartley, D., Rosen, R. and Rasmussen, R. (1992). Global average concentration and trend for hydroxyl radicals deduced from ALE/GAGE trichloro-ethane (methyl chloroform) data for 1978–1990. *J. Geophys. Res.,* **97**, 2445–2461.

Raper, O.F., Farmer, C.B., Zander, R. and Park, J.H. (1987). Infrared spectroscopic measurements of halogenated sink and reservoir gases in the stratosphere with the ATMOS instrument. *J. Geophys. Res.,* **92**, 9851–9858.

Reber, C.A., Trevathen, C.E., McNeal, R.J. and Luther, M.R. (1993). The Upper Atmosphere Research Satellite (UARS) mission. *J. Geophys. Res.,* **98**, 10,643–10,647 (pp. 10,643–108,14 comprise a special section on UARS in this issue).

Reihs, C.M., Golden, D.M. and Tolbert, M.A. (1990). Nitric acid uptake by sulfuric acid solutions under stratospheric conditions: determination of Henry's law solubility. *J. Geophys. Res.,* **95**, 16,545–16,550.

Ridley, B.A., McFarland, M., Schmeltekopf, A.L., Proffitt, M.H., Albritton, D.L., Winkler, R.H. and Thompson, T.L. (1987). Seasonal differences in the vertical distributions of NO, NO$_2$, and O$_3$ in the stratosphere near 50° N. *J. Geophys. Res.,* **92**, 11,919–11,929.

Rinsland, C.P., Levine, J.S., Goldman, A., Sze, N.D., Ko, M.K.W. and Johnson, D.W. (1991). Infrared measurements of HF and HCl total column abundances above Kitt Peak, 1977–1990: seasonal cycles, long-term increases, and comparisons with model calculations. *J. Geophys. Res.,* **96**, 15,523–15,540.

Ritzhaupt, G. and Devlin, J.P. (1991). Infrared spectra of nitric and hydrochloric acid-hydrate thin films. *J. Phys. Chem.,* **90**, 90–95.

Roche, A.E., Kumer, J.B., Mergenthaler, J.L., Ely, G.A., Uplinger, W.G., Potter, J.F., James, T.C. and Sterritt, L.W. (1993). The Cryogenic Limb Array Etalon Spectrometer (CLAES) on UARS: experiment description and performance. *J. Geophys. Res.,* **98**, 10,763–10,775.

Roscoe, H.K., Kerridge, B.J., Gray, L.J., Wells, R.J. and Pyle, J.A. (1986). Simultaneous measurements of stratospheric NO and NO_2 and their comparison with model predictions. *J. Geophys. Res.,* **91**, 5405–5419.

Roscoe, H.K. (1991). Review and revision of measurements of stratospheric N_2O_5. *J. Geophys. Res.,* **96**, 10,879–10,884.

Rowland, F.S. and Molina, M.J. (1975). Chlorofluoromethanes in the environment. *Rev. Geophys. Space Phys.,* **13**, 1–35.

Rowland, F.S., Sato, H., Khwaja, H. and Elliott, S.M. (1986). The hydrolysis of chlorine nitrate and its possible atmospheric significance. *J. Phys. Chem.,* **90**, 1985–1988.

Russell, J.M. III, Farmer, C.B., Rinsland, C.P., Zander, R., Froidevaux, L., Toon, G.C., Gao, B., Shaw, J. and Gunson, M. (1988). Measurements of odd nitrogen compounds in the stratosphere by the ATMOS experiment on spacelab 3. *J Geophys. Res.,* **93**, 1718–1736.

Sanders, R.W., Solomon, S., Carroll, M.A. and Schmeltekopf, A.L. (1989). Visible and near-ultraviolet spectroscopy at McMurdo station, Antarctica 4. Overview and daily measurements of NO_2, O_3, and OClO during 1987. *J. Geophys. Res.,* **94**, 11,381–11,391.

Schiller, C., Wahner, A., Platt, U., Dorn, H.-P., Callies, J. and Ehhalt, D.H. (1990). Near UV atmospheric measurements of column abundances during airborne Arctic stratospheric expedition, January–February 1989: 2. OClO observations. *Geophys. Res. Lett.,* **17**, 501–504.

Schoeberl, M.R. and Hartmann, D.L. (1991). The dynamics of the stratospheric polar vortex and its relation to springtime ozone depletions. *Science,* **251**, 46–52.

Smith, R.H., Leu, M.-T. and Keyser, L.F. (1991). Infrared spectra of solid films formed from vapors containing water and nitric acid. *J. Phys. Chem.,* **95**, 5924–5930.

Solomon, P., de Zafra, R., Parrish, A. and Barrett, J.W. (1984). Diurnal variation of stratospheric chlorine monoxide: a critical test of chlorine chemistry in the ozone layer. *Science,* **224**, 1210–1214.

Solomon, S. (1990). Progress towards a quantitative understanding of Antarctic ozone depletion. *Nature,* **347**, 347–354.

Solomon, S. and Albritton, D.L. (1992). Time-dependent ozone depletion potentials for short- and long-term forecasts. *Nature,* **357**, 33–37.

Solomon, S., Garcia, R.R., Sherwood, F.S. and Wuebbles, D.J. (1986). On the depletion of Antarctic ozone. *Nature,* **321**, 755–758.

Solomon, S., Miller, H.L., Smith, J.P., Sanders, R.W., Mount, G.H., Schmeltekopf, A.L. and Noxon, J.F. (1989a). Atmospheric NO_3 1. Measurement technique and the annual cycle at 40° N. *J. Geophys. Res., 94,* 11,041–11,048.

Solomon, S., Mount, G.H., Sanders, R.W. and Schmeltekopf, A.L. (1987) Visible spectroscopy at McMurdo Station, Antarctica, 2. Observations of OClO. *J. Geophys. Res., 92,* 8329–8338.

Solomon, S., Sanders, R.W., Carroll, M.A. and Schmeltekopf, A.L. (1989b). Visible and near-ultraviolet spectroscopy at McMurdo station, Antarctica 5. Observations of the diurnal variations of BrO and OClO. *J. Geophys. Res., 94,* 11,393–11,403.

Solomon, S., Sanders, R.W., Garcia, R.R. and Keys, J.G. (1993) Increased chlorine dioxide over Antarctica caused by volcanic aerosols from Mount Pinatubo. *Nature, 363,* 245–248.

Stachnik, R.A., Hardy, J.C., Tarsala, J.A., Waters, J.W. and Erickson, N.R. (1992). Submillimeterwave heterodyne measurements of stratospheric ClO, HCl, O_3 and HO_2: first results. *Geophys. Res. Lett., 19,* 1931–1934.

Stimpfle, R.M., Wennberg, P.O., Lapson, L.B. and Anderson, J.G. (1990). Simultaneous, in situ measurements of OH and HO_2 in the stratosphere. *Geophys. Res. Lett., 17,* 1905–1908.

Stockwell, W.R. and Calvert, J.G. (1983). The mechanism of the $HO–SO_2$ reaction. *Atmos. Environ., 17,* 2231–2235.

Stolarski, R.S. and Cicerone, R.J. (1974). Stratospheric chlorine: a possible sink for ozone. *J. Can. Chem., 52,* 1610–1615.

Stolarski, R., Bojkov, R., Bishop, L., Christos, Z., Staehelin, J. and Zawodny, J. (1992). Measured trends in stratospheric ozone. *Science, 256,* 342–349.

Tabazadeh, A. and Turco, R.P. (1993). Stratosphere chlorine injection by volcanic eruptions: HCl scavenging and implications for ozone. *Science, 260,* 1082–1086.

Tolbert, M.A. and Middlebrook, A.M. (1990). Fourier transform infrared studies of model polar stratospheric cloud surfaces: growth and evaporation of ice and nitric acid/ice. *J. Geophys. Res., 95,* 22,423–22,431.

Tolbert, M.A., Rossi, M.J. and Golden, D.M. (1988a). Heterogeneous interactions of chlorine nitrate with hydrogen chloride, and nitric acid with sulfuric acid surfaces at stratospheric temperatures. *Geophys. Res. Lett., 15,* 847–850.

Tolbert, M.A., Rossi, M.J. and Golden, D.M. (1988b). Antarctic ozone depletion chemistry: reactions of N_2O_5 with H_2O and HCl on ice surfaces. *Science, 240,* 1018–1021.

Tolbert, M.A., Rossi, M.J., Malhotra, R. and Golden, D.M. (1987). Reaction of chlorine nitrate with hydrogen chloride and water at Antarctic stratospheric temperature. *Science, 238,* 1258–1260.

Toohey, D.W., Anderson, J.G., Brune, W.H. and Chan, K.R. (1990). In situ measurements of BrO in the Arctic stratosphere. *Geophys. Res. Lett., 17,* 513–516.

Toon, G.C. and Farmer, C.B. (1989). Detection of HOCl in the Antarctic stratosphere. *Geophys. Res. Lett., 16,* 1375–1377.

Toon, G.C., Farmer, C.B., Lowes, L.L., Schaper, P.W., Blavier, J.-F. and Norton, R.H. (1989). Infrared aircraft measurements of stratospheric composition over Antarctica during September 1987. *J. Geophys. Res., 94,* 16,571–16,596.

Toon, O.B., Growell, E., Gary, B., Bait, L., Newman, P., Pueschel, R., Russell, P., Schoberl, M., Toon, G., Traub, W., Valero, F., Selkirk, H. and Jordan, J. (1993). Heterogeneous reaction probabilities, solubilities, and physical state of cold volcanic aerosols. *Science,* **261**, 1136–1140.

Toon, O.B., Hamill, P., Turco, R.P. and Pinto, J. (1986). Condensation of HNO_3 and HCl in the winter polar stratospheres. *Geophys. Res. Lett.,* **13**, 1284–1287.

Toumi, R., Kerridge, B.J. and Pyle, J.A. (1991). Highly vibrationally excited oxygen as a potential source of ozone in the upper stratosphere and mesosphere. *Nature,* **351**, 217–219.

Traub, W.A., Johnson, D.G. and Chance, K.V. (1990). Stratospheric hydroperoxyl measurements. *Science,* **247**, 446–449.

Tung, K.-K., Ko, M.K.W., Rodriguez, J.M. and Sze, N.D. (1986). Are Antarctic ozone variations a manifestation of dynamics or chemistry. *Nature,* **322**, 811–814.

Turco, R.P., Toon, O.B. and Hamill, P. (1989). Heterogeneous physicochemistry of the polar ozone hole. *J. Geophys. Res.,* **94**, 16,493–16,510.

Turnipseed, A.A., Birks, J.W. and Calvert, J.G. (1991). Kinetics and temperature dependence of the BrO + ClO reaction. *J. Phys. Chem.,* **95**, 4356–4364.

Van Doren, J.M., Watson, L.R., Davidovits, P., Worsnop, D.R., Zahniser, M.S. and Kolb, C.E. (1991). Uptake of N_2O_5 and HNO_3 by aqueous sulfuric acid droplets. *J. Phys. Chem.,* **95**, 1684–1689.

Wahner, A., Callies, J., Dorn, H.-P., Platt, U. and Schiller, C. (1990a). Near UV atmospheric absorption measurements of column abundances during airborne Arctic stratospheric expedition, January–February 1989: 1. Techniques and NO_2 observations. *Geophys. Res. Lett.,* **17**, 497–500.

Wahner, A., Callies, J., Dorn, H.-P., Platt, U. and Schiller, C. (1990b). Near UV atmospheric absorption measurements of column abundances during airborne Arctic stratospheric expedition, January–February 1989: 3. BrO observations. *Geophys. Res. Lett.,* **17**, 517–520.

Wahner, A., Jakoubek, R.O, Mount, G.H., Ravishankara, A.R. and Schmeltekopf, A.L. (1989a). Remote sensing observations of nighttime OClO column during the airborne Antarctic ozone experiments, September 8, 1987. *J. Geophys. Res.,* **94**, 11,405–11,411.

Wahner, A., Jakoubek, R.O., Mount, G.H., Ravishankara, A.R. and Schmeltekopf, A.L. (1989b). Remote sensing observations of daytime column NO_2 during the airborne Antarctic ozone experiments August 22 to October 2, 1987. *J. Geophys. Res.,* **94**, 16,619–16,632.

Waters, J.W., Froidevaux, L., Read, W.G., Manney, G.L., Elson, L.S., Flower, D.A., Jarnot, R.F. and Harwood, R.S. (1993). Stratospheric ClO and ozone from the microwave limb sounder on the upper atmosphere research satellite. *Nature,* **362**, 597–602.

Waters, J.W., Hardy, J.C., Jarnot, R.F. and Pickett, H.M. (1981). Chlorine monoxide radical, ozone, and hydrogen peroxide: stratospheric measurements by microwave limb sounding. *Science,* **214**, 61–64.

Waters, J.W., Stachnik, R.A., Hardy, J.C. and Jarnot, R.F. (1988). ClO and O_3 stratospheric profiles: balloon microwave measurements. *Geophys. Res. Lett.,* **15**, 780–783.

Webster, C.R. and May, R.D. (1987). Simultaneous in situ measurements and diurnal variations of NO, NO_2, O_3, jNO_2, CH_4, H_2O, and CO_2 in the 40- to 26-km region using an open path tunable diode laser spectrometer. *J. Geophys. Res.,* **92**, 11,931–11,950.

Webster, C.R., May, R.D., Toumi, R. and Pyle, J.A. (1990). Active nitrogen partitioning and the nighttime formation of N_2O_5 in the stratosphere: simultaneous in situ measurements of NO, NO_2, HNO_3, O_3 and N_2O using the BLISS diode laser spectrometer. *J. Geophys. Res., 95*, 13,851–13,866.

Weinstock, E.M., Phillips, M.J. and Anderson, J.G. (1981). In situ observations of ClO in the stratosphere: a review of recent results. *J. Geophys. Res., 86*, 7273–7278.

Wennberg, P.O., Stimpfle, R.M., Weinstock, E.M., Dessler, A.E., Lloyd, S.A,. Lapson, L.B., Schwab, J.J. and Anderson, J.G. (1990). Simultaneous in situ measurements of OH, HO_2, and H_2O: A test of modeled stratospheric HO_x chemistry. *Geophys. Res. Lett., 17*, 1909–1912.

WMO (World Meteorological Organization) (1986). Atmospheric ozone 1985. *WMO Global Ozone Research and Monitoring Project, Report No. 16.*

WMO (1990a). Report of International Ozone Trends Panel 1988. *WMO Global Ozone Research and Monitoring Project, Report No. 18.*

WMO (1990b). Scientific assessment of stratospheric ozone: 1989. *WMO Global Ozone Research and Monitoring Project, Report No. 20.*

WMO (1992). Scientific assessment of stratospheric ozone: 1991. *WMO Global Ozone Research and Monitoring Project, Report No. 25.*

Worsnop, D.R., Fox, L.E., Zahniser, M.S. and Wofsy, S.C. (1993). Vapor pressures of solid hydrates of nitric acid: Implications for polar stratospheric clouds. *Science, 259*, 71–74.

Wuebbles, D.J. (1983). Chlorocarbon emission scenarios: potential impact on stratospheric ozone. *J. Geophys. Res., 88*, 1433–1433.

Zander, R., Gunson, M.R., Foster, J.C., Rinsland, C.P. and Namkung, J. (1990). Stratospheric $ClONO_2$, HCl, and HF concentration profiles derived from Atmospheric Trace Molecule Spectroscopy Experiment Spacelab 3 observations: an update. *J. Geophys. Res., 95*, 20,519–20,525.

Zhang, R., Wooldridge, P.J., Abbatt, J.P.D. and Molina, M.J. (1993a). Physical chemistry of the H_2SO_4/H_2O binary system at low temperatures: stratospheric implications. *J. Phys. Chem., 97*, 7351–7358.

Zhang, R., Wooldridge, P.J. and Molina, M.J. (1993b). Vapor pressure measurements for the $H_2SO_4/HNO_3/H_2O$ and $H_2SO_4/HCl/H_2O$ systems: incorporation of stratospheric acids into background sulfate aerosols. *J. Phys. Chem., 97*, 8541–8548.

12

Acidic precipitation

Jeremy M. Hales

12.1. INTRODUCTION AND SCOPE

"Acid rain", first noted in the middle part of the 19th century, is the popular term for the wet and dry deposition of acidic substances. Over the past two decades the phenomenon of man-induced acid deposition has evolved into an issue of considerable public concern. This complex phenomenon is the end-product of a series of processes involving emissions of precursor chemicals (especially sulfur and nitrogen oxides), their transport and chemical transformation to acidic substances, and eventual wet or dry deposition. Concern about the effects of acid deposition have included damage to crops, forests, and aquatic life.

Acidic precipitation is a commonly used term[1] that is applied to describe the chemical nature of rain, snow, and cloud-water, most often under the influence of human-induced atmospheric pollution. This term is somewhat misleading, because the acidity (i.e., the free hydrogen ion content) of precipitation results from a chemical balance that typically involves multiple species—some of which are acidic, while others are basic. While chemically contaminated rain-water is indeed acidic in most cases, it is not uncommon to encounter conditions involving very "dirty" rain that contains considerable amounts of basic material and thus has low acidity. Consequently, acidity is of limited usefulness as a general index of precipitation quality. While there is no doubt that any highly acidic precipitation sample is chemically contaminated, there is no guarantee that samples having near-neutral acidity levels are particularly clean, without verification by more complete chemical analysis.

This relationship between acidity and chemical make-up can be expressed in terms of an ion balance:

$$\Sigma \text{anion equivalents} = \Sigma \text{ cation equivalents} \tag{12.1}$$

For example, if a rain-water sample contains only SO_4^{2-}, SO_3^{2-}, NO_3^-, Cl^-, HCO_3^-, and CO_3^{2-} as its externally donated anions, and Na^+, NH_4^+, and Ca^{2+} as its corresponding cations, then an appropriate ion balance would be

$$[H^+] + [Na^+] + [NH_4^+] + 2[Ca^{2+}] = 2[SO_4^{2-}] + 2[SO_3^{2-}] + [NO_3^-]$$
$$+ [Cl^-] + [OH^-] + [HCO_3^-]$$
$$+ 2[CO_3^{2-}] \tag{12.2}$$

where the bracketed quantities denote molar concentrations. In this manner hydrogen ion concentration is dictated by the composite electroneutrality of all principal ions existing in the sample. Usually acidity is expressed as the pH, which is the base-10 logarithm of the molar hydrogen ion concentration, i.e.,

$$\text{pH} = \log_{10}[H^+] \tag{12.3}$$

---

[1] A somewhat more general interpretation of this term, which includes dry as well as wet deposition processes, is also applied extensively in the literature. While the present treatment will be confined to wet removal phenomena, it is emphasized that dry pathways, on average, are about equally important and must be included whenever atmospheric modeling and/or material balances are considered. This is reflected in the discussion of lifetimes, distance scales, and microscopic balances presented in Section 12.2.2. Detailed treatments of dry deposition are available elsewhere, e.g., Hicks (1990).

Water itself ionizes to a limited extent, resulting in hydrogen ion concentrations for absolutely pure water close to 10^{-7} molar (pH 7), a value that is generally (although somewhat inaccurately) considered as "neutral". Atmospheric carbon dioxide dissolves and partially ionizes in water to form additional hydrogen ion, resulting in a pH in the range of 5.5. Other naturally occurring acidic and basic species, including organics, further alter this value (see Keene and Galloway, 1984), leading to some debate regarding "typical" acidity values for pristine, background precipitation pH (see Charlson and Rhode, 1982). It is not unusual, on the other hand, to observe rain pH to be as low as 4, or even in the high 3s, in highly polluted regions of the globe. Table 12.1 gives examples of the chemical compositions of typical polluted (Pennsylvania) and relatively pristine (Bermuda) rain-water samples. Important things to note here are the seasonal differences, the obvious contribution of oceanic constituents to samples at the Bermuda site, and the shifts in relative importances of the various ions between the sites.

In view of these multiple chemical contributions, it is generally preferable to denote this general subject as *precipitation chemistry*—a term which denotes both the chemical make-up of atmospheric water and the chemical conversion processes that occur within these aqueous media. A companion term, *precipitation scavenging*, characterizes the natural process by which atmospheric pollutants are attached to cloud and precipitation particles and are consequently delivered to the Earth's surface, i.e., the phenomenon leading to the atmospheric chemistry that we observe through network measurements.

TABLE 12.1   Example Monthly Precipitation-Weighted Averages of Precipitation Chemistry Measurements at a Polluted Continental Site and an Oceanic Site

Ion	Monthly Averaged Concentration (μmol l^{-1}) State College, Pennsylvania July 1979/December 1979	Monthly Averaged Concentration (μmol l^{-1}) St Georges Harbor, Bermuda July 1980/December 1980
H^1	85/34	21.6/10.8
Na$^+$	2.1/2.1	37.8/286
K$^+$	1.2/0.69	1.30/6.93
Mg^{2+}	0.68/2.5	4.28/35.7
Ca^{2+}	2.7/5.6	12.0/7.5
NH$_4^+$	18/7	3.21/2.14
Cl$^-$	5.7/5.1	44.6/342
SO$_3^2$	0/2.5	-/-[a]
SO$_4^{2-}$	36/16	12.9/20.9 (9.7/3.65)[b]
NO$_3^-$	30/23	4.26

[a]Not measured.
[b]Compensated for sea-salt contribution.

Source: MAP35 (1982) and Church et al. (1982).

The goal of this chapter is to provide an introductory overview of both precipitation scavenging and precipitation chemistry, thus enabling the reader to appreciate the general extents of chemical perturbations to precipitation quality and the processes by which these consequences occur. This is accomplished by viewing the phenomena on a descending series of spatial scales, starting from a global-scale standpoint and ending with a microscopic analysis of how molecules and particles become incorporated with hydrometeors[2] and fall to the surface. A final section addresses the large-scale spatial distributions of atmospheric chemistry that result from these scavenging processes, and provides further indication of the impact of mankind on the precipitation environment. This chapter makes liberal use of a variety of currently available literature resources on precipitation scavenging and precipitation chemistry. In particular, two National Acid Precipitation Assessment Program (NAPAP) reports (Hicks, 1990; Sisterson, 1990) provide much of the underlying basis for this treatment, and the reader is referred to these references for more in-depth discussion of these topics.

12.2. MACROSCOPIC CYCLES AND BALANCES

12.2.1. Overview of the Scavenging Sequence

The precipitation-scavenging process can be visualized conceptually in terms of Figure 12.1, which represents the pathways of pollutants as they are released from their sources and delivered to their receptors at the Earth's surface. This overall relationship is often referred to as the *source–receptor sequence* and its general features have been discussed at length elsewhere (e.g., Butcher et al., 1992). Precipitation scavenging can be considered as a subset of this composite sequence, and in viewing the depiction of Figure 12.1, it is evident that three steps must be accomplished in order for atmospheric wet removal to occur. Specifically,

1. the pollutant particles or molecules must be brought into the presence of condensed atmospheric water—in the form of clouds, precipitation, or both;
2. the pollutant must somehow become attached to the local hydrometeors;
3. the pollutant-containing hydrometeors must be delivered to the Earth's surface.

---

[2] The term *hydrometeor* is generally defined as any airborne element of condensed water, either suspended or falling; thus cloud droplets, ice particles, and raindrops are all considered to be hydrometeors.

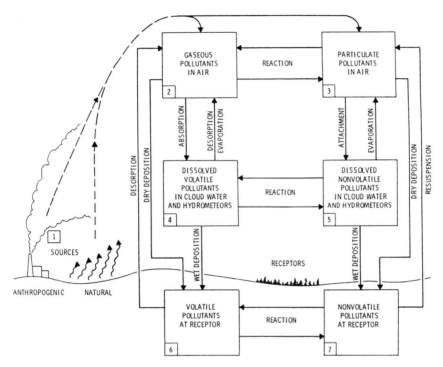

FIGURE 12.1. Schematic depiction of the atmospheric source–receptor sequence.

Furthermore, the pollutant may undergo one or more chemical reaction steps once it is in physical contact with the condensed atmospheric water. While such reactions are not absolutely necessary for scavenging to occur, they often enhance the process by stabilizing the pollutant in the hydrometeors, thus inhibiting its escape back to the gaseous phase. A prime example of this effect is the aqueous phase oxidation of dissolved SO_2 (which is volatile and desorbs easily from aqueous solution) to SO_4^{2-} (which is non-volatile and tends to remain associated with the aqueous phase).

Figure 12.1 illustrates several important features of precipitation scavenging. First, the arrows indicate that both forward and reverse processes may take place, and competitions for pollutant may occur among alternative removal pathways. Pollution may be captured from the gaseous phase by a hydrometeor, for example, but it may also be delivered back from this aqueous phase by evaporation of the hydrometeor's water or by revolatilization of the pollutant itself. Similarly, reversible chemical reactions may occur within the various condensed aqueous phases, and wet-deposited material may be re-emitted from the Earth's surface via a variety of volatilization and resuspension processes. As a consequence, precipitation scavenging and its companion steps in the atmospheric source–receptor sequence are more appropriately viewed in terms of set of composite *cyclical* process than as a series of unidirectional pathways between the source and the receptor.

12.2.2. Lifetimes, Distance Scales, and Macroscopic Balances

Figure 12.1 may be considered as the conceptual basis for a pollutant material balance as the pollutant is emitted to, processed within, and removed from the atmosphere. Although the task of supplying the appropriate numbers characterizing the various input and output rates from the component boxes is beyond the scope of this discussion, it is informative to consider briefly a simplified material balance of this type wherein the total emission–deposition sequence is represented by a single box, representing a volume V of atmosphere as shown in Figure 12.2. If Q_o is the total emission rate of pollutant to that volume and $F_o = f_w + f_d + f_r$ represents the pollutant's total removal rate (made up of wet, dry, and chemical reaction removal components), then the rate of change of pollutant mass within the volume is

$$\frac{dM_o}{dt} = V \frac{dC}{dt} = Q_o - F_o \qquad (12.4)$$

(C denotes a volume-averaged pollutant concentration). Furthermore, if steady state conditions exist such that $Q_o = F_o$, then the rate of pollutant change within the volume is zero and the mean lifetime of pollutant molecules (the average time a molecule takes between its release at the source and its removal from the system) is given (see Bolin and Rodhe, 1973) by

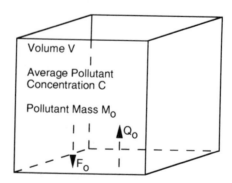

FIGURE 12.2.   Schematic of atmosphere volume applied for residence time estimation.

$$\tau = \frac{M_o}{F_o} \tag{12.5}$$

Similarly, the lifetimes for pollutants being removed by wet, dry, and chemical conversion processes are given by

$$\tau_w = \frac{M_o}{f_w}, \qquad \tau_d = \frac{M_o}{f_d}, \qquad \tau_c = \frac{M_o}{f_c} \tag{12.6}$$

leading to the relationship

$$\frac{1}{\tau} = \frac{1}{\tau_w} + \frac{1}{\tau_d} + \frac{1}{\tau_c} \tag{12.7}$$

It is important to note several of the idealizations used in this simple material balance development. First, giving the atmosphere a volume V implies that some sort of "top" is presumed to exist on the atmosphere; this is obviously an assumption of questionable validity. Furthermore, the neglect of loss terms from the box by advective outflow presumes the box is sufficiently large that all pollution emitted in the volume V is either deposited or reacted within its boundaries. This implies that, for many of the more long-lived pollutants, the box must encompass the entire planet. Moreover, the steady state assumption is absolutely necessary for equations (12.4)–(12.7) to be valid, and the reader

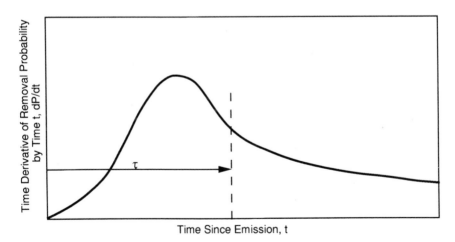

FIGURE 12.3. Hypothetical plot of a typical residence time probability density function.

should use extreme caution in applying these forms for non-steady conditions.[3] Finally, one should note that the average lifetimes presented above are single quantities that represent large and often complex distributions of pollutant lifetimes, and more complete depiction requires expression in terms of lifetime density functions such as that shown in Figure 12.3.

Regardless of these approximations and imperfections, it is of some interest to examine reported values of pollutant lifetimes. Table 12.2 summarizes typical values which are based on estimates of atmospheric loadings, removal rates, and other parameters by various authors. Combining these values with typical transport winds of a few meters per second provides an appreciation for the distances that pollution can travel before removal at a receptor. Applying an assumed $5 \, m/s^{-1}$ mean wind with Table 12.2's reported sulfur lifetime of 4.7 days, for example, suggests source–receptor distance scales of the order of $2\,000 \, km$. Owing to the large uncertainties associated with these estimates, the reader should view the values in Table 12.2 with some caution and consider them as semiquantitative indicators only. These lifetimes will be re-examined later in the context of observed spatial wet deposition distributions.

12.3. STORM-SCALE PHENOMENA

12.3.1. Cloud Formation Processes

Section 12.2.1 has noted that the first essential step of the precipitation-scavenging process involves the introduction of hydrometeors and pollution particles into common airspace. This step may occur by one (or a combination) of the following two types of processes:

A. by condensation of water vapor in the vicinity of pollution elements, or
B. by relative movement (e.g., advection, diffusional mixing) of the hydro-meteors and the pollution.

Both these processes depend strongly on storm development characteristics, and because of this, it is appropriate here to provide a brief overview of storm formation processes and storm features.

Clouds and precipitation occur because water vapor contained within the atmosphere condenses to form liquid (or solid) water. As can be noted from Figure 12.4, substantial quantities of water vapor can exist in air at warmer temperatures, but these allowable levels reduce rapidly as temperature decreases. If cooling of a moist air parcel proceeds to the point of water vapor saturation,

---

[3] Hales and Renne (1990) have extended this treatment to non-steady-state systems. As expected, the formulations are somewhat more complex under such conditions.

TABLE 12.2   Some Representative Atmospheric Residence Times of Selected Trace Contaminants

Chemical Species	Estimated Residence Time (days)	Source
Combined wet and dry deposition, and reactive depletion processes		
Total sulfur	4.7	Application of equation (12.5) to M_o and F_o estimates from Charlson et al. (1992)
Total sulfur	2.1	Global/regional budgets; Rodhe (1978)
SO_2	1.0	Global/regional budgets; Rodhe (1978)
SO_2	0.7	Marine boundary layer; Charlson et al. (1992)
SO_4^{2-}	3.3	Global/regional budgets; Rodhe (1978)
Dimethyl sulfide	1.5	Marine boundary layer; Charlson et al. (1992)
Methane sulfonic acid	0.5	Marine boundary layer; Charlson et al. (1992)
Total reactive nitrogen[a]	6.9	Application of equation (12.5) to M_o and F_o estimates from Jaffe (1992)
NO_y[b]	16	Application of equation (12.5) to M_o and F_o estimates from Jaffe (1992)
$NH_3 + NH_4^+$	2.9	Application of equation (12.5) to M_o and F_o estimates from Jaffe (1992)
Wet deposition only		
Total sulfur	3.8	Global/regional budgets; Rodhe (1978)
SO_2	4.2	Global/regional budgets; Rodhe (1978)
SO_4^{2-}	3.3	Global/regional budgets; Rodhe (1978)
Total reactive nitrogen[a]	9.2	Application of equation (12.5) to M_o and F_o estimates from Jaffe (1992)
NO_y[b]	26	Application of equation (12.5) to M_o and F_o estimates from Jaffe (1992)
$NH_3 + NH_4^+$	3.7	Application of equation (12.5) to M_o and F_o estimates from Jaffe (1992)
Dry deposition only		
Total sulfur	4.2	Global/regional budgets; Rodhe (1978)
SO_2	2.5	Global/regional budgets; Rodhe (1978)
SO_4^{2-}	Large	Global/regional budgets; Rodhe (1978)
Total reactive nitrogen[a]	27	Application of equation (12.5) to M_o and F_o estimates from Jaffe (1992)
NO_y[b]	43	Application of equation (12.5) to M_o and F_o estimates from Jaffe (1992)
$NH_3 + NH_4^+$	15	Application of equation (12.5) to M_o and F_o estimates from Jaffe (1992)
Chemical reaction only		
SO_2	3.3	Global/regional budgets; Rodhe (1978)

[a]Includes $NO_y + NH_3 + NH_4^+$.
[b]Includes all odd-nitrogen oxides, including organic nitrogen compounds.

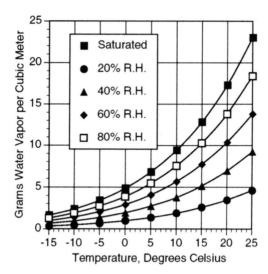

FIGURE 12.4.   Moisture content of air as a function of relative humidity and temperature.

then hydrometeor formation will occur in a manner such as to maintain the system in an approximately saturated state. Minor but significant deviations from saturation sometimes occur in cloud systems owing to transient effects of the evaporation/condensation process. This will be discussed in somewhat more detail in Section 12.4, where cloud physics processes are considered. Typical concentration ranges of the cloud-water produced by this process are indicated in Table 12.3.

Air cooling can occur by conduction of heat to cold surfaces and by radiant energy loss, but in so far as cloud and storm formation is concerned, a third cooling mechanism—that associated with expansion of air as it ascends to elevations of lower pressure—is predominant. This mechanism is a direct consequence of the first law of thermodynamics, which for present purposes can be stated as

$$\delta T = \frac{RT}{pc_p} \delta p \qquad (12.8)$$

where δ is the incremental operator, T is the absolute temperature, R is the gas law constant, p is atmospheric pressure, and c_p is the specific heat of air at constant pressure. In essence, equation (12.8) states that any parcel of air that experiences a small pressure change δp will exhibit a corresponding temperature change δT. Since air pressure diminishes with height, any ascension will result in a direct reduction of the air parcel's temperature and will lead to condensation if

TABLE 12.3   Typical Measured Values of Cloud-Water Characteristics

Cloud Type	Liquid Water Content ($g\,m^{-3}$)	Mean Droplet Radius (μm)	Droplet Concentration (droplets/ml)
Small continental cumulus			
Australia	0.4		420
USA	1.0	9	300
England	0.45	6	210
Russia	0.15	4.7	310
Cumulus congestus			
USA	2.0	24	64
Russia	0.45		95
Cumulonimbus			
USA	2.5	20	72
Small maritime cumulus			
Hawaii	0.5	15	75
Caribbean	0.4		45
Altostratus			
Russia	0.6	6.6	220
Stratus			
Hawaii	0.35	6	260
Orographic			
Hawaii	0.3		45

Source: Mason (1971).

saturation conditions are exceeded. The storm classes discussed immediately below differ from one another primarily in their various mechanisms for elevating air and thus accomplishing the cooling–condensation process.

12.3.2. Cyclonic Storm Systems

Cyclonic storm systems are so named because of their occurrence in regions of large-scale counter-clockwise or "cyclonic" airflow and can be divided roughly into two primary categories. The first of these contains the so-called *extratropical cyclones*, which arise as a consequence of discontinuities (fronts) between adjacent air masses. These are also commonly known as *frontal storms*. The second category refers to *tropical cyclones*, which are driven primarily by latent heat exchanges associated with the condensation of water in moist oceanic air. Hurricanes are tropical cyclones having surface wind speeds exceeding 120 miles h^{-1}. As their name implies, extratropical cyclones occur primarily in the mid- and high-latitudes, whereas tropical cyclones predominate to the south. Tropical cyclones generally lose energy if and when they migrate off the ocean and onto land surfaces. On occasion, however, a tropical cyclone can move northward— over land or sea—and merge with a frontal system, spawning an extratropical cyclone of unusually high intensity.

Because of preferred patterns of extratropical cyclones over the polluted regions of North America, Europe, and Asia, this storm type will be given primary emphasis here. Tropical cyclones account for substantial precipitation in areas such as India, southern Asia, and southern North America, however, and are important secondary contributors in this regard.

Figure 12.5 is an idealized schematic of a frontal system. Here the warm sector to the south can be envisioned as a wedge of warm air that has penetrated into cold northern air, resulting in sharp horizontal temperature gradients. In the northern hemisphere such systems generally move in an easterly direction, with relatively rapid movement by the westward cold front. This causes the cold front to gradually overtake and merge with the warm front, leading to the development of a so-called *occluded front* to the north as the system ages.

Idealized vertical cross-sections of warm, cold, and occluded fronts are shown in Figures 12.6(a)–12.6(c), respectively. As can be noted from the wind vectors, eastward-moving air from the warm sector is forced to move up and over the cold air to the east, resulting in ascension, cooling, and water condensation as discussed in Section 12.3.1. From a scavenging perspective one can note that

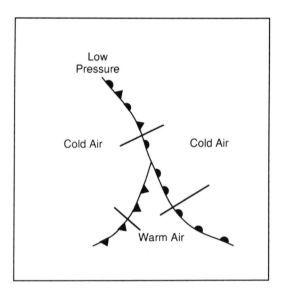

FIGURE 12.5.   Schematic of a northern, mid-latitude frontal system.

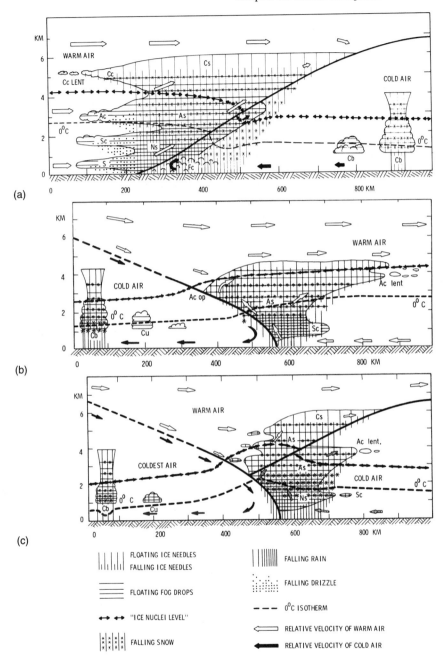

FIGURE 12.6. Schematic depictions of frontal storm cross-sections: (a) warm front; (b) cold front; (c) occluded front. (Adapted from Godske et al., 1957.)

warm sector pollution is advected to the condensation region *in the same air parcels* as the condensing water vapor, thus effecting Step 1 of the scavenging sequence (a Type A process). In contrast, pollution in lower layers of the eastward cold air move relative to the front and experience precipitation from aloft that falls *through* the pollution (a Type B process); thus Step 1 is accomplished here mainly by relative movement of the pollution and the hydrometeors. Figures 12.6(b) and 12.6(c) show similar idealizations of cold and occluded fronts, which can be used for further insights regarding Step 1. Each of these figures indicates regions of unstable air, leading to convective disturbances, as well as the motions associated directly with the frontal boundaries. This additional class of storm systems is discussed immediately below.

12.3.3. Convective Storm Systems

In its simplest possible manifestation a convective storm is initiated in an isolated air mass having an unstable temperature/humidity structure. Because of this instability, any small vertical motion results in an increase in buoyancy, which induces additional motion. The pressure drop associated with this ascension leads to cooling and ultimately to the condensation of water vapor. Release of latent heat by this condensation adds new sensible energy to the system, thus further increasing buoyancy and encouraging even greater vertical velocities and further

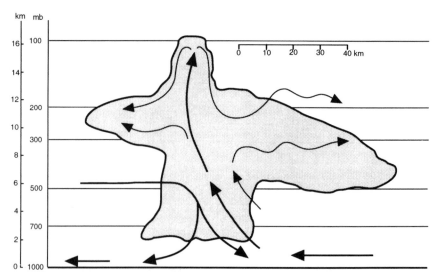

FIGURE 12.7.   Schematic vertical cross-section of a mature convective storm, showing primary flow patterns. (Adapted from Court and Griffiths, 1985.)

development of the cloud. Simultaneously, various cloud physics processes operate to produce precipitation-size particles. Buoyed by the updraft in early development stages, these particles ultimately grow sufficiently large that they fall, helping to create a storm downdraft region that partially compensates for the air proceeding upward elsewhere. Convective storms can be sufficiently energetic to reach large heights, and they actually penetrate into the stratosphere on occasion. Figure 12.7 provides a schematic cross-section of a simple convective storm at a relatively mature development stage. It should be noted that vertical air motions associated with such storms are capable of elevating boundary layer pollutants to great heights, provided that scavenging processes do not remove these pollutants en route. In the context of Step 1 of the scavenging sequence, convective storms introduce pollutants and hydrometeors to the same airspace by both Type A and Type B processes, i.e., both by local water condensation and by relative movement of the hydrometeors and pollutant.

Usually convective storms do not occur as simple isolated entities as implied by Figure 12.7, but instead appear in groups such as small clusters, lines of cells, or as massive cell complexes. Larger groups of cells, known generally as mesoscale convective complexes (MCCs), can occupy areas of several thousand square kilometers and be responsible for large amounts of precipitation.

12.3.4. Additional Cloud and Storm Types

Additional processes for cloud and storm formation include air parcel lifting by terrain (orographic clouds), land water transitions (generating "lake effect" clouds in central North America), advection of moist air over cooler land or water surfaces, and radiative cooling—either of land surfaces or of the cloud itself. Often these phenomena act as enhancements of synoptic-scale disturbances (particularly frontal storms) rather than as isolated precipitation-forming entities, although exceptions do exist. Near-surface radiation fogs and orographic clouds, for example, can deliver significant amounts of water (and water-borne pollutants) to the surface by droplet impaction and sedimentation, even in the absence of rain or large-scale weather disturbances. In a final analysis, however, each of the above cloud/storm classes provides some sort of mechanism (or combination of mechanisms) to effect Step 1 of the scavenging sequence, either through process Type A or B. Just how efficient a given storm is in accomplishing this effect is a strong determinant of its precipitation-scavenging efficiency.

12.4. MICROSCOPIC INTERACTIONS

Once a pollutant particle or molecule is transported into a region of the storm where condensed water is present it may attach to cloud-water, ice, and precipitation by a variety of mechanisms. As indicated in Figure 12.1, these processes can often be reversible, in the sense that both attachment and detachment can occur. The rates at which these interactions take place depend on the abundance of the condensed water, the interconversion processes that are occurring locally between various classes of water (e.g., ice, vapor, cloud-water, etc.), and the affinity for water of the particular pollutant in question. Physical combination of dry pollutant with condensed water is of course absolutely necessary for the scavenging process to occur, and the rate of this attachment has a large bearing on the overall wet removal rate. Often the mechanisms for pollutant attachment differ substantially, depending on whether the pollutant in question is a gas or an aerosol particle; consequently, these two classes of pollutants are discussed separately below.

12.4.1. Aerosol Attachment

Aerosol attachment to hydrometeors can be envisioned to occur in two classes of processes, depending upon

- whether the process involves migration of water vapor to (and subsequent condensation on) the particle (Class I), or
- whether it occurs by contact of the particle to an existing hydrometeor (Class II).

The rates of both Class I and Class II processes depend strongly on particle size and chemical composition; thus scavenging tends to be a highly selective phenomenon, removing some particles with high efficiency and others relatively slowly, if at all.

Aerosol particles can become attached to condensed water phases by a number of individual mechanisms; these include:

- *nucleation* (Class I): the migration of water vapor molecules to the surface of an aerosol particle, with subsequent condensation in sufficient quantities to form a liquid drop;
- *Brownian diffusion* (Class II): the diffusional transport of aerosol particles to an existing hydrometeor;
- *diffusiophoresis* (Class II): the transport of aerosol particles to the surface of a hydrometeor—a sweeping action induced by the flux of water vapor molecules migrating to the same surface;
- *thermophoresis* (Class II): the thermally induced transfer of aerosol particles to the surface of any hydrometeor that is cooler than its surroundings;
- *electrophoresis* (Class II): the migration of particles to a hydrometeor surface caused by an EMF gradient;
- *impaction and interception* (Class II): the inertial collection of aerosol particles by falling hydrometeors.

Figure 12.8 provides some conceptual appreciation for these processes.[4] Figure 12.8(a) depicts an initially dry aerosol particle, which absorbs water molecules to nucleate a cloud droplet through a Class I interaction. The particle shown here is represented schematically as an agglomerate of a variety of substances ("internally mixed" aerosol particle). While many particles are complex multicomponent conglomerates of this type, others are composed of relatively pure substances. Regardless of its degree of compositional heterogeneity, however, the nucleating capability of any particle depends strongly on its chemical affinity for water; thus strongly hydrophilic substances, such as soluble salts and acids, nucleate much more readily than hydrophobic substances, such as pure soot. As indicated by the schematic, aerosol particles may be reinjected back to the aerosol phase via cloud evaporation. Although Figure 12.8(a) indicates evaporative ejection of the very same particle that formerly nucleated into a droplet, it should be recognized that dissolution, as well as a number of other possible mechanisms, is likely to modify the form of particles during the uptake/ reinjection process.

Various Class II attachment mechanisms for cloud droplets are indicated schematically in Figure 12.8(b). Without going into large detail, one should note that two of these mechanisms (diffusiophoresis and thermophoresis) depend strongly on heat and water vapor transfer being experienced by the droplet. Diffusiophoretic deposition is affected directly by the flux of water vapor molecules migrating to and condensing on the droplet. Thermophoretic deposition depends on the droplet temperature being lower than the surrounding air, allowing the droplet to act as a miniature thermal precipitator. Latent heat of condensation/evaporation raises/lowers the droplet temperature; thus a strong interplay (and partial mutual cancellation) of diffusiophoretic and thermophoretic effects is expected to occur in most cloud environments. Electrophoresis obviously depends strongly on relative charging of the aerosol particles and the cloud droplets, while Brownian diffusion—particle motion caused through random collisions with gas molecules—varies sharply and inversely with particle size.

Figure 12.8(c) depicts the process of cloud droplet growth by coagulation of smaller droplets and by further condensation. It is important to note here that the pollution burdens of the coagulating droplets are consolidated in the product drop, and if the product drop evaporates, it will probably eject a single particle which is some clustered combination of the original nuclei. If the product droplet

---

[4] It should be noted here that the elements in these figures are not necessarily to scale. In general the aerosol particles are depicted at a somewhat larger scale than the cloud particles, which are in turn drawn at a somewhat larger scale than the precipitation elements. This is necessary because of the large size range of the entities involved: raindrops and snowflakes as large as a few milimeters, and aerosol particles as small as a few nanometers.

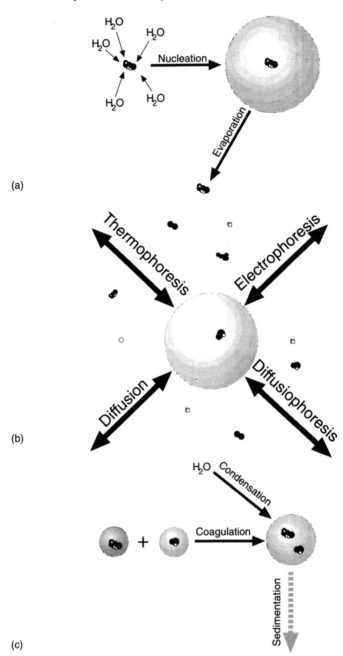

FIGURE 12.8.   Microscopic pollutant attachment mechanisms: (a) cloud-droplet nucleation; (b) diffusive and phoretic mechanisms; (c) condensation and coagulation (autoconversion).

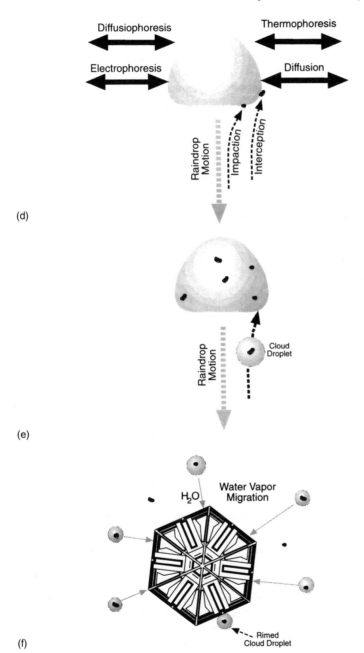

FIGURE 12.8 (*Continued*)   (d) Capture mechanisms of raindrops; (e) raindrop growth by cloud-water accretion; (f) Bergeron–Findeison and riming processes.

grows by these processes to sufficient size, then it will acquire sufficient gravitational velocity to become classified as a precipitation particle. Cloud physicists often refer to this droplet coagulation process as "autoconversion".

As indicated in Figure 12.8(d), hydrodynamic drag deforms precipitation-size water drops, causing them to assume "inverse bullet" forms. Also as indicated by this figure, diffusiophoresis, thermophoresis, electrophoresis, and Brownian diffusion operate to various degrees for precipitation-size drops as well. In addition, however, inertial collection—caused by the relative velocities and inertia of the raindrop and particle—arises as an important additional Class II scavenging mechanism. In most of the classical literature, inertial collection is subdivided into two categories: interception (which would not occur if the aerosol particle did not have a finite radius) and impaction (which would not occur if the particle did not have finite mass). From this it is obvious that the efficiency of these processes should increase with increasing particle mass and volume.

A special category of inertial collection, known as accretion, is depicted in Figure 12.8(e). This is similar to the situation depicted in Figure 12.8(d), except that the inertially collected particle is a cloud droplet rather than a pollutant aerosol particle. As can be appreciated from this figure, accretion tends to accumulate pollution loadings as the cloud droplets are collected. Cloud-water accretion can also occur when the falling hydrometeor is an ice particle through a process known as riming, as shown in Figure 12.8(f). In such situations the cloud droplet usually freezes immediately on contact, resulting in agglomerated ice particles such as graupel and hail.

Figure 12.8(f) also depicts an additional cloud physics mechanism that is of profound importance. Known as the Bergeron–Findeisen effect, this process occurs whenever water droplets and ice crystals coexist locally in the same cloud. Supercooled water droplets often coexist with ice crystals in sub-freezing clouds. The vapor pressure of water above such droplets, however, exceeds the corresponding vapor pressure associated with ice. This results in a migration of water to the ice surface with a resulting tendency for droplet evaporation— essentially de-scavenging the pollutant originally held by the sacrificial cloud drops. For this reason, snowstorms whose precipitation is generated primarily by the Bergeron–Findeisen process are particularly inefficient in pollution removal.

As noted above, most salts and acid-forming aerosol particles are usually so hygroscopic that the nucleation mode of attachment dominates whenever cloud formation processes are active. Under such conditions other mechanisms are usually important only in special circumstances and are typically ignored in acid-rain-related calculations. In the context of present diagnostic and regional source–receptor models the usual procedure is to consolidate hydrometeor- and aerosol-size spectra into representative single sizes and to apply estimated attachment parameters for these "lumped" models of natural behavior. It is also usual practice to presume that nucleation dominates in-cloud attachment of aerosol particles and to treat below-cloud scavenging in terms of a capture efficiency that is derived from best estimates of Brownian and inertial phenomena.

12.4.2. Pollutant Gas Attachment

Owing to the high diffusive mobility of gases, advective/diffusive attachment dominates other mechanisms for pollutant gas uptake by all classes of hydrometeors. Well-established mathematical formulations exist to describe gas phase transport under such circumstances; several additional features become active in such situations, however, and this lends substantial complexity to the composite attachment process.

One such feature (or complex of features) involves the solubility of the gas in the condensed aqueous media, which is associated with an inherent reversibility in the attachment process. Gas molecules can migrate to a water droplet surface and be absorbed within its interior, but desorption back to the gas phase and subsequent migration away from the drop are also possible. This is in direct contrast with the behavior of non-volatile aerosol particles, which, once captured, can be reasonably expected to remain associated with the aqueous phase unless total evaporation occurs. At a state of diffusional equilibrium the net gas uptake by the aqueous phase is zero, with absorption and desorption of molecules forming a dynamic balance.

The question of gas solubility does not pose any large practical problem for acid precipitation assessment as long as numerical values for pertinent solubility parameters are known. Fortunately, solubility parameters for most of the common acid precursor gases have been measured. For example, the water solubilities of sulfur dioxide, nitric acid, peroxyacetyl nitrate, ozone, hydrogen peroxide, hydrogen chloride, ammonia, formaldehyde, and carbon dioxide have all been quantified to a point where they can be employed reliably in direct scavenging calculations. It is important to note here, however, that many of these gases participate in ionic dissociation reactions upon contact with water and that such behavior often imparts a significant degree of non-linearity on the scavenging process.

In a related area, work is progressing to measure attachment equilibria of gases to ice surfaces. It is presently unclear whether these interactions can be classified in terms of solubility, surface adsorption, or some combination of the two, but they have proven to be of obvious importance as scavenging pathways for some strongly hydrophilic gases, such as HNO_3.

A second set of features complicating the question of pollutant gas uptake involves the interfacial and aqueous phase transport of the absorbing material. As noted above, molecular interchange across the gas–liquid interface is determined partly by gas phase transport to the aqueous surface, but transport across any surface film (such as an organic monolayer) and mass transfer within the liquid are potentially important influencing factors as well. In addition, there is usually a non-zero probability that pollutant molecules approaching the liquid surface may be reflected on contact rather than being accommodated by the aqueous medium. This latter feature has been generally treated mathematically using the concept of an "accommodation coefficient" whose value ranges between 0 (total reflection of approaching molecules by the surface) and 1 (total accommodation). In aggregate each of these features can be visualized as a series "resistance" to pollutant uptake; comparatively high resistances are considered to be "rate-influencing", while particularly low resistances are unimportant as controlling features of the pollutant capture process.

12.4.3. Aqueous Phase Transformations

As noted above, once a pollutant molecule has come into contact with a condensed aqueous phase, it may or may not react to form a different chemical species. Obviously such reaction is not essential for the wet removal process, since the pollutant may be brought to the Earth's surface in either reacted or unreacted form; chemical reaction is an important facilitator of acidic precipitation formation and the scavenging process, however, since it often stabilizes pollutants in precipitation, thus preventing their escape back to the gaseous medium. Chemical reaction is also often effective in changing the acidity of the precipitation elements and in affecting the solubility of various pollutant gases.

12.4.3.1. Ionic Reactions and Acid–Base Equilibria

Many chemical species of importance to acidic rain formation dissociate at least partly to form ionic species when in aqueous solution. Sulfur dioxide, for example, can be absorbed from the gas phase to exist as a whole molecule in aqueous solution, which is capable of adding a water molecule and dissociating to form bisulfite ion, which in turn can dissociate to form sulfite ion:

$$SO_2(g) \leftrightarrow SO_2(aq) \tag{12.9}$$

$$SO_2(aq) + H_2O \leftrightarrow HSO_3^- + H^+ \tag{12.10}$$

$$HSO_3^- \leftrightarrow SO_3^{2-} + H^+ \tag{12.11}$$

Here the double-headed arrows denote that the reactions may proceed in both the forward and reverse directions; if the forward and reverse rates of reaction are equal, then a state of *dynamic equilibrium* is said to exist.

It is standard practice in physical chemistry to express individual chemical relationships such as (12.9)–(12.11) in mathematical form using measured constants for the various equilibria. Solubility equations, expressing the aqueous phase concentrations of the pollutants in terms of their gaseous phase values, can be derived by combination of these individual expressions. When, as in the case for SO_2 given above, the dissolution process involves several stages of equilibria, the aqueous phase concentrations are not directly proportional to their gaseous phase counterparts. This non-proportionality is reflected in the corresponding solubility equation and can be an important source of non-linearity in the source–receptor sequence.

Ionization reactions generally proceed sufficiently rapidly that their rates are not of concern as rate-influencing steps in the scavenging sequence; thus most aspects of ionic phenomena in the scavenging process can be expressed solely in terms of equilibria. In particular, hydrogen ion concentrations are typically computed in terms of the ion balance, i.e., equation (12.1), on the basis of the other ionic constituents. Comparison of pH measurements in precipitation chemistry samples with values computed on the basis of ion balances has demonstrated the general validity of this approach.

12.4.3.2. Sulfur Chemistry in Cloud and Precipitation Systems

Most cloud- and precipitation-borne sulfur occurs in the (IV) (sulfite ion) and (VI) (sulfate ion) oxidation states. Aqueous phase sulfur in the (IV) oxidation state occurs mainly through absorption of gaseous sulfur dioxide according to equations (12.9)–(12.11). As noted previously, this absorption process is generally limited by solubility equilibrium, which is pH-dependent. In addition to the acid–base equilibria indicated in Section 12.4.3.1, sequestering by organic complexing agents is possible; for example, dissolved formaldehyde is known to complex S(IV), stabilizing it in a form that is immune to oxidation and enhancing its total solubility as well.

Aqueous phase sulfur in the (VI) oxidation state is incorporated both by scavenging of pre-existing sulfate particles and by aqueous phase oxidation of S(IV). The relative importance of the direct and reactive mechanisms appears to depend strongly on storm conditions and chemical mix and has been the subject of continued conjecture. Scott (1982), for example, has predicted that in-cloud oxidation mechanisms account for 55–70% of precipitation sulfate in summertime convective storms. The fraction of SO_2 converted to sulfate in clouds in the winter is estimated to be similar, but the total precipitation-borne sulfur burden is only about one-third of that observed in the summer (see Table 12.1). High regional conversion amounts have also been estimated based upon the calculated abundance of aerosol and precipitation sulfate (McNaughton and Scott, 1980). Measurements made upwind and downwind of individual clouds have in most cases indicated aqueous phase formation of S(VI) (Hegg and Hobbs, 1982), while measurements upwind and downwind of urban areas appear to indicate substantial in-cloud production of S(VI) as well (Hales and Dana, 1979; Patrinos, 1985). The acidity of cloud-water has been observed to be greater than can be explained by the composition of aerosols and gases in nearby clear air (Lazrus et al., 1983; Daum et al., 1984).

Measurements of oxidants and the distribution of sulfur pollutants between the gaseous and aqueous phases are consistent with a large fraction of SO_2 converted to aqueous phase S(IV) in the summer. A different situation, in which pre-existing aerosol could account for most sulfate in precipitation, has also been observed in winter cases, such as in western Michigan (Scott and Laulainen, 1979). These measurements, combined with enhanced understanding of oxidant behavior that has been acquired during recent years, suggest a substantial attenuation of reactive sulfur scavenging during winter months.

Reaction rate coefficients for S(IV) oxidation by different chemical pathways have been measured in a number of laboratories. Pursuant to assumptions on atmospheric composition, and reaction rates with H_2O_2, O_3, HNO_2, NO_2, and O_2 as a function of pH, these results are summarized below.

- Reaction with O_2 in the absence of a catalyst is negligibly slow.
- Reaction with H_2O_2 is nearly pH-independent over the pH range of cloud-water and precipitation . At ambient gas phase H_2O_2 levels in the vicinity of 1 ppb, conversion rates of SO_2 of several percent per minute can occur. H_2O_2 mixing ratios of a few parts per billion are common in the eastern US during summer months.
- Reaction with O_3 is expected to be important at pH levels above 4.5. The inverse relation between SO_2 solubility and H^+ concentration causes this reaction (and other non-acid-catalyzed reactions) to strongly decrease in importance at lower pH.
- At the high end of the plausible range for transition metal ion concentrations, catalyzed reactions with O_2 may be significant. Transition metal ion concentrations typical of the industrial coastline of Los Angeles (0.5 μg m^{-3} of Fe^{3+}, 0.02 μg m^{-3} of Mn^{2+}) were calculated to cause approximately one-third of the S(VI) concentrations existing in a fog (Jacob and Hoffman, (1983).

The inverse dependence between hydrogen ion concentration and S(IV) ozone oxidation rates indicated above should be interpreted with some caution owing to point-to-point variability of acidity within a storm. Diagnostic storm models, for example, suggest that many storms contain very acidic cloud water in inflow regions, but comparatively neutral water at other points within the cloud. These model calculations also suggest that ozone oxidation processes often proceed efficiently in these cleaner regions, even though the precipitation produced by the storm may be strongly acidic.

Regardless of these ozone-based considerations, current evidence suggests that H_2O_2 is probably the most important oxidizing agent for SO_2 in the aqueous phase, at least for the pH conditions (4.0–4.5) typical of summer precipitation in the eastern US. The observed non-coexistence of SO_2 and H_2O_2 in clouds is consistent with a rapid reaction forming S(VI), proceeding until the limiting reagent is depleted (Daum et al., 1984; Kelly et al., 1985). Therefore a reduction in SO_2 emissions would be proportional to S(VI) production when this production is controlled by the H_2O_2 reaction and H_2O_2 exists in sufficient quantities to ensure depletion of SO_2.

In-cloud oxidation of SO_2 by H_2O_2 can lead to non-linear relations between ambient SO_2 and precipitation S(VI). The non-linearity arises because there will be circumstances in which locally in summer, ambient SO_2 levels exceed those of H_2O_2 (probably always in winter). There is observational evidence that under these circumstances the H_2O_2 oxidizes an equivalent amount of SO_2. The remaining SO_2 is not readily oxidized because of the low pH and consequent low SO_2 solubility. Clouds are generally more acidic in the summertime, and as SO_2 solubility also decreases with increasing temperature, dissolved SO_2 concentrations in the summertime are considerably lower than in the wintertime. Accordingly, if SO_2 concentrations are reduced to a value equal to the concentration of H_2O_2, it is expected that there would be little change in the amount of S(VI) formed (Chameides, 1984). The prevalence and regional-scale implications of the above local imbalance between oxidant and SO_2 have not, however, been quantified. Non-linear relations between SO_4^{2-} formation and SO_2 are expected also for other oxidation mechanisms, as discussed above in conjunction with the pH dependence of SO_2 solubility.

12.4.3.3. Nitrogen Chemistry

Nitrate in precipitation can result from the scavenging of gas phase HNO_3, specific gas phase nitrogen oxides, and aerosol NO_3^- (see also Chapter 6). In contrast with sulfur chemistry, there is no evidence that significant amounts of acid are formed from aqueous phase oxidation of the emitted pollutants NO and NO_2 (Lee and Schwartz, 1981). According to recent measurements of the solubility and kinetics of peroxyacetyl nitrate and HONO, these compounds are likewise not expected to be significant sources of NO_3^- in precipitation.

Measurements of cloud-water and precipitation composition during a study of a frontal storm suggest the continued formation of HNO_3, consistent with an in-cloud aqueous phase source (Lazrus et al., 1983). In the absence of sunlight it is expected that HNO_3 can be produced from the gas phase reaction sequence

$$O_3 + NO_2 \longrightarrow O_2 + NO_3 \qquad (12.12)$$

$$NO_3 + NO_2 \longrightarrow N_2O_5 \qquad (12.13)$$

followed by

$$N_2O_5 + H_2O \longrightarrow HNO_3 + O_2 \qquad (12.14)$$

with conversion rates of NO_2 of order 10% h^{-1} (Schwartz, 1986). During the day, NO_3 is rapidly destroyed by photolysis or by reaction with NO. However, even in the presence of sunlight or NO, the above reaction sequence could be driven toward formation of HNO_3 if mass transfer of N_2O_5 to water (currently unknown) can compete effectively with the gas phase back reactions. The solubility and kinetics of NO_3 also are not well understood. Continued production of gas phase HNO_3 in air within clouds during the daytime can occur also from the usual gas phase photochemical route

$$NO_2 + OH \longrightarrow HNO_3 \qquad (12.15)$$

The rate of conversion of NO_2 by this process can be several per cent per hour (see e.g., Calvert and Stockwell, 1983), but will differ in the cloud owing to changed chemical environment and solar intensity. The corresponding sulfur reaction

$$SO_2 + OH \rightarrow HSO_3 \xrightarrow{H_2O} H_2SO_4 \qquad (12.16)$$

occurs with a rate constant an order of magnitude lower than is the case for NO_2. As a consequence, an unrealistically long cloud lifetime would be required for this gas phase reaction to yield an appreciable apparent in-cloud acid source.

Other, as yet unidentified, sources of aqueous phase NO_3^- may exist. Gas phase measurements at Niwot Ridge, Colorado and Point Arena, California, indicate that nitrogen pollutant concentrations detected by a non-selective chemiluminescent NO_y detector are often more than twice those that could be identified on the basis of measurements of individual compounds (Fahey et al., 1986).[5]

12.4.3.4. Aqueous Phase Photochemistry

Recent calculations have explored the possibilities of an aqueous phase "photochemistry" involving interactions with radicals produced *in situ* or transferred from the gaseous to the aqueous phase (e.g., Chameides and Davis, 1982). These studies have a much shorter history than their gas phase counterparts. Many interesting mechanisms have been suggested, but the

---

[5] More recent measurements of this type conducted by several authors are reported in the June 30, 1993 issue of the *Journal of Geophysical Research*.

quantitative evaluation of rates depends upon further direct measurements in laboratory investigations.

As in the gas phase, the OH radical can oxidize hydrocarbons and recombination reactions of HO_2 radical (or its reaction products with H_2O) can form H_2O_2. OH and HO_2 are rapidly interconverted, as in the gas phase, via reactions with O_3 and hydrocarbons. Formation of O_3 via reactions of NO, which is a prominent feature of gaseous phase photochemistry, is not expected to occur to any significant extent in the aqueous phase owing to the very low solubility of NO.

Obviously, both primary and secondary photochemical phenomena occurring in cloud environments depend strongly on local actinic flux. Although model estimates of such flux levels have been made, the questions of exact values and spatial variabilities of these values within clouds remain major factors that limit quantitative knowledge in this area.

12.5. SPATIAL DISTRIBUTIONS AND TEMPORAL TRENDS

In concordance with the stepwise characterization of precipitation scavenging given in Section 12.2, it is obvious that the processes described in Sections 12.3 and 12.4 control, and result in, wet delivery of pollutant to the Earth's surface— provided, of course, that storm formation proceeds to the point where hydrometeors make contact with the ground. This final section discusses the manifestation of this process in terms of observed precipitation chemistry spatial distributions and temporal trends.

12.5.1. Spatial Distributions in North America and Europe

Figure 12.9 shows spatial distributions of precipitation chemistry observed in North America for the selected major ions SO_4^{2-}, NO_3^-, NH_4^+, and H^+. These are annual precipitation-weighted averages, corresponding seasonally stratified contour plots typically show marked differences owing to meteorological (in some situations) source variability. Figure 12.10 shows corresponding contour plots for Europe. The wet-deposition patterns on these plots generally reflect the pollutant source areas; moreover, the dispersion of the patterns appears visually consistent with the distance scales and lifetimes discussed in Section 12.2.

Numerous attempts have been made to account quantitatively for these deposition patterns on a source–receptor basis using several approaches, ranging from large computer-modeling applications through more simple, macroscopic material balances. The computer-modeling analyses have often been moderately

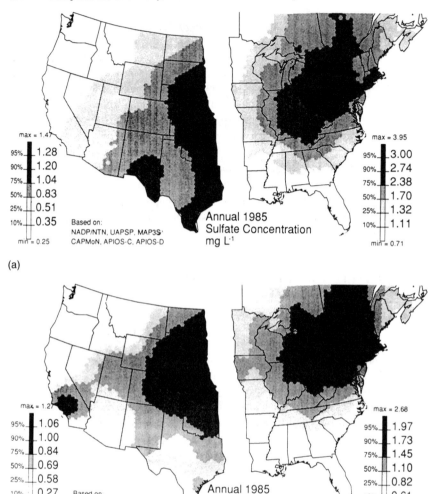

FIGURE 12.9. Observed 1985 annual average precipitation chemistry for North America: (a) sulfate ion concentration, mg l^{-1}; (b) nitrate ion concentration, mg l^{-1}.

successful in producing deposition fields that are similar to the observations (e.g., Dennis 1990). Because of system complexity, however, their predictions of more detailed phenomena, such as source attribution, have been somewhat doubtful. Substantial effort continues to refine model elements in key areas to increase overall model proficiency in these areas.

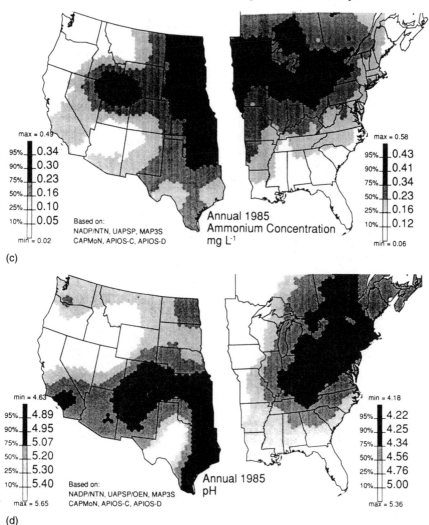

FIGURE 12.9 (*Continued*)   (c) Ammonium ion concentration, mg l⁻¹; (d) pH. (From Sisterson, 1990.)

Hilst and Chapman (1990) provide a more recent example of a macroscopic material balance approach, wherein they superimpose the North American emission inventories for sulfur and nitrogen oxides with 4 year averages of wet deposition patterns for SO_4^{2-} and NO_3^- and then perform a variability analysis to estimate source influences associated with each precipitation chemistry monitoring site. Specific steps in the Hilst–Chapman analysis are as follows.

1. Select a simple semi-empirical equation (such as a linear slope–intercept expression, $D = b + mE$) to depict the relationship between precipitation-borne pollutant concentration D and surrounding emission strength E.
2. Define a tentative source region by drawing a circle of radius a around each monitoring station on the map. Apply the emission inventory to calculate the emission rates E within these circles.

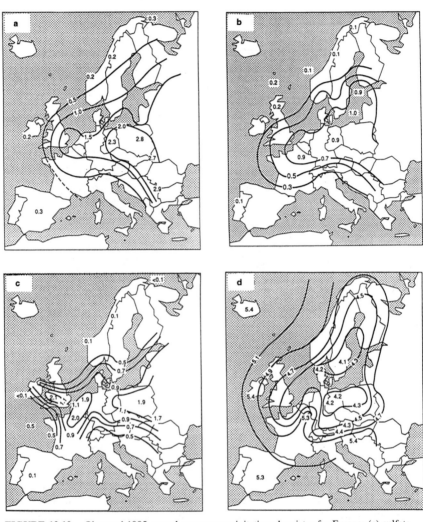

FIGURE 12.10.   Observed 1985 annual average precipitation chemistry for Europe: (a) sulfate ion concentration, $mgS\,l^{-1}$; (b) nitrate ion concentration, $mgN\,l^{-1}$; (c) ammonium ion concentration, $mgN\,l^{-1}$; (d) pH. (From Sisterson, 1990.)

3. Perform a least-squares fit of the results to determine the parameters (e.g., slope m, intercept b) associated with the equation chosen in Step 1. Also calculate the statistical variance (r^2) associated with the fit of this equation to these data.
4. Repeat Step 1 for a number of representative values of circle radius a.
5. Plot r^2 versus a, and note the region where additional increases in a do not contribute significantly to further reduction of r^2. This value of a is taken to be the approximate source–receptor length scale for wet deposition.

Two noteworthy conclusions were made as a result of this analysis. First, North American areas having relatively low emission densities exhibit weak associations between source strength and wet deposition, even with source areas as large as 560 km in radius. This implies that external (or otherwise unaccounted) emissions dominated the observed precipitation chemistry and that characteristic source–wet deposition distances are long in these regions. Weak association also was demonstrated in the polluted regions of the US and southern Canada during winter months. On the other hand, the associations were moderately strong in these polluted areas during summer, leading to the conclusion that characteristic wet deposition length scales were on the order of 400–600 km under summertime polluted conditions.

The Hilst–Chapman analysis is limited by several idealizations, including its tacit presumption of isotropic pollutant dispersion (reflected by the choice of circular source regions) and its limited ability to deal with external and extraneous pollutant contributions. On the other hand, it provides an encouragingly consistent semiquantitative picture when compared with the residence time estimates discussed in Section 12.2 and with the basic modeling results obtained to date. From these there is little doubt that source–receptor distance scales for wet removal of sulfur oxides and reactive nitrogen can range from several hundred to thousands of kilometers and that seasonal differences in these scales can be profound. There is no doubt whatsoever that long-range transport is a key element of the acidic deposition phenomenon.

12.5.2. Temporal Trends in North America and Europe

Temporal behaviors of the major ions SO_4^{2-}, NO_3^-, NH_4^+, and H^+, expressed as aggregate averages of selected North American stations, are shown in Figure 12.11. Noteworthy in these plots are the large station-to-station variability for any given year and the indications of small, if any, trends during this period of record. Hydrogen and ammonium ions have no discernible trends within the variability of the data; SO_4^{2-} and NO_3^- show barely discernible downward and upward trends, respectively. The short period of record for these data (1979–1988) should be

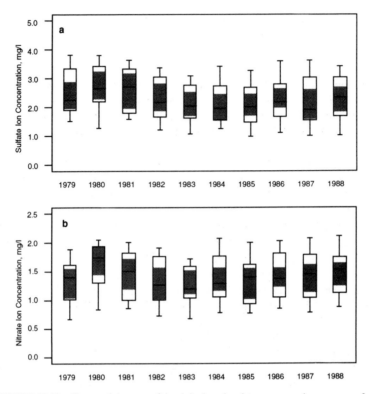

FIGURE 12.11.   Temporal patterns of precipitation chemistry, expressed as averages from selected North American stations; box plots are the 10th, 25th, 50th, 75th, and 90th percentiles of the station set: (a) sulfate ion; (b) nitrate ion.

noted with the expectation that earlier years, involving lower emission densities, probably resulted in lower precipitation-borne ion concentrations. Unfortunately, few continuous precipitation chemistry stations operated prior to this time, thus precluding meaningful station-aggregated data. The results of the Hubbard Brook, New Hampshire station represent the longest available North American record in this regard, and its results are presented in Figure 12.12 to provide the reader with some insight of possible regional behavior during earlier time periods. Unfortunately, this record begins in 1964, which is still well into the period of high North American NO_x and SO_x emissions. It does, however, give some insight into longer-term variability experienced by a single station.

European sampling stations tend to exhibit more year-to-year variability than experienced in North America, possibly because of the location of many European sources to the west side of the continent (influenced by the prevailing winds from the Atlantic Ocean), in conjunction with the more pronounced

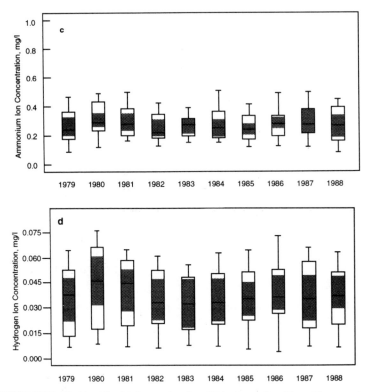

FIGURE 12.11 (*Continued*)   (c) Ammonium ion; (d) hydrogen ion. (From Simpson et al., 1992.)

interannual variability of weather patterns. A typical European site, located at Forschult in central Sweden, provides an example of this variability (Figure 12.13). Combined with station-to-station differences, such year-to-year variability complicates European trend analysis and has resulted in varying trend estimates depending on the periods of record and the station ensembles chosen for analysis (e.g., Granat, 1978; Rodhe and Granat, 1984).

12.6. CONCLUSIONS

This introductory overview of precipitation chemistry is intended to provide a conceptual appreciation of how acidic precipitation is formed, to give a basic understanding of the chemical basis for rain acidity, and to demonstrate features of the acid precipitation phenomenon from both temporal and spatial perspectives. Several important conclusions can be made from this overview.

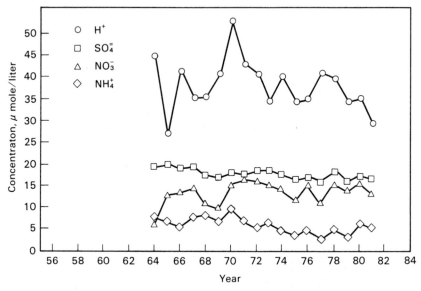

FIGURE 12.12.   Temporal patterns of annual average precipitation chemistry for the Hubbard Brook, New Hampshire sampling site. (From measurements of Likens et al. 1984.)

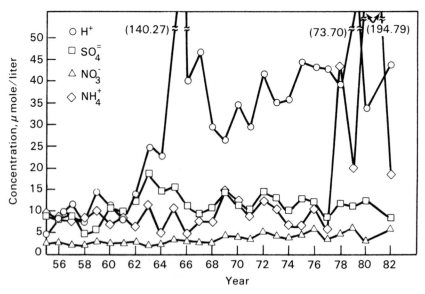

FIGURE 12.13.   Temporal patterns of annual average precipitation chemistry for the Forschult, Sweden sampling site. (Adapted from Hicks, 1990.)

- Precipitation acidity is a direct and calculable consequence of the presence of ions existing in precipitation. Several of these ions are primarily anthropogenic in origin, while others originate mainly from natural sources.

- There is no doubt whatsoever that man's contribution to the pollution burden can and does have a significant effect on precipitation acidity as well as on the general chemical make-up of wet deposition. This is especially apparent in the spatial distributions of precipitation chemistry with respect to major pollution sources, but is reflected in modeling analyses as well. On the other hand, natural sources can contribute significantly to the ionic content of precipitation, as evidenced by measurements taken at remote sites.

- Atmospheric lifetimes of key acidifying pollutants prior to wet deposition are sufficiently long to render long-range transport an important determinant of spatial wet deposition patterns. Characteristic source–receptor length scales, while quite uncertain in most cases, range from several hundred to thousands of kilometers. For many pollutants these length scales have a pronounced seasonal dependence, with a tendency for increase during winter months.

- The pollutant-scavenging process leading to acidic precipitation proceeds in a series of steps, including introduction of the pollutant to a storm region, its physical attachment to hydrometeors, and its wet delivery to the Earth's surface. Chemical conversion processes in both the gaseous and aqueous phases are often strong determinants of overall wet removal efficiency. Moreover, many of these steps are reversible, complicating the source–receptor pathway and adding substantial uncertainties to any quantitative depiction of the process.

- General semiquantitative agreement between theory and measurement exists in so far as gross features of the source–receptor process are concerned. Limitations in our ability to quantitatively depict the transport–attachment–reaction steps of the scavenging sequence, however, severely limit our ability to attribute specific sources responsible for deposition at a given site or to project future trends in any confident manner. Significant additional knowledge regarding these processes will be necessary before realistic projections of this type will be possible.

References

Bolin, B. and Rodhe, H. (1973). A note on the concepts of age distribution and transit time in natural reservoirs. *Tellus*, **XXV**, 58–62.

Butcher, S.S., Charlson, R.J., Orians, G.H. Wolfe, G.V. (1992). *Global Biogeochemical Cycles*, Academic Press, New York, NY.

Calvert, J.G. and Stockwell W.R. (1983). Acid generation in the troposphere by gas-phase chemistry, *Environ. Sci. Technol.*, **17**, 428A.

Chameides, W.L. (1984). The photochemistry of a remote marine stratiform cloud. *J. Geophys. Res.*, **89**, 4739–4755.

478   Composition, Chemistry, and Climate of the Atmosphere

Chameides, W.L. and Davis, D.D. (1982). The free-radical chemistry of cloud droplets and its impact on the composition of rain. *J. Geophys. Res.*, **87**, 4863.

Charlson, R.J., Anderson, T.L. and McDuff, R.E. (1992). The sulfur cycle. In Butcher, S.S., Charlson, R.J., Orians, G.H. and Wolfe, G.V., Eds, *Global Biogeochemical Cycles*, Academic Press, New York, NY, Chap. 13.

Charlson, RJ. and Rhode, H. (1982). Factors controlling the acidity of natural rainwater. *Nature*, **295**, 683–685.

Church, T.M., Galloway, J.N., Jickells, T.D. and Knapp, A.H. (1982). The chemistry of western Atlantic precipitation at the mid Atlantic coast and on Bermuda. *J. Geophys. Res.*, **87**, 11,013–11,018.

Court, A. and Griffiths, J.F. (1985). Thunderstorm climatology. In Kessler, E., Ed., *Thunderstorm Morphology and Dynamics*, University of Oklohoma Press, Norman, OK.

Daum, P.H., Schwartz, S.E. and Newman, L. (1984). Acidic and related constituents in liquid water stratiform clouds. *J. Geophys. Res.*, **89**, 1447.

Dennis, R.L. (1990). Evaluation of regional acidic deposition models and selected applications of RADM. *NAPAP State of Science and Technology Report 5*, National Acid Precipitation Assessment Program, Washington, DC.

Fahey, D.W., Hübler, G., Parrish, D.D., Williams, E.J., Norton, R.B., Ridley, B.A., Singh, H.B., Liu, S.C. and Fehsenfeld, F.C. (1986). Reactive nitrogen species in the troposphere: measurements of NO, NO_2, HNO_3, particulate nitrate, peroxyacetyl nitrate (PAN), O_3, and total reactive odd nitrogen at Niwot Ridge, Colorado. *J. Geophys. Res.*, **91**, 9781–9793.

Godske, C.L., Bergeron, T., Berkness, J. and Bundgaard, R.C. (1957). *Dynamic Meteorology and Weather Forecasting*, American Meteorological Society, Boston, MA.

Granat, L. (1978). Sulfate in precipitation as observed by the European atmospheric chemistry network. *Atmos. Environ.*, **12**, 413–424.

Hales, J.M. and Dana, M.T. (1979). Precipitation scavenging of urban pollutants by convective storm systems. *J. Appl. Meterol.*, **18**, 294.

Hales, J.M. and Renne, D.S. (1990). Atmospheric source–receptor relationships: concepts and terminology. *Report PNL-7470*, Pacific Northwest Laboratory, Richland, WA.

Hegg, D.A. and Hobbs, P.V. (1982). Measurements of sulfate production in natural clouds. *Atmos. Environ.*, **16**, 2663.

Hicks, B.B. (1990). Atmospheric processes research and process model development. *NAPAP State of Science and Technology Report 2*, National Acid Precipitation Assessment Program, Washington, DC.

Hilst, G.R. and Chapman, E.G. (1990). Source–receptor relationships for wet $SO_4^=$ and NO_3^- production. *Atmos. Environ.*, **24A**, 1889–1901.

Jacob, D.J. and Hoffman, M.R. (1983). A dynamic model for the production of H^+, NO_3^-, and $SO_4^=$ in urban fog. *J. Geophys. Res.*, **88**, 6611.

Jaffe, D.A. (1992). The nitrogen cycle. In Butcher, S.S., Charlson, R.J., Orians, G.H. and Wolfe, G.V., Eds, *Global Biogeochemical Cycles*, Academic Press, New York, NY, Chap. 14.

Keene, W.C. and Galloway, J.N. (1984). Organic acidity in precipitation of North America. *Atmos Environ.*, **18**, 2491–2497.

Kelly, T.J., Daum, P.H. and Schwartz, S.E. (1985). Measurement of peroxides in cloudwater and rain. *J. Geophys. Res.,* **90**, 7861.

Lazrus, A.L. et al. (1983). Acidity in air and water in the case of warm frontal precipitation. *Atmos. Environ.,* **17**, 581.

Lee, Y.N. and Schwartz, S.E. (1981). Reaction kinetics of nitrogen dioxide with liquid water at low partial pressure. *J. Phys. Chem.,* **85**, 840.

Likens, G.E., Bormann, F.H., Pierce, R.S., Eaton, J.S. and Munn, R.E. (1984). Long-term trends in precipitation chemistry at Hubbard Brook, New Hampshire. *Atmos. Environ.,* **18**, 2641–2647.

McNaughton, D.J. and Scott, B.C. (1980). Modeling evidence of in-cloud transformations of sulfur dioxide to sulfate. *J. Air Pollut. Control Assoc.,* **30**, 272.

MAP3S (1982). The MAP3S/RAINE precipitation chemistry network: statistical overview for the period 1976–1980. *Atmos Environ.,* **16**, 1603–1631.

Mason, B.J. (1971). *The Physics of Clouds,* Clarendon Press, Oxford.

Patrinos, A.A.N. (1985). The impact of urban and industrial emissions on mesoscale precipitation quality. *J. Air Pollut. Control Assoc.,* **35**, 719.

Rodhe, H. (1978). Budgets and turn-over times of atmospheric sulfur compounds. *Atmos. Environ.,* **12**, 671–680.

Rodhe, H. and Granat, L. (1984). An evaluation of sulfate in European precipitation 1955–1982. *Atmos. Environ.,* **18**, 2627–2639.

Schwartz, S.E. (1986). Chemical conversion in clouds. In Lee, S.D., Schneider, T., Grant, L.D. and Verker P.J., Eds, *Aerosols: Research, Risk Assessment and Control Strategies,* Lewis Publishers, Chelsea, MI, pp. 349–375.

Scott, B.C. (1982). Predictions of in-cloud conversion rates of SO_2 to SO_4 based on a simple chemical and dynamic model. *Atmos. Environ.,* **16**, 1735.

Scott, B.C. and Laulainen, N.S. (1979). On the concentration of sulfate in precipitation. *J. Appl. Meterol.,* **18**, 136.

Simpson, J.C., Olsen, A.R. and Bittner, E.A. (1992). 1988 wet deposition temporal and spatial patterns in North America. *PNL-8049.* Pacific Northwest Laboratory, Richland, WA.

Sisterson, D.L. (1990). Deposition monitoring: methods and results. *NAPAP State of Science and Technology Report 6,* National Acid Precipitation Assessment Program, Washington, DC.

13

Air pollution and climate change

Donald J. Wuebbles

13.1. INTRODUCTION AND OVERVIEW

One of the indisputable facts about climate is that it is always changing. However, over the last 10,000 years the globally averaged surface temperature has varied over a range of less than 2 K. Humanity is now undertaking an experiment, by emitting long-lived radiatively important gases into the atmosphere in unprecedented amounts, that could result in a warming of the global temperatures well beyond the natural variability. Numerical models of the climate system suggest that surface temperatures could increase by as much as 3 K (best estimated range of 2–5 K) over the next century (IPCC, 1990, 1992).

Of particular concern for the potential changes to climate are the growing atmospheric concentrations of gases that absorb terrestrial infrared radiation, termed greenhouse gases. Much of this concern has centered around carbon dioxide (CO_2) because of its importance as a greenhouse gas and also because of the rapid rate at which its atmospheric concentration has been increasing. However, in the last decade it has been shown that other greenhouse gases are contributing to about half of the overall increase in the greenhouse effect on climate. In addition to these direct effects, research studies have shown that chemical interactions in the atmosphere can lead to additional "indirect" effects on climate. As an example, changes in stratospheric ozone have received much attention because of concerns about ultraviolet radiation (see Chapter 11 for more details), but ozone is also a greenhouse gas and changes in its distribution can affect climate. Production of sulfuric aerosols and other particles in the troposphere and stratosphere may also be having a significant influence on climate.

At present there remain many uncertainties in understanding the effects of the changing atmospheric composition on climate. Because of potential environmental and socio-economic impacts from climate change, it is necessary to gain a much fuller understanding of the forces and interactions that affect climate. The purpose of this chapter is to describe the greenhouse gases and other influences affecting climate, the processes affecting climate, plus the past climate record and the future possibilities for climate change. Potential impacts from climate change will also be touched on.

13.2. CLIMATE AND THE GREENHOUSE EFFECT

13.2.1. What is Climate?

Climate is generally defined as a description of the average (or typical) behavior of the atmosphere. Therefore climate is the aggregation of the weather; it is usually expressed in terms of mean (or average) conditions and variances, including the probability of extremes and space and time covariance properties. An immediate question arises as to the nature of the averaging process used in defining climate. Traditionally 30 years has generally been used as the averaging time for climate (Leith, 1993). Shorter averaging times are more associated with weather-sampling fluctuations; longer times interfere with the concept of climate change. The time scales of interest for climate change are on the order of decades to centuries.

One of the fundamental climate variables is the planetary annual average surface temperature. Other variables of interest include averages of regional surface temperatures, the frequency and amount of rainfall, and the extent and frequency of floods and droughts. Concerns about climate change, including those related to the greenhouse gases, therefore relate to the evolution of the mean behavior and variances in such variables.

13.2.2. The Greenhouse Effect

The "greenhouse effect" is a natural feature of the world that refers to a physical property of the Earth's atmosphere; without the greenhouse effect we would not be here. If the Earth had no atmosphere (but the same reflectivity to solar radiation, or albedo, as it has now), its average surface temperature would be about 255 K rather than the comfortable 288 K found today. The difference in temperature is due to the presence of an atmosphere; in particular, to a suite of gases called greenhouse gases. In its existing state the Earth–atmosphere system balances absorption of solar radiation by emission of infrared radiation to space. The atmosphere absorbs more infrared energy than it re-radiates to space, resulting in a net warming of the Earth–atmosphere system and of surface temperature. This phenomenon, demonstrated in Figure 13.1, is referred to as the greenhouse effect.

As a means to understanding the greenhouse effect, it is useful to examine the annual and global average radiative energy budget of the combined Earth–

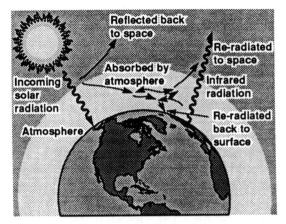

FIGURE 13.1.   Schematic diagram of the greenhouse effect. Energy radiated by the Earth at infrared wavelengths is absorbed by greenhouse gases in the atmosphere and is re-radiated; some of the energy is re-radiated downward, leading to a net warming of the lower atmosphere.

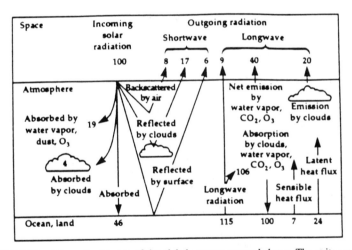

FIGURE 13.2.   Schematic diagram of the global average energy balance. The units are percentage of incoming solar radiation. The solar fluxes are shown on the left-hand side and the long-wave (IR) fluxes are on the right-hand side. Of the 100 units of incoming solar radiation (each equivalent to about $3.4 \, W \, m^{-2}$), about 23 are absorbed in the atmosphere, 46 are absorbed by the surface, and 31 are reflected to space. The planetary energy balance is achieved by the emission of 69 units to space as IR radiation. The solar energy absorbed by the surface is used in part to directly heat the atmosphere (sensible heat flux) and to evaporate moisture (latent heat flux). Atmospheric emission of IR radiation downward to the surface is about equal to the solar radiation reaching the top of the atmosphere and more than twice as large as the amount of solar radiation absorbed at the surface. This greenhouse energy permits the surface to warm significantly more than would be permitted by the solar radiation alone. (Based on MacCracken and Luther, 1985.)

atmosphere system. A schematic diagram of the global average energy balance is shown in Figure 13.2. The Sun emits most of its energy at wavelengths between 0.2 and 4.0 μm, primarily in the ultraviolet (UV), visible, and near-infrared (IR) wavelength regions. This is illustrated in Figure 13.3. A very small fraction of this energy is intercepted by the Earth as it orbits the Sun. The atmosphere absorbs approximately 23% (MacCracken and Luther, 1985) of the incoming solar radiation, principally by ozone (O_3) in the UV and visible ranges and by water vapor in the near IR.

The Earth re-emits the energy it absorbs back to space in order to maintain an energy balance. Because the Earth is much colder than the Sun, the bulk of this emission takes place at longer wavelengths than those for incoming solar radiation. Most of this radiation is emitted in the wavelength range from 4 to 100 μm, which is the region generally referred to as long-wave or IR radiation (see Figure 13.3). Although water vapor, carbon dioxide, and other greenhouse gases are relatively inefficient absorbers of solar radiation, these gases are strong absorbers of IR radiation. Clouds also play an important role in determining the

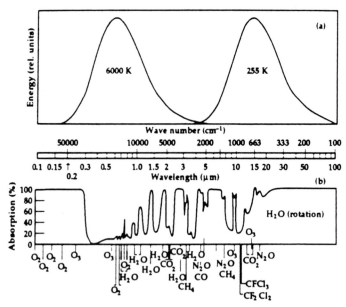

FIGURE 13.3. (a) Blackbody curves showing the variation in emitted energy with wavelength for temperatures typical of the Sun (6000 K) and the Earth (255 K), respectively, and (b) percentage of atmospheric absorption for radiation passing from the top of the atmosphere to the surface. Note the comparatively weak absorption of the solar spectrum and the region of weak absorption from 8 to 12 μm in the long-wave spectrum, referred to as the "window region". (Based on MacCracken and Luther, 1985.)

energy balance; one of the largest uncertainties in determining the climate change expected from greenhouse gases is the current limitation in understanding cloud processes.

The greenhouse gases (and clouds) re-emit the absorbed long-wave radiation based on their local atmospheric temperature, which tends to be cooler than the Earth's surface temperature. The absorbed energy is re-radiated in all directions, such that some of the radiation is lost to space. Some of the radiation, however, is emitted downward, leading to a net trapping of long-wave radiation and a warming of the surface.

13.2.3. Infrared Absorption by Greenhouse Gases

The greenhouse gases absorb the Earth-emitted infrared radiation in specific energy (or wavelength) bands determined by the quantum mechanical properties of the specific molecule. Water vapor is the single most important greenhouse gas in the Earth's atmosphere. The vibrational–rotational bands of water vapor absorb most of the radiation at wavelengths less than $8 \, \mu m$ and the rotational bands effectively absorb radiation at wavelenghts greater than $18 \, \mu m$ (see Figure 13.3). Atmospheric water vapor content is in approximate equilibrium between evaporation, condensation, and transport; changes in the tropospheric concentration of water vapor are thought to be primarily linked to climatic processes, while human activities are thought to have little direct impact on its concentrations. As a result, water vapor is considered when evaluating climatic feedback processes but is not one of the greenhouse gases included in concerns about future climate change. Carbon dioxide is the second most effective greenhouse gas, with its $15 \, \mu m$ band dominating the infrared absorption from 12 to $18 \, \mu m$. Other important greenhouse gases include ozone (O_3), methane (CH_4), nitrous oxide (N_2O), and the chlorofluorocarbons (CFCs, particularly $CFCl_3$ and CF_2Cl_2).

The spectral region from about 8 to $12 \, \mu m$ is referred to as the "window" region because of the relative transparency to radiation over these wavelengths. Nearly 80% of the radiation emitted by the surface in the window region escapes to space (Ramanathan, 1988). Most of the non-CO_2 greenhouse gases with the potential to affect climate, including O_3, CH_4, N_2O, and the CFCs, all have strong absorption bands in the atmospheric window region. Relatively small changes in the concentrations of these gases can produce a significant change in the net radiative flux affecting the Earth–troposphere climatic system.

As the concentration of a greenhouse gas becomes high, it can absorb most of the radiation in its energy bands; once any of its absorption wavelengths becomes saturated, it is unable to absorb more energy at that specific wavelength, and further increases in its concentration have a diminishing effect on climate. This is called the band saturation effect. For example, the radiative forcing from further increases in carbon dioxide concentration in the current atmosphere will increase as the natural logarithm of its concentration because of this effect. For carbon dioxide, parts of its absorption spectrum (particularly in the 15 μm band) are already so opaque that additional CO_2 molecules are less effective. Methane and nitrous oxide also exist in sufficient quantities that significant absorption is occurring; it is found that their forcing is approximately proportional to the square root of their concentration. Also, at the wavelengths where water vapor and carbon dioxide strongly absorb IR radiation, the greenhouse effect of other gases will be minimal. On the other hand, absorption by other gases at wavelengths that are not saturated, such as the chlorofluorocarbons or other halocarbons, varies linearly with concentration.

Another important consideration in radiative absorption is the band overlap effect. If a gas absorbs at wavelengths that are also absorbed by other gases, then the effect on radiative forcing of increasing its concentration can be diminished. For example, there is significant overlap between some of the absorption bands of methane and nitrous oxide that need to be carefully considered in determining their radiative forcing on climate. Other gases with primary absorption in the window region are generally unaffected by band overlap effects.

13.2.4. Factors Affecting Climate

Any factor which alters the radiation received from the Sun or that radiation lost to space, or which alters the redistribution of energy within the atmosphere and between the atmosphere, land, and ocean, can affect climate. Changes in the solar energy output reaching the Earth constitute the primary external forcing on the climate system. The Sun's output of energy is known to vary by small amounts over the 11 year cycle associated with sunspots, and there are indications that the solar output may vary by larger amounts over longer time periods. Slow variations in the Earth's orbit, over time scales of multiple decades to thousands of years, have led to changes in the solar radiation reaching the Earth and are thought to have played an important role in affecting the past climate, including the formation of the ice ages.

As the concentrations of greenhouse gases increase, the net trapping of IR radiation is enhanced, leading to an enhancement of the greenhouse effect and a potential warming of the lower atmosphere and the Earth's surface. The amount of warming from greenhouse gases will depend on several factors, including the amount of increase in concentration of each greenhouse gas, the radiative properties of the gases involved, the concentration of other greenhouse gases already present in the atmosphere, the resulting interactions with other radiatively important atmospheric constituents (such as through chemistry, as discussed later), and the effects of climatic feedbacks (also discussed later). The amount of warming can also depend on local effects such as the variation in the density of the greenhouse gas with altitude.

Aerosols, the small particles in the atmosphere, can also influence climate by absorbing and reflecting solar radiation. Moreover, changes in aerosol concentrations can alter cloudiness and change cloud reflectivity through their effects on cloud properties. Sources of atmospheric aerosols that can influence climate include dust and sulfur dioxide from volcanic eruptions, sulfates of industrial origin, and carbonaceous aerosols from biomass burning. In a general sense, aerosols tend to cool climate.

Although greenhouse-gas- or aerosol-induced changes in the radiative balance of the Earth will tend to alter atmospheric and oceanic temperatures and associated circulation and weather patterns, one should also recognize that climate varies naturally over all time scales. If the human-induced effects on climate are to be distinguished from natural variations, it is necessary to understand and identify the human-induced "signal" in climate against the background "noise" of natural climate variability.

13.3. THE GREENHOUSE GASES AND CLIMATE FORCING

13.3.1. The Greenhouse Gases

A necessary starting point in understanding the current and potential future effects of human activities on climate is to determine the changes that have occurred in concentrations of greenhouse gases and aerosols and to evaluate and estimate the future changes that may occur. Understanding the potential for future changes require understanding the strength of their sources, both natural and anthropogenic, and the strength of the processes determining their removal from the atmosphere.

The chemical formulas of many of the gases important to climate change are shown in Table 13.1. Also shown in this table are the known direct and indirect ways in which these gases can influence climate forcing. Most of the gases are themselves absorbers of long-wave terrestrial radiation, i.e., they are greenhouse gases. Several of the gases, such as hydroxyl (OH) and carbon monoxide (CO), do not have a direct influence in climate, but because of their importance in atmospheric chemical processes, these gases still have a strong influence on the rate of climate change. The atmospheric concentrations of a number of globally important trace gases are increasing; there is ample evidence suggesting that surface emissions, largely from anthropogenic sources, are primarily responsible for the increase in atmospheric concentrations of such gases as carbon dioxide, methane, nitrous oxide, and the chlorofluorocarbons.

Earlier chapters dealt extensively with the budgets and changing concentrations of greenhouse gases and aerosols in the atmosphere. The reader is referred to those chapters for further discussion on these gases and aerosols. Chapter 3

TABLE 13.1   Key Greenhouse Gases and Their Relevance to Climate

Trace Constituent	Common Name	Importance for Climate
CO_2	Carbon dioxide	Absorbs infrared radiation; affects stratospheric O_3
O_3	Ozone	Absorbs ultraviolet and infrared radiation
CH_4	Methane	Absorbs infrared radiation; affects tropospheric O_3 and OH; affects stratospheric O_3 and H_2O; produces CO_2
N_2O	Nitrous oxide	Absorbs infrared radiation; affects stratospheric O_3
$CFCl_3$	CFC-11	Absorbs infrared radiation; affects stratospheric O_3
CF_2Cl_2	CFC-12	Absorbs infrared radiation; affects stratospheric O_3
$C_2F_3Cl_3$	CFC-113	Absorbs infrared radiation; affects stratospheric O_3
C_2F_5Cl	CFC-115	Absorbs infrared radiation; affects stratospheric O_3
CHF_2Cl	HCFC-22	Absorbs infrared radiation; affects stratospheric O_3
$C_aH_bF_cCl_d$	Other HCFCs, HFCs	Absorb infrared radiation; HCFCs can affect stratospheric O_3
CCl_4	Carbon tetrachloride	Absorbs infrared radiation; affects stratospheric O_3
CH_3CCl_3	Methyl chloroform	Absorbs infrared radiation; affects stratospheric O_3
C_2H_2, etc.	Non-methane hydrocarbons	Absorbs infrared radiation; affects tropospheric O_3 and OH
OH	Hydroxyl	Scavenger for many atmospheric pollutants. including CH_4, CO, CH_3CCl_3, and CHF_2Cl
CO	Carbon monoxide	Affects tropospheric O_3 and OH cycles; produces CO_2
NO_x	Nitrogen oxides	Affects tropospheric O_3 and OH cycles; precursor of acidic nitrates; stratospheric emission affects O_3
CH_3Br	Methyl bromide	Affects stratospheric ozone
CF_2ClBr	Ha-1211	Absorbs infrared radiation; affects stratospheric O_3
CF_3Br	Ha-1301	Absorbs infrared radiation; affects stratospheric O_3
SO_2	Sulfur dioxide	Forms aerosols which scatter solar radiation
COS	Carbonyl sulfide	Forms aerosol in stratosphere which scatter solar radiation

Source: Based on Wuebbles and Edmonds (1991).

describes the changing concentrations of carbon dioxide, methane, nitrous oxide, and other important greenhouse gases. Chapter 7 discusses the chlorofluorocarbons and other halogens. Chapters 10 and 11 discuss the changes occurring in tropospheric and stratospheric ozone, respectively. Chapter 5 gives a more general discussion of aerosols in the atmosphere, while Chapter 8 discusses the sources of sulfuric aerosols in the troposphere and stratosphere.

13.3.2. The Role of Atmospheric Chemistry

Atmospheric chemistry is playing several important roles in determining current climate and the potential extent of future climate change. With the exception of carbon dioxide, the atmospheric concentrations of the trace constituents of concern to climate change are all influenced by chemical and photochemical processes in the atmosphere. The amount of time a greenhouse gas exists in the atmosphere, its atmospheric lifetime, is largely determined, for most greenhouse gases, by chemical sink processes. One of the important greenhouse gases, ozone, is primarily produced and destroyed through atmospheric chemistry. In addition, chemical interactions between greenhouse gases and other radiatively important atmospheric constituents provide important climatic feedback mechanisms that need to be understood and accounted for. The following discussion will focus on several special gases that are particularly influenced by chemical interactions.

13.3.2.1. Ozone

Ozone plays several important roles in affecting climate. Although the direct radiative effect on climate from CO_2 and the other greenhouse gases in Table 13.1 depends primarily on their concentration and distribution in the troposphere, the climatic effect of ozone depends on its distribution throughout the troposphere and stratosphere. Ozone is a primary absorber of solar ultraviolet (UV) and visible radiation in the atmosphere. It is the effect of ozone in determining the amount of ultraviolet radiation reaching the Earth's surface that has largely resulted in the national and international policy actions (on production of CFCs and other halocarbons) to protect the amount of stratospheric ozone. However, ozone is also a greenhouse gas, with a large infrared absorption band at 9.6 μm.

It is the balance between these radiative processes that determines the net effect of changes in the ozone distribution on climate (Lacis et al., 1990; Wang et al., 1993). Increases in ozone above roughly 28 km tend to decrease the surface temperature as a result of the increased absorption of solar radiation before it reaches the troposphere. Increases in ozone in the lower stratosphere and upper troposphere tend to increase the surface temperature because of the relative importance of the infrared greenhouse effect in this region. This is because the greenhouse effect produced by ozone is directly proportional to the temperature contrast between the radiation absorbed and the radiation emitted; this contrast is greatest near the tropopause where temperatures are at a minimum compared with the surface temperature.

As discussed in earlier chapters, emissions of greenhouse gases, such as methane and the CFCs, can affect concentrations of tropospheric and stratospheric ozone, leading to a radiative influence on climate in addition to the direct radiative effects of the gas emitted.

The absorption of solar radiation by ozone also explains the increase in temperature with altitude in the stratosphere. The thermal structure of the stratosphere is sensitive to changes in ozone and, because of its well-mixed concentration and its strong infrared properties, to changes in carbon dioxide. Changes in stratospheric temperatures can then effect stratospheric chemistry because of the strong temperature dependence of many important stratospheric processes. Resulting effects on stratospheric chemistry can further affect climatic forcing.

Because of its strong UV absorption properties, a decrease in stratospheric ozone can increase photolysis of tropospheric constituents, resulting in changes in the chemistry of the troposphere and additional possible effects on climatic radiative forcing.

13.3.2.2. Hydroxyl

Hydroxyl radical (OH) is not a greenhouse gas and does not have a direct radiative effect on climate. Nonetheless, OH is extremely important because of its chemical reactivity and the role it plays in determining the atmospheric lifetimes of many greenhouse gases and other radiatively important atmospheric constituents. Reaction with OH is the primary chemical sink for such gases as CH_4, CO, CH_3CCl_3, CH_3Cl, CH_3Br, SO_2, and other hydrocarbons and hydrogen-containing halocarbons. Most of the compounds being used or considered as replacements for CFCs have reaction with OH as their primary loss process in the atmosphere (see Chapter 7 for more details). Reaction of OH with methane and the halocarbons in the troposphere also limits the amount of these gases reaching the stratosphere, where these species can affect concentrations of stratospheric ozone.

Because its production is dependent on the amount of water vapor and ozone, changes in climate can influence the amount of tropospheric ozone. The primary destruction of OH occurs through reactions with CO and CH_4; as a result, levels of tropospheric OH depend on emissions and concentrations of these gases. Increases in the concentrations of CO and CH_4 can lead to decreased concentrations of OH, with a subsequent positive feedback on the atmospheric lifetimes, and consequently the concentrations, of CO, CH_4, and other molecules scavenged by OH.

13.4. RADIATIVE FORCING ON CLIMATE

A perturbation to the atmospheric concentration of an important greenhouse gas (or in the distribution of aerosols) induces a radiative forcing that can affect climate. In keeping with the definition adopted by the IPCC (1990, 1992) and WMO (1986, 1992), radiative forcing of the surface–troposphere system is defined as the change in net radiative flux at the tropopause after allowing for stratospheric temperatures to readjust to radiative equilibrium. The tropopause is chosen because it is considered in a global annual mean sense that the surface and troposphere are closely coupled. This definition is based on earlier climate-modeling studies which indicated an approximately linear relationship between the global mean radiative forcing at the tropopause and the resulting global mean surface temperature change (e.g., Hansen et al., 1981; Ramanathan et al., 1985, 1987).

Table 13.2 shows the radiative forcing per molecule for several of the important greenhouse gases relative to the radiative forcing for CO_2, assuming a background atmosphere corresponding to concentrations measured in 1990 (based on IPCC, 1990). As mentioned earlier, radiative forcing for these gases, all of which have absorption bands in the window region, are calculated to be much greater than the radiative forcing for CO_2.

Recent climate model studies (e.g., Wang et al., 1991, 1992, 1993; Cox et al., 1994) suggest that the relationship between global mean radiative forcing and global mean surface temperature change is not as simple as thought previously and may be dependent on the forcing mechanism. In particular, changes in radiative forcing with strong horizontal or vertical structure may produce different effects on climate than the same globally averaged radiative forcing for well-mixed gases like CO_2. For example, recent studies suggest that such effects may be important in cases where there are changes in the vertical structure of ozone (Wang et al., 1993) or where there are non-uniform changes in the geographical distribution of aerosols. In addition, the prior climate model studies have often expressed the total forcing on climate from all greenhouse gases in terms of the amount of CO_2 that would give this amount of forcing. Wang et al. (1991, 1992) indicate that such an assumption is invalid, even if all the gases are well mixed in the troposphere; however, this conclusion was based on calculations assuming instantaneous radiative forcing (not allowing the stratosphere to respond) and may be less of a problem if stratosphere adjustment were included.

TABLE 13.2   Radiative Forcing (ΔF) Relative to CO_2 per Unit Molecule Change in the Atmosphere for 1990 Background Atmosphere

Gas	ΔF per Molecule Relative to CO_2
CO_2	1
CH_4	21
N_2O	206
CFC-11	12400
CFC-12	15800
HCFC-22	10660
CH_3CCl_3	2730
CF_3Br	16000

Source: Based on IPCC (1990).

13.4.1. Past Trends in Radiative Forcing

13.4.1.1. Changes in Greenhouse Gases

Atmospheric concentrations of important greenhouse gases have increased substantially since the beginning of the industrial period (from the late 1700s). For example, concentrations of CO_2 have increased from about 280 ppm in the pre-industrial era (according to ice core records and other data sets) to 355 ppm in 1992. Methane concentrations have more than doubled, from about 0.7 to 1.75 ppm in 1992. There were no production and emissions of CFCs in the pre-industrial era.

Published analyses of the direct radiative forcing due to the changes in greenhouse gas concentrations since the late 1700s are generally in good agreement, determining an increase in radiative forcing of about $2.1–2.6\,W\,m^{-2}$ ($2.3\,W\,m^{-2}$ from IPCC, 1990, and Shi and Fan, 1992; $2.1\,W\,m^{-2}$ from Hansen et al., 1993, and Kiehl and Briegleb, 1993; $2.6\,W\,m^{-2}$ from Hauglustaine et al., 1993). Table 13.3 shows the change in direct radiative forcing determined by IPCC (1990) for this period and for the last decade (1980–1990). Of the $2.3\,W\,m^{-2}$ change in radiative forcing from greenhouse gases over the last two centuries, approximately $0.5\,W\,m^{-2}$ occurred within the last decade. By far the largest effect on radiative forcing has been the increasing concentration of carbon dioxide.

13.4.1.2. Changes in Ozone

A significant decease in stratospheric ozone has occurred over the last two decades, with much of the decrease occurring in the lower stratosphere at middle to high latitudes (Stolarski et al., 1991; WMO, 1992; Reinsel et al., 1994). Available evidence indicates that decrease in ozone is primarily related to the increasing abundance of chlorine and bromine in the stratosphere from emissions

TABLE 13.3   **Calculated Change in Radiative Forcing for Changes in Atmospheric Concentrations of Greenhouse Gases from Periods from the Beginning of the Industrial Period**

Greenhouse gas	Radiative Forcing ($W\,m^{-2}$)	
	1765–1990	1980–1990
CO_2	1.50	0.30
CH_4	0.42	0.06
N_2O	0.10	0.03
CFCs, other halocarbons	0.30	0.13
Total	2.32	0.52

Includes direct radiative effects only.

Source: Based on IPCC (1990, 1992).

of halocarbons (WMO, 1992). Analyses of the resulting radiative forcing effects have generally determined a cooling tendency from this decrease in ozone (WMO, 1992; IPCC, 1992; Ramaswamy et al., 1992; Schwarzkopf and Ramaswamy, 1993; Wang et al., 1993). These analyses show a large dependence of the radiative forcing on the latitudinal variations and assumed altitude dependence of the ozone loss. The globally averaged change in radiative forcing due to this ozone decrease is approximately -0.1 to $-0.2 \, \text{W m}^{-2}$.

The modeling analysis of Hauglustaine et al. (1994) finds a net warming rather than cooling; however, their model determined less ozone change in the lower stratosphere than is observed, and their derived forcing is dominated by changes in upper stratospheric ozone, where a decrease in ozone leads to a positive radiative forcing (see, e.g., Lacis et al., 1990).

Early studies (see, e.g., IPCC, 1992; WMO, 1992) described the globally averaged radiative forcing from the decrease in stratospheric ozone as being of the same magnitude but opposite in sign to the radiative forcing from the CFCs. However, this overly simplistic conclusion did not account adequately for the contribution of all of the chlorine- and bromine-containing halocarbons to the ozone loss. Daniel et al. (1994), by accounting for the relative effects of different halocarbons affecting ozone, have partitioned the direct and indirect (due to effects on ozone) radiative forcing from halocarbons. As shown in Figure 13.4, the direct radiative forcing from halocarbons (since pre-industrial times) is

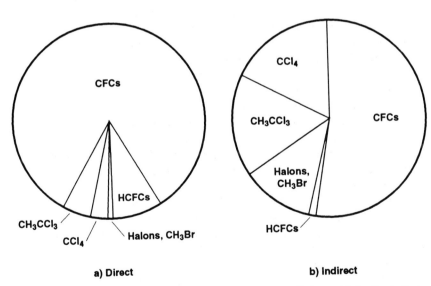

a) Direct   b) Indirect

FIGURE 13.4.   Relative partitioning of the radiative forcing in 1990 due to (a) the direct effects of CFCs and other halocarbons and (b) the indirect effects from CFCs and halocarbons on observed changes in stratospheric ozone over recent decades. (Based on Daniel et al., 1994.)

primarily due to CFCs; however, the cooling effect from the changes in ozone is more evenly divided between CFCs, bromine compounds, and other halocarbons. The net radiative forcing from direct plus indirect effects implies that CFCs account for roughly half of the forcing they would have if the indirect effects were not included.

Changes in tropospheric ozone, above the boundary layer, can also affect climate. Changes in ozone in this region are much more uncertain, although calculations indicate that an increase in ozone over recent decades (as suggested by some measurements) should add to the warming effect in radiative forcing from greenhouse gases (Wang et al., 1993; Hauglustaine et al., 1994).

13.4.1.3. Effects from Aerosols

Changes in concentrations of tropospheric aerosols represent the largest source of uncertainty in determining the effect of human activities on past climate forcing. Human-related aerosols may play a significant role in explaining the discrepancy between observed temperature changes and that expected from greenhouse gases. The best understood mechanism by which changes in concentrations of aerosols can influence climate is by scattering of incoming solar radiation. In addition, aerosols can act as cloud condensation nuclei, thus increasing the number concentration of cloud droplets and increasing cloud albedo. Corresponding decrease in cloud droplet size may inhibit promotion of precipitation, extend cloud lifetime, and increase atmospheric water vapor content.

Past changes in radiative forcing from aerosols are thought to have a cooling effect, thus mitigating some of the effect from greenhouse gases. Changes in amount of sulfate, nitrate, and carbonaceous aerosols may have contributed to effects over the last century. However, the geographical distribution and the diurnal and seasonal patterns of the radiative forcing from sulfate aerosols are quite different from that for greenhouse gases. Radiative forcing from sulfate aerosols is centered primarily over the industrialized regions in mid-latitudes of the northern hemisphere. Forcing from greenhouse gases is operative diurnally and seasonally, while aerosols have their primary effects during daylight (owing to the solar scattering effect) and during the summer.

Current estimates for the globally averaged direct solar effect on radiative forcing from changes in sulfate aerosols during the industrial period range from -0.2 to $-0.9\,W\,m^{-2}$ (Charlson et al., 1991; Kiehl and Briegleb, 1993; Hansen et al., 1993; Taylor and Penner, 1994). The northern hemispheric averaged effect would be roughly double these values. Carbonaceous aerosols from biomass burning may contribute another -0.1 to $-0.5\,W\,m^{-2}$ during this period (Penner et al., 1992; Andrae, 1993; Hansen et al., 1993). Indirect effects on clouds and cloud albedo are much more uncertain, and the radiative forcing from these effects may vary from zero up to as much as $-1\,W\,m^{-2}$ (Charlson et al., 1992).

Stratospheric sulfate aerosols can result in a large change in radiative forcing for several years following a major volcanic eruption. For example, the June 1991 eruption of Mount Pinatubo in the Philippines is estimated to have resulted in a maximum radiative forcing effect of $-4\,\mathrm{W\,m^{-2}}$ and as much as $-1\,\mathrm{W\,m^{-2}}$ up to 2 years after the eruption (Hansen et al., 1992).

13.4.1.4. Combined Effects

Based on the above discussion, the globally averaged radiative forcing on climate has increased by about $1.3 \pm 0.6\,\mathrm{W\,m^{-2}}$ over the last two centuries. However, this does not consider changes in tropospheric ozone that have likely increased the radiative forcing further. Changes in stratospheric water vapor (from oxidation of increasing levels of methane) and other indirect effects resulting from atmospheric chemistry are also not included. Effects on cloud albedo from aerosols could reduce the warming effect further. Because of the hemispheric and other inhomogeneous variations in concentrations of aerosols, the overall change in radiative forcing could be much greater or much smaller at specific locations about the globe, with the largest increase in radiative forcing expected in the southern hemisphere (where aerosol content is smallest).

13.4.2. Projections of Future Radiative Forcing

Future changes in radiative forcing are quite uncertain owing to uncertainties in future emissions and concentrations of greenhouse gases and aerosols. As an example of individual responses, Table 13.4 shows model-derived radiative forcing for changes in concentrations of several different gases and changes in the solar flux. The resulting change in surface temperature shown in Table 13.4

TABLE 13.4 Radiative Forcing at the Tropopause (ΔF) and Derived Surface Temperature Changes Assuming No Climatic Feedbacks (ΔT_0) for Selected Changes in Forcing Mechanisms

Forcing Mechanism		ΔF ($\mathrm{W\,m^{-2}}$)	ΔT_0 (K)
CO_2	Increase from 300 to 600 ppmv	4.35	1.31
CH_4	Increase from 1.6 to 3.2 ppmv	0.52	0.16
N_2O	Increase from 0.28 to 0.56 ppmv	0.92	0.27
CFC-11	Increase from 0 to 1 ppbv	0.22	0.07
CFC-12	Increase from 0 to 1 ppbv	0.28	0.08
O_3	Decrease by 50%	-1.2	-0.38
Solar constant	Increase by 2%	4.58	1.35

If climatic feedbacks were included, climate models suggest that temperatures would increase by a factor of 1–3 (IPCC, 1990, 1992).

Source: Based on results of Rind and Lacis (1993).

assumes that no climatic feedback processes occur in the atmosphere. These results are based on the one-dimensional radiative–convective model of Rind and Lacis (1993), although similar results have been found with other models (Cess et al., 1985; Ramanathan et al., 1987). However, in another study, Cess et al. (1993) examined the results for a doubling of the CO_2 concentration from 15 different climate models and found a range in the calculated radiative forcing of $3.3–4.8\,W\,m^{-2}$; their analysis indicated that the cause was model-to-model differences in the infrared radiation codes for treating CO_2, showing the importance of careful modeling of radiative transfer in these models.

A range of scenarios has been developed by IPCC (1992) for future emissions of greenhouse gases. None of these scenarios should be considered as a prediction of the future, but they do illustrate the effects of various assumptions about economics, demography, and policy on future emissions. Overall, the six scenarios developed indicate that greenhouse gas emissions could significantly increase over the next century in the absence of special measures to reduce emissions. The emissions from two of the six scenarios for 1990–2100 developed by IPCC (1992) are given in Table 13.5. These scenarios will be investigated as examples of the possible effect of greenhouse gases on climate.

Scenario IS92a is similar to the earlier scenario for Business as Usual developed for IPCC (1990). Scenario IS92a assumes moderate population growth (11.3 billion people by 2100), moderate economic growth, and partial compliance with controls on CFCs. In this scenario, CO_2 produced from fossil fuel burning and cement production increases by a factor of 3 from 1990 to 2100, whereas CO_2 emissions from deforestation are significantly reduced. With this scenario, concentrations of atmospheric CO_2 would be expected to be greater than 900 ppm by 2100 (Jain et al., 1994). Human-related emissions of CH_4 and N_2O approximately double. CFCs and halons are gradually phased out, but at a much slower rate than now expected. The other scenario shown in Table 13.5, the IPCC (1992) low emission scenario IS92c, assumes population grows and then declines by the middle of the next century, assumes low economic growth, and assumes severe constraints on fossil fuel supplies and use. In this scenario, CO_2 emissions eventually decline below 1990 levels. Resulting CO_2 concentrations reach 600 ppm in 2100 (Jain et al., 1994). Methane and nitrous oxide emissions only increase slightly.

Figure 13.5 shows the derived globally averaged radiative forcing as a function of time for these scenarios relative to the radiative forcing for the pre-industrial background atmosphere (using the LLNL model of Jain et al., 1994). Calculated radiative forcing increases to $8\,W\,m^{-2}$ by 2100 for scenario IS92a and to $4.8\,W\,m^{-2}$ for scenario IS92c. Both cases are a sizable increase over the $2.4\,W\,m^{-2}$ derived for 1990, implying a significant warming tendency (Section 13.5 examines the resulting effects on climate). Effects of aerosols and indirect chemical effects were not included.

TABLE 13.5   Current and Projected Annual Greenhouse Gas Emission Estimates for the Scenarios (a) IS92a and (b) IS92c

Gas	1990	2000	2005	2025	2050	2075	2100	Change (%) 1990–2100
(a)								
CO_2 (Gt C)								
Industrial	6.1	7.2	8.0	11.1	13.7	16.9	20.4	+230
Land use	1.3	1.3	1.2	1.1	0.8	0.4	−0.1	−108
N_2O (Tg N)	12.9	13.8	14.1	15.8	16.6	16.7	17.0	+32
CH_4 (Tg)	506	545	568	659	785	845	917	+61
CFC-11(kt)	289	168	137	94	85	16	2	−99
CFC-12 (kt)	362	200	161	98	110	22	1	−100
CFC-113 (kt)	147	29	22	21	24	0	0	−100
CFC-114 (kt)	13	4	3	3	3	0	0	−100
CFC-115 (kt)	7	5	4	1	1	0	0	−100
CCl_4 (kt)	119	34	15	19	21	0	0	−100
CH_3CCl_3 (kt)	738	353	137	97	110	0	0	−100
HCFC-22 (kt)	138	275	329	568	1058	1232	1225	+788
CF_3Br (kt)	4	4	4	2	1	1	0	−100
CO_2 Equiv. (Gt C)[a]	20.3	19.4	20.2	24.2	28.9	30.6	35.3	+74
(b)								
CO_2 (Gt C)								
Industrial	6.1	6.2	6.5	7.1	6.8	5.4	4.8	−23
Land use	1.3	1.3	1.3	1.1	0.7	0.2	−0.2	−115
N_2O (Tg N)	12.9	13.6	13.8	15.0	15.0	14.2	13.0	+6
CH_4 (Tg)	506.0	526.0	540.0	589.0	613.0	584.0	546.0	+8
CFC-11 (kt)	289	168	137	94	85	16	2	−99
CFC-12 (kt)	362	200	161	98	110	22	1	−100
CFC-113 (kt)	147	29	22	21	24	0	0	−100
CFC-114 (kt)	13	4	3	3	3	0	0	−100
CFC-115 (kt)	7	5	4	1	1	0	0	−100
CCl_4 (kt)	119	34	15	19	21	0	0	−100
CH_3CCl_3 (kt)	738	353	137	97	110	0	0	−100
HCFC-22 (kt)	138	275	329	568	1058	1232	1225	+788
CF_3Br (kt)	4	4	4	2	1	1	0	−100
CO_2 equiv. (Gt C)[a]	20.3	18.3	18.4	19.9	19.7	16.3	14.9	−26

[a] Equivalent CO_2 emission rate defined as the product of the 100 year global warming potential (IPCC, 1992; see also Section 13.8.1 and Table 13.6) and the mass (human-related) emission rate of the trace gases.

Future emissions of SO_2 and the resulting sulfate aerosols are projected to increase (as much as double) over the next century (IPCC, 1992), primarily as a result of energy production and use. However, because of pressures to control emissions, there are quite significant uncertainties in the extent of this increase. Increases are largely projected for developing countries, while downward trends

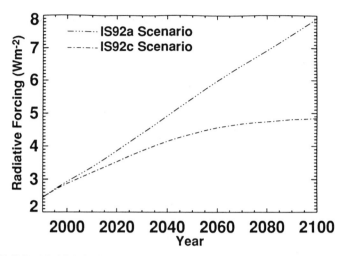

FIGURE 13.5. Model-derived change in radiative forcing for 1990–2100 (relative to pre-industrial atmosphere) for greenhouse gas emissions from IPCC scenarios IS92a and IS92c. (Based on Jain et al., 1994.)

are projected for North America and Europe. The possible changes in the geographical distribution of the emissions add to the difficulty of projecting future effects of aerosols on climate. Other factors affecting future radiative forcing, such as changes in tropospheric ozone or the forcing from volcanic eruptions, are extremely difficult to characterize.

A 1% change in total solar irradiance reaching the Earth is equivalent to a radiative forcing of $2.4\,W\,m^{-2}$ at the tropopause, comparable with the radiative forcing from increasing concentrations of greenhouse gases over the last few centuries. The 11 year sunspot cycle results in a change in the total irradiance of about 0.1 % (Lean, 1991). The Maunder Minimum, a period about 1 K colder than present from about 1645 to 1715, is associated with a reduction in solar irradiance of $(0.25 \pm 0.1)\%$ (Lean et al., 1992; Baliunas and Jastrow, 1990). If a similar event were to begin now or in the near future, it would partially offset the anticipated increase in radiative forcing from greenhouse gases (Wigley and Kelly, 1990).

13.5. PROJECTIONS OF FUTURE CLIMATE

13.5.1. Methods for Predicting Climate

The climate system consists of the coupling between several important elements, including the atmosphere, the oceans, the cryosphere, the biosphere, and the geosphere. Figure 13.6 gives a schematic illustration of the many components that compose the coupled climatic system. As mentioned earlier, the fundamental process driving the climate system is heating by incoming solar radiation and cooling by infrared radiation to space. The heating is strongest at tropical latitudes, while cooling predominates at polar latitudes in the winter hemisphere. The latitudinal gradient in heating drives the large-scale circulation in the atmosphere and the ocean. This provides the basic heat transfer necessary to balance the system.

Two approaches are being used by researchers for estimating the effect of changes in radiative forcing on the climate system. The method most accepted by researchers is the use of numerical models of the atmospheric general circulation (referred to as global circulation models or GCMs) that are akin to the models used for numerical weather forecasting. For climate studies, these models generally include representations of other elements of the climate system (using ocean models, land surface models, etc.) with varying degree of sophistication.

Most investigations of climate change with these models have generally used GCMs coupled with simple representations of the ocean and relatively simple schemes for land surface processes. Recent model studies have included more detailed representation, but low resolution, models of ocean processes.

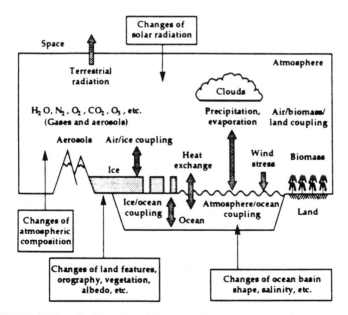

FIGURE 13.6. Schematic illustration of the components of the coupled atmosphere–ocean–ice–land climatic system. The solid (line) arrows represent external processes and the shaded arrows represent internal processes in climatic change. (Based on IPCC, 1990.)

Interpretation of climate change results from these models has primarily focused on the overall global effects of climate change. The resolution of existing climate models is still too coarse, and overall uncertainties are too great, to allow more than a limited regional interpretation of results.

The alternative approach being used for projecting climate change is based on reconstructions of past climates using paleoclimatic data (e.g., Budyko et al., 1987; Budyko and Izrael, 1987; MacCracken et al., 1990). The usefulness of past climates for projecting future climate has been questioned because of limited quality and quantity of paleoclimatic data over sufficient areas of the globe and because of questionable applicability of past climate forcing to projected forcing from future scenarios (IPCC, 1990). Nonetheless, paleoclimatic data provide valuable information about climate variability and the spectrum of possible climatic change, while also providing valuable data for evaluating climate models over different climatic regimes.

Analyses of past and current data on climate along with analyses of climate from numerical models are being used in combination to determine what future climate may be like if humans continue to affect radiative forcing. The rest of this section focuses on the processes that influence the response of climate and the models being used to analyze climate change. Section 13.5.3 will look at the past climate record and what it can tell us about the future.

13.5.2. Climatic Feedbacks

Many aspects of the climate system and its response to a change in radiative forcing are still not well understood. A significant number of the uncertainties are associated with understanding the interactive climate feedback processes that are key to determining the extent of climate response from a given change in climate forcing. The feedback processes determine whether the climate responds to amplify (through positive feedbacks) or reduce (through negative feedbacks) the expected warming from an increase in radiative forcing (Cess and Potter, 1988; Mahlman, 1991). Current models of climatic processes indicate that the net effect of the feedbacks being considered in these models produces an amplification in surface temperature ranging from a factor of 1 (i.e., no major amplification) to a factor of 3 (IPCC, 1990, 1992). Another way of representing this uncertainty in the response of the climate system to changes in radiative forcing is to base it on the equilibrium surface temperature for doubling of atmospheric CO_2 concentration calculated by climate models. Recent climate model calculations estimate this range at 1.5–4.5 °C (IPCC, 1990, 1992), with a best case estimate of 2.5 °C (IPCC, 1992; Hoffert and Covey, 1992). This uncertainty range in the climate sensitivity will be considered in later sections evaluating the effects of changes in surface temperature expected from past and projected future changes in radiative forcing.

Some of the most important feedbacks that have been identified and are considered in such models are described below.

13.5.2.1. H_2O Greenhouse Feedback

As the lower atmosphere (the troposphere) warms, it can hold more water vapor. The enhanced water vapor traps more IR radiation and amplifies the greenhouse effect. Ramanathan (1988) indicates that, based on studies with one-dimensional climate models, this feedback amplifies the air temperature by a factor of about 1.5 and the surface warming by a factor of about 3. The recent report by the IPCC (1992) determined a surface temperature amplification factor of 1.6 for water vapor feedback (see also Cess, 1992).

13.5.2.2. Ice Albedo Feedbacks

Warming induced by the global greenhouse effect melts sea-ice and snow cover. Whether it is ocean or land, the underlying surface is much darker (i.e., it has a lower albedo) than the ice or snow; therefore it absorbs more solar radiation, thus amplifying the initial warming. According to Ramanathan (1988), ice albedo feedback amplifies the global warming by 10–20%, with larger effects near sea-ice margins and in polar oceans. A comparison of 17 climate models generally agrees with this analysis, although a few models did get much larger albedo feedback effects (Cess et al., 1991).

13.5.2.3. Cloud Feedback

Cloud feedback mechanisms are extremely complex and are still poorly understood. Changes in cloud type, amount, altitude, and water content can all affect the extent of the climatic feedback. The sign of the feedback is also not understood, although current climate models generally find this to be a positive feedback. The treatment of clouds in GCMs is getting more and more complex; however, remaining uncertainties are so great that it is premature to attempt reliable conclusions about the magnitude of cloud feedback processes (IPCC, 1992).

13.5.2.4 Ocean–Atmosphere Interactions

The oceans influence the climate in several fundamentally important ways. First, because of the importance of the water vapor greenhouse feedback, the air and land temperature response is affected by the warming of the ocean surface. If the oceans do not respond to the greenhouse heating, the H_2O feedback will be turned off, because increased evaporation from the warmer ocean is the primary source of increasing atmospheric water vapor. Second, oceans can sequester the radiative heating into the deeper layers, which, because of their enormous heat capacity, can significantly delay the overall global warming effect over land surfaces. Current climate models suggest this delaying effect may cause a lag in the expected temperature response of at least a few decades, with the best estimate being about 50 years (Schlesinger and Jiang, 1990).

13.5.2.5. Other Feedbacks

Other feedbacks exist, such as changes in albedo related to changes in land features and biomass, but many of these feedbacks are not well understood. It is the combined effect of the many uncertainties in feedback processes that has resulted in the factor-of-3 (1.5–4.5 °C) difference in the amount of equilibrium warming estimated for a radiative equivalent doubling of CO_2 concentrations that was mentioned earlier.

13.5.3. Climate Model Projections

The GCM-based climate models attempt to synthesize the existing knowledge of the physical and dynamical processes in the overall climate system and try to account for the complex interactions between the various processes. These models incorporate well-established scientific laws (e.g., conservation of mass, energy, and momentum), empirical knowledge, and implicit representations. A typical GCM involves hundreds of thousands of equations and many variables. Climate models simulate the global distributions of the quantities thought to be important to climate, including temperature, winds, cloudiness, and rainfall. These models can represent on average most of the features of the current climate, including the large-scale distributions of pressure, temperature, winds and precipitation, which are well represented spatially and seasonally. However, on regional scales these variables are not determined as well. Errors in surface air temperature of 2–3 °C were found in selected regions in a comparison of models for IPCC (1990). In their current state of development the representations of many of the processes have major uncertainties that affect the predictability of climate change.

Historically, the standard calculation for predicting climate change effects has been based on a doubling of CO_2 concentrations. These equilibrium calculations have derived the range of 1.5–4.5 °C warming effect mentioned earlier. In addition to the warming effect from increased concentrations of greenhouse gases, current climate models predict that average precipitation will increase and areas of sea-ice and seasonal snow cover will decline. Air temperatures are projected to increase more over land than over the oceans, and soil moisture is calculated to decline in some mid-latitude continental areas. The models also suggest that global warming will intensify extreme events such as droughts and storms.

Simpler models, which simulate the behavior of GCMs, are often used to examine the time-dependent changes in globally averaged surface temperature for a variety of emissions scenarios. Figure 13.7 shows model-calculated changes

in global mean surface temperature for the IPCC IS92a and IS92c scenarios
described in Section 13.4.2. The Jain et al. (1994) model used in this analysis is
a globally averaged energy balance climate model that depends on results from
the GCM-based climate models for representation of the climate sensitivity
effects resulting from climatic feedbacks. The IS92a scenario results in a degree
or more warming than the IS92c scenario by 2100. Evaluated changes in surface
temperature for these scenarios are highly dependent on the assumed climate
sensitivity. The best-case estimate for the climate sensitivity of 2.5 °C (for a
doubling of CO_2) results in a warming relative to 1990 of about 1.1 °C by 2100
for scenario IS92c and about 2.1 °C for scenario IS92a. In scenario IS92a a
warming of 2 °C relative to the pre-industrial atmosphere is exceeded as early as
2030, 2055, and 2095 for the climate senstitivites of 4.5, 2.5, and 1.5 °C,
respectively. Slightly higher temperature changes were determined for these
scenarios in the Annex of IPCC (1992).

Recently, more GCM climate models have been applied to transient analyses
of climate change (IPCC, 1992; Manabe and Stouffer, 1993). Internal model
variability obscures geographical patterns of climate change for the first few
decades of the scenarios examined, but once established, the patterns are
generally similar to results from the equilibrium calculations (IPCC, 1992). For
instance, for increased greenhouse gases (often represented in terms of increasing
CO_2 only): surface air temperatures increase more over land than over the
oceans; precipitation increases on average at high latitudes, in the monsoon

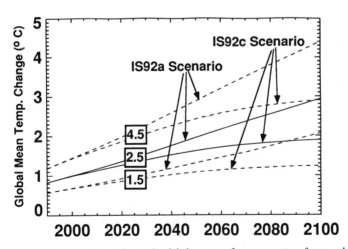

FIGURE 13.7.   Model-derived change in global mean surface temperature for greenhouse gas
emissions from IPCC scenarios IS92a and IS92c, including effects of the uncertainty range in
climate sensitivities (1.5, 2.5, and 4.5 °C for doubling of CO_2 concentration). (Based on Jain et
al., 1994.)

region of Asia, and in the winter at mid-latitudes; and soil moisture decreases on average in some mid-latitude continental areas in summer. However, the transient CO_2 calculations show less warming over some ocean areas as compared with the equilibrium calculations.

13.6. A LOOK AT THE PAST

13.6.1. The Temperature Record

The instrumental record of surface temperature is fragmentary until the mid-19th century, after which it improves slowly (IPCC, 1990). Global temperature data are available from about 1890. For the data since about 1900, similar temperature changes are seen in three independent data sets: one collected over land and two over the oceans (IPCC, 1990, 1992). Analyses of these data sets have attempted to negate known complicating factors such as biases introduced by local warming due to urban growth near some stations.

Figure 13.8 shows the record of the global mean surface air temperature compiled by Jones et al. (1991) from the available observations, given as anomalies from the mean temperature for 1950–1979. Since the late 19th century, the data indicate a surface temperature warming of $0.45 \pm 0.15\,°C$ (IPCC, 1990, 1992). Long-term warming trends in the northern and southern hemispheres are similar (IPCC, 1992).

Model calculations from Jain et al. (1994) based on the radiative forcing from greenhouse gases are also shown in Figure 13.8. These energy balance model

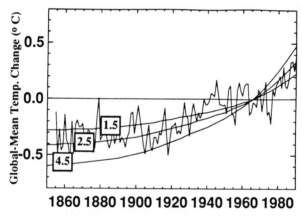

FIGURE 13.8. Observed and calculated mean global surface temperature anomalies plotted in reference to the 1950–1979 period. The jagged line reflects the temperature record constructed from measurements by Jones et al. (1991). The smooth curves are based on the model of Jain et al. (1994) for the radiative forcing from greenhouse gases and examining the uncertainty range in climate sensitivities.

calculations account for the uncertainty range in climate sensitivity discussed earlier. The model-calculated surface temperature changes in 1990 relative to the 1950–1979 reference period are 0.23, 0.34, and 0.48 °C for the $2 \times CO_2$ temperature sensitivities of 1.5, 2.5, and 4.5 °C, respectively. This compares with the observed temperature anomaly of 0.38 °C in 1990 (Jones et al., 1991). Overall, the calculated temperature changes from greenhouse gases only crudely approximate those observed. However, the calculations do not consider other effects on climate over this period, including the effects of aerosols, solar variability, volcanic eruptions, and natural climate variability. Also, while the observed warming is roughly consistent with increased concentrations of greenhouse gases, it is also within the range of natural variability.

The observed surface temperatures in the 1980s through 1991 are the warmest this century. Surface and satellite data indicate that surface and atmospheric temperatures declined during 1992 (Dutton and Christy, 1992; Hansen et al., 1993), with a mean surface cooling of 0.3–0.4 °C. Summer 1992 cooling was greatest in the continental interiors. This temperature response appears to be consistent with the temperature change expected from sulfuric aerosol effects following the Mount Pinatubo volcanic eruption (Hansen et al., 1993).

Recent analyses (Karl et al., 1991; Kukla and Karl, 1993; IPCC, 1992) indicate that much of the temperature increase over the last few decades has occurred because of an increase in night-time temperatures rather than daytime temperatures. What causes this observed trend? Greenhouse gases should be just as effective during any time of day. However, the radiative effects of sulfuric and other aerosols should primarily be during the daytime. Increased cloudiness would keep night-time minimum temperatures higher and daytime maximum temperatures lower. Cloudiness could increase naturally, as a response to increased aerosols, or as a response to climate change. A combination of all of these effects may be responsible for the observed diurnal trend in temperatures.

Various analyses have attempted to examine the combined effects of natural and human-related changes in radiative forcing to explain the observed temperature record. Analyses by Schlesinger and Ramankutty (1992) and Kelly and Wigley (1992) indicate that there is strong circumstantial evidence for solar variations contributing to the observed temperature changes over the last century, but these studies also indicate that greenhouse gases have been the dominant contributor to the observed temperature increase over this period. Assumed trends in sulfate aerosols included in these studies are also important for achieving the best match to the observed temperature record. Schlesinger and Ramankutty (1992) identify a temperature oscillation with a period of 65–70 years in the temperature record, apparently resulting from predictable internal variability in the ocean–atmosphere system. This oscillation may also contribute to obscuring the greenhouse warming signal in the temperature record.

13.6.2. Paleoclimates

Climate varies naturally on all time scales, ranging from a few years to millions of years. Figure 13.9 illustrates the temperature record over the last million years. Isotopic composition in ice cores and other proxy data for surface temperature were used to derive this record. Over this period, glacial (ice ages) and interglacial cycling have occurred on a time scale of 100,000 years with temperature variations of 5–7 °C. Since the end of the last ice age, about 10,000 years ago, global mean surface temperatures have fluctuated over a range up to 2 °C on time scales of centuries. As seen in the results from climate models,

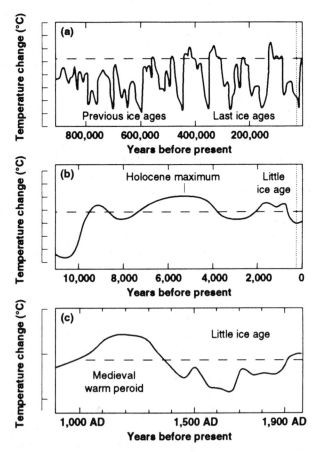

FIGURE 13.9. Schematic diagrams of the global temperature variations since the Pleistocene on three time scales: (a) the last million years, (b) the last ten thousand years; (c) the last thousand years. The dashed line nominally represents conditions near the beginning of the 20th century. (Based on IPCC, 1990.)

representative projections of future greenhouse gas emissions and concentrations could lead to global mean surface temperatures in the next century that are greater than have occurred in the last 200,000 years.

As mentioned in Section 13.5.1, analyses of past climates are being used as analogs for projecting future states of climate. For example, Hansen et al. (1993) use paleoclimatic data to estimate that the climate sensitivity for ice age cooling is $3 \pm 1\,°C$ for equivalent radiative forcing change to a doubling of atmospheric CO_2.

Budyko and Izrael (1987) suggest three periods of the paleoclimatic record as analogues of a future warmer climate. These periods are: (1) the climate optimum (approximately $3-4\,°C$ warmer than present at northern mid-latitudes) of the Pliocene, 3.3–4.3 million years before present; (2) the Eemian interglacial optimum ($1-3\,°C$ warmer than present) at 125,000–130,000 years before present; and (3) the mid-Holocene maximum (about $1\,°C$ warmer on average) at 5000–6000 years before present. These analogues imply that warmer temperatures at mid-latitudes and increased precipitation over the continents could be expected from increased levels of greenhouse gases. However, there are many uncertainties in the interpretation of these climatic reconstructions that limit their usefulness as analogs for future climate.

13.7. IMPACTS OF CLIMATE CHANGE

Many questions about how the changing composition of the atmosphere will affect climate remain unanswered. None the less, a number of research studies are attempting to determine what sorts of impacts on our environment may result from global warming. This section attempts to highlight existing understanding of the concerns regarding the consequences of climate change (see e.g., IPCC, 1990, 1992; NAS, 1991; US Department Of Energy (DOE) Multi-Laboratory Climate Change Committee, 1990; MacCracken, 1988).

At this point there is much less certainty about the climate changes expected in local regions compared with the expected overall global warming. Computer models of the atmosphere and climate generally agree that an overall global warming will occur because of increasing radiative forcing resulting from increasing concentrations of greenhouse gases, but they are much less certain in their representation of the spatial distribution of this warming. There is even less certainty about regional changes in precipitation. Some of the potential impacts of climate change are described below.

13.7.1. Sea-Level Rise

Sea-levels are expected to rise as a response to global warming, but the rate and timing remain uncertain. Over the last century the global mean sea-level has risen about 10–20 cm. Over the next century current models project a further increase

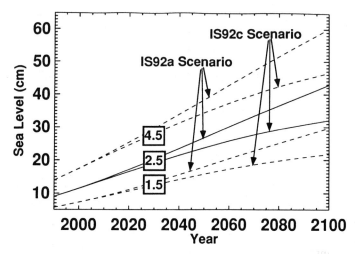

FIGURE 13.10. Model estimated changes in sea-level for IPCC (1992) scenarios IS92a and IS92c, including effects of the uncertainty range in climate sensitivities. (Based on Jain et al., 1994.)

in global mean sea-level of 10–100 cm for typical scenarios of greenhouse gas emissions and resulting climate effects (IPCC, 1990, 1992; NAS, 1991). Figure 13.10 shows the Jain et al. (1994) model-derived estimated effect on sea-level based on the IPCC (1992) scenarios IS92a and IS92c. For the climate sensitivity range of 1.5–4.5 K (for a doubling of CO_2 concentration) the model-derived sea-level rise for 1990 ranges from 5.6 to 13.6 cm relative to the pre-industrial period, well within the uncertainty range of analyses of sea-level changes over this time period (IPCC, 1990). Projections for these scenarios from 1990 to 2100 determine an increase in sea-level of 20–46 cm for scenario IS92a and an increase of 16 to 32 cm for scenario IS92c.

Even if further increases in greenhouse forcing were halted by 2030, sea-level would continue to rise from 2030 to 2100 by as much again as from 1990 to 2030. Most of the contribution to sea-level rise is expected to derive from thermal expansion of the oceans and the increased melting of mountain glaciers. The prospect of such an increase in the rate of sea-level rise is of concern to low-lying coasts. The most severe effects of the sea-level rise may result from extreme events (e.g., storm surges), the incidence of which may also be affected by climatic change.

13.7.2. Effects on Human Health

Direct effects on human health of the emitted greenhouse gases are believed to be small. Other links between climate change and human health have not been well established, although modest effects could result. For example, temperature increases could stress human health (e.g., heat stress), but in cold areas they could reduce stress; some diseases and pests could be more prevalent. However, good predictions of effects on health cannot be made without good predictive data on changes in local temperatures, humidities, and levels of precipitation. Stratospheric O_3 depletion could affect health because of the corresponding increase in UV radiation.

13.7.3. Agriculture and Food Supplies

Increased atmospheric CO_2 could potentially enhance growth of some food crops as well as increase water use efficiency. However, changing climatic patterns could require changes in cropping patterns and consequently in infrastructures and costs, perhaps bringing benefits to some regions while negatively impacting others. According to the DOE Multi-Laboratory Climate Change Committee, it appears likely that the future rate of growth in the worldwide demand for food and fiber can be met assuming typical projections of climate change. However, rapid changes in climate or more severe climate change could make adaptation more difficult.

13.7.4. Ecosystems

Ecosystem structure and species distribution are thought to be particularly sensitive to the rate of change of climate. Local changes in temperature, precipitation, and soil moisture and the occurrence of extreme events are all thought to exert stress on ecosystems. The globally-averaged rate of temperature change calculated by the Jain et al. (1994) model is shown for the IPCC IS92a and IS92c scenarios in Figure 13.11. Depending on the climate sensitivity, the IS92a scenario is likely to result in a rate of change equal to or much greater than 0.1 K per decade. It is uncertain whether this rate of change is sufficient to allow for adequate adaptation by ecosystems. Changes in precipitation or in frequency of events such as droughts or storms may be more directly related to stress on ecosystems, but there remains greater uncertainty in estimating these effects of climate change.

Increasing concentrations of CO_2 increase plant growth. However, rapid changes in climate threaten a reduction in biodiversity. Some existing species of plants and animals might be unable to adapt, being insufficiently mobile to migrate at the rate required for survival. Many uncertainties about the adaptability of natural ecosystems to climate change still remain.

13.8. POLICY CONSIDERATIONS

There are many possible policy responses to the prospects of human-related effects on climate (see, e.g., NAS, 1991; Pearman, 1992; Goss Levi et al., 1992;

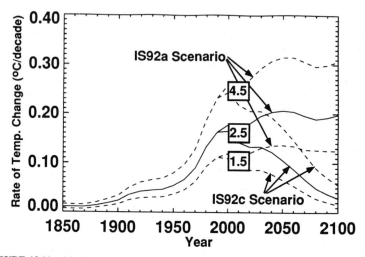

FIGURE 13.11. Model derived rate of change in global mean temperature for IPCC (1992) scenarios IS92a and IS92c, including effects of the uncertainty range in climate sensitivities. (Based on Jain et al., 1994.)

Schlesinger, 1993). In particular, for concerns about global warming related to emissions of greenhouse gases, these responses include: (1) options that reduce greenhouse gas emissions; (2) options that offset the effects of greenhouse gas emissions by either removing the gases directly from the atmosphere or altering their effect on climate; and (3) options that help human and ecologic systems adjust or adapt to new climatic conditions and events. The first two categories of these options are referred to as mitigation options and the third are adaptation options.

While it is not the intention in this chapter to consider or examine the range of possible policy options, it is interesting to consider the reduction in human-related emissions that would be required to stabilize the concentration of greenhouse gases at current concentrations. IPCC (1990) estimated it would require greater than a 60% reduction in human-related emissions of CO_2, 15–20% reduction in emissions of methane, 70–80% in emissions of N_2O, 70–85% in emissions of CFCs, and 40–50% in emissions of HCFC-22. In attempts to reduce the changes in radiative forcing on climate, by far the greatest challenge confronting policy makers is determining means of reducing emissions of CO_2. Societal dependence on fossil fuels will make it extremely difficult to reduce emissions sufficiently to stabilize the current concentration of CO_2.

Tropospheric aerosols have lifetimes of about 1 week or less. Radiative forcing from these aerosols is due to about 1 week's emissions, while forcing from greenhouse gases is generally the result of decades of emissions. As a result, any policy to control SO_2 emissions or other sources of tropospheric aerosols will have an immediate impact on radiative forcing from aerosols. Likewise, a policy considering the use of aerosols to compensate for greenhouse gas effects (ignoring the problems with distribution, diurnal and seasonal effects, effects of aerosols on atmospheric chemistry, etc.) would require a long term commitment to generate aerosols.

13.8.1. Global Warming Potentials

Simplified measures for representing human-related effects on climate can provide important insights to scientific and policy analyses. The concept of global warming potentials (GWPs) was developed as a means to provide a simple representation of the relative effects on climate resulting from a unit mass emission of a greenhouse gas (IPCC, 1990, 1992). Global warming potentials are expressed as the time-integrated radiative forcing from the instantaneous release of 1 kg (i.e., a small mass emission) of a trace gas expressed relative to that of 1 kg of the reference gas, CO_2:

$$GWP(x) = \frac{\int_0^n a_x[x(t)] \, dt}{\int_0^n a_{CO_2}[CO_2(t)] \, dt}$$

where n is the time horizon over which the calculation is considered, a_x is the climate-related radiative forcing response due to a unit increase in atmospheric concentration of the gas in question, $[x(t)]$ is the time-decaying concentration of that gas, and the corresponding quantities for the reference gas are in the denominator. The radiative forcing responses a, are derived from radiative transfer models. The trace gas concentrations $[x(t)]$ remaining after time t are based upon the atmospheric lifetimes of the gas in question. The reference gas has been taken generally to be CO_2, since this allows a comparison of the radiative forcing role of the emission of the gas in question with that of the dominant greenhouse gas that is emitted as a result of human activities. GWPs can account for both the direct radiative effects from a gas and the indirect effects resulting from chemical effects on other greenhouse gases.

The best choice of integration time horizon in evaluating GWPs has been the subject of much discussion and controversy (IPCC, 1990; WMO, 1992). The complexities of treating CO_2 and the carbon cycle prevent integration of GWPs to steady state. There is, however, no given value of integration time for determining GWPs that is ideal over the range of uses of this concept. GWPs are generally calculated over three time horizons, for 20, 100, and 500 years. It is believed that these three time horizons provide a practical range for policy applications (IPCC, 1992). GWPs determined for the longer integration period provide a measure of the cumulative chronic effects on climate, while the integration over the shorter period is representative of near-term effects. GWPs evaluated over the 100 year period appear generally to provide a balanced representation of the various time scales for climatic response (WMO, 1992). The best choice of time horizon will depend on the specific analysis being considered. One needs to balance the effects of near-term responses in comparing greenhouse gases with consideration of the long-term persistence of any environmental effects from long-lived gases.

The availability of an index that places the greenhouse gases on an equivalent scale for their effects on climate has considerable practical value. Such an index can be used in a variety of different analyses. Some examples of such uses are: (1) assessing the relative contributions of the many human activities contributing to greenhouse gas emissions; (2) comparing (and ranking) greenhouse gas effects from competing technologies and energy uses, including consideration of different energy policies; (3) developing approaches to minimize the impact of human activities on the climate system; (4) ranking the emissions from various countries; (5) providing a basis for comparing reductions in the total climate-forcing effects from various countries; and (6) functioning as a signal to policy makers for encouraging some activities and discouraging others. By using such indices, industries and governments can also determine approaches most appropriate for them to meet commitments to help reduce the changes in radiative forcing from emissions of greenhouse gases. At the same time it should

TABLE 13.6   Selected Global Warming Potentials (GWPs) Evaluated for the Direct Radiative Forcing from Greenhouse Gases

Gas	Time Horizon		
	20 years	100 years	500 years
CO_2	1	1	1
CH_4	35	11	4
N_2O	260	270	170
CFC-11	4500	3400	1400
CFC-12	7100	7100	4100
CFC-113	4600	4500	2500
CFC-114	6100	7000	5800
CFC-115	5500	7000	8500
CH_3CCl_3	360	100	34
HCFC-22	4200	1600	540
HFC-134a	3100	1200	400
CF_3Br	5600	4900	2300

Source: Based on IPCC (1992).

TABLE 13.7   **Estimated Human-Related Emissions for 1992 (in Tg year^{-1}) Multiplied by GWPs from Table 13.6**

Gas	Emissions (Tg year^{-1})	Time Horizon 20 years	100 years	500 years
CO_2	22700[a]	22700 (1)	22700 (1)	22700 (1)
CH_4	340	11900[b] (0.52)	3740[b] (0.16)	1360[b] (0.06)
N_2O	4.7	1220 (0.05)	1270 (0.06)	800 (0.04)
CFC-11	0.20	918 (0.04)	690 (0.03)	290 (0.01)
CFC-12	0.33	2340 (0.10)	2340 (0.10)	1340 (0.06)
CFC-113	0.17	800 (0.04)	780 (0.03)	430 (0.02)
CFC-114	0.005	32 (0.001)	36 (0.001)	30 (0.001)
CFC-115	0.013	69 (0.003)	88 (0.004)	110 (0.005)
CH_3CCl_3	0.59	210 (0.009)	59 (0.003)	20 (0.001)
HCFC-22	0.22	900 (0.04)	340 (0.01)	120 (0.005)
CF_3Br	0.007	39 (0.002)	34 (0.001)	16 (0.001)

[a]For CO_2, fossil fuel sources only (= 6.2 GtC).
[b]Indirect chemical effects could approximately double (or perhaps more than double) the values calculated.

be noted that no single indexing approach or tool is likely to meet all the needs of policy makers. Policy analyses often need a variety of tools, including results from climate-modeling studies, along with careful consideration of the state of the science.

Table 13.6 gives values of GWPs for a few select gases based on the IPCC (1992) analysis (these values are currently being revised for the 1994 IPCC interim assessment). For all integration time horizons the GWPs for all greenhouse gases are greater than 1, the GWP for CO_2 by definition. Because of their strong infrared absorption and their long atmospheric lifetimes, the GWPs for CFCs are quite large.

Table 13.7 shows an example of using GWPs in which the mass emission rates from human-related emissions for 1992 are multiplied by the GWPs over the integration time horizons from Table 13.6. In the near term (20 year integration), current emissions of methane (especially if indirect effects are included) are nearly as important as CO_2 in terms of their effect on radiative forcing, while the cumulative effect over long time horizons (500 years) is totally dominated by the effects of current carbon dioxide emissions. Another example, in defining equivalent CO_2 emissions, is given in Table 13.5.

13.9. CONCLUSIONS

There are several aspects of the concerns about climate change for which there is a high level of certainty. Absorption properties of the relevant greenhouse gases are generally well understood. Global concentrations of important greenhouse gases, namely, CO_2, CH_4, N_2O, and CFCs, are known to have been increasing for many decades. Available evidence indicates that these increasing concentrations are largely related to human activities. Climate models accurately show that these greenhouse gases are increasing the radiative forcing on climate, producing a tendency for global warming.

There are other aspects of climate change that are less certain but highly probable. For example, changes in the distributions of ozone and aerosols are likely affecting radiative forcing but need further clarification. Also, climate models indicate that global mean surface temperature will likely warm by several degrees Kelvin over the next century as a result of increasing radiative forcing from greenhouse gases (estimated at 2–5 K by IPCC, 1992). Such warming should be accompanied by an increase in tropospheric water vapor and an increase in globally averaged precipitation. Stratospheric temperatures should decrease as a result of the increasing CO_2.

There are other aspects that are still highly uncertain. In particular, the regional changes in climate expected from global warming are poorly understood. Likewise, the impacts on humanity and the biosphere from changes in climate are poorly understood. The present state of knowledge of tools for analyzing the consequences of climate change is such that, even if climate change could be forecast with certainty, the relationship between climate change and societal damage or benefit could not be well determined. There is an acute need to systematically expand our understanding of the consequences of climate change.

In looking at the broad perspective of understanding climate change, Schneider (1994) suggests that the scientific community needs to work across many scales and disciplines to understand physical, chemical, biological, and relevant societal processes, their interactions, the heterogeneity in net radiative forcing they imply, the implications of transient, heterogeneous forcing for the Earth systems response, and the synergisms that will be discovered from coupled Earth systems research.

Acknowledgments

This work was performed under the auspices of the U.S. Department of Energy by Lawrence Livermore National Laboratory under Contract No. W-7405-ENG-48 and was supported in part by the Department of Energy's Environmental Sciences Division and by the U.S. Environmental Protection Agency.

References

Andreae, M.O. (1993). Climatic effects of changing atmospheric aerosol levels. In Henderson-Sellers, A., Ed., *World Survey of Climatology, Volume XX: Future Climates of the World*.

Baliunas, S. and Jastrow, R. (1990). Evidence for long-term brightness changes of solar-type stars. *Nature*, **348**, 520–523.

Budyko, M.I. and Izrael, Yu. A., Eds (1987). *Anthropogenic Climatic Changes*, Springer Verlag, Berlin.

Budyko, M.I., Ronov, A.B., and Yanshin, A.L. (1987). *History of the Earth's Atmosphere*, Springer Verlag, Berlin.

Cess, R.D. (1992). Comparison of general circulation models. In Levi, B.G., Hafemeister, D. and Scribner, R., Eds, *Global Warming: Physics and Facts*, American Institute of Physics, New York, NY.

Cess, R.D. and Potter, G.L. (1988). A methodology for understanding and intercomparing atmospheric climate feedback processes in general circulation models. *J. Geophys. Res.*, **93**, 8305–8314.

Cess, R.D., Potter, G.L., Ghan, S.J. and Gates, W.L. (1985). The climatic effects of large injections of atmospheric smoke and dust: a study of climatic feedback mechanisms with one and three dimensional climate models. *J. Geophys. Res.*, **90**, 12,937–12,950

Cess, R.D., Potter, G.L., Zhang, M.-H., Blanchet, J.-P., Chalita, S., Colman, R., Dazlich, D.A., Del Genio, A.D., Dymnikov, V., Galin, V., Jerrett, D., Keup, E., Lacis, A.A., Le Treut, H., Liang, X.-Z., Mahfouf, J.-F., McAvaney, B.J., Meleshko, V.P., Mitchell, J.F.B., Morcrette, J.-J., Norris, P.M., Randall, D.A., Rikus, L., Roeckner, E., Royer, J.-F., Schlese, U., Sheinin, D.A., Slingo, J.M., Sokolov, A.P., Taylor, K.E., Washington, W.M., Wetherald, R.T. and Yagai, I. (1991). Interpretation of snow–climate feedback as produced by 17 general circulation models. *Science*, **253**, 888–892.

Cess, R.D., Zhang, M.-H., Potter, G.L., Barker, H.W., Colman, R., Dazlich, D.A., Del Genio, A.D., Esch, M., Fraser, J.R., Galin, V., Gates, W.L., Hack, J.J., Ingram, W.J., Kiehl, J.T., Lacis, A.A., Le Treut, H., Liang, X.-Z., Mahfouf, J.-F., McAvaney, B.J., Meleshko, V.P., Morcrette, J.-J., Randall, D.A., Roeckner, E., Royer, Sokolov, A.P., Sporyshev, P.V., Taylor, K.E., Wang, W.-C. and Wetherald, R.T. (1993). Uncertainties in carbon dioxide radiative forcing in atmospheric circulation models. *Science,* **262**, 1252–1255.

Charlson, R.J., Langner, J., Rodhe, H., Leovy, C.B. and Warren, S.G. (1991). Perturbation of the Northern Hemisphere radiative balance by backscattering from anthropogenic sulfate aerosols. *Tellus,* **43AB**, 152–163.

Charlson, R.J., Schwartz, S.E., Hales, J.M., Cess, R.D., Coakley, J.A., Jr., Hansen, J.E. and Hofmann, D.F. (1992). Climate forcing by anthropogenic aerosols. *Science*, **255**, 423–430.

Cox, S.J., Wang, W.-C. and Schwartz, S.E. (1994). Different climate response to radiative forcings by sulphate aerosols and greenhouse gases. *Nature*, in press.

Daniel, J.S., Solomon, S. and Albritton, D.L. (1994). On the evaluation of halocarbon radiative forcing and global warming potentials. *J. Geophys. Res.*, in press.

DOE Multi-Laboratory Climate Change Committee (1990). *Energy and Climate Change,* Lewis Publ., Chelsea, MI.

Dutton, E.G. and Christy, J.R. (1992). Solar radiative forcing at selected locations and evidence for global lower tropospheric cooling following the eruptions of El Chichon and Pinatubo. *Geophys. Res. Lett.*, **19**, 2313–2316.

Goss Levi, B., Hafemeister, D. and Scribner, R., Eds (1992). *Global Warming: Physics and Facts*, American Institute of Physics, New York, NY.

Hansen, J., Johnson, D., Lacis, A., Lebedeff, S., Lee, P., Rind, D. and Russel, G. (1981). Climate impacts of increasing carbon dioxide. *Science,* **213**, 957–966.

Hansen, J.E., Lacis, A., Ruedy, R. and Sato, M. (1992). Potential climate impact of Mount Pinatubo eruption. *Geophys. Res. Lett.,* **19**, 215–218.

Hansen, J., Lacis, A., Ruedy, R., Sato, M. and Wilson, H. (1993). How sensitive is the world's climate. *Natl. Geogr. Res. Explor.,* **9**, 142–158.

Hauglustaine, D.A., Granier, C., Brasseur, G.P. and Megie, G. (1994). The importance of atmospheric chemistry in the calculation of radiative forcing on the climate system. *J. Geophys. Res.,* **99**, 1173–1186.

Hoffert, M.I. and Covey, C. (1992). Deriving global climate sensitivity from paleoclimate reconstructions. *Nature,* **360**, 573–576.

IPCC (Intergovernmental Panel on Climate Change) (1990). Houghton, J.T., Jenkins, G.J. and Ephraums, J.J., Eds, *Climate Change: The IPCC Scientific Assessment,* Cambridge University Press, Cambridge.

IPCC (1992). Houghton, J.T., Jenkins, G.J. and Ephraums, J.J., Eds, *1992 IPCC Supplement,* Cambridge University Press, Cambridge.

Jain, A.K., Kheshgi, H.S. and Wuebbles, D.J. (1994). Integrated science model for assessment of climate change. *87th Annual Meeting Report of the Air and Waste Management Association, Cincinnati, OH, June 19–24,* Paper 94-TP59.08.

Jones, P.D., Wigley, T.M.L. and Farmer, G. (1991). Marine and land temperature data sets: a comparison and a look at recent trends. In Schlesinger, M.E., Ed., *Greenhouse-Gas-Induced Climate Change: A Critical Appraisal of Simulations and Observations,* Elsevier, Amsterdam.

Karl, T.R., Kukla, G., Razuvayev, V.N., Changery, M.J., Quayle, R.G., Helm, R.R., Jr., Easterling, D.R. and Fu, C.B. (1991). Global warming: evidence for asymmetric diurnal temperature change. *Geophys. Res. Lett.,* **18**, 2253–2256.

Kelly, P.M. and Wigley, T.M.L. (1992). Solar cycle length, greenhouse forcing and global climate. *Nature,* **360**, 328–330.

Kiehl, J.T. and Briegleb, B.P. (1993). The relative roles of sulfate aerosols and greenhouse gases in climate forcing. *Science,* **260**, 311–314.

Kukla, G. and Karl, T.R. (1993). Nighttime warming and the greenhouse effect. *Environ. Sci. Technol.,* **27**, 1468–1474.

Lacis, A.A., Wuebbles, D.J. and Logan, J.A. (1990). Radiative forcing of climate by changes in the vertical distribution of ozone. *J. Geophys. Res.,* **95**, 9971–9981.

Lean, J. (1991). Variations in the sun's radiative output. *Rev. Geophys.,* **29**, 505–535.

Lean, J., Skumanich, A. and White, O. (1992). Estimating the sun's radiative output during the Maunder minimum. *Geophys. Res. Lett.,* **19**, 1591–1594.

Leith, C.E. (1993). Numerical models of weather and climate. *Lawrence Livermore National Laboratory Report UCRL-JC-114919.*

MacCracken, M.C. (1988). Greenhouse warming: what do we know? UCRL-99998, Lawrence Livermore National Laboratory, Livermore, CA.

MacCracken, M.C., Budyko, M.I., Hecht, A.D. and Izrael, Y.A., editors (1990). Prospects for Future Climate. Lewis Publishers, Chelsea, MI.

MacCracken, M.C. and Luther, F.M. (1985). *Projecting the climatic effects of increasing carbon dioxide. DOE Report DOE/ER-0237,* US Department of Energy, Washington, DC.

Mahlman, J.D. (1991). Assessing global climate change: when will we have better evidence. In *Climate Change and Energy Policy*, American Institute of Physics, New York, NY.

Manabe, S. and Stouffer, R.J. (1993). Century-scale effects of increased atmospheric CO_2 on the ocean–atmosphere system. *Nature*, **364**, 215–218.

NAS (National Academy of Sciences) (1991). *Policy Implications of Greenhouse Warming*, Policy Implications of Greenhouse Warming—Synthesis Panel, National Academy Press, Washington, DC.

Pearman, G.I., Ed. (1992). *Limiting Greenhouse Effects: Controlling Carbon Dioxide Emissions*, Wiley, New York, NY.

Penner, J.E., Dickinson, R.E. and O'Neill, C.A. (1992). Effects of aerosol from biomass burning on the global radiation budget. *Science*, **256**, 1432–1434.

Ramanathan, V. (1988). The greenhouse theory of climate change: a test by an inadvertent global experiment. *Science*, **240**, 293–299.

Ramanathan, V., Callis, L., Cess, R., Hansen, J., Isaksen, I., Kuhn, W., Lacis, A., Luther, F., Mahlman, J., Reck, R. and Schlesinger, M. (1987). Climate–chemical interactions and effects of changing atmospheric trace gases. *Rev. Geophys.*, **25**, 1441–1482.

Ramanathan, V., Cicerone, R.J., Singh, H.B. and Kiehl, J.T. (1985). Trace gas trends and their potential role in climate change. *J. Geophys. Res.*, **90**, 5547–5557.

Ramaswamy, V., Schwarzkopf, M.D. and Shine, K.P. (1992). Radiative forcing of climate from halocarbon-induced global stratospheric ozone loss. *Nature*, **355**, 810–812.

Reinsel, G.C., Tiao, G.C., Wuebbles, D.J., Kerr, J.B., Miller, A.J., Nagatani, R.M., Bishop, L. and Ying, L.H. (1994). Seasonal trend analysis of published ground-based and TOMS total ozone date through 1991. *J. Geophys. Res.*, in press.

Rind, D. and Lacis, A. (1993). The role of the stratosphere in climate change. *Surv. Geophys.*, **14**, 133–165.

Schlesinger, M.E. (1993). Greenhouse policy. *Nat. Geogr. Res. Explor.*, **9**, 159–172.

Schlesinger, M.E. and Jiang, X. (1990). Simple model representation of atmosphere-ocean GCMs and estimation of the time scale of CO_2-induced climate change. *J. Climate.*, **3**, 1297–1315.

Schlesinger, M.E. and Ramankutty, N. (1992). Implications for global warming of intercycle solar irradiance variations. *Nature*, **360**, 330–333.

Schneider, S.H. (1994). Detecting climatic change signals: are there any fingerprints? *Science*, **263**, 341–347.

Schwarzkopf, M.D. and Ramaswamy, V. (1993). Radiative forcing due to ozone in the 1980s: dependence on altitude of ozone change. *Geophys. Res. Lett.*, **20**, 205–208.

Shi, G. and Fan, X. (1992). Past, present and future climatic forcing due to greenhouse gases. *Adv. in Atmos. Sci.*, **9**, 279–286.

Stolarski, R.S., Bloomfield, P., McPeters, R.D. and Herman, J.R. (1991). Total ozone trends deduced from Nimbus 7 TOMS data. *Geophys. Res. Lett.*, **18**, 1015–1018.

Taylor, K. and Penner, J.E. (1994). Anthropogenic aerosols and climate change. *Nature*, **369**, 734–737.

Wang, W.-C., Dudek, M.P. and Liang, X.-Z. (1992). Inadequacy of effecive CO_2 as a proxy in assessing the regional climate change due to other radiatively active gases. *Geophys. Res. Lett.*, **19**, 1375–1378.

Wang, W.-C., Dudek, M.P., Liang, X.-Z. and Kiehl, J.T. (1991). Inadequacy of effective

CO_2 as a proxy in simulating the greenhouse effect of other radiatively active gases. *Nature,* **350**, 573–577.

Wang, W.-C., Zhuang, Y.-C. and Bojkov, R.D. (1993). Climate implications of observed changes in ozone vertical distributions at middle and high latitudes of the Northern Hemisphere. *Geophys. Res. Lett.,* **20**, 1567–1570.

Wigley, T.M.L. and Kelly, P.M. (1990). Holocene climatic change, ^{14}C wiggles and variations in solar irradiance. *Philos. Trans. R. Soc. A,* **330**, 547–560.

(World Meteorological Organization) WMO (1986). Atmospheric ozone: 1985. *Global Ozone Research and Monitoring Project, Report No. 16,* Geneva.

WMO (1992). Scientific assessment of stratospheric ozone depletion: 1991. *Global Ozone Research and Monitoring Project, Report No. 25,* Geneva.

Wuebbles, D.J. and Edmonds, J. (1991). *Primer on Greenhouse Gases,* Lewis Publ., Chelsea, MI.

Appendix: Useful Numerical Values and Symbols

This appendix provides useful numerical values and symbols that are commonly employed in atmospheric sciences and are also used in this book.

A. Common Units, Symbols and Relationships

Length	meter (m) or centimeter (cm)	$1\,m = 10^2\,cm$
Mass	kilogram (kg) or gram (g)	$1\,kg = 10^3\,g$
Time	second (s)	
Frequency	hertz (Hz)	s^{-1}
Energy	joule (J)	$kg\,m^2\,s^{-2}$
Power	watt (W)	$J\,s^{-1}$ (or $kg\,m^2\,s^{-3}$)
Pressure	pascal (Pa)	$kg\,m^{-1}\,s^{-2}$
	millibar (mb)	$10^2\,Pa$ (hPa)
Temperature	degree Kelvin (K)	
	degree Celsius (°C)	$°C = K - 273.15$
Volume	liter (l)	$10^{-3}\,m^3$; $10^3\,cm^3$
Mixing ratios	parts per million (ppm or ppmv)	10^{-6} v/v
	parts per billion (ppb or ppbv)	10^{-9} v/v
	parts per trillion (ppt or pptv)	10^{-12} v/v
Wavelength	Ångström (Å)	$10^{-8}\,cm$; $10^{-4}\,\mu m$

B. Common Prefixes

Prefix	Multiple	Symbol	Prefix	Multiple	Symbol
deci	10^{-1}	d	deca	10^{1}	da
centi	10^{-2}	c	hecto	10^{2}	h
milli	10^{-3}	m	kilo	10^{3}	k
micro	10^{-6}	μ	mega	10^{6}	M
nano	10^{-9}	n	giga	10^{9}	G
pico	10^{-12}	p	tera	10^{12}	T
femto	10^{-15}	f	peta	10^{15}	P
atto	10^{-18}	a	exa	10^{18}	E

C. Universal Constants

Avogadro's number (N_A) 6.0220×10^{23} molecules gmole^{-1}
Boltzmann's constant (k) 1.3807×10^{-23} J K^{-1} molecule^{-1}
Planck's constant (h) 6.6262×10^{-34} J s
Speed of light in vacuum ($c*$) 2.9979×10^{8} m s^{-1}
Stefan–Boltzmann constant (σ) 5.6696×10^{-8} W m^{-2} K^{-4}
Universal gas constant (R) 8.3144 J K^{-1} gmole^{-1}
(or 0.0821 l atm K gmole^{-1})

D. Mean Properties of the Earth and its Atmosphere

Earth

Mean radius of the Earth 6371 km
Total volume of the Earth 1.083×10^{12} km^3
Total surface area of the Earth 5.10×10^{8} km^2 (oceans 70.8%; continents 29.2%)
Total mass of the Earth 5.976×10^{24} kg
Mass of the biosphere 1.148×10^{16} kg
Estimated age of the Earth 5×10^{9} years
Mean height of continents above sea level 0.875 km
Mean depth of world oceans 3.794 km
Average distance from the Sun to Earth 1.50×10^{8} km
Acceleration due to gravity at the Earth's surface 9.81 m s^{-2}
Mean rotational velocity of the Earth 29.8 km s^{-1}
Solar irradiance on a perpendicular plane outside the Earth's atmsphere 1.38×10^{3} W m^{-2}

Atmosphere

Mean composition of dry air 78.084% N_2; 20.946% O_2; 0.946% Ar; 0.032% CO_2
Mean mass of the atmosphere 5.14×10^{18} kg
Mean mass of the troposphere 4.23×10^{18} kg
Mean mass of the lower stratosphere (below 30 km) 8.48×10^{17} kg
Mean mass of the upper stratosphere (30–50 km) 5.80×10^{16} kg
Mean surface temperature 288.15 K
Mean surface pressure 1013.25 mb (or hPa); 760 torr
Mean molecular weight of dry air 28.97 g/gmole
Density of dry air at 0 °C and 1013 mb 1.293 kg m^{-3}

Specific heat of dry air at constant pressure
(C_p) $1005 \, \mathrm{J\,kg^{-1}K^{-1}}$
Specific heat of dry air at constant volume
(C_v) $718 \, \mathrm{J\,kg^{-1}\,K^{-1}}$
Ratio of specific heats of dry air
($\gamma = C_p/C_v$)   1.40
Volume of ideal gas at 0 °C and 1013 mb $22413.6 \, \mathrm{cm^3\,gmole^{-1}}$

TABLE A-1 **Temperature, pressure, density, and number density of air as a function of altitude (geometric) from U. S. Standard Atmosphere (1976)**

Altitude (km)	Temperature (K)	Pressure (mb)	Density (kg m^{-3})	Number density (molec. cm^{-3})
0	288.15	1.01(3)*	1.23	2.55(19)
2	275.15	7.95(2)	1.00	2.09(19)
4	262.17	6.17(2)	8.19(−1)	1.70(19)
6	249.19	4.72(2)	6.60(−1)	1.37(19)
8	236.22	3.57(2)	5.26(−1)	1.09(19)
10	223.25	2.65(2)	4.14(−1)	8.60(18)
12	216.65	1.94(2)	3.12(−1)	6.49(18)
14	216.65	1.42(2)	2.28(−1)	4.74(18)
16	216.65	1.04(2)	1.67(−1)	3.46(18)
18	216.65	7.57(1)	1.22(−1)	2.53(18)
20	216.65	5.53(1)	8.89(−2)	1.85(18)
22	218.57	4.05(1)	6.45(−2)	1.34(18)
24	220.56	2.97(1)	4.69(−2)	9.76(17)
26	222.54	2.19(1)	3.43(−2)	7.12(17)
28	224.53	1.62(1)	2.51(−2)	5.21(17)
30	226.51	1.20(1)	1.84(−2)	3.83(17)
32	228.49	8.89	1.36(−2)	2.82(17)
34	233.74	6.63	9.89(−3)	2.06(17)
36	239.28	4.99	7.26(−3)	1.51(17)
38	244.82	3.77	5.37(−3)	1.12(17)
40	250.35	2.87	4.00(−3)	8.31(16)
42	255.88	2.20	3.00(−3)	6.23(16)
44	261.40	1.70	2.26(−3)	4.70(16)
46	266.94	1.31	1.71(−3)	3.57(16)
48	270.65	1.03	1.32(−3)	2.74(16)
50	270.65	8.00(−1)	1.03(−3)	2.14(16)
60	247.02	2.19(−1)	3.10(−4)	6.44(15)
70	219.58	5.22(−2)	8.28(−5)	1.72(15)
80	198.64	1.04(−2)	1.85(−5)	3.84(14)
90	186.87	1.84(−3)	3.42(−6)	7.12(13)

*Power of 10 in parentheses: 1.01(3) = 1.01 × 10³.

Index